D1560226

The Chemistry of Indoles

ORGANIC CHEMISTRY

A SERIES OF MONOGRAPHS

1. Wolfgang Kirmse. CARBENE CHEMISTRY, 1964
2. Brandes H. Smith. BRIDGED AROMATIC COMPOUNDS, 1964
3. Michael Hanack. CONFORMATION THEORY, 1965
4. Donald J. Cram. FUNDAMENTAL OF CARBANION CHEMISTRY, 1965
5. Kenneth B. Wiberg (Editor). OXIDATION IN ORGANIC CHEMISTRY, PART A, 1965; PART B, *In preparation*
6. R. F. Hudson. STRUCTURE AND MECHANISM IN ORGANO-PHOSPHORUS CHEMISTRY, 1965
7. A. William Johnson. YLID CHEMISTRY, 1966
8. Jan Hamer (Editor). 1,4-CYCLOADDITION REACTIONS, 1967
9. Henri Ulrich. CYCLOADDITION REACTIONS OF HETEROCUMULENES, 1967
10. M. P. Cava and M. J. Mitchell. CYCLOBUTADIENE AND RELATED COMPOUNDS, 1967
11. Reinhard W. Hoffman. DEHYDROBENZENE AND CYCLOALKYNES, 1967
12. Stanley R. Sandler and Wolf Karo. ORGANIC FUNCTIONAL GROUP PREPARATIONS, 1968 VOLUME I; VOLUME II, *In preparation*
13. Robert J. Cotter and Markus Matzner. RING-FORMING POLYMERIZATIONS, PART A, 1969; PART B, *In preparation*
14. R. H. DeWolfe. CARBOXYLIC ORTHO ACID DERIVATIVES, 1970
15. R. Foster. ORGANIC CHARGE-TRANSFER COMPLEXES, 1969
16. James P. Snyder (Editor). NONBENZENOID AROMATICS, I, 1969
17. C. H. Rochester. ACIDITY FUNCTIONS, 1970
18. Richard J. Sundberg. THE CHEMISTRY OF INDOLES, 1970

In preparation

A. R. Katritzky and J. M. Lagowski. CHEMISTRY OF THE HETEROCYCLIC *N*-OXIDES

Ivar Ugi. ISONITRILE CHEMISTRY

G. Chiurdoglu. CONFORMATIONAL ANALYSIS

The Chemistry of Indoles

RICHARD J. SUNDBERG

Department of Chemistry
University of Virginia
Charlottesville, Virginia

1970
ACADEMIC PRESS
New York and London

ACADEMIC PRESS, INC.
111 Fifth Avenue, New York, New York 10003

United Kingdom Edition published by
ACADEMIC PRESS, INC. (LONDON) LTD.
Berkeley Square House, London W1X 6BA

LIBRARY OF CONGRESS CATALOG CARD NUMBER: 71-107554

PRINTED IN THE UNITED STATES OF AMERICA

CONTENTS

PREFACE ix

LIST OF ABBREVIATIONS xi

I. ELECTROPHILIC SUBSTITUTION REACTIONS ON THE INDOLE RING

A. Introduction to the Structure and Reactivity of Indoles 1
B. Protonation of Indoles 3
C. Nitration 11
D. Halogenation 14
E. Azo Coupling Reactions and Reactions with Azides 17
F. Sulfonation 18
G. Alkylation of Indoles and Metalloindoles with Alkyl Halides and Related Alkylating Agents 19
H. Reactions with Carbenes 31
I. Acylation of Indoles and Metalloindoles 33
J. Reactions with Ketones and Aldehydes 39
K. Reactions with Iminium Bonds 56
L. Reactions with Electron Deficient Olefins, Acetylenes, and Quinones 67
M. Miscellaneous Electrophilic Substitutions 77
N. The Mechanism of Electrophilic Substitution in 3-Substituted Indoles 78
O. Theoretical Treatment of Indole Reactivity and Substituent Effects 83
References 85

II. GENERAL REACTIONS OF FUNCTIONALLY SUBSTITUTED INDOLES

A. Introduction 93
B. Substitution and Elimination Reactions at Indolylcarbinyl Carbon Atoms 94
C. Hydrogenolysis of Indolylcarbinyl Substituents 108
D. Cleavage of Substituents from the Indole Ring. Decarboxylations and Related Reactions 114
E. Reactions of 3*H*-Indole Derivatives 120
F. Reactions of Vinylindoles 125
G. Reduction of the Indole Ring and Interconversion of Indoles and Dihydroindoles (Indolines) 129

H. The Indoline–Indole Synthetic Method 134
I. Metalation of Indoles 135
References 137

III. SYNTHESIS OF THE INDOLE RING

A. The Fischer Indole Synthesis 142
B. Synthesis of Indoles from Derivatives of α-Anilinoketones: The Bischler Synthesis 164
C. Indoles from the Reactions of Enamines with Quinones: The Nenitzescu Synthesis 171
D. Reductive Cyclizations 176
E. Electrophilic Cyclization of Styrene Derivatives 183
F. The Madelung Synthesis 189
G. Oxidative Cyclizations 191
H. Miscellaneous Reactions Leading to Indole Ring Formation 193
References 206

IV. SYNTHETIC ELABORATION OF THE INDOLE RING

A. Synthetic Approaches to Indolealkanoic Acids 214
B. Synthesis of Derivatives of Tryptamine 218
C. Synthesis of Tryptophan and Its Derivatives 230
D. Synthesis of Derivatives of β-Carboline 236
E. Synthesis of Indole Alkaloids 251
References 275

V. OXIDATION, DEGRADATION, AND METABOLISM OF THE INDOLE RING

A. The Reactions of Indoles with Molecular Oxygen and Peroxides 282
B. Ozonolysis of Indoles 295
C. Oxidation of Indoles with Other Oxidizing Agents 296
D. Oxidation of Indole Derivatives in Biological Systems and Chemical Models of Biological Systems 308
References 312

VI. REARRANGEMENT, RING EXPANSION, AND RING OPENING REACTIONS OF INDOLES

A. Simple Substituent Migrations 316
B. Rearrangements of Indolinones and Indolinols 322
C. Ring Expansions and Ring Opening Reactions of Indoles 331
References 339

VII. HYDROXYINDOLES AND DERIVATIVES INCLUDING OXINDOLE, 3-INDOLINONE, AND ISATIN

A. Reactions of Oxindoles 341
B. Synthesis of the Oxindole Ring 357
C. Derivatives of 3-Hydroxyindole and 3-Indolinone (Indoxyls) 364
D. Synthesis of 3-Indolinone Derivatives 367
E. Indoles with Carbocyclic Hydroxyl Substituents 368
F. 1-Hydroxyindoles 372
G. 2,3-Indolinedione (Isatin) Derivatives 377
H. 3*H*-Indole-3-One-1-Oxides (Isatogens) 382
References 388

VIII. AMINOINDOLES

Text 393
References 399

IX. KETONES, ALDEHYDES, AND CARBOXYLIC ACIDS DERIVED FROM INDOLE

A. Ketones and Aldehydes Derived from Indole 401
B. Synthesis of Indolyl Ketones and Indolecarboxaldehydes 412
C. Indolecarboxylic Acids 421
References 426

X. NATURALLY OCCURRING DERIVATIVES OF INDOLE AND INDOLES OF PHYSIOLOGICAL AND MEDICINAL SIGNIFICANCE

A. Introduction 431
B. Indole-Derived Antibiotics 431
C. Indoles of Physiological Significance: Tryptophan, Serotonin, and Melatonin 438
D. Indole Derivatives as Hallucinogens 440
E. Medicinally Significant Indoles 440
F. Luciferins Derived from Indole 442
G. Miscellaneous Indole Conjugates and Metabolites 443
References 445

AUTHOR INDEX 449

SUBJECT INDEX 477

PREFACE

A great deal of the chemistry of indole and its derivatives has been developed since the comprehensive reviews of Sumpter and Miller ("The Chemistry of Heterocyclic Compounds," Vol. 8, A. Weissberger, Ed., Interscience, New York, 1954) and of Julian, Meyer, and Printy ("Heterocyclic Compounds," Vol. 3, R. C. Elderfield, Ed., Wiley, New York, 1952, pp. 1–274) appeared in the early 1950's.

This monograph surveys the chemistry of indole derivatives reported in the literature from the early 1950's through 1967. A number of papers which appeared during 1968 and early 1969 are also included. The impetus for continued research into the chemistry of indoles during this period has come largely from interest in the potent biological activity of many indole derivatives and from efforts to elucidate the structure of indole alkaloids and to synthesize them. Much of the work reviewed in this volume has thus been the result of effort toward synthetic objectives, and the book reflects this synthetic shading. More studies of a mechanistic nature appeared, along with continued synthetic work, in the 1960's, and such studies are beginning to furnish clearer insight into the mechanisms of reaction of indole derivatives.

In the first two chapters I have attempted to correlate some of the most important types of reactions of the indole ring on a mechanistic basis. Chapters III and IV review the methods of synthesizing indoles. Chapters V and VI, respectively, describe oxidations and rearrangements of indole derivatives. The special features of the reactivity and synthesis of hydroxyindoles, aminoindoles, and acylindoles are discussed in Chapters VII, VIII, and IX, respectively. Finally, in Chapter X, certain classes of indoles found in nature are discussed. I have attempted to prepare tables which include representative examples of some of the more general reactions and syntheses of indole derivatives; they are not intended to be exhaustive.

I express my appreciation to Dr. M. Akram Sandu (University of Minnesota), Mr. James A. Janke (University of Minnesota), Mrs. Richard H. Smith, Jr. (Charlottesville), and Professor J. B. Patrick (Mary Baldwin College), for reading portions of the manuscript, and also to Mr. Woodfin V. Ligon, Jr. (University of Virginia), who read the entire manuscript. Special thanks are due to my friend and teacher Professor Wayland E. Noland for his interest and encouragement in the preparation of the manuscript as well as for reading portions of the manuscript. My wife Lorna is due thanks not only for patient encouragement but also for typing the first draft of the manuscript; I thank Mrs. Stuart R. Suter for typing the final manuscript. Finally, my daughters Kelly Kay and Jennifer Leigh deserve credit for letting this effort infringe on many of their plans and projects.

Abbreviations Used in Tables and Structural Formulas

Ac	CH_3CO
Bz	Benzene
DMF	Dimethylformamide
Et	CH_3CH_2
Me	CH_3
Ph	C_6H_5
Pr	$CH_3CH_2CH_2$
THF	Tetrahydrofuran

The Chemistry of Indoles

I
ELECTROPHILIC SUBSTITUTION REACTIONS ON THE INDOLE RING

A. INTRODUCTION TO THE STRUCTURE AND REACTIVITY OF INDOLES

Indole is a planar heteroaromatic molecule. The ten-electron pi system is formed by a pair of electrons from the nitrogen atom and eight electrons from the eight carbon atoms. Figures 1 through 3 depict the total atomic electron densities, the sigma atomic electron densities, and the pi electron densities, respectively, as obtained from self-consistent field molecular orbital calculations (193) which give dipole moments in excellent agreement with experimental values (23).

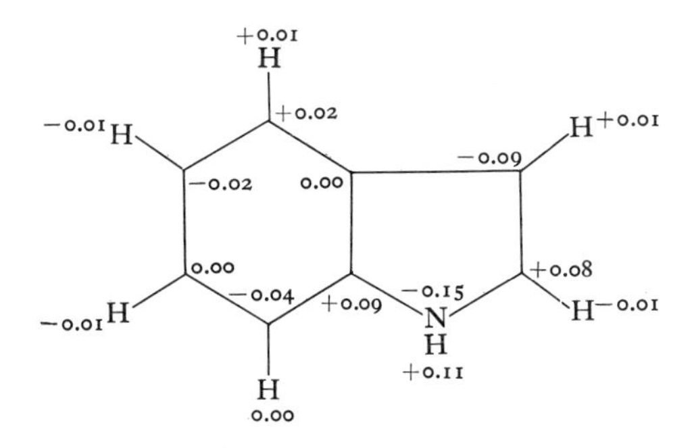

Fig. 1. Total net atomic electron densities of indole (23).

The data in Fig. 3 are in accord with the familiar description of indole in terms of the resonance concept, in that there is significant delocalization of electron density from nitrogen to the carbon atoms of the ring, particularly to C-3. The most fundamental chemical properties of the indole ring are

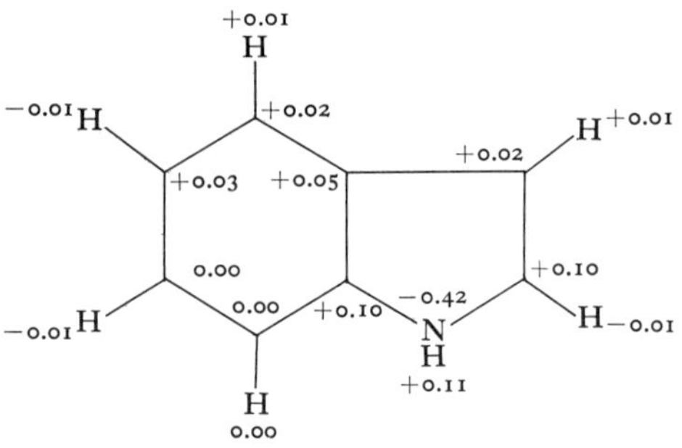

Fig. 2. Net sigma atomic electron densities of indole (23).

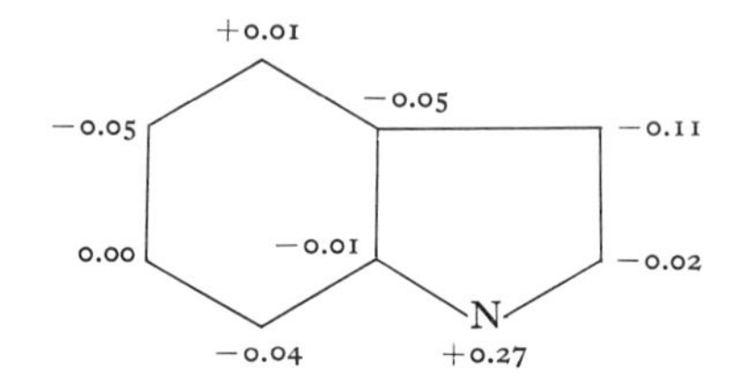

Fig. 3. Net pi atomic electron densities of indole (23).

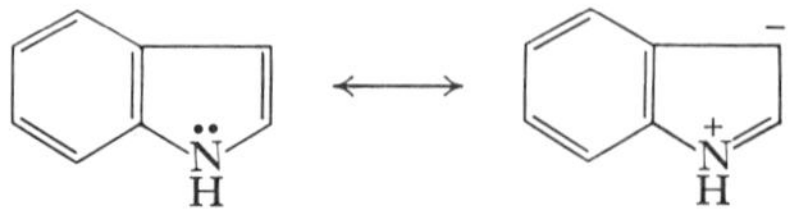

Fig. 4. Simple resonance description of indole.

correctly predicted by the structural picture in Fig. 4. The heterocyclic nitrogen atom is very weakly basic because of delocalization of its unshared pair of electrons into the pi system. A pK_a of –2.5 has been quoted (18) and protonation is at C-3, not nitrogen (96, 99). Acyl substituents on nitrogen are relatively readily cleaved by hydrolysis, as is typical of acyl derivatives of nitrogen atoms having delocalized unshared pairs (241, 242). The electron-rich aromatic ring tends to form complexes with acceptors such as polynitrobenzenes (74, 278) or tetracyanoethylene (54, 75). Association constants (54, 75) of such complexes have been studied and the crystal structure (88) of the indole-trinitrobenzene and 3-methylindole-trinitrobenzene complexes have been determined by X-ray methods.

Self-consistent field molecular orbital calculations have been used to calculate bond lengths of indole as well as the resonance energy (63). Lack of experimental structural data prevents a direct comparison with theory in the case of indole, but agreement with experimental data is satisfactory for the same theoretical treatment of related heterocyclic systems (63). Figure 5 shows experimental bond distances for 3-methylindole in the 3-methyl-indole-trinitrobenzene complex, and Fig. 6 shows the theoretical indole bond distances calculated by Dewar and Gleicher (63).

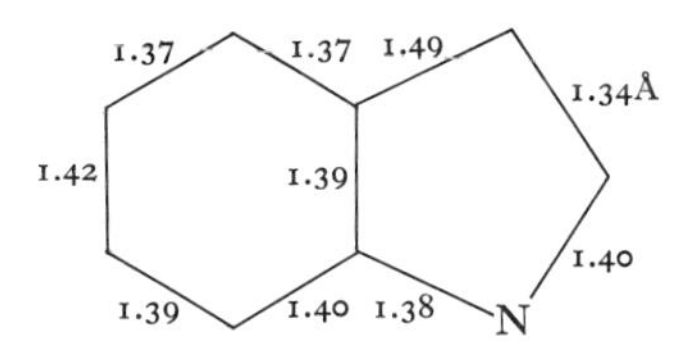

Fig. 5. Bond lengths in the 3-methylindole-trinitrobenzene complex (88).

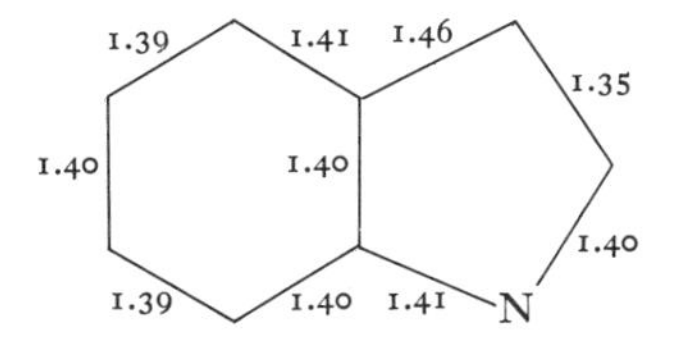

Fig. 6. Calculated bond lengths for indole (63).

The resonance energy of indole has been estimated at 41–58 kcal/mole on the basis of thermochemical data (290).

The indole ring is reactive toward electrophilic substitution, the 3-position being the most reactive site for substitution. The high reactivity of the 3-position can be predicted on the basis of pi electron density (34), localization energy (34), or frontier electron density (79) information obtained from molecular orbital calculations. Attempts to order the relative reactivity of the positions on the carbocyclic ring have been less successful (34). A more thorough discussion of theoretical attempts to predict the reactivity of the indole ring is deferred until Section L. Electrophilic substitution has been the most completely studied reaction of the indole ring and it is considered from an experimental point of view in the following sections of this chapter.

B. PROTONATION OF INDOLES

The delocalization of electron density from the nitrogen atom to the ring pi system diminishes the electron density at the nitrogen atom and indole is therefore a very weak base. By working in strongly acidic solutions it has, however, been possible to determine the structure of the conjugate acids of simple indoles and to study the relative basicities of substituted indoles (18, 96, 99, 121). Interpretation of the ultraviolet (uv) and nuclear magnetic resonance (nmr) spectra of indoles has established that the indole ring is protonated in the 3-position (96, 99, 121). The appearance of the nmr signal of the methyl group in the conjugate acid of 3-methylindole as a doublet because of spin–spin splitting with the added proton provides the most direct evidence for 3-protonation (96, 99). The possibility that a small amount of N-protonation would escape detection by nmr measurements has been rendered unlikely by examination of the uv spectrum of the quaternary indolium ion **B1** which is an appropriate model for the N-protonated

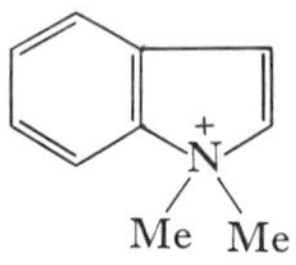

B1

conjugated acid (97a). Hinman and Lang (96) have concluded that there is no indication in the uv spectrum of partially protonated indoles of significant N-protonation. Kamlet and Dacons (121) have also concluded from the uv spectra of 2-arylindoles in 85% phosphoric acid that protonation must occur predominantly at the 3-position.

Because of their weak basicity, the quantitative basicity of indoles must be determined with reference to acidity functions, and ambiguities in the numerical values of the pK values are thereby introduced (96, 99). Nevertheless, the relative basicity of a number of indoles have been determined and certain substituent effects have been established. Acidity constants for certain substituted indoles are recorded in Table I. On the basis of their data on 25 indole derivatives, Hinman and Lang (96) have tabulated the effect of certain substituents on the pK of the ring. These effects are shown in Table II. Hinman and Lang have presented reasonable explanations for the effects. The base-weakening effect of 3-alkyl substituents is particularly noteworthy and has been attributed in part to decreased hyperconjugative stabilization of the conjugate acid.

Reversible protonation permits isotopic exchange at certain positions on the indole ring. Early work in this area was done by Koizumi (131). Hinman and co-workers (95, 96, 99) have provided examples of such exchange processes. With 2-methylindole, rapid N–H exchange occurs in dioxane solutions 5×10^{-5} M in sulfuric acid. The exchange rate is considerably greater than that of the comparable process in indole. Refluxing a solution

TABLE I □ DISSOCIATION CONSTANTS OF THE CONJUGATE ACIDS OF SOME SUBSTITUTED INDOLES

Substituent	pK_a[a]	pK_a[b]
1,2-diMe	0.30	0.34
2-Me	−0.28	−0.10
1,2,3-triMe	−0.66	—
2,3-diMe	−1.49	−1.10
1-Me	−2.32	−1.80
1,2-Dimethyl-5-nitro	−2.94	—
2-Methyl-5-nitro	−3.58	—
None	−3.62	−2.4
3-Me	−4.55	−3.35

[a] From ref. (96) and based on the H_I function described therein.
[b] From ref. (18) and based on the H_0 acidity function.

TABLE II □ SUBSTITUENT EFFECTS[a]

Substituent	ΔpK
1-Me	$+0.7 \pm 0.11$
2-Me	$+2.9 \pm 0.16$
3-Me	-1.1 ± 0.16
3-Pr	-1.2 ± 0.09
5-NO_2	-3.3 ± 0.03
5-Me	$+0.5 \pm 0.09$

[a] From ref. (96). Used with the permission of the American Chemical Society.

of indole in deuterium oxide containing 5×10^{-3} M sulfuric acid results in exchange at both the 1- and 3-positions. The N–D can be re-exchanged by refluxing with water, a process which makes 3-deuteroindole available. The exchange of N–H in neutral solution is facilitated by electronegative substituents, suggesting that such substituted indoles may undergo exchange via the conjugate base formed by removal of the proton from nitrogen (111). The protonation of several ionic derivatives of indole has been examined. Indolylmagnesium iodide, when treated with tetrahydrofuran containing deuterium oxide and then heated, gives indole deuterated to the extent of about 50% at both the nitrogen atom and C-3 (111). Powers, Meyer, and Parsons (200) have reported the most complete investigation of protonation of metallo derivatives of indole to date. In contrast to the

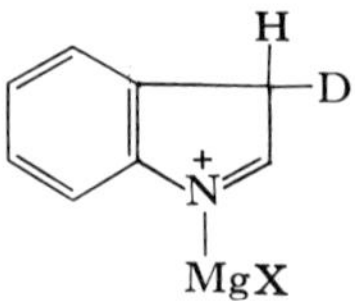

B2

sodium and lithium salts of indole, which give only small amounts of indole-3-*d* on reaction with deuterium oxide, solutions of indolylmagnesium iodide in ether give extensive amounts of indole-3-*d*. The extent of deuterium incorporation into C-3 ranges up to 70% and is highly dependent on the amount of deterium oxide used in the reaction. The suggestion that the 3*H*-indole-magnesium complex **B2** formed by 3-protonation of indolylmagnesium iodide may undergo exchange at the 3-position at a rate competitive with tautomerization to 1*H*-indole has been advanced to explain the extensive introduction of deuterium (200).

The detailed mechanisms of the exchange processes in indole and certain of its methyl derivatives have been discussed by Hinman and Whipple (99) and by Challis and Long (44). Exchange at the nitrogen atom can take place via reversible N-protonation or by 3-protonation followed by deprotonation of the resulting indolenine. Slow exchange of the 2-proton in 3-methylindoles is attributed to competitive protonation at the 2-position. The loss of tritium from 2-methylindole-3-*t* has been shown to be catalyzed by general acids (44) and also by hydroxide ion. The detritiation of 1,2-dimethylindole-3-*t* is not appreciably catalyzed by hydroxide ion, suggesting that base-

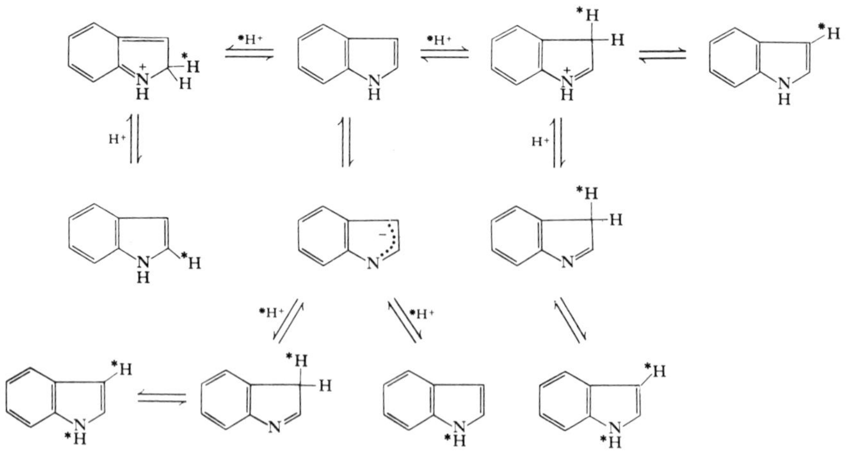

Scheme I. Exchange mechanisms for indole.

catalyzed exchange involves anion formation by loss of the proton on nitrogen. The range of mechanisms available for exchange on the indole ring is summarized in Scheme I.

Indoles with 5-hydroxyl substituents rapidly and selectively exchange the 4-proton in deuterium oxide containing catalytic amounts of trimethylamine (58). An explanation of this selectivity probably lies in the greater stability of the conjugated system **B3** over **B4**. In **B4** the aromaticity of the system is destroyed whereas in **B3** a conjugated pyrrole ring remains. Intermediate **B3** would arise if the mechanism of exchange involves C-protonation of a phenoxide ion. Similar selectivity is noted in certain other reactions of these hydroxy derivatives (Chapter VII, E).

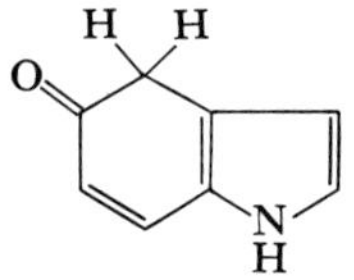

H H O N H

B3

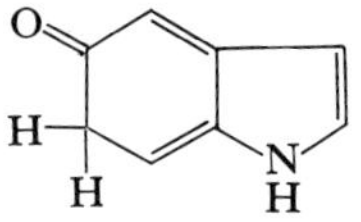

O H H N H

B4

Incomplete protonation of indoles in concentrated solution initiates dimerization and trimerization of certain indoles. Both a dimer (125, 182), and trimer (125) of indole have been characterized. A brief review of the area has appeared (235). The structures of the dimer and trimer were correctly formulated by Smith (236) in 1954 on the basis of mechanistic consideration of the process, the observation that triindole is a primary amine, and earlier structural investigations (228, 229). The structures **B5**

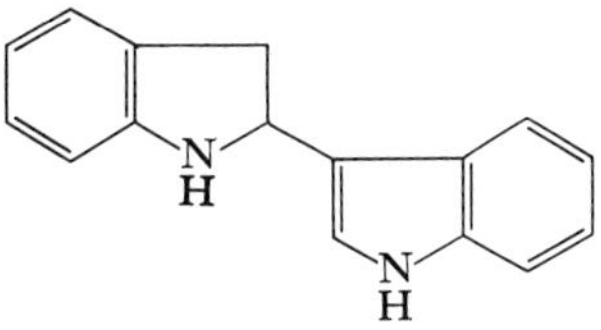

N H N H

B5

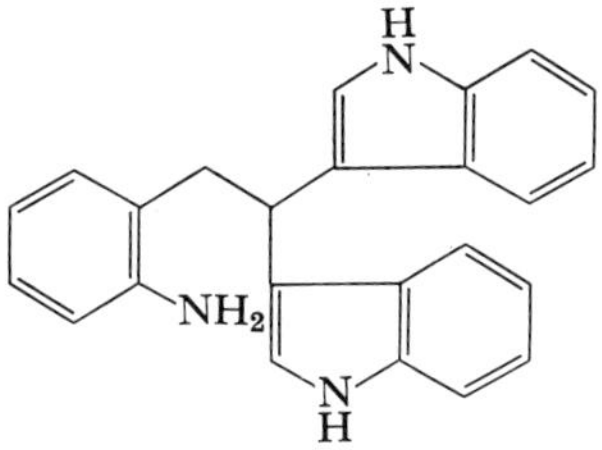

H N NH_2 N H

B6

and **B6**, respectively, were proposed by Smith as the correct formulations of diindole and triindole. Subsequently, Hodson and Smith (102) confirmed the diindole structure by a degradative sequence. Noland and Kuryla (168) shortly thereafter confirmed the structure of triindole by its independent synthesis in two steps from *o*-nitrophenylacetaldehyde and indole.

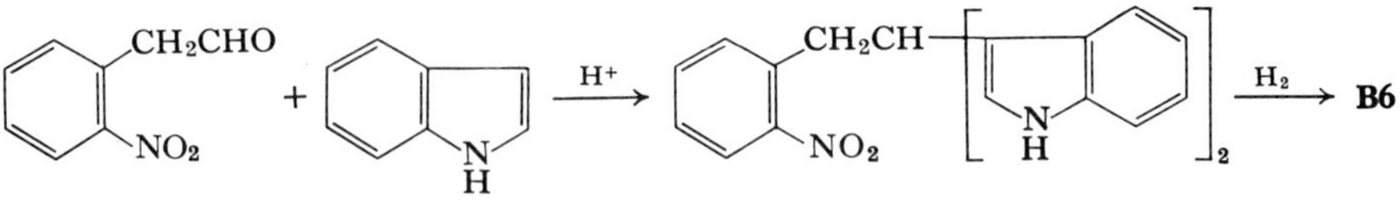

B7 B8 B9

The dimerization is initiated by 3-protonation of indole generating the electrophilic cation **B10**. Electrophilic attack on the 3-position of an unprotonated indole ring generates diindole. As subsequent discussion will show, the high reactivity of the 3-position of indole toward electrophilic attack is a very general feature of indole chemistry. The formation of

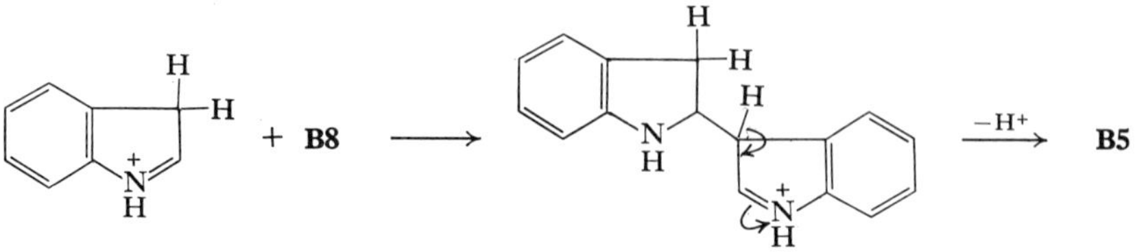

B10 B11

triindole occurs when the experimental conditions permit subsequent reaction of diindole. Protonation of the basic indoline nitrogen of **B5** generates a potential leaving group. Ionization of the C–N bond in **B12** is favored by the ability of the indole ring to delocalize the resulting positive charge. A new electrophile, the 3-alkylideneindolenine cation **B13** is generated and further electrophilic substitution on unprotonated indole occurs.

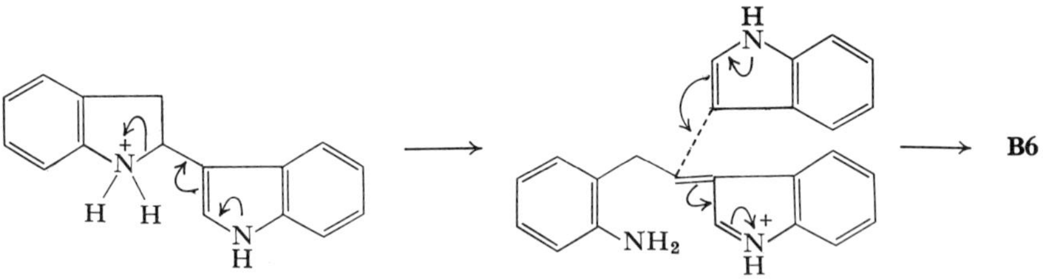

B12 B13

The scope of the dimerization reactions is readily interpreted in terms of these mechanisms. 2-Methylindole fails to dimerize (163, 228, 229). The reduced electrophilicity of the cation **B14** generated by protonation of 2-methylindole causes the failure of the dimerization in this case. Both the electronic and steric effect of the 2-methyl group in **B14** will reduce the reactivity relative to the unsubstituted analog **B10**. However, when an

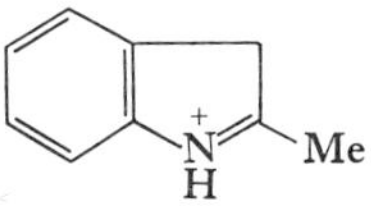

B14

equimolar mixture of indole and 2-methylindole is subjected to dimerization conditions (163), a quantitative yield of the mixed dimer **B15** can be obtained as the hydrochloride. The course of this reaction demonstrates

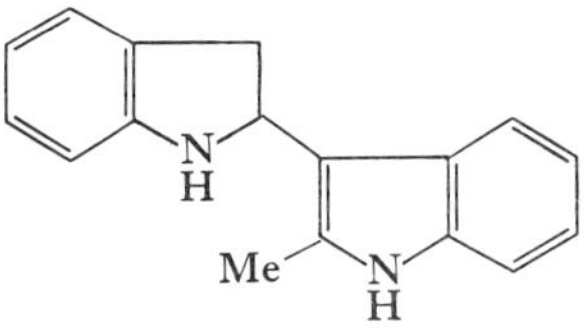

B15

the high electrophilicity of **B10** relative to **B14** as well as the fact that 2-methylindole is even more effective than indole as the nucleophilic component in dimerization. The enhanced nucleophilicity of 2-methylindole is no doubt the result of the electron-releasing effect of the 2-methyl group. Studies of dimerization and trimerization reactions of mixtures of indoles have been generalized to include not only substituted indoles but also pyrroles (163). The general conditions for a successful mixed dimerization include the presence of one indole which is capable of acting as the electrophilic component in dimerization (A component) and a second indole which is more reactive as a nucleophilic component (B component). If the B component is less reactive as nucleophile than the A component, homodimerization of the A component will dominate.

Three groups (16, 98, 237) established independently that the dimer of 3-methylindole has structure **B16**. The dimerization is again initiated by

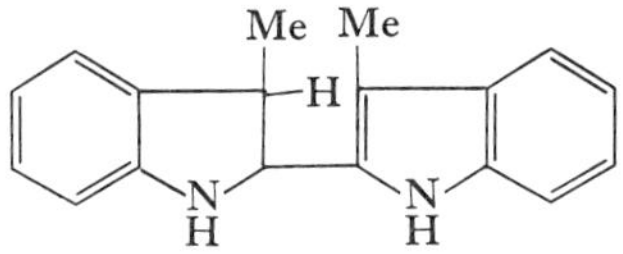

B16

3-protonation, but since electrophilic aromatic substitution at the 3-position is blocked, the product results from electrophilic substitution at the 2-position. Attempted mixed dimerization of indole and skatole (163) gives instead diindole and **B16**, demonstrating that indole is more reactive than 3-methylindole as both electrophile (as the conjugate acid) and nucleophile.

It seems appropriate at this stage to point out certain general features of indole chemistry which are illustrated by various reactions described in the preceding paragraphs. The preference of a proton for the 3-position of the indole ring is only one example of the preference for attack by electrophiles at C-3. The stability of the generalized cation **B17** relative to **B18** must chiefly reflect the retention of styrene-type conjugation in **B17** versus the *o*-quinoid conjugation of **B18**. The preference for structures of type **B17** over **B18** is one of the important unifying features of indole chemistry.

B17 **B18**

The electrophilicity of structures such as **B17**, as illustrated by the dimerization reactions, is also a general and useful concept which will recur in later discussion. A key step in the trimerization of indole, the conversion of **B12** to **B13** (page 8), illustrates still another important general feature of the chemistry of indole rings. Electronegative groups can readily depart from a carbon atom attached to C-3 of the indole ring; the electron-rich heteroaromatic ring is effective in stabilizing the positive charge which remains. This stabilization is the result of delocalization involving the ring and in some cases also involves subsequent loss of the positive charge as a proton and generation of reactive intermediates having the general structure **B21**.

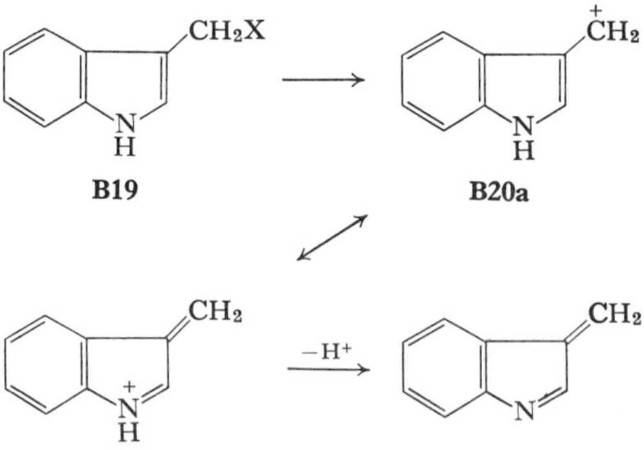

B19 **B20a**

$-H^+$

B20b **B21**

The protonation of tryptamine (**B22**) and systems of the physostigmine type (**B24**) illustrates the interplay of several potential protonation sites on reactivity (106). The exocyclic nitrogen atom in tryptamine is far more basic than the indole ring, and the positive charge introduced on the exocyclic nitrogen has a strong electrostatic destabilizing effect on further protonation. The pK for protonation of the indole ring of tryptamine is -6.31 compared to -4.55 for 3-methylindole (96). Spectroscopic evidence permits assign-

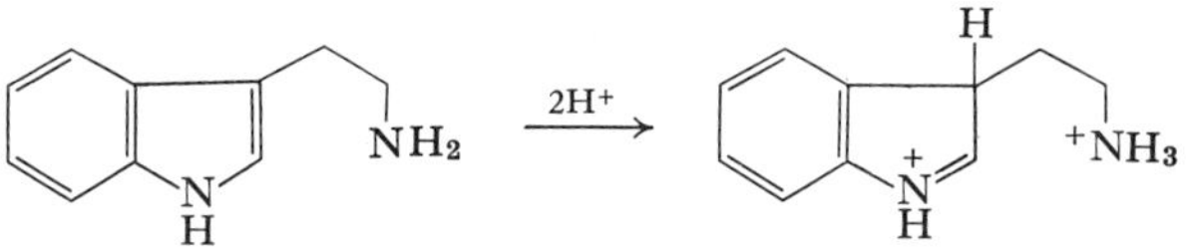

B22 $\xrightarrow{2H^+}$ **B23**

ment of structure **B23** to the diprotonated species. In neutral solution the ring system **B24** exists as shown. At moderate acidity, protonation at the trialkylnitrogen atom occurs. As the acidity is increased a second proton is added, but the second protonation is accompanied by ring opening and dispersal of the charge. Or, considering the reverse of the process **B24** → **B26**, generation of a nucleophilic site in the side chain by deprotonation

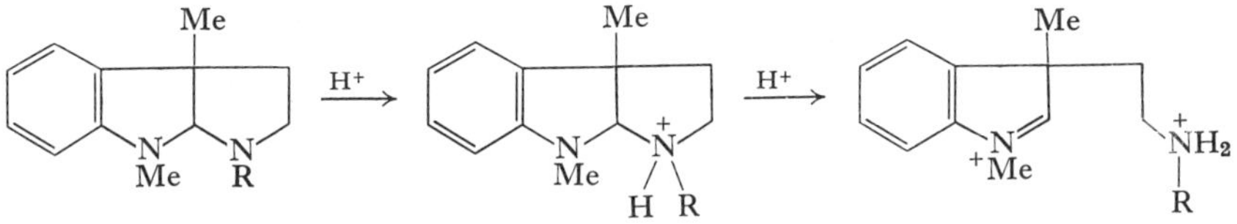

B24 $\xrightarrow{H^+}$ **B25** $\xrightarrow{H^+}$ **B26**

results in addition of the nucleophilic nitrogen to the neighboring electrophilic iminium bond.

C. NITRATION

The course of nitration of simple derivatives of indole has received considerable recent attention, principally from Noland and co-workers (171, 174–177) and Berti, Da Settimo, and co-workers (15, 17, 59–61). Some typical results obtained by these workers are recorded in Table III. Generalizations drawn from these data have greatly clarified the matters of orientation and extent of nitration under various reaction conditions. A summary of data in this area published in 1950 attests to the multiplicity of product types found under various nitration conditions (226).

A clear-cut result which emerges from the data in Table III is the fact that the orientation of nitration in simple 2-substituted indoles depends

TABLE III □ NITRATION OF REPRESENTATIVE INDOLES[a]

Entry	Substituents	Nitrating[a] media	Position(s) of nitration (ratio) (% yield)	Ref.
1	None	A	[b]	(176)
2	1-Me-2-Ac	A	5, ?	(64)
3	1,2-diMe	A	5, 82%	(176)
		B	3,6,[c] 28%	(177)
4	1-Ac-2,3-diMe	A	5, ?	(191)
		B	6, 14%	(191, 227)
5	1,2,3,3-tetra-Me-3*H*-indolium ion	A	5, 77%	(33)
6	1-Me-3-formyl	A	5:6 (1.5:1), 93%	(61)
7	2-Me	A	5, 84%	(176)
		B	3,6,[c] 39%	(177)
8	2-Ph	A	5, 87%	(175)
		B	3,6,[c] 40%	(175)
9	2-Me-3-formyl	A	5:6 (1.5:1), 92%	(61)
10	2-Carboxy-3-nitro	B	5, 54%	(174)
11	2,3,3-triMe indolium ion	A	5, 81%	(177)
12	3-Ac	B	6:4 (16:1), 35%	(174)
13	3-Cyano	B	6:4 (1.7:1), 32%	(174)
14	3-Formyl	A	5:6 (1.5:1), 85%	(17)
		B	6, 16%	(17)

[a] Nitrating media: A = Nitrate salts in concentrated sulfuric acid. B = Concentrated nitric acid or nitric acid in acetic acid.
[b] Product is an amorphous solid.
[c] Dinitration occurs.

upon the reaction medium. Nitration in concentrated sulfuric acid gives high yields of 5-nitro derivatives, but nitration in nitric acid or nitric acid–acetic acid mixtures gives lower yields of 3,6-dinitro derivatives. The first observation is striking. The 3-position of the indole ring, normally the most reactive toward electrophilic substitution, has been bypassed. The proposals of Berti and Da Settimo (17) and Noland, Smith, and Johnson (176) that *5-nitration is the result of nitration via the conjugate acid* represented a significant advance in formulating a comprehensive picture of indole nitrations. The proposal is supported by spectroscopic data which indicate that 2-methylindole is completely protonated in concentrated sulfuric acid (97). 3,3-Dialkyl-3*H*-indolium ions (**C1**) are suitable structural models for the conjugate acids of 2-alkylindoles. Noland, Smith, and Rush (177) as well as Brown and Katritzky (33) have observed 5-nitration in the 3,3-dialkyl-3*H*-indolium system. The iminium grouping in **C1** or in the con-

C1 $\xrightarrow[NHO_3]{H_2SO_4}$ C2

R = H 81,88%
R = Me 77% Refs. (33, 177)

jugate acid of a 2-substituted indole thus acts as a para-directing group (33). There is at least one example of such a directing effect in a noncyclic system. Nitration of the conjugate acid of the benzylidene derivative of aniline gives a 94% yield of the para product (6, 60). There does not seem to be data available to indicate the extent to which the iminium linkage is activating or deactivating as a substituent.

As is apparent from Table III, nitrations in nitric acid and nitric acid–acetic acid mixtures give 3,6-dinitro derivatives from simple 2-substituted indoles. Studies on the polynitration of 2-alkylindoles and several derivatives in such media have established 3,6,4 or 3,4,6 as the preferred order of orientation (177). Nitrations in these media generally require thermal initiation and apparently proceed concurrently with oxidation-reduction reactions which produce oxides of nitrogen (177). The present view (177) of nitration in nitric acid suggests that 3-nitration involves the nitration (or nitrosation) of a small concentration of the unprotonated indole in equilibrium with its conjugate acid. The 3-nitro derivative is less basic than its precursor and subsequent nitration proceeds rapidly on the unprotonated indole to the dinitro stage. Nitration of 2-methyl-3-nitroindole (**C3**) to 3,6-dinitro-2-methylindole (**C4**) has been shown (177) to be an efficient process (67% yield at 28°C, 30 min) in concentrated nitric acid.

C3 $\xrightarrow{HNO_3}$ C4 (67%) + C5 (3%)

As shown in Table III, nitration of certain derivatives of indole-3-carboxaldehyde in concentrated sulfuric acid gives high yields of mixtures of 5- and 6-nitration products (17, 61). The full interpretation of this result must take into account several pieces of data which are not available at this point. The formyl group no doubt decreases the basicity of the indole ring and also introduces the possibility of O-protonation. Either or both of these factors could be related to the increased tendency toward 6-nitration.

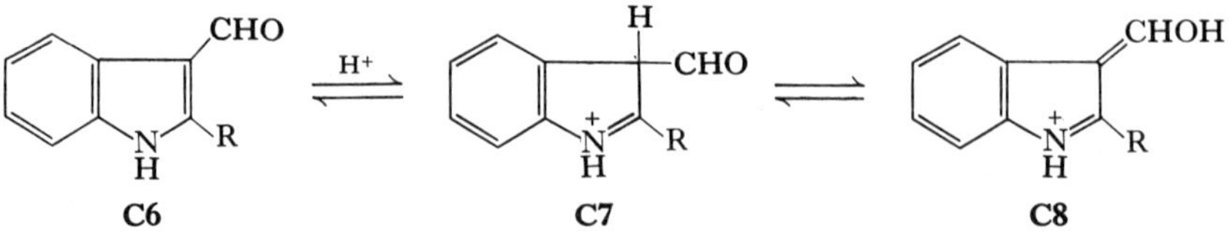

C6 ⇌ (H+) C7 ⇌ C8

Loss of 3-formyl and 3-acetyl substituents have been observed (174). Entry 10 in Table III represents an anomolous orientation. An explanation has been advanced (174) in terms of the existence of the substrate in a cyclic arrangement possessing structural features similar to 3*H*-indolium ions.

In summary, it is now well established that the reaction medium used for nitration of indoles, particularly as it affects the indole-3*H*-indolium ion equilibrium, plays a very important part in determining both nitration rate and position of substitution. The data available indicate that nitrations in concentrated sulfuric acid often proceed cleanly, although there are exceptions among indoles which are especially susceptible to oxidative degradation (176). The general order $3 > 6 > 4$ for reactivity of unprotonated indoles is now supported by numerous examples. In general, the synthetic success of nitration of unprotonated indoles seems to depend on the susceptibility of the indole to oxidative attack. Nitration is most likely to proceed satisfactorily when the ring carries electron withdrawing substituents which stabilize the ring toward oxidation.

D. HALOGENATION

Halogenation of indoles has not been studied extensively from the mechanistic point of view. The results of various synthetic studies, however, show that halogenations follow the normal pattern of electrophilic substitution. The 3-position is the preferred site of attack. If this position is blocked, substitution occurs at the 2-position or on the benzene ring. Table IV records the results of typical indole halogenations.

Pyridinium bromide perbromide in pyridine has been found to be a superior (189) brominating agent for indole. Halogenations are complicated by the fact that 2- and 3-haloindoles are unstable, especially in the presence of acid. 3-Iodoindole, for example, is solvolyzed to indoxyl acetate (**D2**) in glacial acetic acid containing silver acetate (7).

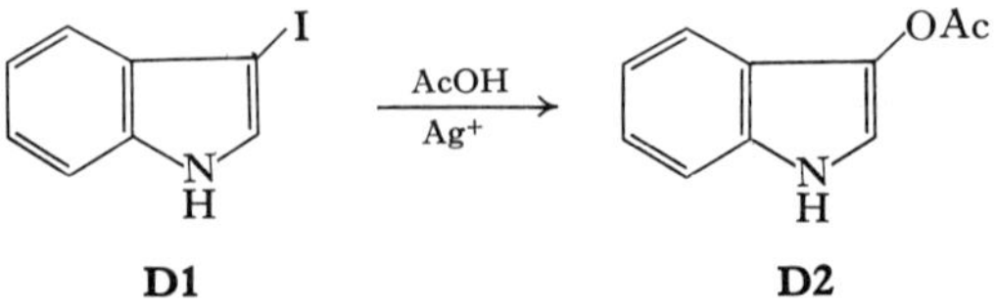

D1 → (AcOH, Ag^+) D2

TABLE IV □ HALOGENATION OF REPRESENTATIVE INDOLES

Substitution	Halogenating agent	Position of halogenation (% yield)	Ref.
None	Pyridinium bromide perbromide	3 (64)	(189)
None	Iodine-potassium iodide	3	(7)
	Iodine monochloride	3 (76)	(251)
	Dioxane dibromide	3 (60)	(251)
1-Benzoyl	Bromine	3 (82)	(277)
	Chlorine	3 (80)	(277)
2-Bromo-3-Me	*N*-Bromosuccinimide	6 (48)	(95a)
2-Carbethoxy-3-bromo	Bromine	5 (83)	(135)
2-Carbethoxy-3-chloro	Bromine	5	(135)
	Phosphorus pentachloride	6	(135)
2-Carboxy	Bromine	3 (88)	(136)
2-(*o*-Iodophenyl)	Iodine monochloride	3	(35)
2-Me	*N*-Chlorosuccinimide	3 (78)	(198)
2,3-diMe	Bromine, silver sulfate in concentrated sulfuric acid	5 (75)	(132)
3-Carbethoxy	Bromine	5 and 6	(140, 144)
3-Me	*N*-Bromosuccinimide	2 (45)	(95a)

D3 —H^+→ D4 —H_2O→ D5 —$-HCl$→ D6 → D7

D8 → D9 → D7

3-Chloroindole is converted to oxindole in aqueous acidic methanol (198). 1-Benzyl-2-chloroindole is hydrolyzed to 1-benzyloxindole under similar conditions (198). The hydrolytic reactions of halo derivatives are probably initiated by 3-protonation. The hydrolysis of 3-chloroindole can be rationalized in this way (198). Protonation of 2-haloindoles at C-3 results in the generation of an imido halide and high reactivity of the halide toward nucleophilic displacement is then expected. Halogenation in hydroxylic solvents often leads directly to oxindoles. Such reactions are considered in Chapter V, C.

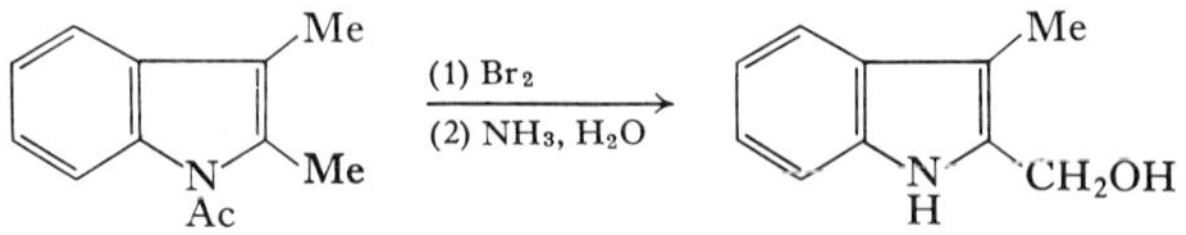

Me, N, Ac, Me — (1) Br_2, (2) NH_3, H_2O → Me, N, H, CH_2OH

D10 **D11**

3-Haloindoles are apparently somewhat more stable to alkaline media since 3-bromoindole and 3-chloroindole can be obtained by alcoholysis of corresponding benzoyl derivatives in ethanol containing sodium ethoxide (187, 277) and 2-bromo-3-methylindole can be recovered unchanged from hot ethanolic potassium hydroxide solution (95a).

1,2,3-Trisubstituted indoles can undergo halogenation on the 2-substituent under certain conditions. Thus, bromination of 1-acetyl-2,3-dimethylindole in acetic acid, followed by treatment with aqueous ammonia gives 2-hydroxymethyl-3-methylindole (**D11**) (30, 191). This transformation can be formulated as occurring through a bromine addition product. A similar mechanism can be formulated for the conversion of 1,2,3-trimethylindole to the pyridinium salt **D16** (143). Bromination of 2,3-dimethylindole and

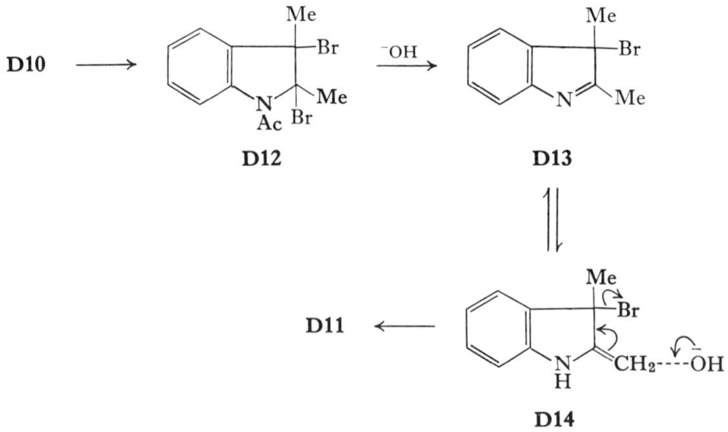

D10 → (Me, Br, N, Ac, Me, Br) **D12** —^-OH→ (Me, Br, N, Me) **D13**

⇅

D11 ← (Me, Br, N, H, CH_2---$\bar{O}H$) **D14**

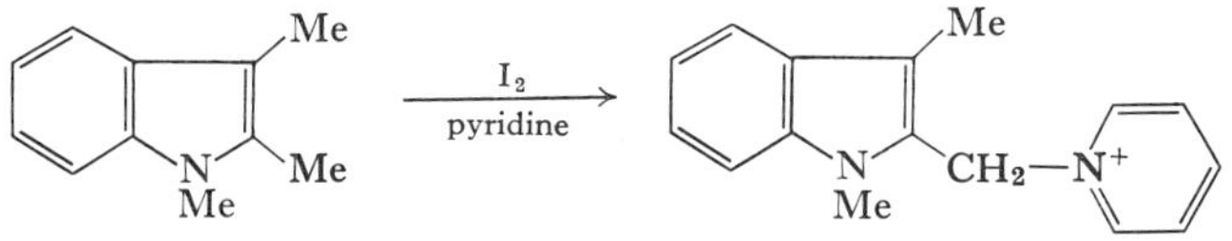

D15 $\xrightarrow[\text{pyridine}]{I_2}$ D16

tetrahydrocarbazole by *N*-bromosuccinimide in pyridine gives **D17** and **D18**, respectively (219). An addition-elimination-displacement sequence again would seem to be the most likely formulation for this reaction.

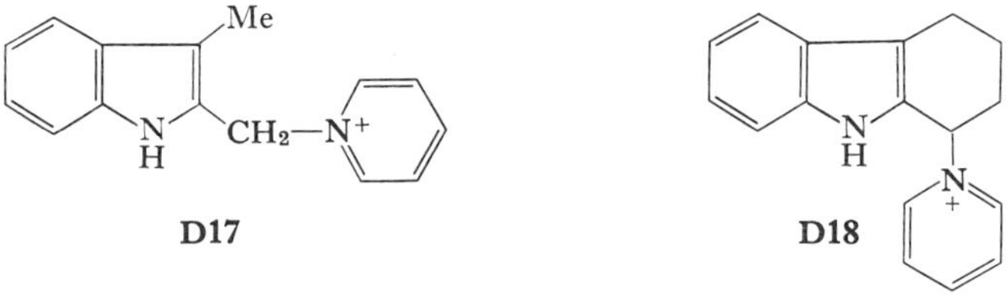

D17 D18

E. AZO COUPLING REACTIONS AND REACTIONS WITH AZIDES

The indole ring is sufficiently reactive at the 3-position to be substituted by diazonium ions. The kinetics of this reaction have been investigated (19) in dilute aqueous solution. *p*-Nitrobenzenediazonium ion showed first-order behavior in the presence of excess indole but other diazonium ions exhibit more complex behavior. Relatively few azo derivatives have been isolated from such coupling reactions (19, 41). Because of the facility of following azo coupling reactions by spectrophotometric methods, this coupling reaction would seem to be potentially useful as a probe of substituent effects on relative reactivity at the 3-position.

2-Methylindole reacts with picryl azide to give 3,3′-azobis(2-methylindole) (9). Other indoles react similarly. *p*-Toluenesulfonyl azide effects

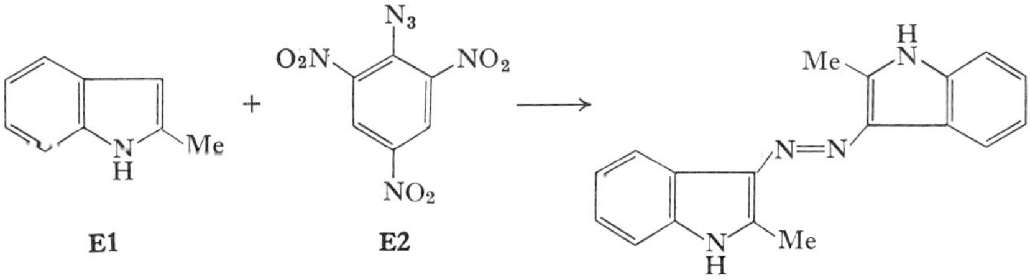

E1 + E2 ⟶ E3

an analogous transformation, but some 2-methyl-3-(*p*-toluenesulfonamido)-indole is also formed. It has been proposed that a common intermediate,

E4 E5 E8

E6 E9

E7

the dipolar ion **E5**, can account for both types of product, but other mechanisms can be envisaged.

F. SULFONATION

The sulfonation of indole with pyridine-sulfur dioxide at 120°C is reported to give indole-2-sulfonic acid, which is isolated as the barium salt in 95% yield (252). At lower temperatures prior 1-sulfonation apparently occurs. 3-Methylindole is reported to react similarly but 2-methylindole fails to give characterizable products (252). Successful sulfonations of 2-phenylindole and indole-3-acetic acid have also been reported (256). Mechanistic discussion seems unjustified on the basis of the sparse data available.

G. ALKYLATION OF INDOLES AND METALLOINDOLES WITH ALKYL HALIDES AND RELATED ALKYLATING AGENTS

In this section the alkylation of indoles by such agents as alkyl halides, alkyl sulfonates, epoxides, and carbonium ions will be considered. Such alkylating agents normally react at the 1- or 3-position of the indole ring. Prior to detailed consideration of the mechanisms of such alkylations, the acidity of indole and the structures of the metallo derivatives of indole must be considered. Indole is a weak acid. The ionization constants (as weak acids) of a number of indoles have been measured on an acidity scale developed for strongly basic solutions (286a). A value of 16.97 ± 0.10 is quoted for the pK_a of indole. The acidity is then roughly comparable to that of an aliphatic alcohol but considerably greater than aniline (29). Substituents play their

TABLE V □ ALKYLATIONS OF ANIONS OF INDOLES

Entry	Indole substituent-metal ion	Solvent	Alkylating agent	Position of alkylation or (ratio 1:3)	Total yield (%)	Ref.
1	None-Na	NH_3	MeI	1	85–95	(196)
2	None-Na	NH_3	Benzyl chloride	1	45	(192)
3	None-Na	THF	MeBr	(24:1)	—	(42)
4	None-Na	THF	EtI	(3:1)	—	(42)
5	None-Na	THF	Benzyl bromide	(1:1)	—	(42)
6	None-Na	THF	Allyl bromide	(1.2:1)	—	(42)
7	None-Na	DMF	Allyl bromide	(20:1)	—	(42)
8	None-Mg	Ether-Bz	Allyl bromide	3	70	(32)
9	None-Mg	Anisole	Propynyl bromide	3	39	(283)
10	None-Mg	Bz	*n*-Butyl chloride	3	42	(80)
11	2-Me-Na	NH_3	MeI	1	90	(195)
12	2,3-diMe-Na	NH_3	Allyl bromide	(~1:1)	76	(154)
13	2,3-diMe-Na	Toluene	MeI	(~1:1)	41	(155)
14	2,3-diMe-Na	Toluene	Benzyl chloride	3	52	(155)
15	2,3-diMe-Mg	Ether	Allyl bromide	3[a]	47	(154)
16	2,3-diMe-Mg	Ether	Chloroacetonitrile	3	28	(156)
17	3-Me-Mg	Bz	EtI	3	38	(108)
18	3-Me-Mg	Ether-Bz	Allyl bromide	[b]	76	(107)
19	3-Me-Mg	Bz	*i*-Propyl (bromide) ?	3	12	(156)
20	3-Et-Mg	Bz	EtI	3	33	(108)
21	3-(2-Aminoethyl)-Na	NH_3	MeI	1	60	(195)
22	diNa of tryptophan	NH_3	MeI	1	96	(287)
23	diNa of tryptophan	NH_3	Allyl bromide	1	70	(287)

[a] Extent of 1-alkylation not determined.
[b] 3:2:1 ratio = 11:7:1.

normal role, with electron withdrawing substituents increasing the acidity of the N–H group. The N–H group is, of course, acidic toward organometallic reagents. The complete characterization of the structure of the metallic derivatives of indole and of their reactions with alkylating agents would require information on factors such as the ionic versus covalent character of the nitrogen-metal bond, the extent of ion aggregation, and, in the case of covalent bonding, the position of bonding. Although the magnesium derivative of indole was once (119) formulated as the covalent C-Mg derivative, this idea appears to be firmly ruled out by nmr studies which show no evidence of metal bonding at C-3 (206). One is left with the problem of determining whether, under given conditions, the indolyl anion exists free or paired with a metallic ion. The tightness of the coordination of the metal ion at nitrogen will strongly affect the reactivity of the anion. In Table V typical results of the alkylation of various metallo derivatives of indole are summarized. Useful generalizations can be drawn from these data.

The metallo derivatives of indole considered in Table V have generally been prepared in one of three ways. When the sodium derivative is to be used in liquid ammonia, sodium amide is generated in excess ammonia. The indole is added, forming the salt, and the alkylation is carried out using the resulting solution (196). For alkylation in other solvents, the solvent can be added and ammonia removed by evaporation (155), or sodium hydride can be used to form the sodium salt with liberation of hydrogen (42). The magnesium salts are prepared by reaction of the indole with methyl or ethyl magnesium halides. The hydrocarbon is evolved and the magnesium salt of indole remains. Of prime synthetic and mechanistic importance is the question of the extent of 3 versus 1 alkylation. Though alkylation is sometimes observed at C-2 when the 3-position is blocked (see entry 18, Table V and a more complete discussion in Section N), the competition between positions 1 and 3 is the common one. One of the most complete studies of the factors affecting 1 versus C-3 alkylation is that of Cardillo *et al.* (42). This study reports rather precise ratios of 3 versus 1 alkylation as a function of solvent and alkylating agent for the sodium salt of indole. Product ratios were determined by gas chromatography. Most of the other yields recorded in Table V are isolated yields and subject to the losses involved in isolation and purification. Primary bromides were found to give predominantly 1-alkyl derivatives although detectable amounts (2–13%) of 3-alkyl and 1,3-dialkyl products were also found. Secondary and tertiary bromides give similar results, although the usual tendency of branched halides to give elimination products reduced the efficiency of the alkylation reaction. Significantly, the proportion of 3-alkylation increased with allyl bromide and with benzyl bromide. The shift toward 3-alkylation with the reactive allyl and benzyl halides is a general phenomenon. Compare, for example, entries 13 and 14, Table V. A second very significant feature of the data of Cardillo *et al.* (42) is seen by comparison of entries 6 and 7. Use of dimethylformamide instead of tetrahydrofuran as solvent under otherwise comparable conditions greatly favors 1-alkylation. A final generalization which can be drawn from Table V, and that is well documented by other data, is the shift toward 3-alkylation when a magnesium, rather than a sodium derivative, is alkylated. Compare, for example, entries 6 and 8, and 12 and 15.

The extent of ion aggregation in solutions of indolyl anions will depend upon the identity of the metal ion. In general, a protic solvent such as ammonia is capable of solvating both ions and one expects extensive dissociation to relatively free ions. Solvents such as dimethylformamide, which can effectively solvate cations, also favor dissociation of ion aggregates. In nonpolar solvents such as ethers and hydrocarbons one expects to find a maximum amount of ion aggregation. The degree of association of the ions will also depend upon the strength with which a metal ion acts as a Lewis acid in coordinating with the indolyl anion. Small, highly charged magnesium is expected to be more tightly associated with the indolyl anion than a sodium ion would be. In nonpolar solvents very tight coordination is to be expected. The species being alkylated may be one of a continuum ranging from a solvated indolyl anion free of its counter ion to a species in which the attraction between the cation and anion may be so tight that it can usefully be considered to be covalently bonded. It is clear that these different species may be expected to react differently toward a given alkylating agent. The "free" indolyl anion appears to have a strong preference for reaction at nitrogen. (See entries 1, 7, 11, 21, 22, and 23, Table V.) The free anion bears a full negative charge and this charge is normally thought to be localized in an sp^2 orbital on nitrogen. Nitrogen alkylation can occur by S_N2 type reactions with the electron-rich nitrogen acting as nucleophile. If the pair of electrons generated by anion formation is extensively involved with coordination to a metal atom, its availability to act as the nucleophile in an S_N2 reaction is greatly reduced. The pi system may become the preferred site of attack and C-alkylation can occur. The effect of metal cations in coordinating with the halogen atom and assisting its ionization may also be important. Such coordination is expected to increase the reactivity of the halide and shift the reaction in the direction of becoming a more S_N1-like process. The effects largely parallel (42) results obtained with other ambident anions such as phenolate (56) and pyrryl anions (101). Because of the tendency to produce 1-alkylation in high yield, alkylation of sodium derivatives of indole in liquid ammonia has become a widely used synthetic procedure. Alkylation of the magnesium derivative is the preferred method if 3-alkylation is desired. Additional examples of preparatively significant alkylation of metal derivatives of substituted indoles are included in Table VI.

TABLE VI □ ADDITIONAL EXAMPLES OF ALKYLATIONS OF INDOLE ANIONS

Indole substituent-metal ion	Solvent	Alkylating agent	Position of alkylation	Yield (%)	Ref.
3-(2-Dimethylamino-ethyl)-5-benzyloxy-Na	NH_3	MeI	1	92	(14)
5-Benzyloxy-Mg	Eth	Chloroacetonitrile	3	—	(240)
None-Mg	Eth	2-Chloromethyl-pyridine	3	66	(62)
None-Mg	Eth	4-Chloromethyl-pyridine	3	29	(62)
2-Me-Mg	Eth-Bz	β-Dimethylamino-ethyl chloride	3	25	(81)
None-Na	NH_3	1,4-Dibromobutane	1[a]	62	(117)
None-Na	NH_3	1,6-Dibromohexane	1[b]	60	(117)

[a] Product is 1-[4-(1-indolyl)butyl]indole.
[b] Product is 1-[6-(1-indolyl)hexyl]indole.

A synthetically useful alternative to alkylation in liquid ammonia involves heating indoles with alkylating agents in organic solvents (91, 141, 142, 192, 250a). Either 1- or 3-alkylation can occur. Nitrogen alkylation, however, appears to predominate except when benzyl (and probably allyl) compounds are employed. Table VII records examples of such alkylations. Excess alkylating agent can accomplish polymethylation (141).

1,2-Dimethylindole when heated with allyl bromide in dilute phosphoric acid gives two alkylation products, **G3** and **G4** (76). The mechanism of this interesting reaction has not been defined. In particular, it is not clear if the acid is an essential component of the system. It is stated that use of more dilute acid favors the formation of **G4**. There are several paths that can be visualized for the formation of the products, including paths involving Cope

G1 + $CH_2{=}CHCH_2Br$ (**G2**) ⟶

G3 (1-Me, 3,3-bis($CH_2CH{=}CH_2$), 2-${=}CH_2$ indoline) + **G4** (1-Me, 3-$CH_2CH{=}CH_2$, 2-$CH_2CH_2CH{=}CH_2$ indole)

TABLE VII □ ALKYLATIONS OF INDOLES IN ORGANIC SOLVENTS

Indole substituents	Solvent (base)	Alkylating agent	Position of alkylation	Yield (%)	Ref.
None	Toluene (Na_2CO_3)	Benzyl *p*-toluene sulfonate	3	23	(192)
2-Me	Xylene (Na_2CO_3)	Methyl *p*-toluene sulfonate (excess)	1, 3, 3	70–76	(141)
2-Ph	Chlorobenzene (NaOH)	Dimethyl sulfate	1	97	(142)
2,3-diMe	Chlorobenzene (NaOH)	Dimethyl sulfate	1	100	(142)
5-Benzyloxy	Xylene (K_2CO_3)	Methyl *p*-toluene sulfonate	1	40	(250a)
5-Benzyloxy-3-formyl	Carbitol (K_2CO_3)	Methyl iodide	1	76	(91)
3-(2-Acetamido-2,2-dicarbethoxy)ethyl	Xylene (K_2CO_3)	Methyl *p*-toluene sulfonate	1	70	(137)

type rearrangements. It has been demonstrated (26a) that rearrangements of 3-allyl-2-methyl-3*H*-indoles to 2-(3-butenylindoles) can occur. The conjugate acid of **G3** may be stable to such a rearrangement. This might account for the observed formation of a higher proportion of **G4** in less acidic solutions.

The solvolyses of 2-(3-indolyl)ethyl (tryptophyl) tosylates and bromides have recently attracted attention. Acetolysis of α,α-dideuterotryptophyl tosylate yielded a 1:1 mixture of α,α-dideutero and-β,β-dideuterotryptophyl acetate (47). This result indicates that the indole ring participates in the heterolysis of the C–O bond to give a symmetrical intermediate **G6**. Treatment of the tosylate with base (potassium *t*-butoxide in tetrahydrofuran) gives a compound characterized as **G7**.

CD_2CH_2OTos **G5** → **G6** $\xrightarrow{-OC(Me)_3}$ **G7**

G6 $\xrightarrow{AcOH}$ CD_2CH_2OAc **G8** + CH_2CD_2OAc **G9**

The exceptionally high rate of solvolysis of tryptophyl derivatives is also in accord with participation of the indole nucleus in the ionization reaction, and it has been stated (47) that the reactivity of the indole ring toward such participation probably exceeds that of any other uncharged aromatic ring system investigated to date. Rapid solvolysis of various tryptophyl tosylates has also been observed in Julia's laboratory (118).

Julia and co-workers have observed a number of rearrangements which also indicate that the indole ring acts as a neighboring group in the ionization of leaving groups at the 3-carbon atom. Intramolecular alkylation by the 2-(2-tosyloxy)ethyl substituent has also been proposed to account for the observation of a novel ring expansion to a benzazepine derivative on reaction of the tosylate of 2-(2-indolyl)ethanol with ethyl cyanoacetate in dimethyl sulfoxide (220).

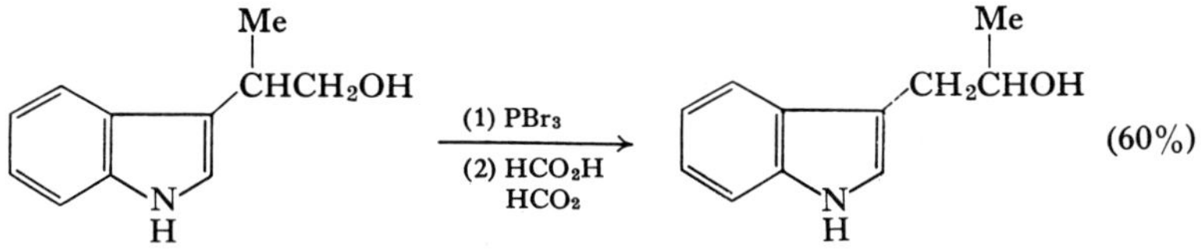

Me
CHCH₂OH
N
H
(1) PBr₃
(2) HCO₂H
HCO₂
Me
CH₂CHOH
N
H
(60%)

G10 G11

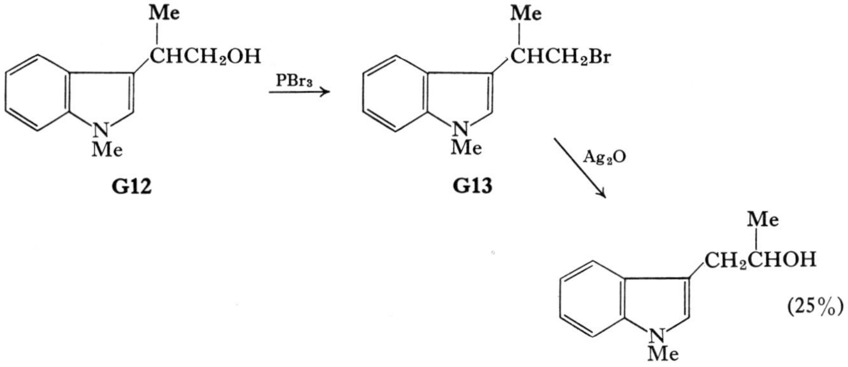

Me
CHCH₂OH
N
Me
PBr₃
Me
CHCH₂Br
N
Me
Ag₂O
Me
CH₂CHOH
N
Me
(25%)

G12 G13 G14

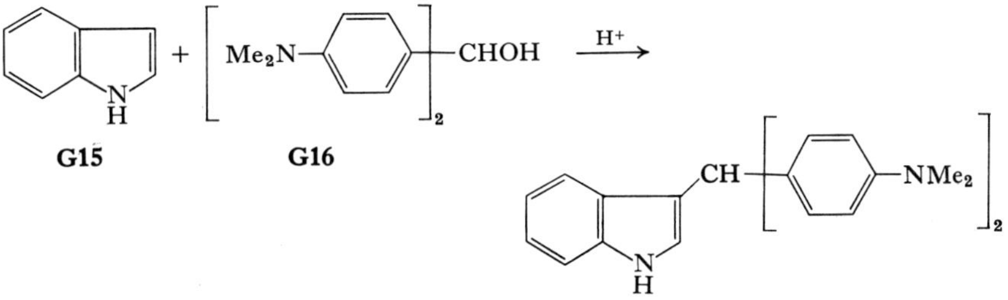

N
H
+
Me₂N
CHOH
2
H⁺
CH
NMe₂
2
N
H

G15 G16 G17

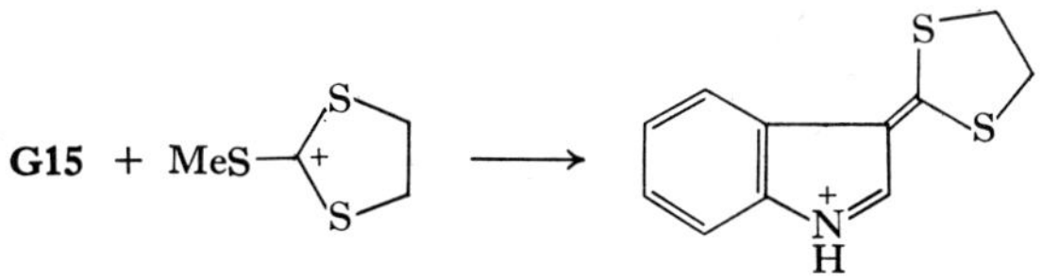

G15 + MeS
S
+
S
S
S
N
H

G18 G19

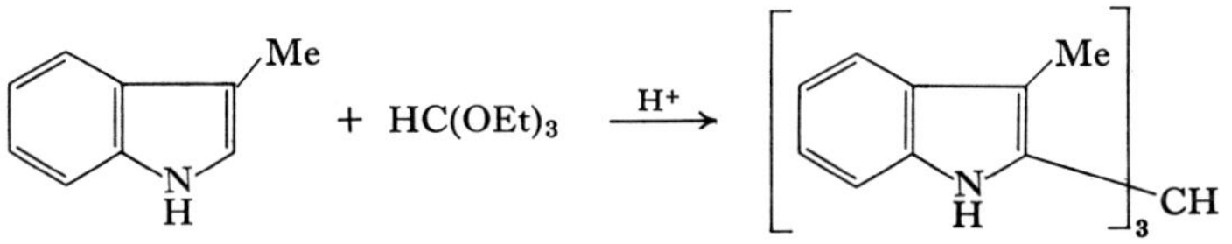

Me
N
H
+ HC(OEt)₃
H⁺
Me
N
H
3
CH

G20 G21 G22

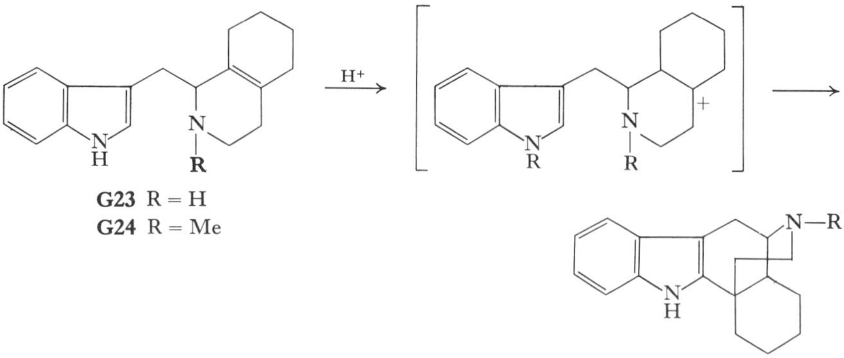

G23 R = H
G24 R = Me

G25 R = H
G26 R = Me

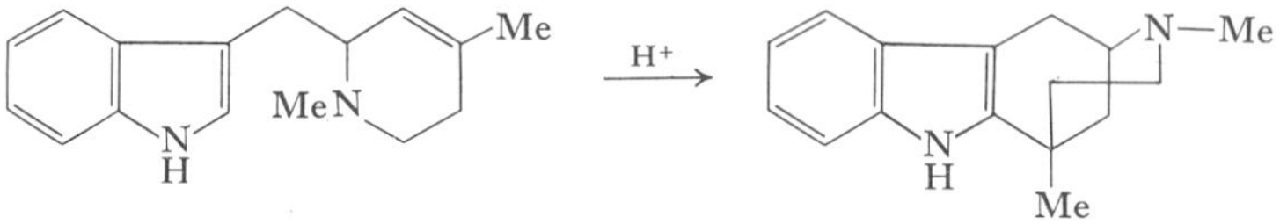

G27 **G28**

Indoles can be alkylated in the absence of basic catalyst under conditions in which carbonium ions are generated from the alkylating agent. Heating indole with bis(*p*-dimethylamino)phenylcarbinol in the presence of an acid catalyst is reported to give a high yield of the 3-alkylation product **G17** (93). Heating 1- or 2-methylindole with triphenylmethyl chloride in refluxing toluene gives high yields of 3-alkylation products (268). Indole and the methylthio-1,3-dithiolium salt **G18** give **G19**, the product of 3-alkylation followed by elimination of methylmercaptan (83). The reaction of

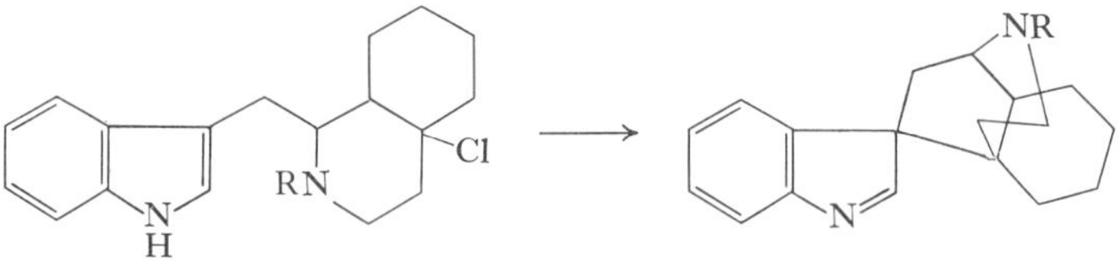

G29 **G30**

3-methylindole with triethyl orthoformate in the presence of acid catalysts (275) is an example of 2-alkylation of a 3-substituted indole and presumably involves alkylation by the diethoxymethyl cation as the first step. Subsequent ionization of C–O bonds in the successive intermediates leads to alkylation of three molecules of 3-methylindole. All of the above reactions would appear to involve generation of a carbonium ion from the alkylating agent followed by Friedel–Crafts alkylation of the indole ring.

Intramolecular alkylation of indoles by olefins has been observed in acidic solution. Morrison and co-workers have reported that the transformations **G23** → **G25** and **G24** → **G26** proceed in 30 and 29% yield, respectively (152). A similar process has been observed on heating of the tetrahydropyridine **G27** with 85% phosphoric acid (151).

The tertiary halide **G29** is converted by methanolic potassium hydroxide to the indolenine **G30** (152), but, since a tertiary halide is involved, one expects that this alkylation proceeds by ionization of the C–Cl bond followed by intramolecular alkylation at C-3. The mechanistic reason for apparent substitution at C-2 in **G24** as opposed to alkylation at C-3 in **G29** is clarified in Section N.

Decomposition of diazoketones and diazoesters in the presence of indole yields 3-alkylation products. These reactions probably involve protonation and decomposition of the diazo compound to a carbonium ion. These reactions are considered in more detail in Chapter IV, A, 2.

There are scattered reports of arylations of indoles, including those shown in the equations below:

G31 + **G32** (Cl, O_2N, NO_2, NO_2) —reflux, xylene→ **G33** (O_2N, NO_2, NO_2) Ref. (268)

G31 + **G34** (Br) —nitrobenzene, CuBr→ **G35** Ref. (201)

G36 (MgBr) + **G37** (Br, Fe) —$CuBr_2$→ **G38** (Fe) + 1-isomer (**G39**) Ref. (158)

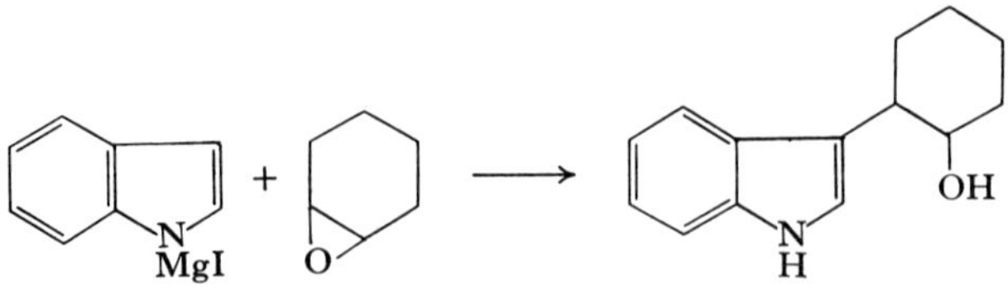

G40 + G41 → G42

Indolylmagnesium iodide reacts with cyclohexene oxide and is alkylated at the 3-position (90). Alkylations of the magnesium derivatives of indole

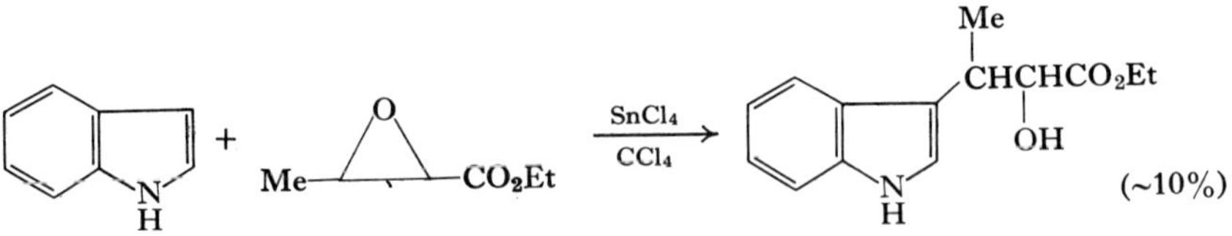

G43 + G44 $\xrightarrow[CCl_4]{SnCl_4}$ G45 (~10%)

with propylene oxide (217) and of 2-phenylindole with ethylene oxide (37) have also been reported. In the latter case use of 2 moles of ethylene oxide gave a 48% yield of 1,3-bis(2-hydroxyethyl)-2-phenylindole. The preparation of tryptophol has been effected by reaction of indole with ethylene oxide in acetic acid and acetic anhydride at 70°C. The crude product mixture, which contains tryptophyl acetate, gives tryptophol (45% yield) after

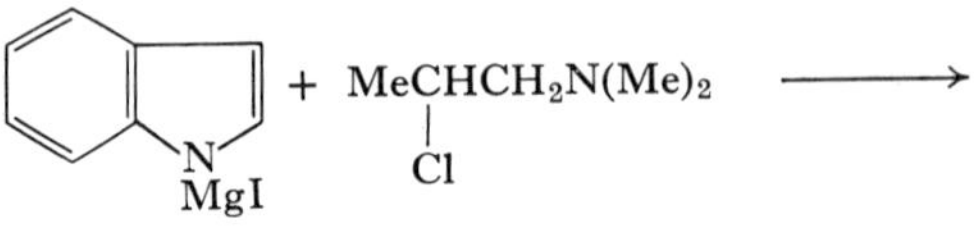

G46 + G47 ($MeCHClCH_2N(Me)_2$) →

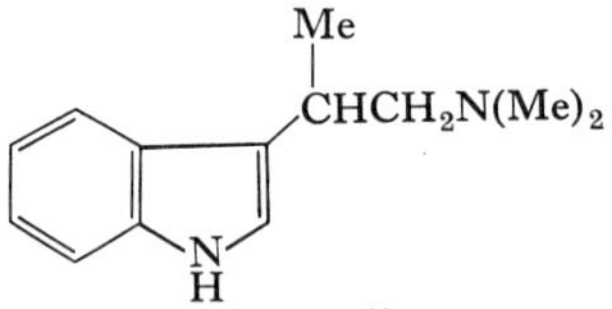

G48 (16%) 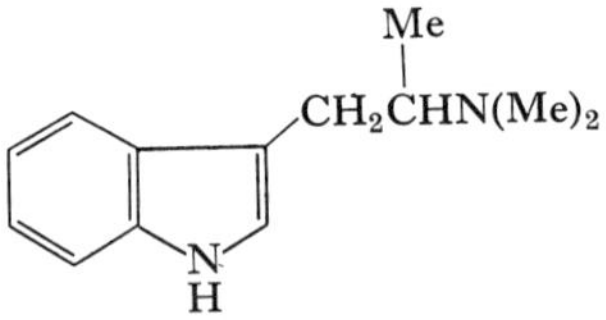 + G49 (5%) 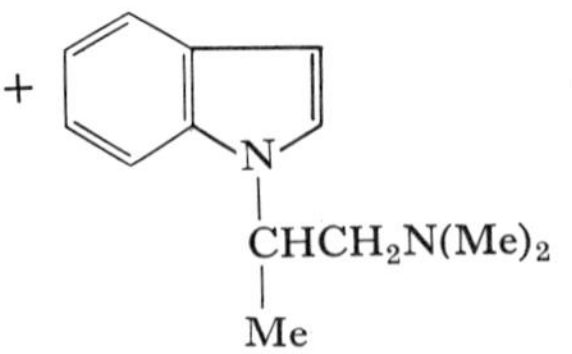 + G50 (<1%) + G51 (<1%)

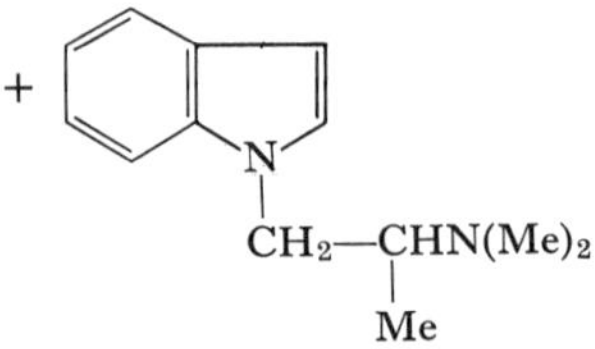

saponification and chromatography (116). The alkylation of indole with ethylene oxide can also be effected using stannic chloride as catalyst (116).

The reaction of indole with the epoxy ester **G44** was investigated during synthetic work involved in structure determination of the antibiotic indolmycin (224). The course of the alkylation reaction under various conditions was investigated and use of stannous chloride at low temperature was found to give the most satisfactory results.

Indoles and indolylmagnesium halides can be alkylated by aziridines in the presence of appropriate catalysts. This reaction is discussed more fully in Chapter IV, B, 6. Aziridinium ions may be involved in alkylation of indole derivatives with β-dialkylaminoethyl halides. Indolylmagnesium iodide gives four products on reaction with 2-chloro-1-dimethylaminopropane (**G47**) (80). Alkylation of the sodium salt of indole with **G53** gives primarily the rearranged product **G54** along with **G55**. These results suggest the intermediacy of an aziridinium ion **G56** in the alkylation reaction.

G52 + $MeCH(Cl)CH_2N(Me)_2$ (G53) $\xrightarrow[\text{benzene}]{\text{NaH}}$ G54 (N-$CH_2CH(Me)N(Me)_2$) + G55 (N-$CH(Me)CH_2N(Me)_2$)

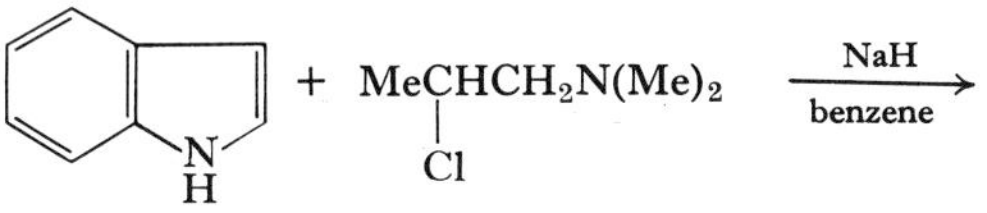

G52

G53

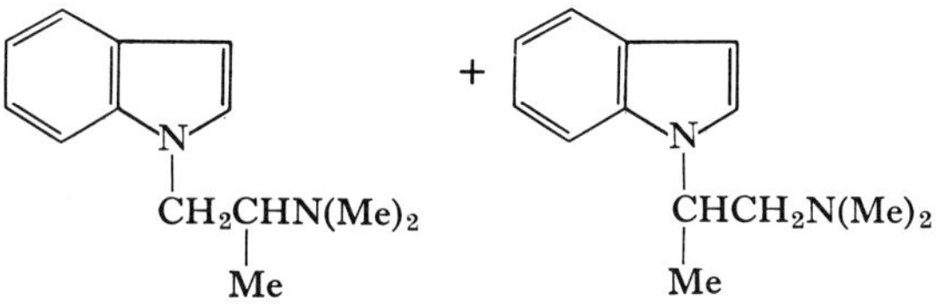

G54

G55

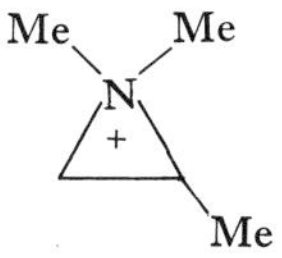

G56

Alkylations with propiolactone (**G58**) and butyrolactone (**G61**) have been reported, but under widely different conditions. Heating several substituted indoles with propiolactone gives 3-alkylation (89), but the potassium salt of indole is reported to react with butyrolactone to give indole-1-butyric acid (**G62**) (208). The magnesium derivative of 6-chloroindole reacts with γ-butyrolactone to give the product of alkylation at C-3 (232). This reaction as well as those cited earlier must be proceeding via alkyl oxygen fission of the lactone.

G57 + G58 $\xrightarrow{110°}$ G59 ($CH_2CH_2CO_2H$, Me)

G60 + G61 $\xrightarrow{200°}$ G62 ($CH_2CH_2CH_2CO_2H$)

The high temperature 3-alkylation of indole by lactones in the presence of potassium hydroxide (77) probably proceeds by an entirely different mechanism (see the discussion which follows).

Indoles are alkylated in the 3-position when heated with primary and secondary alcohols in the presence of base (55, 183, 202). A recent patent describes the extension of this reaction to certain functionalized alcohols (114). Alkylation is believed to be initiated by oxidation of the alcohol to the aldehyde or ketone which then condenses with the indole at C-3. The 3-alkylidene-3*H*-indole **G66** formed in the condensation reaction is then reduced by another molecule of the alcohol. The condensation of indoles

$$\underset{\textbf{G63}}{R_2CHOH} \rightleftharpoons \underset{\textbf{G64}}{R_2CO} + [H_2]$$

G65 + **G64** ⟶ **G66** + H_2O

G66 + **G63** ⟶ **G67** (CHR_2) + **G64**

with ketones and aldehydes to give alkylidene-3*H*-indoles is a very general reaction and is discussed in detail in Section J. Indole reacts with potassium salts of α-hydroxy acids to give 3-indolealkanoic acids (115). The mechanism of this reaction very likely involves condensation with small amounts of the aldehyde or ketone in equilibrium with the hydroxy acid under the conditions of the reaction. The procedure of Fritz for 3-alkylation of indoles with lactones (77) very likely proceeds by a similar mechanism.

A novel alkylation of indole with ethylene and trianilinoaluminum is reported to give 7-substituted products (244). Stroh and Hahn have effected the alkylation of 2-methyl-, 2,3-dimethyl-, 2,3,5-trimethyl-, and 2-phenylindole at C-7. The position of substitution has been established by comparison with authentic samples in several cases. It has been suggested that coordination with the aluminum atom may be involved in determining the position of alkylation (243). Aniline derivatives undergo alkylation in the *ortho* position under similar conditions.

$$\textbf{G68} + CH_2{=}CH_2 \xrightarrow{Al(NHPh)_3} \textbf{G69}$$

G68 G69

H. REACTIONS WITH CARBENES

It has been known for some time that application of the Reimer–Tiemann aldehyde synthesis to indoles gives not only indole-3-carboxaldehydes but also 3-chloroquinolines (205). The results of Marchant and Harvey are illustrative (146). Recognition that dichlorocarbene is an intermediate in

$$\textbf{H1} \xrightarrow[KOH]{CHCl_3} \textbf{H2} + \textbf{H3}$$

H1 H2 H3

$$\textbf{H4} \xrightarrow[KOH]{CHCl_3} \textbf{H5} + \textbf{H6}$$

H4 H5 H6

this reaction has resulted in renewed interest in the mechanism of the reaction and examination of the behavior of indoles toward other carbene sources. Indole, methyllithium, and methylene chloride give quinoline (46).

Tetrahydrocarbazole gives **H8** on reaction with chloroform and potassium butoxide (11). 2,3-Dimethylindole gives an analogous product **H11** along with 2,4-dimethyl-3-chloroquinoline (**H10**) (205). Rees and Smithen (205)

$CHCl_2$

$CHCl_3$ / t-BuO$^-$

N H

H8

H7

as well as Robinson (212) have demonstrated that **H11** is not converted to **H10** with base. It has been proposed that the quinolines arise from unstable

Me Me N H Me Cl Me N Me $CHCl_2$ N Me + EtO$^-$

H9 **H10** **H11**

H9 $:CCl_2$ Me Cl Cl N H Me → **H10**

H12

tricyclic intermediates such as **H12** (205, 212). Rees and Smithen have also examined the reaction of various indoles with dibromocarbene, difluorocarbene, and monochlorocarbene (204, 205). The behavior of dibromocarbene toward 2-methylindole paralleled that found for dichlorocarbene, but difluorocarbene gave only **H13**, and monochlorocarbene gave only a small amount of 2,4-dimethylquinoline. It has been suggested (205) that the

Me N H Me + $:CF_2$ → Me CHF_2 N Me

H9 **H13**

neutral indoles react with carbenes to give primarily unstable adducts such as **H12** whereas dichloromethyl-3*H*-indoles (**H11**) arise from reaction of carbenes with the corresponding anions. The evidence for this proposal is not convincing, especially since 1,2,3-trimethylindole gives products of both types (204).

Thus far, no tricyclic adduct such as **H12** has been fully characterized, although Dobbs (65) has described isolation of an oil by acetylation of the crude product from the reaction of 3-methylindole and chlorocarbene to which he attributed the structure **H14**. Later work ruled out structure **H14** for this product but failed to provide evidence for a definitive structural assignment (65). The existence of an unidentified quinoline precursor does

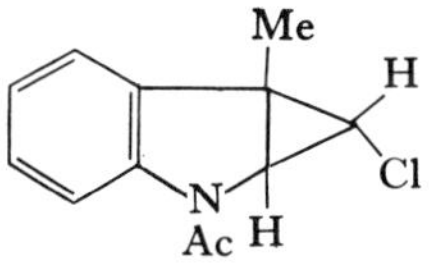

Me H Cl N Ac H

H14

seem to be well established, however. Use of radioactive methylene bromide in a reaction with indole and methylithium gives quinoline containing radioactivity at C-3, as would be expected from the suggested mechanism for the ring expansion (205, 212).

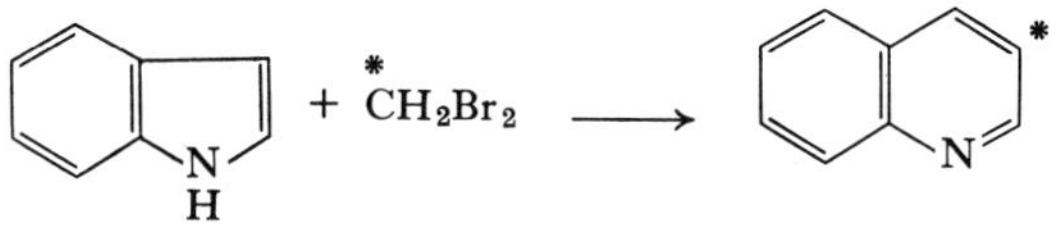

N H + $\overset{*}{C}H_2Br_2$ ⟶ N *

H15 **H16**

Parham and co-workers (187a) have used the reaction of phenyltrichloromercury with indoles having large rings fused to the 2,3 side to prepare quinolines with bridging between C-2 and C-4.

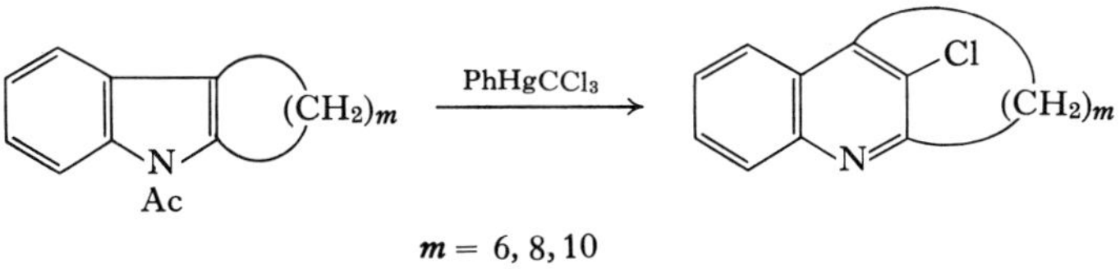

N Ac $(CH_2)_m$ $\xrightarrow{PhHgCCl_3}$ N Cl $(CH_2)_m$

$m = 6, 8, 10$

H17 **H18**

I. ACYLATION OF INDOLES AND METALLOINDOLES

The acylation of indoles and of magnesium derivatives has been a rather widely used synthetic procedure for some time. The 3-position is the normal site of attack although, under certain conditions, acylation may also take place on the nitrogen atom. There are examples of acylation at C-2 when the 3-position is already substituted. Acylations of indole are mechanistically viewed as Friedel–Crafts acylations of a reactive aromatic ring. The acylation of indolylmagnesium reagents must be considered to be the acylation of a species in which there is rather tight coordination between the nitrogen and magnesium atoms. The effect of tight coordination is similar to that described in the earlier discussion of alkylations. Both indole and 2-methylindole are acylated by heating with acetic anhydride. When indole and acetic anhydride are refluxed for 24 hours, 1,3-diacetylindole is formed (223). Alkaline hydrolysis readily removes the *N*-acyl group. A 40% yield of the 3-acetyl derivative has been claimed from the reaction of 2-methylindole with acetic anhydride (24). 3-Acetyl-2-phenylindole has been prepared from 2-phenylindole and acetic anhydride (24). Methyl 1-methylindole-3-acetate gives an excellent yield of the 2-acetyl derivative on reaction with acetyl chloride and zinc chloride (123).

While widely used synthetically, the acylation of magnesium derivatives of indole has not been the subject of detailed mechanistic examination. Kašpárek and Heacock (122) have recently examined carefully the reaction of indolylmagnesium iodide and ethyl chloroformate. Gas chromatography has shown that the products of the reaction are ethyl indole-1-carboxylate (**I2**) and ethyl indole-3-carboxylate (**I3**) and the 1,3-diacylation product **I4**.

$$\text{I1 (N-MgI)} \xrightarrow{ClCO_2Et} \text{I2 (N-}CO_2Et) + \text{I3 (3-}CO_2Et\text{, NH)} + \text{I4 (3-}CO_2Et\text{, N-}CO_2Et)$$

The effect of temperature and molar ratio of the reactants on the relative yields of the three products were reported. A maximum yield of 46% for **I3** was found in contrast to earlier claims of up to 78% (144). 1,3-Bis(chloroacetyl)indole is the only product reported from the reaction of indolylmagnesium bromide with chloroacetyl chloride (4). The 1-chloroacetyl group is easily removed by ethanolic dimethylamine. Successful acylation of indolylmagnesium iodide with 3-carbomethoxypropionyl chloride has been accomplished (10). Yields of 60–70% have been reported for a series of acylations of indolylmagnesium bromide by substituted aroyl halides (40). While this reaction seems generally adaptable to the synthesis of 3-acylindoles, it is likely that careful examination of the products might often reveal the concurrent formation of the easily hydrolyzed 1-acyl derivative. Recent results of Young and Mizianty confirm this expectation (288).

The magnesium salts of 3-substituted indoles can give 1- or 2-acyl derivatives. Oddo (181) reported the formation of 1-acetyl-3-methylindole from the magnesium derivative of 3-methylindole at low temperature, but 2-acetyl-3-methylindole was formed in refluxing ether. Leete and Chen (45) found that the reaction of the magnesium derivative of 3-phenylindole with benzoyl chloride gave 92% 1-benzoyl-3-phenylindole and 2% 2-benzoyl-3-phenylindole. A representative group of synthetically successful acylations can be found in Table LI, Chapter IX.

Acylations of sodium derivatives of indole have been carried out with acid chlorides or *p*-nitrophenyl esters in dimethylformamide (233). Under these conditions the *N*-acyl compound is formed, at least from 2,3-disubstituted indoles. Modest success has been reported for the *N*-acylation of sodio-

MeO, $CH_2CH(Me)CO_2Et$, Me, N–H (**I5**) $\xrightarrow[\text{benzoyl chloride}]{\text{NaH, }Me_2NCHO}$ MeO, $CH_2CH(Me)CO_2Et$, Me, N–COPh (**I6**)

indole under similar conditions using *N*-acyl acid chlorides of amino acids (157).

Although acyl chlorides have been the most widely used acylating agents, other derivatives of carboxylic acids have been employed. The formation of 1-formylindole from indolylmagnesium bromide and ethyl formate has been reported (3). Yields in the range of 50–65% have been reported for the preparation of a number of 1-(β-aminoalkanoyl)indoles (**I9**) from the reaction of indolylmagnesium iodide and ethyl esters of β-aminocarboxylic acids (20). Hishida (100) has reported examples of both 1- and 3-acylation

N–MgI (**I7**) + $EtO_2CCH_2CH(R)NH_2$ (**I8**) ⟶ N–$COCH_2CH(R)NH_2$ (**I9**)

with esters. Generalization of these examples suggests that use of esters as acylating agents may favor 1-acylation. The enol lactones **I11** and **I12** also react to give the 1-acyl derivatives **I13** and **I14** as the principal products (124). No evidence of formation of 3-acyl derivatives is reported. A recent

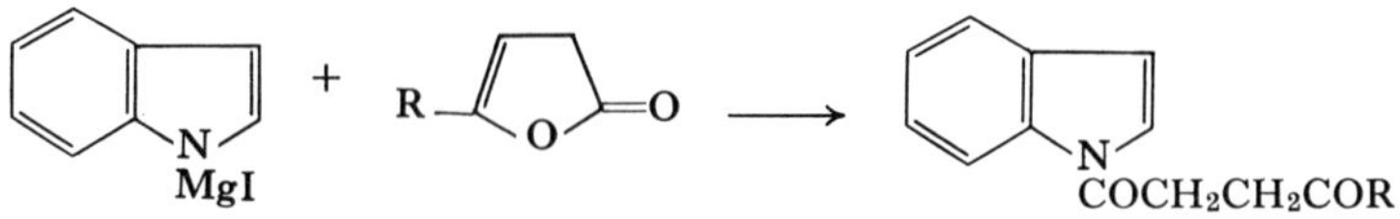

I10 + R = Me I11, R = Ph I12 → R = Me I13 (37%), R = Ph I14 (41%)

patent application (159) reports the acylation of indolylmagnesium iodide in the 3-position by *N*-methylsuccinimide. Diketene reacts with indole in

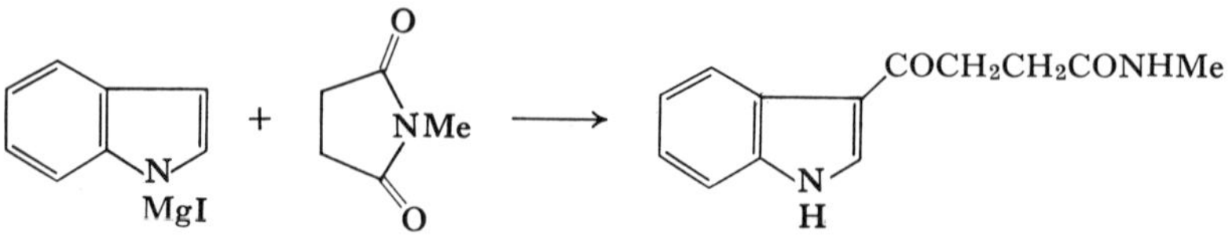

I15 + I16 → I17

benzene containing pyridine to give 1-acetoacetylindole (188).

In the presence of sufficiently reactive acylating agents, 2,3-disubstituted indoles can be acylated in the carbocyclic ring. Heating 2,3-dimethylindole with acetic anhydride and *p*-toluenesulfonic acid gives 1,6-diacetyl-2,3-dimethylindole, but the yield is low (245). Borsche and Groth have carried out the acylation of 1,2-dimethylindole in carbon disulfide with acetyl chloride and aluminum chloride and reported the formation of a ketone which is different from 1,2-dimethyl-3-acetyl indole (24). In the same paper (24), a ketone from 1,2,3-trimethylindole, formed under similar conditions, is assigned the structure 1,2,3-trimethyl-5-acetylindole, but Suvorov and Sorokina's work (245) indicates that this must actually be the 6-acetyl derivative. 1-Acetyl-2,3-dimethylindole is acetylated in good yield in the 6-position under Friedel–Crafts conditions (82). Tetrahydrocarbazoles are also acylated at C-6 of the indole ring (190).

Substitution at C-6 or C-4 and C-6 is the result to be expected on the basis of the previous discussion of orientation effects in indole nitrations (Section C).

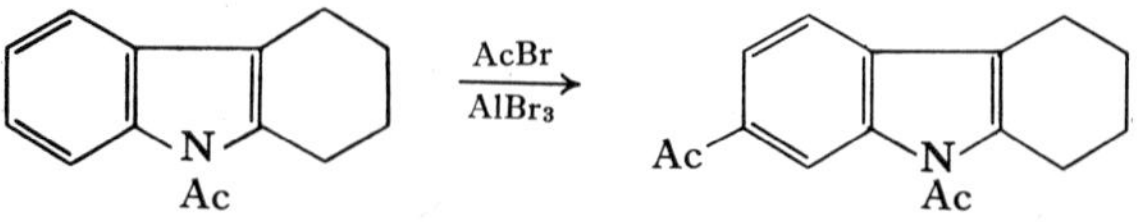

I18 →(AcBr, $AlBr_3$) I19

The 2- and 3-positions of the indole ring are sufficiently reactive that acylations appropriate only for reactive aromatic systems can be successfully applied. The most useful of these are the Vilsmeier–Haack type acylation

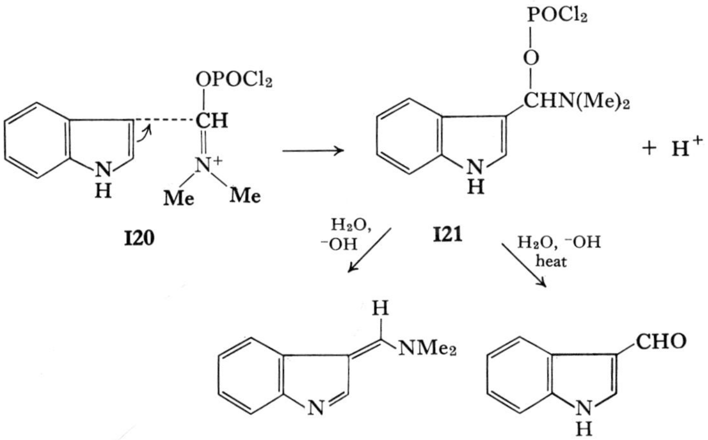

I20 I21 I22 I23

employing amides and phosphoryl trichloride and the Hoesch reaction using nitriles and acid. The former reaction is exemplified by the formylation of indole (234, 271). The mechanism of the reaction has been discussed by Smith. The intermediate **I22** can be isolated under modified conditions (234). Further examples of this extremely useful synthetic procedure are

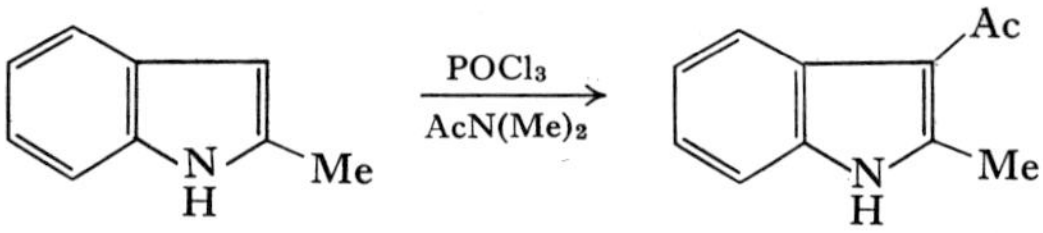

I24 I25

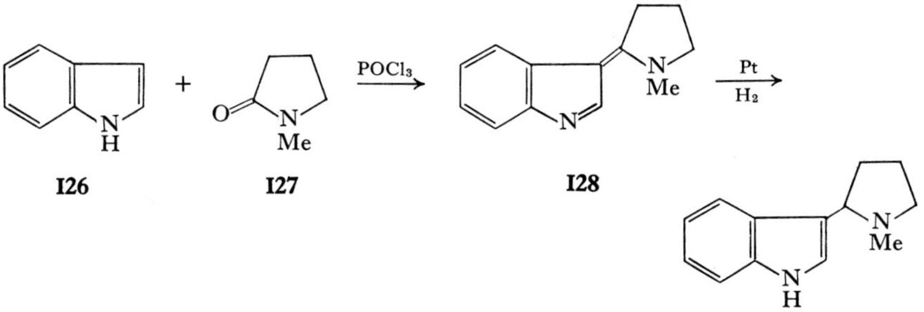

I26 I27 I28 I29

recorded in Table LII, Chapter IX. The reaction proceeds well even when the 2-position is substituted by such electron-withdrawing substituents as carbethoxy (231) or substituted carboxamido (27). While formylation using *N*,*N*-dimethylformamide or *N*-methylformanilide has been the most common application of this reaction, it has been demonstrated by Anthony (5) that acylation can be effected by other amides. The conversion of 2-methylindole to its 3-acetyl derivative and 3-benzoylation of indole using *N,N*-dimethylbenzamide are among the examples reported by Anthony. Other preparations of ketones by this method are recorded in Table LII, Chapter IX. Use of cyclic amides permits the introduction of nitrogen-containing rings at the 3-position (289). As in the case of Friedel–Crafts acylation, if the 1-, 2-, and 3-positions are substituted as in 1-alkyltetrahydrocarbazoles, formylation occurs at C-6 of the indole ring (133). 2,3-Dimethylindole and tetrahydrocarbazole are formylated at nitrogen (133).

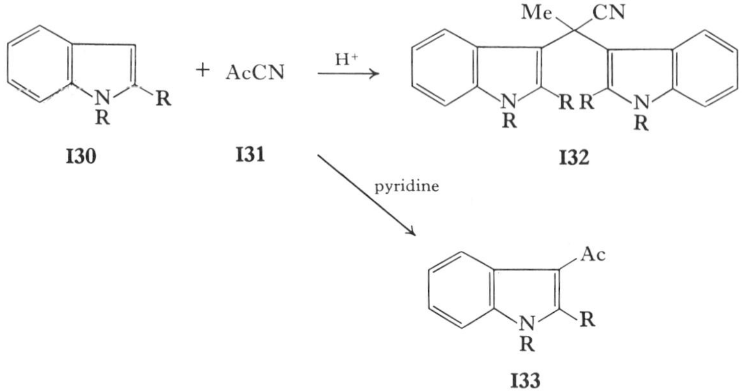

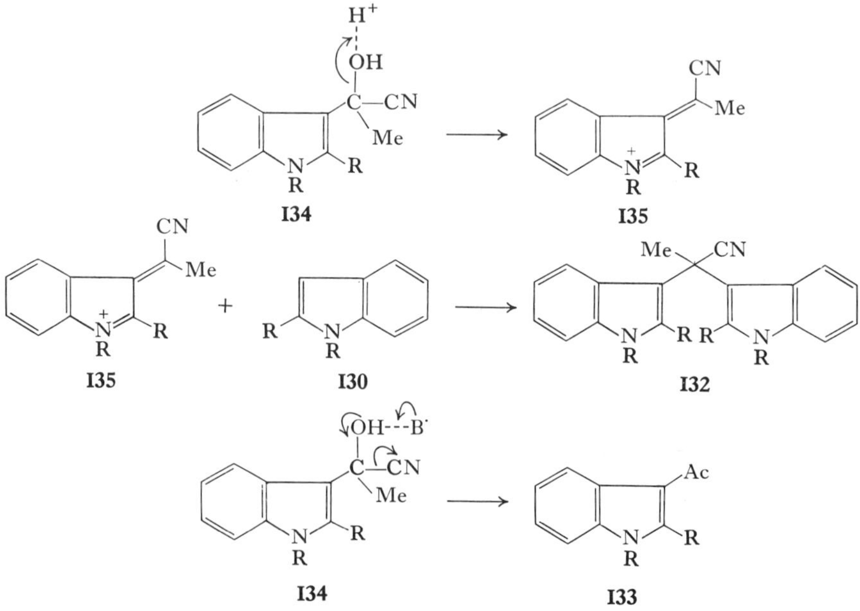

I30 + AcCN (**I31**) —H^+→ **I32**; —pyridine→ **I33**

I34 → **I35**

I35 + **I30** → **I32**

I34 → **I33**

2-Methylindole has been formylated and acylated using hydrogen cyanide or nitriles (10, 230). For example, a 75% yield of 2-methyl-3-benzoylindole has been obtained using benzonitrile and hydrochloric acid. Since indole dimerization and trimerizations (see Section B) occur under similar circumstances, this procedure may be less satisfactory for indoles which are sensitive to such reactions.

The reactions of several indoles with acetyl and benzoyl cyanide have been studied (126, 127). The presence of small amounts of hydrogen chloride in the reaction mixture results in the formation of the substituted acetonitriles **I32**. The presence of pyridine, however, changes the course of the reaction and the 3-acetyl derivatives **I33** are formed. The formation of products such

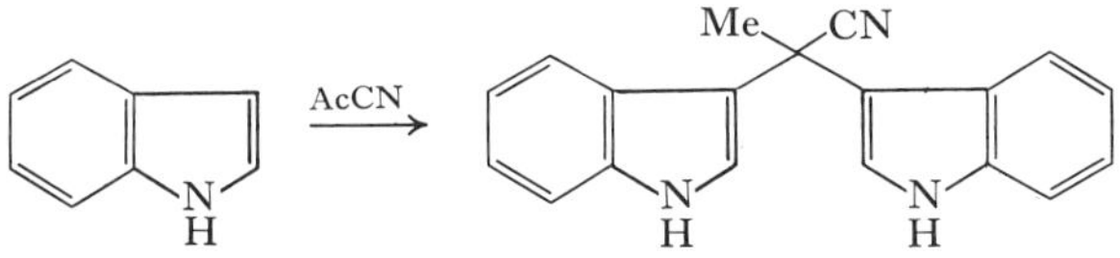

AcCN

I36 **I37**

I32 is characteristic of acid-catalyzed reactions of ketones and indoles and presumably proceeds through the intermediate **I35**. With a base present, **I34** suffers an alternative breakdown with expulsion of cyanide ion. Indole gives mainly **I37** on reaction with acetyl cyanide. Very little 3-acetylindole is formed, even in the presence of basic catalysts (194).

J. REACTIONS WITH KETONES AND ALDEHYDES

Reactions between indoles and aldehydes or ketones are ordinarily initiated by acid-catalyzed attack of the carbonyl compound at the 3-position of the indole ring. Most such reactions, however, then proceed beyond this initial stage. Perhaps the most commonly observed reaction is the reaction of

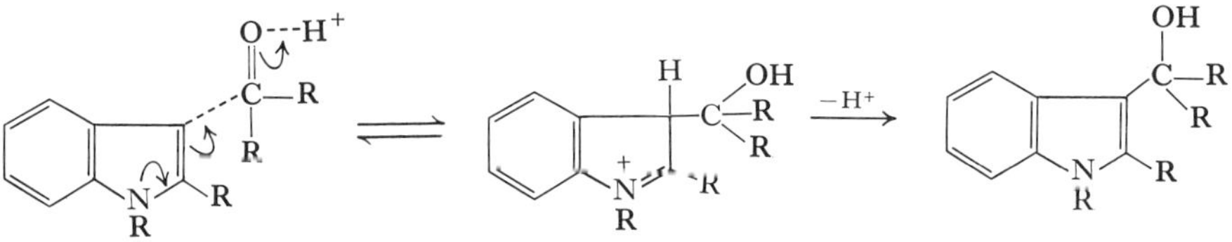

$-H^+$

J1 **J2** **J3**

2 moles of an indole with an aldehyde or ketone to give a diindolylmethane **J5**. This transformation depends upon the instability of the carbinol **J3** in acidic solution. Indolylcarbinols lose water, generating electrophilic

J3 → J4

J4 + J1 → J5

alkylideneindolenines (or the conjugate acid **J4**) which act as electrophiles toward a second molecule of the indole. Table VIII records some typical preparations of diindolylmethanes.

It has been shown (139, 257), that heat, base, or acid convert 3-indolylmethanol into diindolylmethane. These transformations demonstrate the feasibility of the final steps in the mechanism for diindolylmethane formation. Leete (138) has studied the reactions of a number of substituted indolylcarbinols and found that, with the exception of 2-phenyl-3-hydroxymethyl-

TABLE VIII □ REPRESENTATIVE PREPARATIONS OF DIINDOLYLMETHANES

Indole substituent	Aldehyde or ketone	Reaction position	Yield (%)	Ref.
None	Formaldehyde	3	95	(257)
None	2-Pyridinecarboxaldehyde	3	—	(85)
None	4-Pyridinecarboxaldehyde	3	—	(85)
None	Pyruvic acid	3	100	(289a)
None	Acetophenone	3	45	(178)
1-Me	Acetophenone	3	98	(178)
2-Me	Formaldehyde	3	—	(276)
2-Me	Acetone	3	—	(173)
2-Me	Benzaldehyde	3	91	(172)
2-Me	*m*-Nitrobenzaldehyde	3	—	(72)
2-Me	Acetophenone	3	64	(178)
2-Me	2-Pyridinecarboxaldehyde	3	78	(264)
2-Me	3-Pyridinecarboxaldehyde	3	72	(264)
2-Ph	Formaldehyde	3	58	(57)
2,5-diMe	Formaldehyde	3	78	(57)
3-Me	Benzaldehyde	2	44–91	(67, 172)
5-Me	Acetone	3	84	(178)

indole, these carbinols were converted to diindolylmethanes. Biswas and Jackson (21) regard the conversion of 3-indolylcarbinols and related substances to diindolylmethanes as proceeding by the mechanism outlined below. The distinctive feature of this mechanism is that it does not require formation of unsubstituted indole. The isolation of dimeric products having the carbon skeleton of **J10** on diborane reduction of 3-formylindole is in accord with this mechanism. The ease of departure of the hydroxyl group is

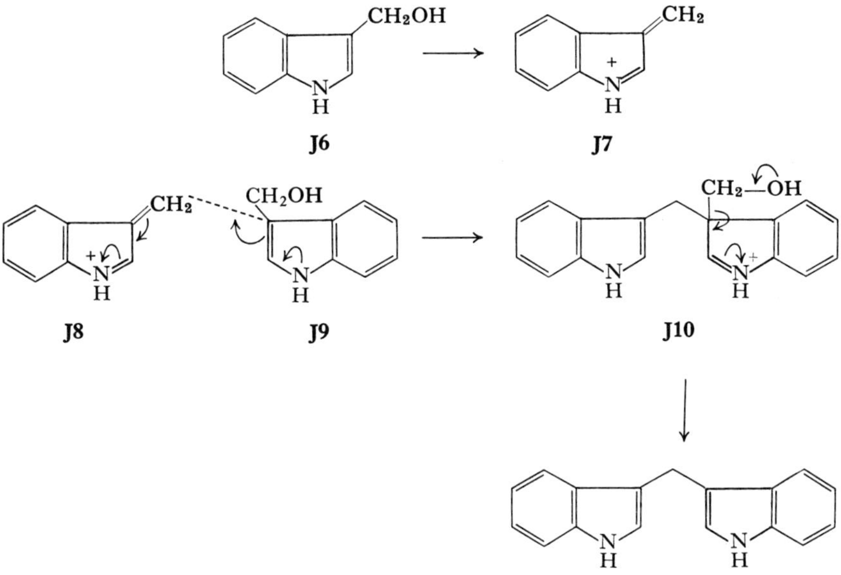

CH_2OH N H **J6** → CH_2 + N H **J7**

CH_2 + N H **J8** CH_2OH N H **J9** → CH_2—OH N H + N H **J10**

↓

N H N H **J11**

notable. This is a general feature of indole chemistry and is the result of the conjugation involving the indole nitrogen. The extreme mechanistic description of the base-catalyzed process would be elimination of a hydroxyl group from the conjugate base (after N–H ionization) of the indole. At the opposite extreme, the process **J12** → **J16** can occur by protonation and ionization of **J12** to the highly stabilized "carbonium ion" **J16**. Intermediate possibilities are no doubt often involved and can perhaps be represented by transition state **J17**. The essential point is that reorganization of the N-(C-2)–(C-3) portion of the pi system of the pyrrole ring can give stable Lewis structures and, thus, ionizations from the "benzylic" carbon at C-3 are much more facile than corresponding ionizations from a simple benzene ring. They are instead comparable to ionizations from *p*-amino- (or *p*-dimethylamino-) substituted benzene rings in which highly favorable conjugation with a nitrogen atom is again possible.

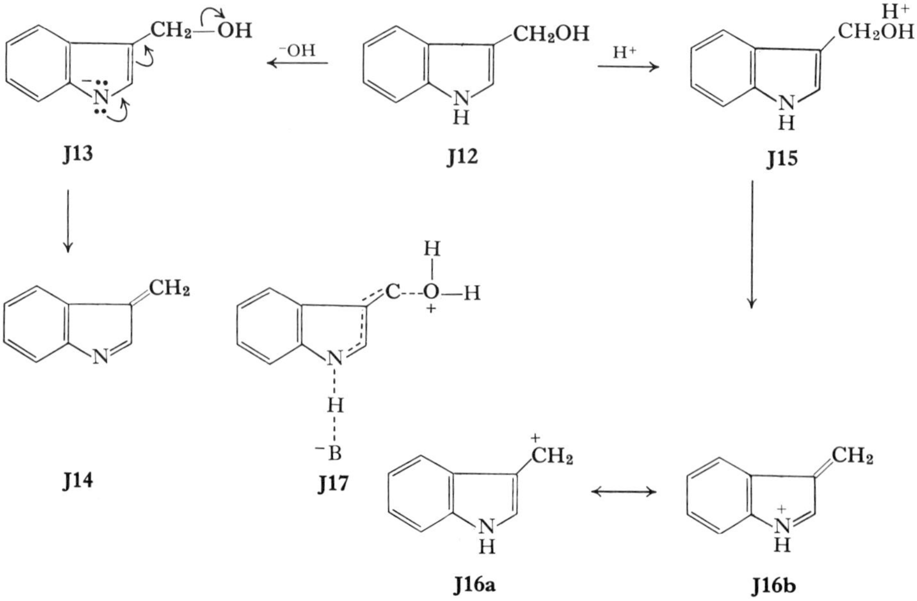

J13 J12 J15

J14 J17 J16a J16b

The structures of the products derived from aldehydes and 3-substituted indoles caused some difficulty prior to the advent of modern spectroscopic methods and an indolenine (**J18**) structure was originally assigned. The presence of typical indole ultraviolet absorption and the presence of N–H absorption bands in the infrared clearly show that the products are actually 2,2′-diindolylmethanes (**J19**) (172).

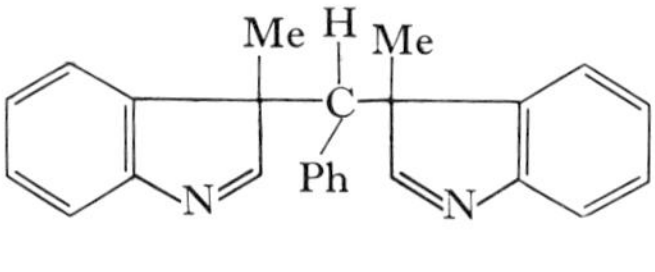

J18 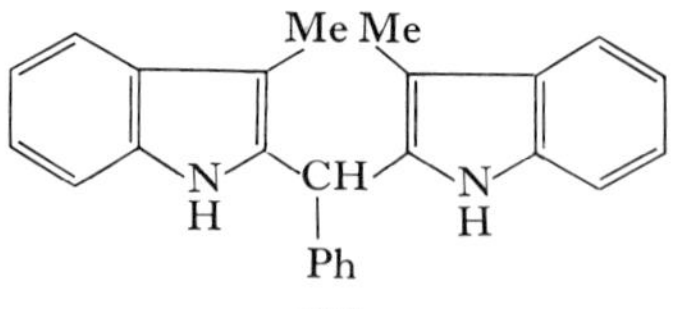 J19

When aromatic carbonyl compounds are condensed with indoles in strongly acidic media, diindolylmethane formation is not always observed. The indole, carbonyl component, diindolylmethane and the intermediates are in equilibrium, and in strongly acidic solution the 3-alkylidene-3*H*-indole may be stable, particularly if the aromatic ring contains electron releasing substituents (39). For example, 2-methylindole and *p*-dimethylaminobenzaldehyde give the alkylidene indolenine **J22** (265, 274).

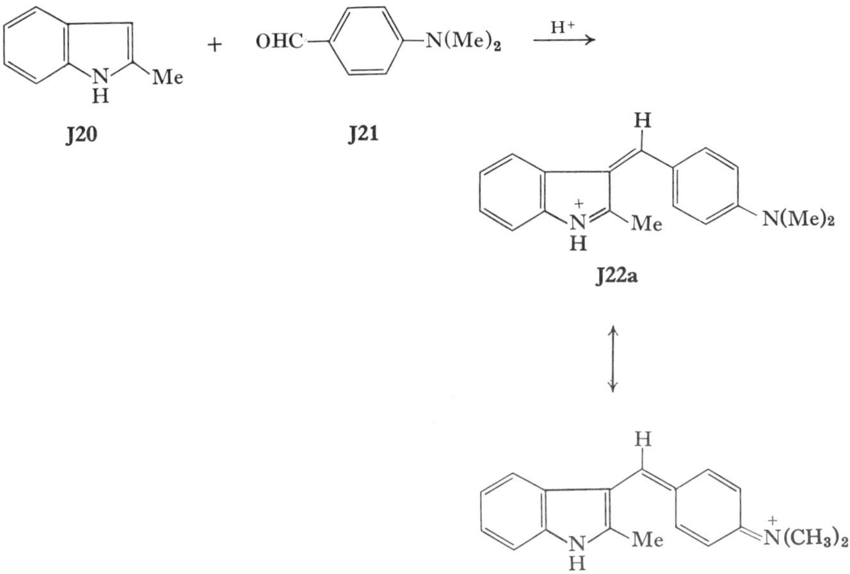

J20 + J21 (OHC–C$_6$H$_4$–N(Me)$_2$) $\xrightarrow{H^+}$ J22a ↔ J22b

J20 **J21** **J22a** **J22b**

The conjugation shown results in decreased electrophilicity of the alkylidene indolenine **J22**. 3-Methylindole gives a similar condensation product with *p*-dimethylaminobenzaldehyde (265). The formation of the highly colored derivatives such as **J22** is the basis of the familiar Ehrlich test for open 2- or 3-positions on the indole nucleus. The reaction has been adapted to quantitative determination of indoles (129).

Burr and Gortner (39) have demonstrated qualitatively that the position of the equilibrium between the diindolylmethane **J24** and the alkylindene-3*H*-indole **J23** is a function of the acidity of the solution, strongly acidic media favoring **J23**. Condensation of aromatic aldehydes with indoles in

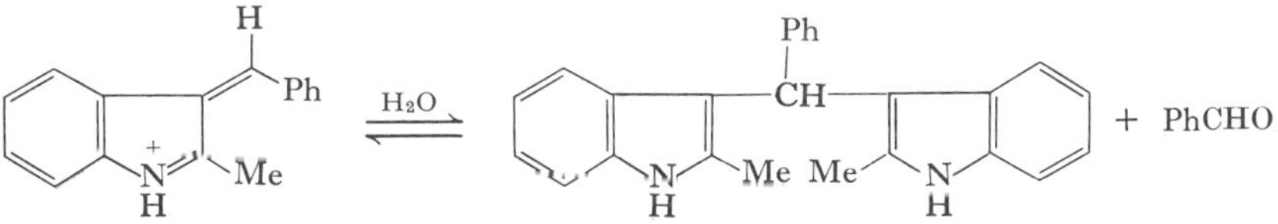

J23 $\underset{}{\overset{H_2O}{\rightleftharpoons}}$ **J24** + PhCHO **J25**

strongly acidic solution can be used as a preparative route to 3-arylidene-3*H*-indoles, the synthesis of **J28** being a recent example (225).

Condensation of indole with either cyclohexanone or 1,3-cyclohexanedione is reported to give the same product which is assigned structure **J31**.

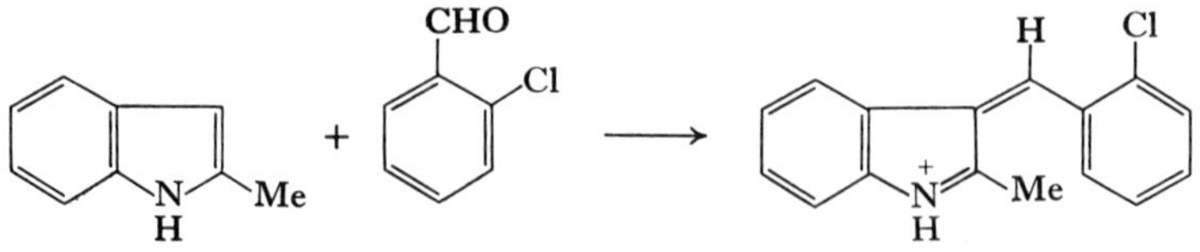

J26 J27 J28

The formation of **J31** from the dione is readily rationalized, but the formation of this product from cyclohexanone requires an oxidation (266).

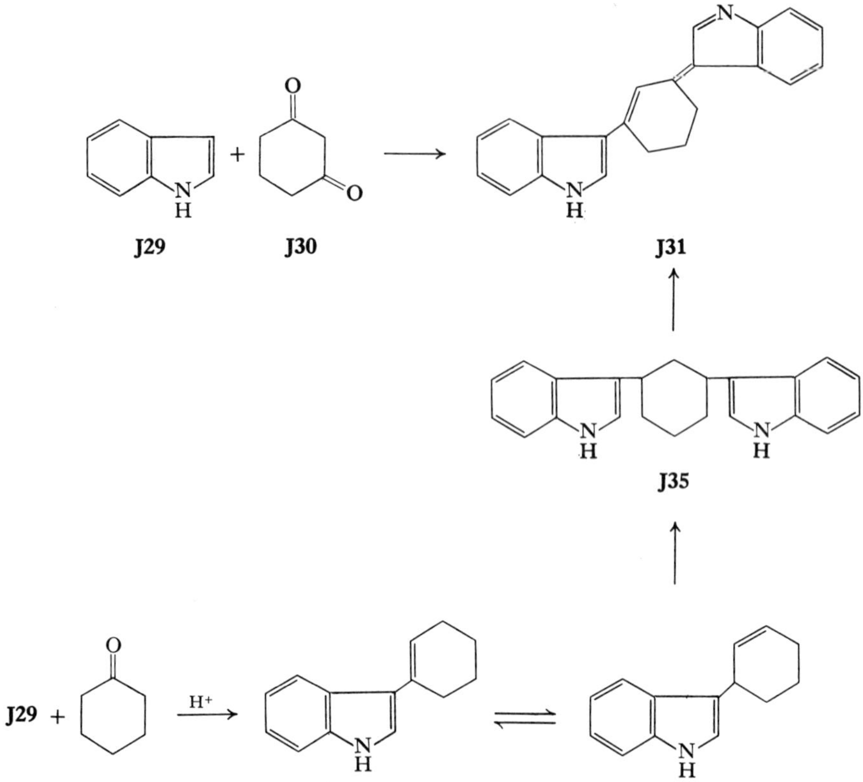

J29 J30 J31

J35

J29 + J32 $\xrightarrow{H^+}$ J33 ⇌ J34

The products of condensation of 3-methylindole with cyclohexanone and of 2-methylindole with cyclohexane-1,4-dione have been assigned structures **J37** and **J39**, respectively, apparently solely on the basis of elemental analyses (266).

A cyclizative version of diindolylmethane formation has been examined. Von Dobeneck and Maas found that the diindolylmethane **J40** reacts with

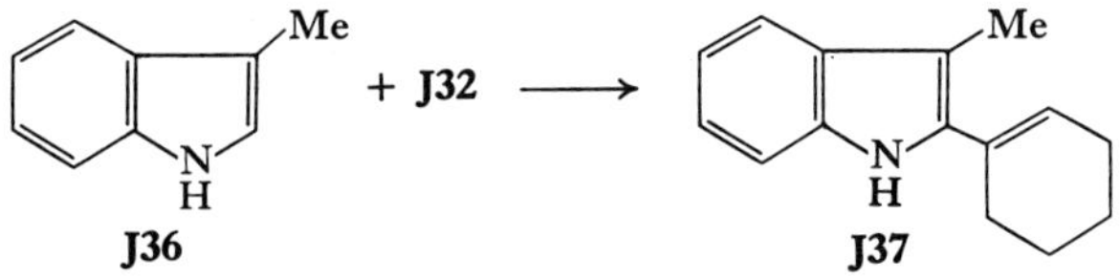

J36 + J32 ⟶ J37

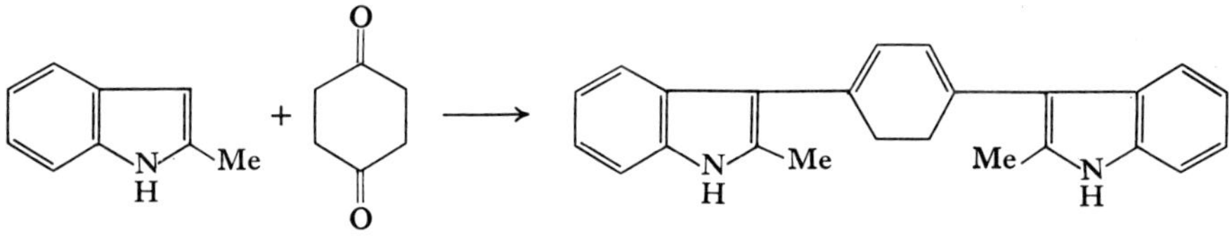

J38 + cyclohexane-1,4-dione ⟶ J39

formaldehyde or aromatic aldehydes to give the cyclic products **J41** (274). Similarly, acetone and indole react in the presence of acid to give **J43** (178). The same product arises from treatment of **J44** with acetone in the presence of acid and **J44** is no doubt an intermediate in the formation of **J43**.

The reaction of indole and formaldehyde terminates at the carbinol stage when sodium methoxide is used as the catalyst (218). There are several examples of reactions of indolylmagnesium derivatives with ketones and aldehydes. Oddo and Toffoli (185) have reported that indolyl- and 2-methylindolylmagnesium bromide give diindolylmethane derivatives with acetaldehyde. Similar results have been reported from the reaction of

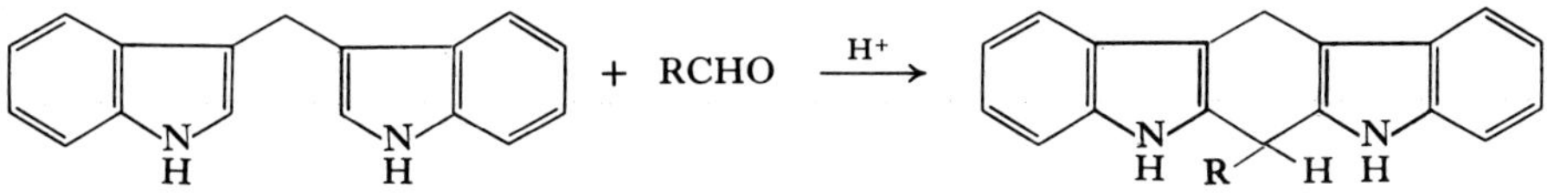

J40 + RCHO $\xrightarrow{H^+}$ J41

J42 + $(Me)_2CO$ $\xrightarrow{H^+}$ J43

J44 + $(Me)_2CO$ $\xrightarrow{H^+}$ J43

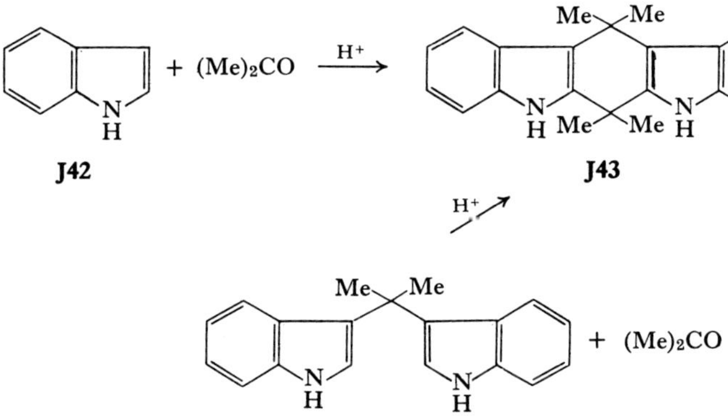

magnesium compounds with acetone (184), acetophenone (174), and a number of aromatic aldehydes (149).

More recently, Bader and Oroshnik (8) investigated the reaction of indolylmagnesium bromide with 2- and 4-pyridinecarboxaldehyde. At –25°C the former aldehyde gives a 50% yield of the carbinol **J47** and 6.5% of the diindolylmethane **J48**. At 0°C the formation of **J48** is favored. Pyridine-4-carboxaldehyde gives a 58% yield of the carbinol **J50** at 0°C, but other products dominate at higher temperature. Hoshino (104) isolated a 1:1

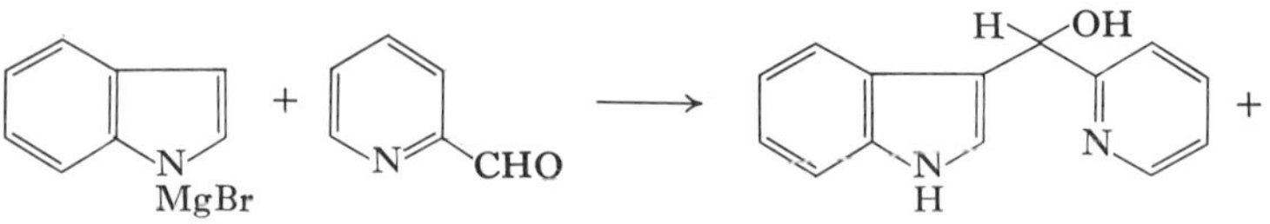

MgBr + N CHO → H OH N N H +

J45 **J46** **J47**

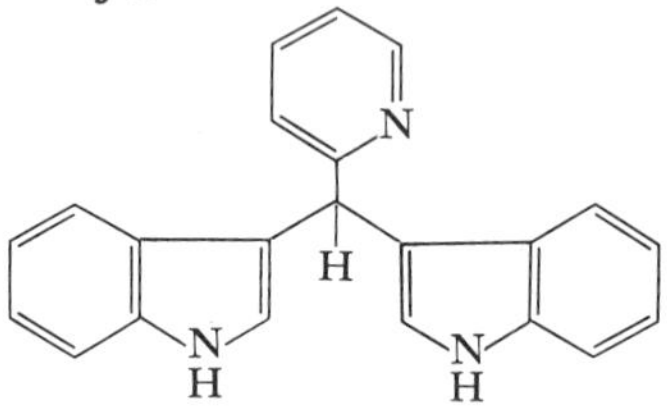

N H N H H N H

J48

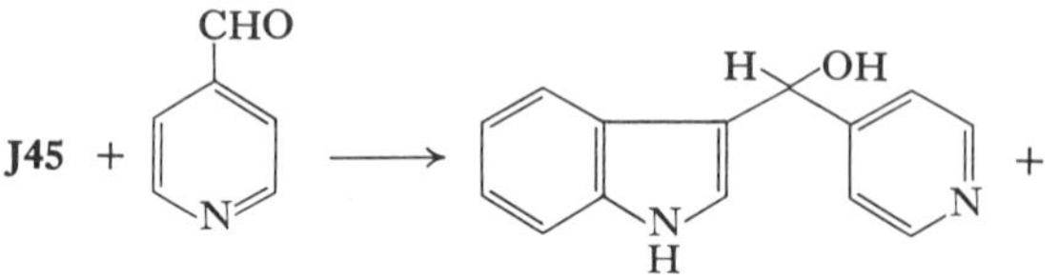

CHO

J45 + N → H OH N H N +

J49 **J50**

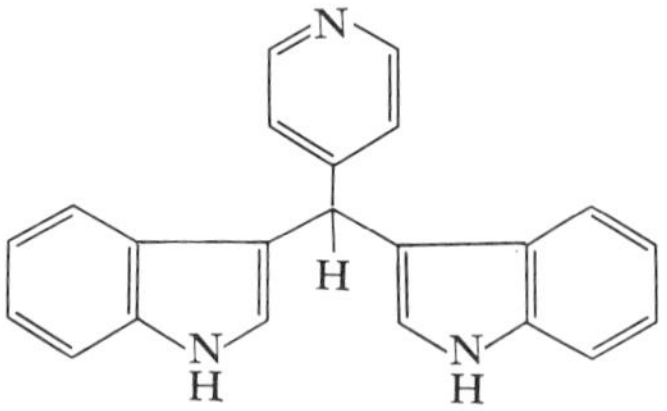

N H N H H N H

J51

N MgI + $(Me)_2CO$ → N MeCOH Me

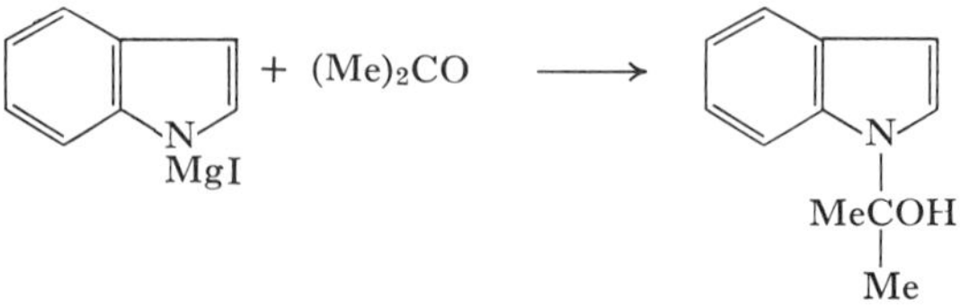

J52 **J53**

addition product from the reaction of indolylmagnesium iodide with acetone. The structure **J53** was assigned on the basis of active hydrogen determinations which indicated a single active hydrogen. The product **J53** is unstable toward decomposition to acetone and indole. It would be interesting to have an nmr spectrum of this product to conclusively rule out the alternative (2 active H's) 3-[2-(2-hydroxy)propyl]indole structure. The inference that can be drawn from these results is that the general instability (138) of 3-indolylcarbinols will permit their isolation from Grignard-type reactions only under carefully controlled conditions.

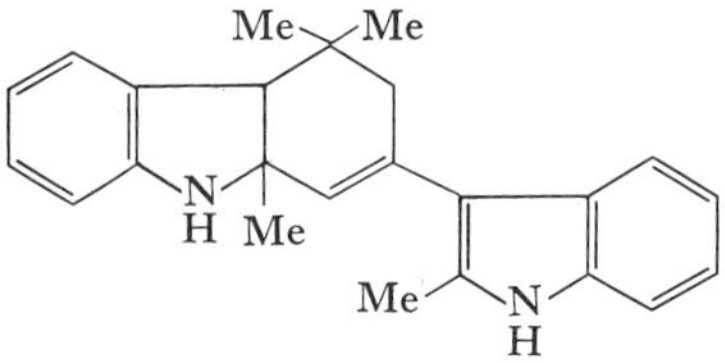

Me Me N H Me Me N H

J54

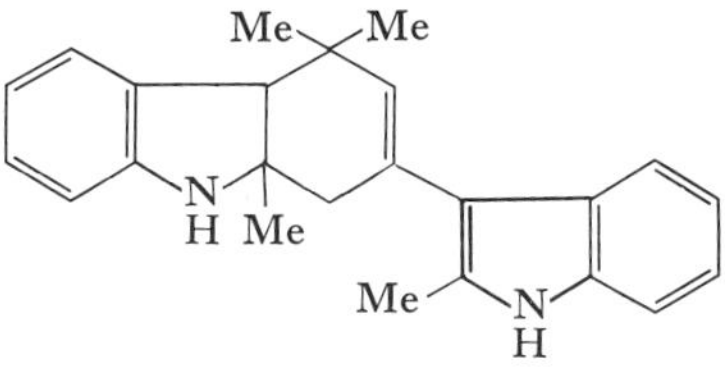

Me Me N H Me Me N H

J55

Noland and co-workers have investigated the acid-catalyzed condensation of 2-methylindole with acetone, mesityl oxide, and phorone (180). Each reaction gives a condensation product which has been shown to have either the structure **J54** or **J55**. The mechanism suggested for this cyclizative condensation involves acid-catalyzed addition of two molecules of **J57** formed by condensation of acetone and 2-methylindole.

Mechanistically related extensions of this reaction are exemplified by the conversions shown below (179). The probable mechanism for these transformations is illustrated for the case of **J68** → **J69**. Indole and 1-methylindole also give cyclizative condensation products similar to those of 2-methylindole along with products of other types (170).

The acid-catalyzed reaction of 2-methylindole and certain 1,3-dicarbonyl compounds has been examined and the products have been found to be 3-vinylindoles such as **J76** and **J78** (173). The formation of products such as **J76** and **J78** may involve a preference for dehydrations of the addition intermediate **J79** to **J80** rather than to **J81**, or **J81** may be formed and go to **J80** faster than it reacts with an indole molecule to give the diindolylmethane **J83**. The conjugation present in the vinylindole **J80** no doubt makes it of lower energy than **J81**. Formation of the vinylindole 2-(1,2-dimethylindole-3-yl)acrylic acid from the reaction of 1,2-dimethylindole and pyruvic acid has also been reported (289a). When indole is condensed with 1,4-diketones, carbazoles are formed from condensation reactions involving both C-2 and C-3 (214).

J56 + $(Me)_2CO$ $\xrightarrow{H^+}$ J57 $\underset{}{\overset{-H^+}{\rightleftharpoons}}$ J58

J57 + J58 ⟶ J59 ⟶ J60 $\xrightarrow{-H^+}$ J54 or J55

Ph
CHCH$_2$COMe
Me
N
H
J60

+

Me
N
H
J61

⟶

Ph
N
H Me
Me
N
H
J62 (34%)

CH$_2$CH$_2$COMe
Me
N
H
J63

+ **J61** ⟶

N
H Me
Me
N
H
J64 (37%)

Me
N
H
J65

+

Me
O
J66

⟶

N
H Me
Me
N
H
J67 (23%)

CH$_2$CH$_2$COMe
N
H
J68

+ **J65** ⟶

N
H H
Me
N
H
J69 (34%)

CH_2CH_2CMe O---H+ Me N H

J68

J65

Me OH CH_2CH_2C Me Me N H

J70

H Me

J71

H+ C H_2 Me

J72

J73

→ J69

+ $MeCOCH_2CO_2Et$ ⟶

J74 J75 J76

J74 + $MeCOCH_2COMe$ ⟶

J77 J78

J79 J80

J81 J82 J83

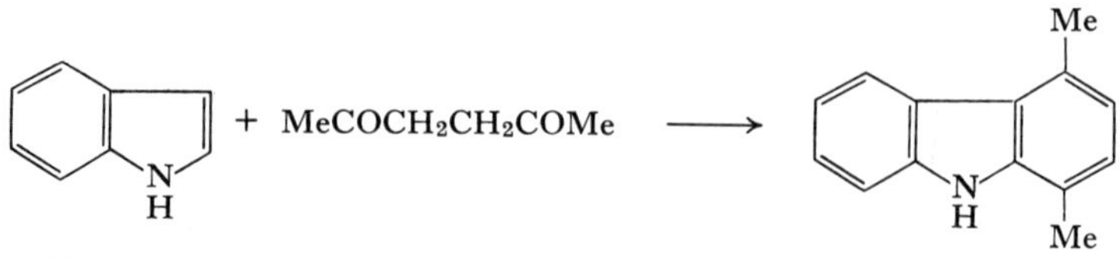

+ $MeCOCH_2CH_2COMe$ ⟶

J84 J85 J86

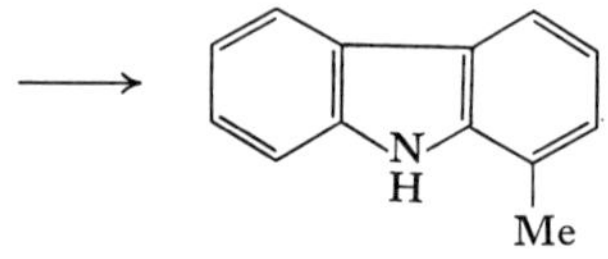

J84 + $MeCOCH_2CH_2CH(OEt)_2$ ⟶

J87

J88

The reaction of 3-methylindole with acetonylacetone results in cyclization to give the pyrido[1,2-a]indole **J91**. Similarly **J89** and levulinaldehyde give **J93** (214, 215). When a substituent is present on nitrogen, thereby

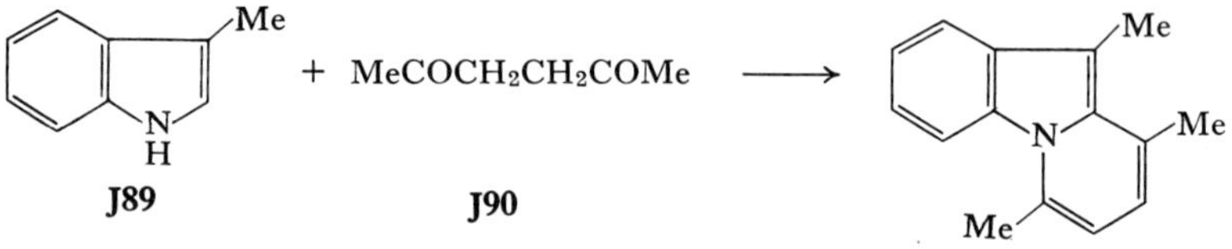

+ $MeCOCH_2CH_2COMe$ ⟶

J89 J90 J91

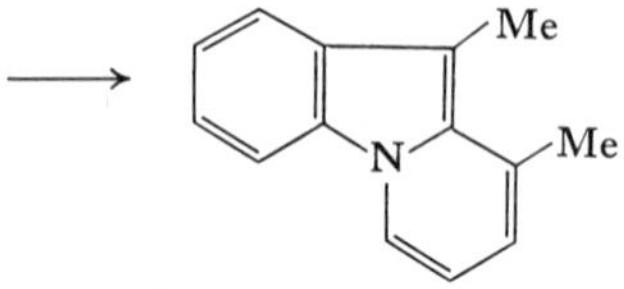

J89 + $MeCOCH_2CH_2CHO$ ⟶

J92

J93

preventing formation of structures such as **J91** and **J93**, an alternative cyclization occurs as illustrated by the reaction of hexane-2,5-dione with 1,3-dimethylindole (216). The most likely mechanistic interpretation of the latter cyclization would appear to involve acid-catalyzed electrophilic attack of the dione on the 3-position of the indole to generate the indolenine **J97**. Cyclization can then proceed via enolization and addition to the indolenine system.

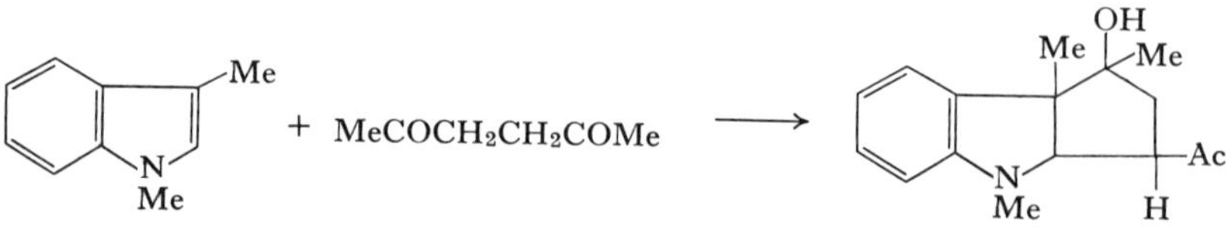

+ $MeCOCH_2CH_2COMe$ ⟶

J94 J95 J96

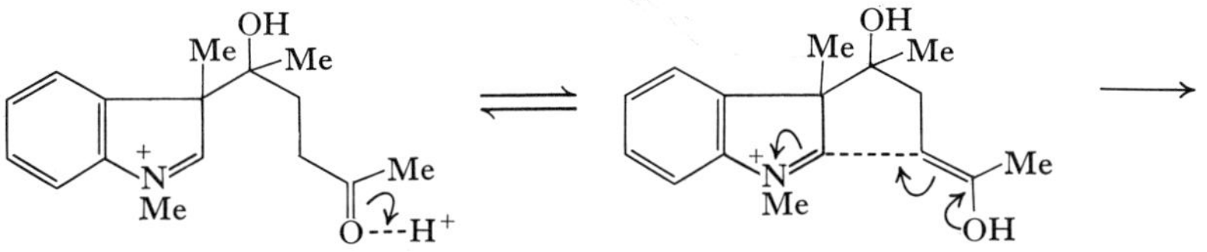

J97 ⇌ J98 ⟶ J96

In the case of the reaction of 3-methylindole and levulinaldehyde (**J92**) the aldehyde function is found bonded to the indole nitrogen. This fact led to the assumption that the cyclization involves initial attack by the dicarbonyl compound at the nitrogen atom. An additional postulate has been advanced that indoles with NH groups and carbonyl compounds may, in general, be in equilibrium with the addition product **J101** and that this equilibrium can control the direction of cyclization when a subsequent irreversible step is available, as in the cyclization to **J93** (172). The reactions

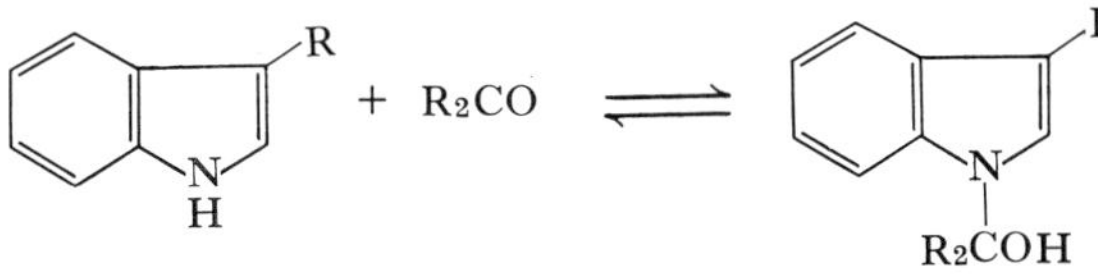

J99 + R_2CO (J100) ⇌ J101

of the 2-pyridylindoles **J102** and **J104** show a similar behavior (26). The *N*-unsubstituted indole undergoes cyclization and dehydration at nitrogen in preference to C-3.

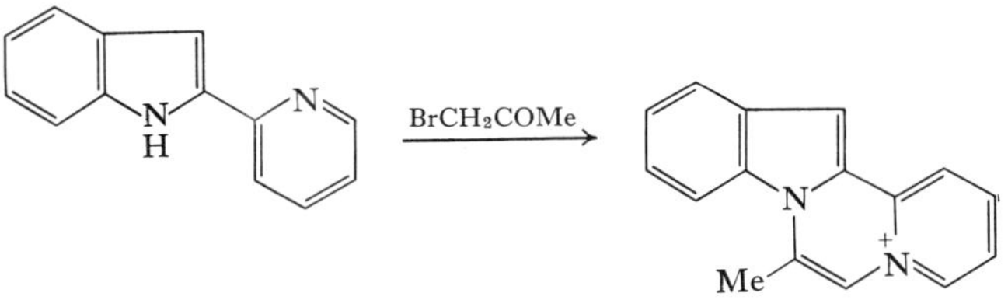

J102 —$BrCH_2COMe$→ J103

J104 —$BrCH_2COMe$→ J105

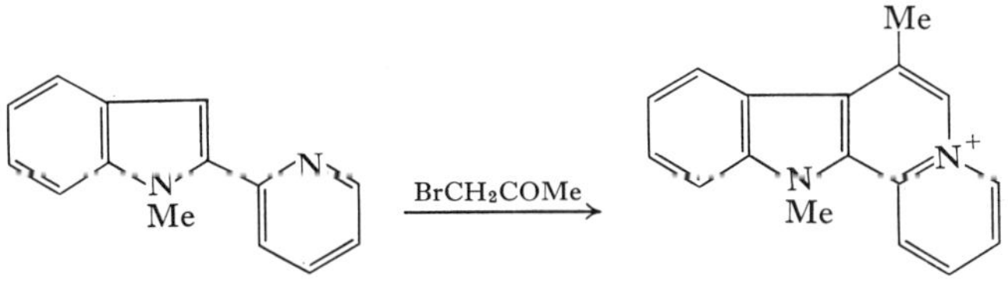

Assuming that kinetic factors determine the product distribution, two possible mechanisms can be invoked, both of which suggest that kinetic preference for addition of the indole NH to a carbonyl is the primary factor determining the direction of cyclization.

2-Methylindole reacts with hydroxymethylene derivatives of ketones to give carbazoles (167). The probable mechanism involves electrophilic attack at C-3 followed by intramolecular condensation involving the methyl substituent.

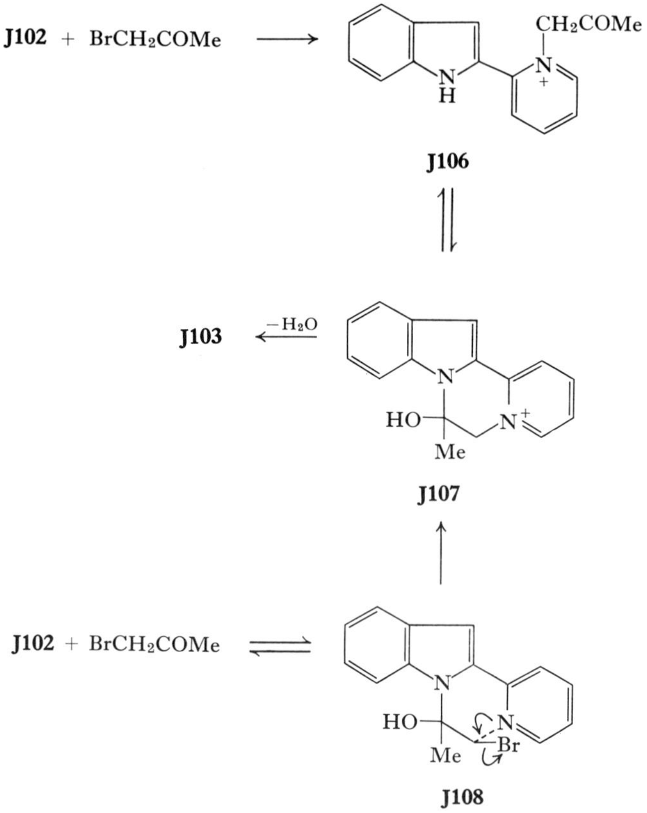

J102 + $BrCH_2COMe$ ⟶ **J106**

J103 $\xleftarrow{-H_2O}$ **J107**

J102 + $BrCH_2COMe$ ⇌ **J108**

Reactions of phthalaldehyde and indoles are interesting in that phthalides formally resulting from lactonization of the intermediate indolylcarbinol are isolated (166, 203). The suggestion has been made (166) that phthalide formation may involve the mechanism below, the key step being nucleophilic addition of the carboxylate group to an alkylidene indolenine intermediate. The work of Rees and Sabet (203) provides convincing evidence that the open-chain form of phthalaldehyde is the electrophile in the reaction. Noland and Johnson's work establishes that ease of phthalide formation for the various reactive positions on the indole ring is $3 > 1 > 2$ (166).

J109 J110 J111

J113 J112

$-2H_2O$

J114

J115 J116 J117

J115 + J116 ⟶ J118 ⟶ J119 ⟶ J120

K. REACTIONS WITH IMINIUM BONDS

Just as the polarized carbonyl bond is sufficiently electrophilic to attack the indole system, the carbon-nitrogen double bond, particularly when the nitrogen atom is protonated or alkylated and bears a full positive charge, is an effective electrophile toward the indole ring. Electrophilic attack by iminium systems is encountered in the widely useful Mannich condensation of formaldehyde and secondary amines with indoles. Electrophilic substitutions involving iminium systems have also been very important in the

$$\underset{\textbf{K1}}{R_2NH} + \underset{\textbf{K2}}{CH_2O} \overset{H^+}{\rightleftharpoons} \underset{\textbf{K3}}{R_2\overset{+}{N}HCH_2OH} \rightleftharpoons \underset{\textbf{K4}}{R_2\overset{+}{N}{=}CH_2} + H_2O$$

$$\underset{\textbf{K4}}{R_2\overset{+}{N}{=}CH_2} + \textbf{K5} \longrightarrow \textbf{K6}$$

synthesis of a number of indole alkaloids and derivatives.

As will be discussed fully in Chapters II and IV, the indolylmethylamines generated by Mannich condensation are very versatile synthetic reagents, and the Mannich condensation has been very widely applied. Table IX

TABLE IX □ TYPICAL MANNICH CONDENSATIONS OF INDOLES

Indole substituents	Aldehyde	Amine	Position of condensation	Yield (%)	Ref.
None	Formaldehyde	diMe (in acetic acid)	3	100	(134)
None	Formaldehyde	diMe (in water)	1	79	(246)
None	Formaldehyde	*t*-Butyl	3	39	(239)
None	Formaldehyde	2,4-Dimethyl-piperidine	3	95	(2)
None	Formaldehyde	Ethyl piperidine-3-carboxylate	3	70	(2)
None	Formaldehyde	*N*-(Dimethylamino)-isopropyl	3	81	(291)
None	Formaldehyde	*N*-(Methylamino)-piperidine	3	49	(291)
None	Acetaldehyde	Isopropyl	3	39	(239)
2-Me	Formaldehyde	Piperidine	3	79	(28)
2-Me	Formaldehyde	*N,N*-Bis(2-hydroxy-ethyl)	3	70	(28)
2-Ph	Formaldehyde	diEt	3	78	(128)
2-Carbethoxy	Formaldehyde	diEt	3	68	(128)
2-Carbethoxy	Formaldehyde	Benzylmethyl	3	93	(28)
2,3-diMe	Formaldehyde	diMe	1	12	(247, 248)
3-Me	Formaldehyde	diMe	1	71	(247, 248)
3-Cyano	Formaldehyde	diMe	1	59	(247, 248)

records some typical examples. Indole, formaldehyde, and dimethylamine give a nearly quantitative yield of gramine (3-dimethylaminomethylindole) in aqueous acetic acid (134). Reaction also occurs in basic solution (134). The reaction proceeds satisfactorily even in the presence of electron-withdrawing substituents in the 2-position, as illustrated by the conversion of ethyl indole-2-carboxylate to its 3-diethylaminomethyl derivative in 68% yield (128).

A recent report (246) describes the reaction of equimolar amounts of indole, formaldehyde, and dimethylamine at 0–5°C. The principal product is a liquid described as 1-dimethylaminomethylindole. Similar results are reported for other secondary amines. The product is reported to have practically no NH absorption in the infrared and to be distinguishable from gramine by gas chromatography. 1-Dimethylaminomethylindole is readily isomerized to gramine. The conditions of the standard preparations (134) of gramine would result in the isomerization of 1-dimethylaminomethylindole to gramine. 3-Methylindole reacts with formaldehyde and dimethylamine in aqueous acetic acid to give a 1-dimethylaminomethyl derivative (247, 248). A later paper reports similar reaction conditions, but describes the product as the 2-derivative (120) without referring to the papers of Swaminathan and workers (247, 248) in which the reaction product was characterized as the 1-dimethylaminomethyl isomer. The weight of the evidence would seem to favor the structural assignment of Swaminathan. Thesing and Binder (258) have described a Mannich base obtained from 3,3′-diindolylmethane, formaldehyde, and dimethylamine. On the basis of lack of NH absorption, it must be **K8**. Similarly, 3-benzylindole gives a

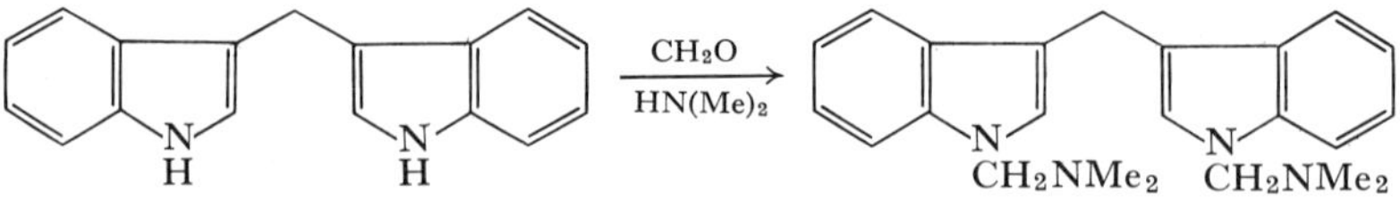

CH_2O / $HN(Me)_2$ → ; N–H, N–H; N–CH_2NMe_2, N–CH_2NMe_2

K7 **K8**

Mannich base lacking NH absorption (258). Nevertheless, a dimethylamino-methyl derivative is also formed by 1,3-dimethylindole and, since this must be **K10**, condensation at C-2 is also possible. Indeed, even 1,2,3-trimethyl-

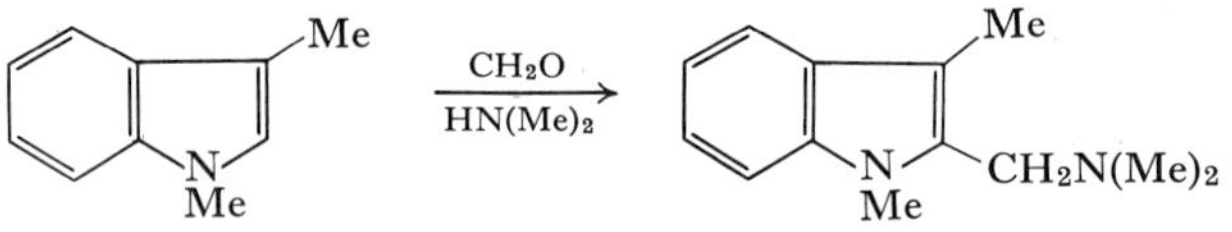

Me; N–Me; CH_2O / $HN(Me)_2$ → Me; N–Me; $CH_2N(Me)_2$

K9 **K10**

indole and *N*-methyltetrahydrocarbazole as well as several other 1,2,3-trialkyl indoles give Mannich bases (261). The formation of **K12** and **K14** must involve the conjugate acids of the indoles. Deprotonation of **K16** can

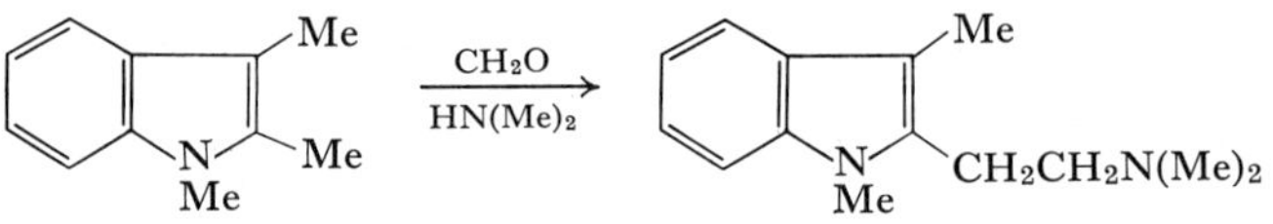

Me; Me; N–Me; CH_2O / $HN(Me)_2$ → Me; N–Me; $CH_2CH_2N(Me)_2$

K11 **K12**

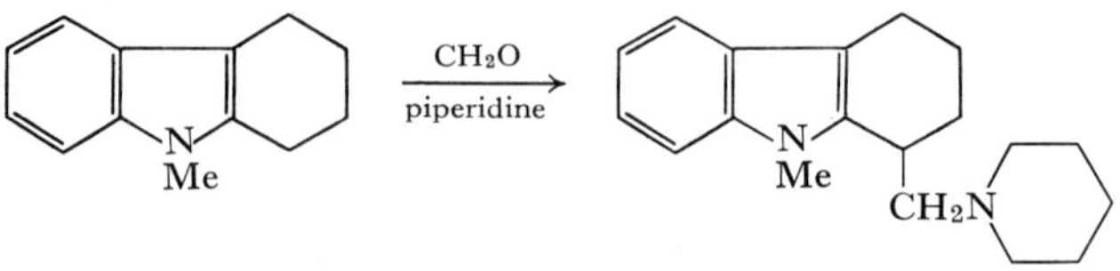

N–Me; CH_2O / piperidine → N–Me; CH_2N

K13 **K14**

then lead to the nucleophilic enamine **K18**. Attack by an iminium ion accounts for the formation of the observed products.

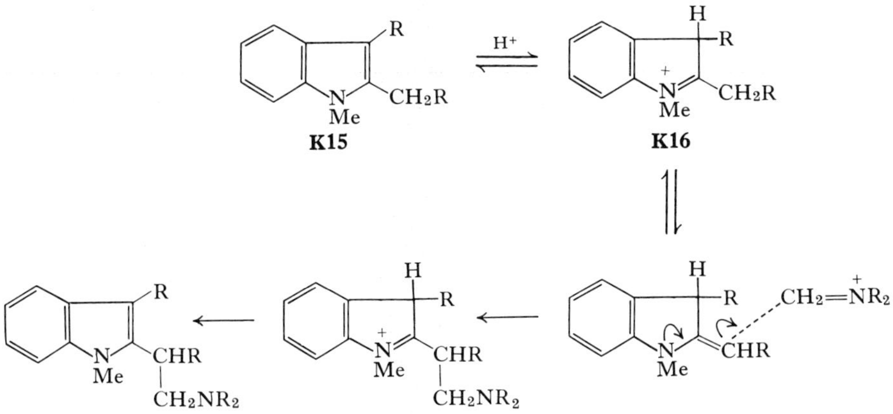

K15 $\xrightleftharpoons{H^+}$ K16

K20 ← K19 ← K18 + $CH_2{=}\overset{+}{N}R_2$ (K17)

An additional interesting variation in the behavior of indole in Mannich condensations has been observed by Thesing and co-workers (260, 263). Both *N*-methylaniline and *N*,*N*-dimethylaniline condense with indole and formaldehyde at C-4 of the aniline ring, giving **K22** and **K23**, respectively.

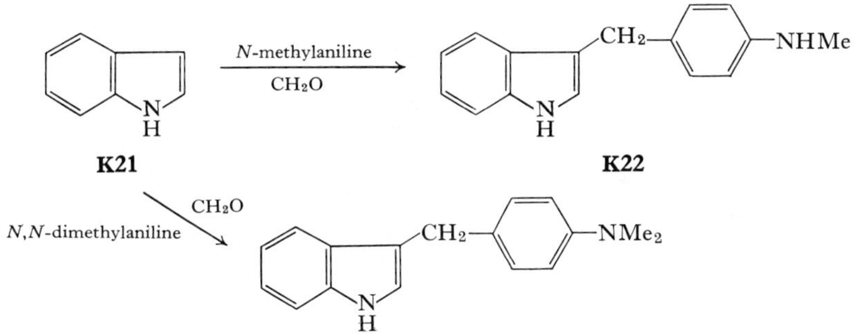

K21 —(*N*-methylaniline, CH_2O)→ K22

K21 —(*N*,*N*-dimethylaniline, CH_2O)→ K23

Rather little is known about the reactions of formaldehyde and primary amines with indoles. Snyder and Matteson (239) have found that *t*-butylamine and formaldehyde give 3-*t*-butylaminomethylindole, and acetaldehyde and isopropylamine give **K25**. Less bulky primary amines are expected to lead to bis condensation (22).

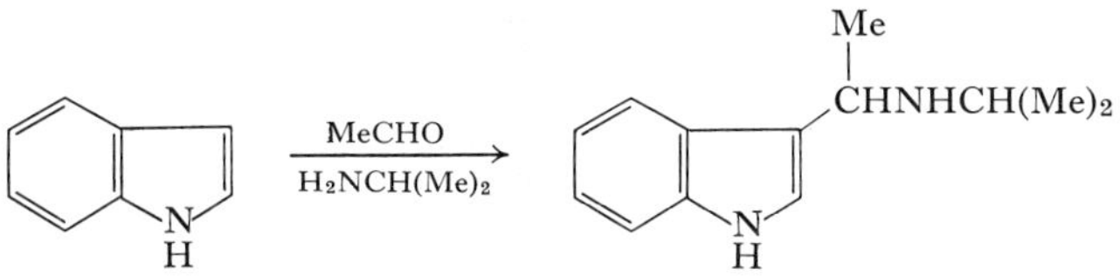

K24 —(MeCHO, $H_2NCH(Me)_2$)→ K25

Imines can be generated by other methods in addition to *in situ* condensation of aldehydes and amines. Snyder and Matteson (239) employed the preformed imines methylene-*t*-butylamine and ethylideneisopropylamine in reactions with indole. They found that when the imines were used in acetic acid the yields of Mannich bases were somewhat higher than when the aldehyde and amine were used directly.

Troxler and co-workers (270) have shown that hydroxyindoles are alkylated in the carbocyclic ring in preference to C-3 under the conditions of the Mannich reaction. It has been shown that 4-hydroxyindole is alkylated at C-5, 5-hydroxyindole at C-4, 6-hydroxyindole at C-7, and 7-hydroxyindole at C-6. The Mannich bases are capable of alkylating such nucleophilic carbon species as nitronate and malonate ions (269).

Condensations of the Mannich type are, in general, reversible. This reversibility leads to reactions in which the indole ring may replace another nucleophilic component of a preformed Mannich base (94, 239). Thus, indole reacts with the tertiary amine **K26** to give **K32** (92). The mechanism of this reaction is presumed to be as illustrated below:

$NCH_2C(CO_2Et)_2(NHCHO)$ (**K26**) ⇌ $^+N{=}CH_2$ (**K27**) + $CH(CO_2Et)_2(NHCHO)$ (**K28**)

K29 + **K27** ⟶ CH_2—N (**K30**)

K30 ⟶ CH_2 (**K31**) $\xrightarrow{^-C(CO_2Et)_2(NHCHO)}$ $CH_2C(CO_2Et)_2(NHCHO)$ (**K32**)

The intermediate **K30** has been isolated from interrupted reactions (92). Similar exchange processes must be involved in the alkylation of indole with the compound **K34** (186). The alkylation of indole with dimethylamino-

K33 + $[(EtO_2C)_2C(NO_2)CH_2]_2NH$ (**K34**) ⟶ $CH_2C(NO_2)(CO_2Et)_2$ (**K35**)

acetonitrile would seem to be best formulated as below and thus represents another example of exchange of Mannich base components (69).

$$Et_2NCH_2CN \longrightarrow Et_2\overset{+}{N}{=}CH_2 + \bar{C}N$$

K36 **K37**

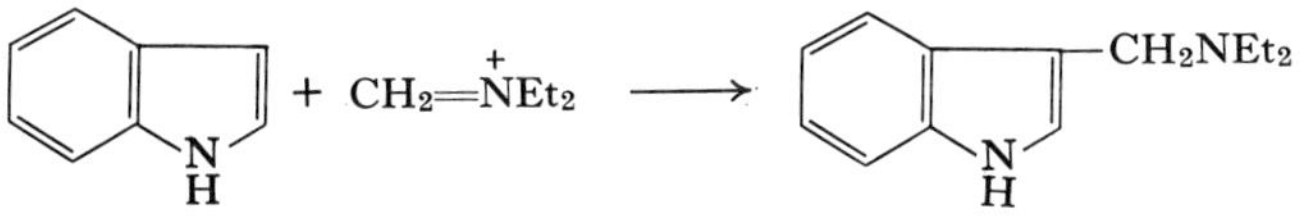

$$\text{indole} + CH_2{=}\overset{+}{N}Et_2 \longrightarrow \text{3-}(CH_2NEt_2)\text{indole}$$

K38 **K39**

$$\textbf{K39} + \bar{C}N \longrightarrow \text{3-}(CH_2CN)\text{indole} + HNEt_2$$

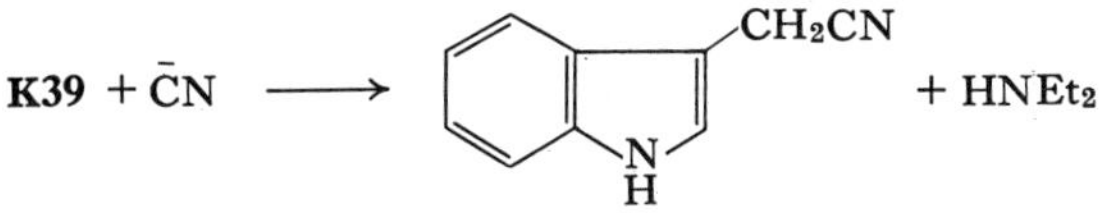

K40

Indole condenses with 1-piperideine in acidic solution (259, 272). The piperideine is generated from its trimer. Marginal success was reported for

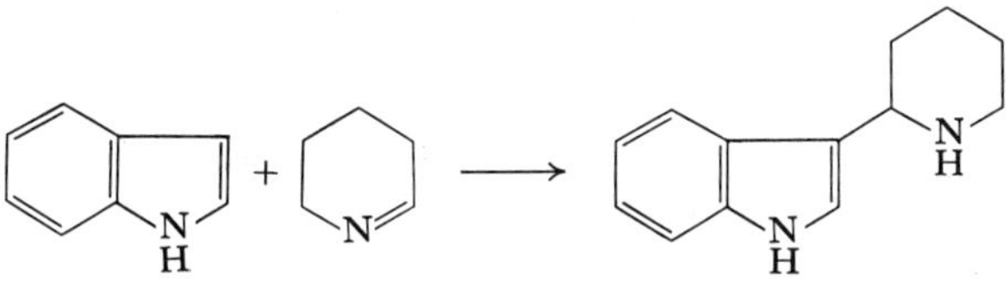

K41 + **K42** ⟶ **K43**

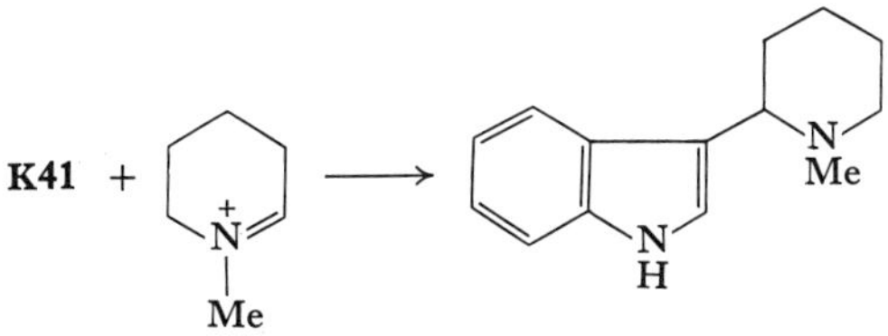

K41 + **K44** ⟶ **K45**

condensation of **K44**, generated by controlled lithium aluminum hydride reduction of *N*-methylpiperidone, and indole (272). 1-Pyrroline, which can be generated by dehydrohalogenation of *N*-chloropyrrolidine, reacts with indole to give a product assumed to be 3-(2-pyrrolidinyl)indole (78). A highly significant new development in the generation of iminium ions involves the reduction of certain substituted pyridines to the 1,4,5,6-tetrahydro stage. In acidic solution most 1,4,5,6-tetrahydropyridines are protonated at C-3, generating iminium salts (199, 280, 281). The versatility and limitations of this reaction have been discussed by Wenkert (279–281).

K46 $\xrightarrow[\text{Pd/C}]{H_2}$ K47 K48 $\xrightarrow[\text{Pd/C}]{H_2}$ K49 Ref. (281)

K50 $\xrightarrow[\text{Pd/C}]{H_2}$ K51 Ref. (281)

K52 + K49 $\xrightarrow{H^+}$ K53 Ref. (281)

K54 + K55 $\xrightarrow{H^+}$ K56 Ref. (281)

K52 + K55 $\xrightarrow{H^+}$ K57 Ref. (199)

K52 + K58 $\xrightarrow{H^+}$ K59 Ref. (280)

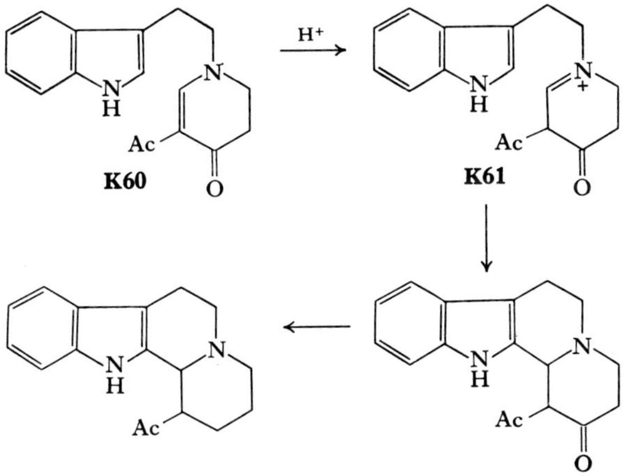

H^+

K60 K61

K63 K62

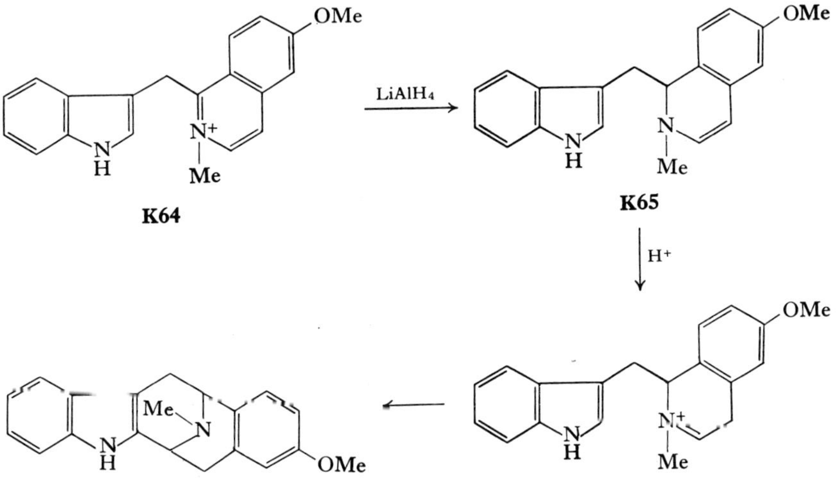

LiAlH$_4$

K64 K65

H^+

K67 K66

K65 ⟶ K68

OMe
N
H
N+
Me

K69 + K70 ⟶ K71

COCH₂Br
N
H
N
N+
O

K71 ⟶ [K72] ⟶ K73

N
H
N
H
OH

K74 + K75 + K76 ⟶ K77

NCOPh
COCl
N
H
N

K74 + K78 ⟶ K79

N+
Me
N
H

The examples shown above illustrate the method. Intramolecular variations have proved very useful in synthesis of certain alkaloid skeletons and these examples are discussed in Section E, Chapter IV.

The success of this procedure depends upon the presence of an electron withdrawing substituent at C-3 on the pyridine ring which permits the catalytic reduction to be terminated at the tetrahydro stage. The ester function has been the one most successfully employed to date. The use of a *t*-butyl ester permits the removal of the C-3 carboxyl group during the course of the alkylation (for example, **K52** → **K59**) via acid-catalyzed hydrolysis of the ester group and decarboxylation (280).

The cyclization and reduction of **K60** to **K63** in acidic solution in the presence of zinc (284) presumably involves an intramolecular cyclization initiated by protonation of the piperidone ring at C-3. The sequence of the reduction and cyclization steps has not been established.

The dihydroisoquinoline **K65** has been prepared by reduction of **K64** with lithium aluminum hydride (292). Protonation at C-3 of the enamine system generates the electrophile **K66** which then cyclizes to **K67**. The yield of **K67** from **K65** is 76% when excess dilute acid is used as the cyclization medium. In the absence of excess acid, **K65** rearranges to **K68** (292). Present mechanistic descriptions (68) of the rearrangement are incomplete.

NCN
+ + BrCN ⟶
N H N N H
K80 **K81** **K82** **K83**
+
CH=CHCH=CHCHO ⟵
N H N CN N H
K85 **K84**

Quaternary pyridinium salts prepared from 2-bromomethyl indolyl ketone cyclize on reduction with lithium aluminum hydride. Generation of an iminium system followed by electrophilic attack at C-3 is involved (197).

The reaction of indoles with quaternized derivatives of pyridine is a related electrophilic substitution (199, 273). Indole reacts with pyridine in the presence of benzoyl chloride to give **K77** and with the methiodide of quinoline to give **K79** (273). Indole, pyridine, and cyanogen bromide react to give **K83** and **K84**, the latter being isolated as the hydrolysis product **K85** (199).

These reactions clearly involve electrophilic attack at C-3 by the pyridinium species at its electron-deficient C-2 and C-4 positions. The reaction of 2-phenylindole with quinoline-1-oxide in the presence of benzoyl chloride must follow a similar pattern (49, 50). Hamana and Kumadaki (87) have

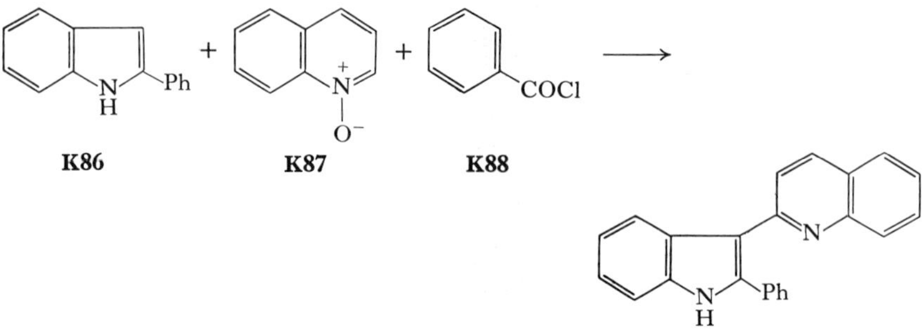

K86 + **K87** + **K88** ⟶ **K89**

recorded several other examples of such reactions. Indole and the *N*-oxide of ethyl nicotinate react to give **K93** in 30% yield. The benzoylated *N*-oxide is presumed to be the electrophilic species, and aromatization proceeds by elimination of benzoic acid.

The reaction of indoles with iminium systems has been of considerable importance in the synthesis of certain indole alkaloid skeletons. These

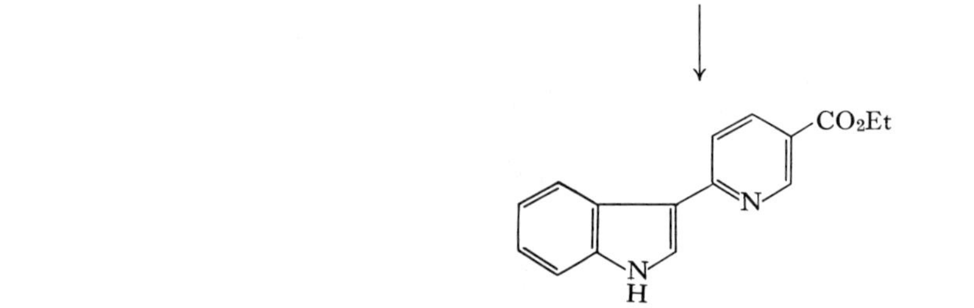

K90 + **K91** ⟶ **K92** + H^+

K92 ⟶ **K93**

applications are discussed in Chapter IV,E. The synthesis of β-carboline derivatives often involves intramolecular reactions of the indole nucleus with iminium systems. The synthesis of β-carboline systems is discussed in Chapter IV,C.

L. REACTIONS WITH ELECTRON DEFICIENT OLEFINS, ACETYLENES, AND QUINONES

The indole nucleus reacts with certain olefins and acetylenes substituted by electron withdrawing substituents. Reactions of this type have been observed under conditions of both acidic and basic catalysis. The acid-catalyzed reactions are considered first. Indole and 2-methylindole give the adducts **L3** and **L4**, respectively, with methyl vinyl ketone in acetic acid solution (249). The yields are excellent. 3-Methylindole gives the corresponding adduct with substitution occurring at the 2-position but the yield is not as satisfactory as in the previous reactions. The mechanism of the

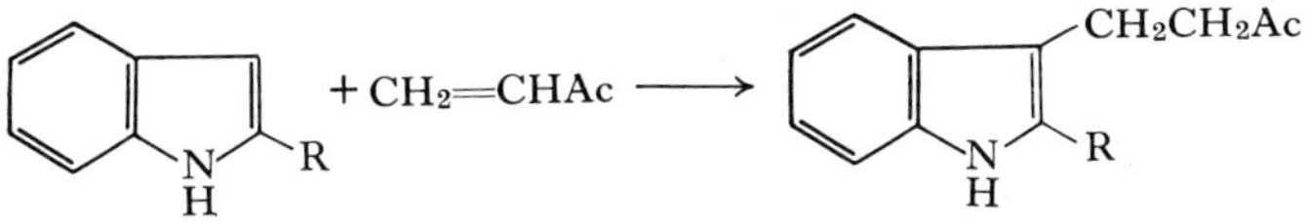

L1 + $CH_2{=}CHAc$ (**L2**) ⟶ **L3**, R = H; **L4**, R = Me (CH_2CH_2Ac at the 3-position)

reaction is considered to involve acid-catalyzed electrophilic attack of the ketone on the 3-position of indole. Robinson and Smith (213) have presented

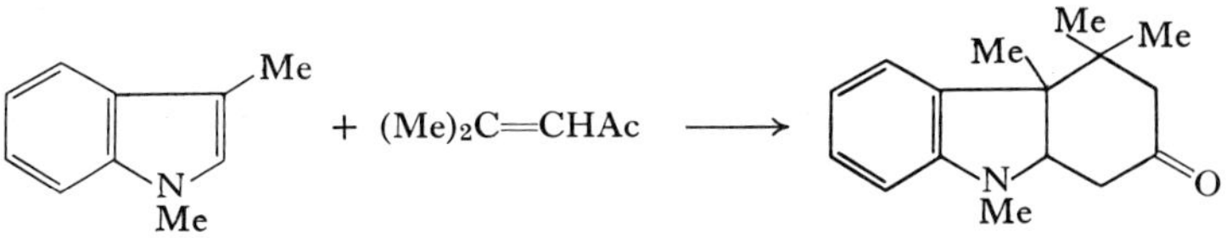

L5 + $(Me)_2C{=}CHAc$ (**L6**) ⟶ **L7**

L5 + **L6** ⟶ **L8** ⇌ **L9** ⟶ **L7**

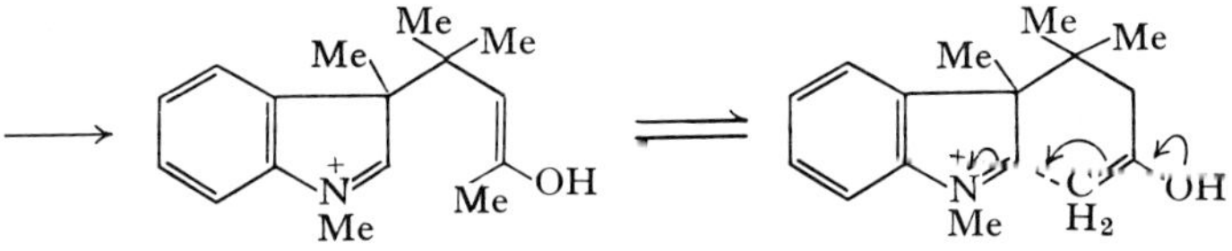

evidence that the product of reaction of mesityl oxide with 1,3-dimethylindole (48, 217) is **L7**. A straightforward rationalization of the formation of **L7** has been presented (213). Evidence against the alternative possibility

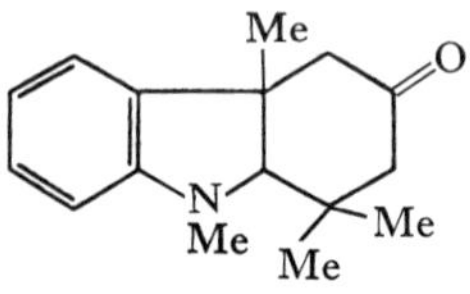

L10

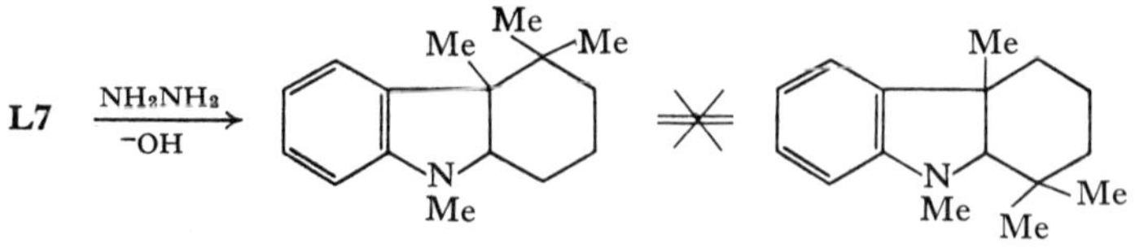

L7 $\xrightarrow[^{-}OH]{NH_2NH_2}$ **L11** ⇸ **L12**

L10 was obtained when the reduction product **L11** was shown to be structurally different from a synthetic sample of **L12**. The reaction of 3-methylcyclohexenone with 1,3-dimethylindole results in the formation of **L15** (48, 217). This transformation can be rationalized by a similar sequence of steps. The orientation of the α,β-unsaturated ketone in the products **L7** and **L15** is correctly predicted if it is assumed that the electron deficient β-carbon atom attacks at C-3 of the indole ring. This feature of this reaction is of significance in connection with the mechanism of substitution in 3-substituted indoles and will be discussed more fully in Section N of this chapter.

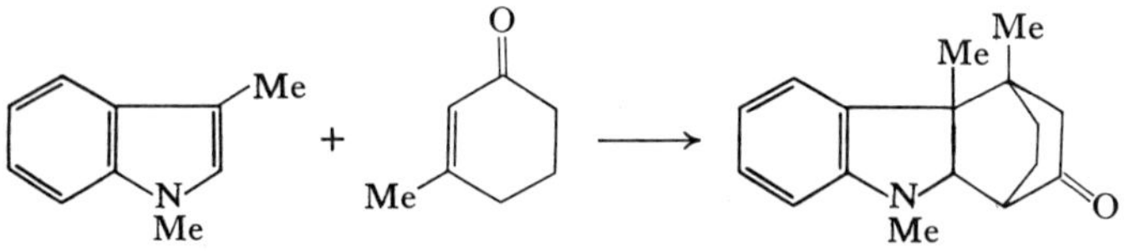

L13 + **L14** ⟶ **L15**

The reaction of 2-methylindole and certain α,β-unsaturated carbonyl compounds in alcoholic hydrogen bromide proceeds by reaction with indole at both the β-carbon and the carbonyl group, followed by oxidation to give alkylidene 3*H*-indolium salts (267). The equations below illustrate these transformations.

2-Methylindole and benzoylacetylene give the adduct **L24** (113). Closely related to this reaction is the alkylation of 2-phenylindole with either

L16 + $CH_2{=}CHCHO$ $\xrightarrow[\text{HBr}]{\text{EtOH}}$ L18

L17

L16 + $CH_2{=}CHAc$ (L19) $\xrightarrow{\text{EtOH, HBr}}$ L21

L16 + $MeCH{=}CHCHO$ (L20) $\xrightarrow{\text{EtOH, HBr}}$ L21

β-chlorovinyl phenyl ketone or β-anilinovinyl phenyl ketone to give **L28** (51). In these reactions addition of the indole to the olefinic linkage is followed by elimination of the β-substituent.

L22 + $PhCOC{\equiv}CH$ (L23) $\longrightarrow$ L24

L25 + $ClCH{=}CHCOPh$ (L26) or $PhNHCH{=}CHCOPh$ (L27) $\longrightarrow$ L28

The 3-substituted indole **L29** has been found to condense with pyruvic acid to give the 1,2-bis(2-indolyl)ethane **L34** (285). This transformation can be satisfactorily rationalized by the steps shown below. The addition of the second indole ring occurs by reaction with the acrylic acid **L32.**

The reactions of indoles with nitroethylenes have received attention because the resulting adducts are potential precursors of tryptamines. This reaction, which was discovered by Noland and co-workers (164), has been extended (161, 165) and applied by others (1). Indole undergoes conjugate addition to nitroethylene in 20% yield. Indolylmagnesium iodide reacts similarly. In addition to nitroethylene, β-nitrostyrene and 1-nitropropene have been found to react with indole or its magnesium derivative. 3-Methylindolylmagnesium iodide gives only a low yield of 3-methyl-1-(2-nitroethyl)indole (1). A 1,3-dialkylated derivative is reported to result from the reaction of 5-benzyloxyindolylmagnesium iodide with nitroethylene (1). The indole itself gives a 45% yield of the 3-nitroethyl adduct **L41** (165).

R + $AcCO_2H$ ⟶ R, COH, Me, CO_2H ⟶

L29 **L30** **L31**

CH_2, CO_2H

L32

L29 + **L32** ⟶ R, R, $CHCH_2$, CO_2H ⟶

L33

R, R, CH_2CH_2

L34

R = —N ═O

Indole and 2-methylindole add to α-acetamidoacrylic acid in acetic acid containing acetic anhydride (238). The role of acetic anhydride in the addition has not been clarified and it is therefore not clear that this reaction is a simple case of acid-catalyzed conjugate addition.

Base-catalyzed addition of indole to acrylonitrile gives the 1-cyanoethyl derivative, as will be discussed shortly. There are, however, reports that in the presence of copper borate (210) or copper acetate (254), 3-(2-cyanoethyl)indole is obtained. An excellent yield of the 3-alkylation product has also been reported for 2-methylindole (254). 1,2-Dimethylindole adds to acrylonitrile to give the 3-alkylated adduct in the presence of copper salts

$$\textbf{L35} + CH_2{=}CHNO_2 \longrightarrow \textbf{L37}\ (3\text{-}CH_2CH_2NO_2)$$

L35 L36 L37

$$\textbf{L35} + PhCH{=}CHNO_2 \longrightarrow \textbf{L39}\ (3\text{-}CH(Ph)CH_2NO_2)$$

L38 L39

$$\textbf{L40}\ (5\text{-}PhCH_2O) + \textbf{L36} \longrightarrow \textbf{L41}\ (5\text{-}PhCH_2O,\ 3\text{-}CH_2CH_2NO_2)$$

L40 L41

but apparently not in their absence (145). 1,2-Dimethyl-5-methoxyindole is also alkylated in excellent yield by acrylonitrile in the presence of cuprous

$$\textbf{L42} + CH_2{=}CHCN \xrightarrow{Cu^+} \textbf{L44}\ (3\text{-}CH_2CH_2CN)$$

L42 L43 L44

chloride (145). The role of the copper catalyst does not seem to have been defined. It may retard competing polymerization of acrylonitrile. Indolylmagnesium bromides are reported to add to methyl α-cyanoacrylate at the 3-position (31).

Indoles react with tetracyanoethylene very rapidly to give charge-transfer complexes but on standing 3-tricyanovinyl derivatives are formed (75, 169, 222) presumably via base-catalyzed elimination of cyanide from the adduct **L47**. 3-Methylindole is converted to 3-methyl-1-tricyanovinylindole in 77% yield under slightly more vigorous conditions than those required with indoles having open 3-positions (169).

The reactions of several indoles with 2- and 4-vinylpyridine have been investigated by Gray and co-workers (85, 86). The electrophilic properties of these olefins, of course, derive from electron withdrawal by the pyridine ring (66). The pyridylethylation procedure gives yields of 50–75% with 2- and 4-vinylpyridine, but lower yields are reported for 5-ethyl-2-vinylpyridine (85). 3-Methylindole gives a low yield of the 1-alkylation product (86).

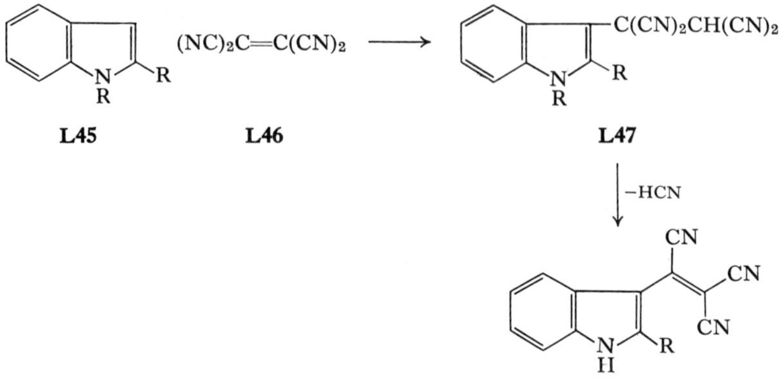

$(NC)_2C{=}C(CN)_2$ ⟶ —$C(CN)_2CH(CN)_2$

L45 L46 L47

$-HCN$

CN CN CN

L48

Indoles with unsubstituted 3-positions react with *o*- and *p*-benzoquinones to give 3-indolylbenzoquinones (36, 38, 150). The reaction is acid-catalyzed and presumably involves addition of the indole to give an indolylhydroquinone. This adduct is subsequently oxidized by excess benzoquinone (38).

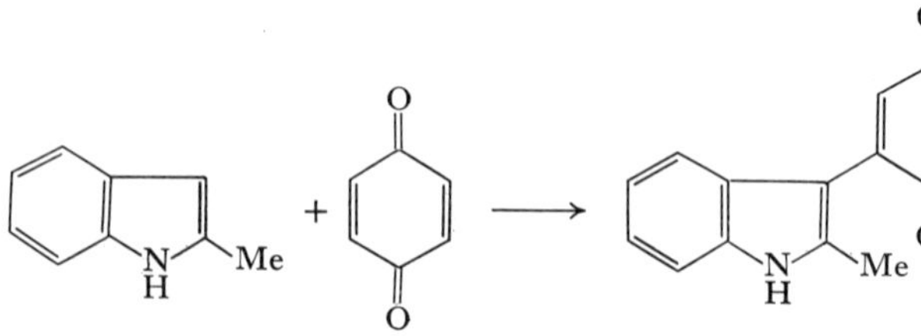

L49 + L50 ⟶ L51 Ref. (38, 150)

L49 (excess) + **L50** ⟶ **L52** Ref. (38)

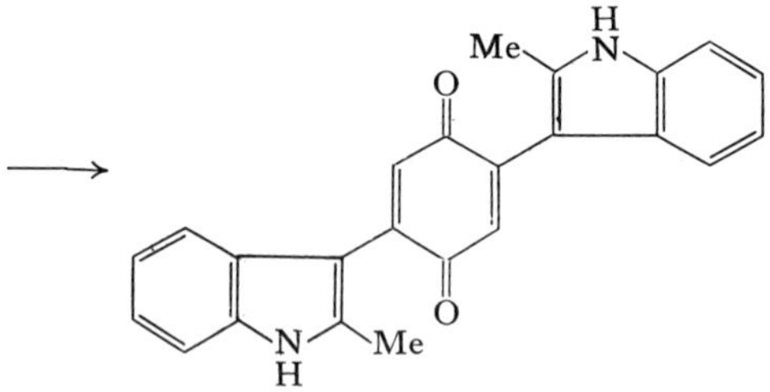

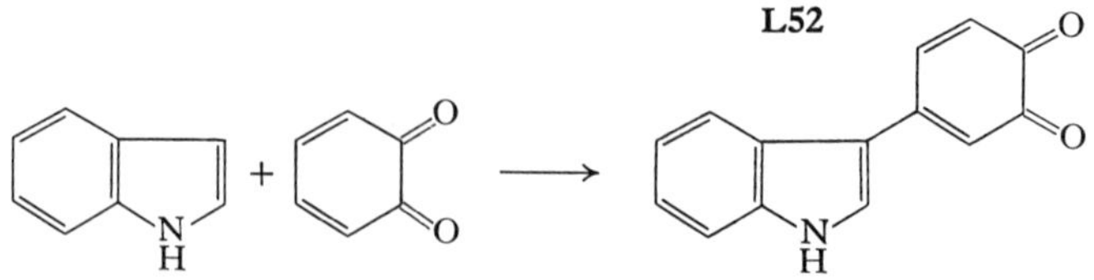

L53 + L54 ⟶ L55 Ref. (38)

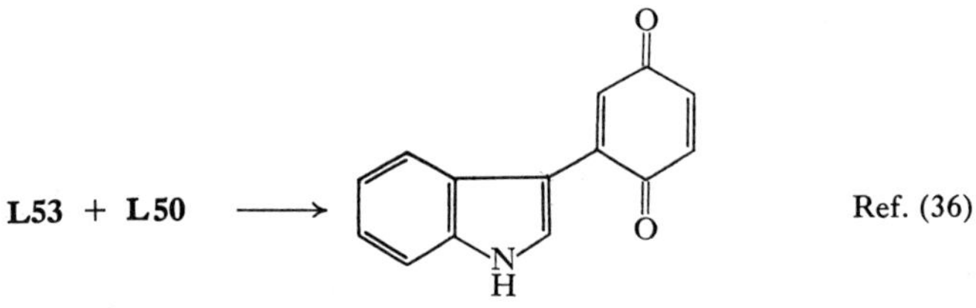

L56

Compound **L51** has been prepared by an alternative synthesis (36) and the other structure assignments (38, 150) thus seem secure. Indole and its 1-methyl and 1-ethyl derivatives react with tetrachloro-*o*-benzoquinone to give adducts assigned structure **L61** (103). The presence of the chlorine

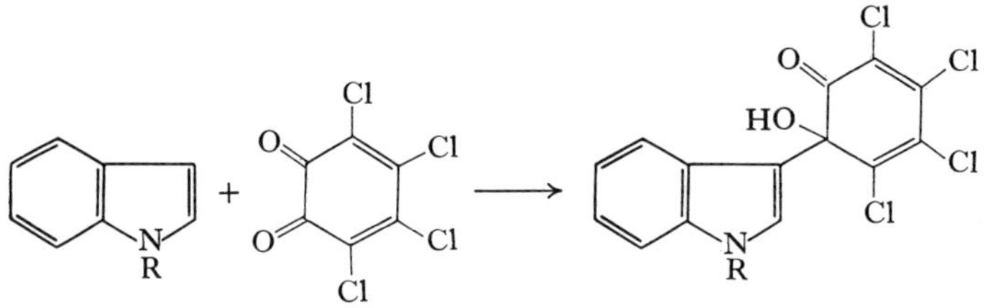

L57 R = H
L58 R = Me
L59 R = Et

L60

L61

substituents prevents the reaction from following the course observed with *o*-benzoquinone. The structural assignment is based on degradative work.

The reaction of 3-alkylindoles with *p*-benzoquinone takes an interesting course (160). High melting adducts containing two molecules of the indole and one of *p*-benzoquinone less a molecule of hydrogen are obtained (38, 160). The reaction is quite general and structural studies have permitted the assignment of structure **L64** to these adducts (160). Several possible mechanisms have been discussed (160). The basic steps must involve electrophilic attack of the quinone at the 3-position of the indole, cyclization

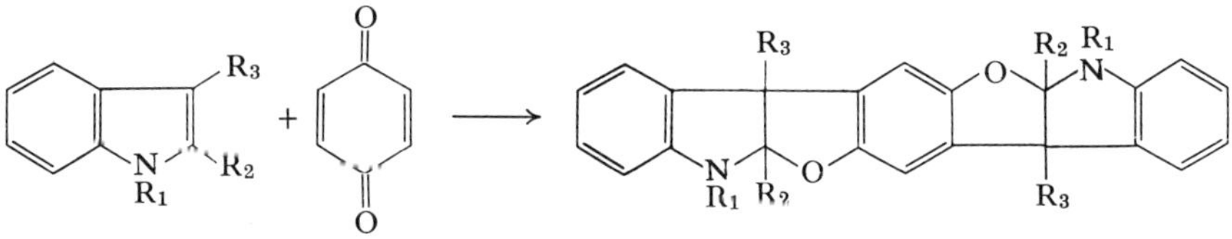

L62 **L63** **L64**

of the hydroquinone adduct (as in **L65** → **L66**) and an oxidation. The sequence of events remains undetermined. The adducts **L67** which are unsubstituted in the 2-position undergo rearrangement to the hydroquinones **L70** via the mechanism shown below (160).

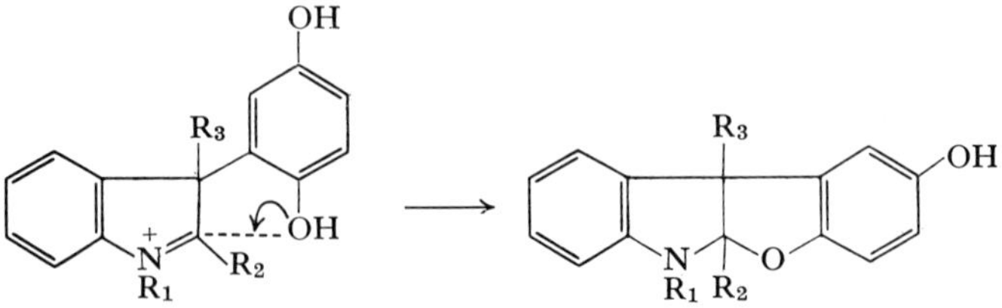

L65 → **L66**

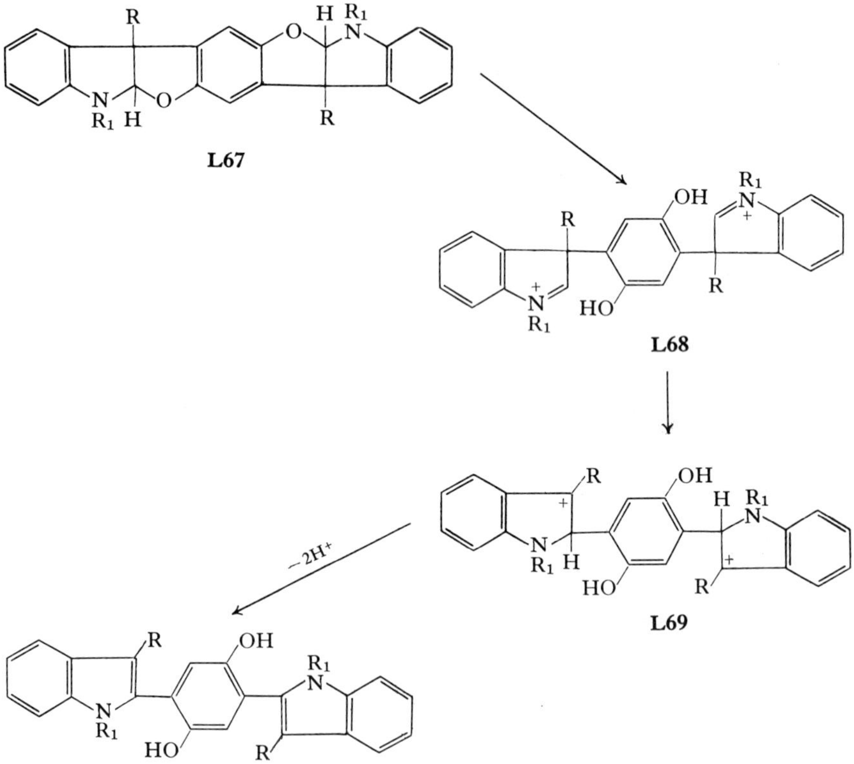

L67 → **L68** → **L69** $\xrightarrow{-2H^+}$ **L70**

TABLE X □ ADDITIONS OF INDOLES TO ELECTRON-DEFICIENT OLEFINS AND ACETYLENES

Indole substituent	Acceptor	Position of reaction	Yield (%)	Ref.
	Unsaturated ketones			
None	Methyl vinyl ketone	3	75	(249)
2-Me	Methyl vinyl ketone	3	84	(249)
2-Me	Benzoylacetylene	3	—	(113)
2-Me	Mesityl oxide	3	55	(48, 216)
2-Me	β-Anilinoacrylophenone	3	—	(51)
2-Me	β-Cyanoacrylophenone	3	—	(52)
3-Me	Methyl vinyl ketone	2	—	(249)
	Nitroethylene derivatives			
None	Nitroethylene	3	20	(161, 164)
None	β-Nitrostyrene	3	78	(161, 164)
Mg deriv.	β-Nitrostyrene	3	39	(161, 164)
2-Me	4-Methoxy-β-nitrostyrene	3	—	(221)
2-Ph	4-Methoxy-β-nitrostyrene	3	—	(221)
5-Benzyloxy	β-Nitrostyrene	3	72	(165)
	Acrylic acid derivatives			
None	α-Acetamidoacrylic acid	3	57	(238)
None	Diethyl β,β-bis(methylthio)-methylenemalonate	3	42	(130)
	Acrylonitrile derivatives			
None	Acrylonitrile	3	71	(254)
None	Tetracyanoethylene	3	76	(222)
None	β,β-Bis(methylthio)methylene-malonitrile	3	50	(130)
1-Me	Tetracyanoethylene	3	59	(169)
1,2-diMe	Acrylonitrile	3	51	(145)
2-Me	Acrylonitrile	3	81	(254)
2-Me	Tetracyanoethylene	3	78	(169)
3-Me	Tetracyanoethylene	1	77	(169)
5-Ethoxy	Acrylonitrile	3	—	(153)
	Quinones			
None	o-Benzoquinone	3	45	(36, 38)
2-Me	p-Benzoquinone	3	90–100	(38, 150)
	Vinylpyridines			
None	4-Vinylpyridine	3	69	(85)
2-Me	4-Vinylpyridine	3	54	(86)
	Vinyl sulfones			
2-Me	p-Methylphenyl β-morpholino-vinyl sulfone	3	—	(147)

N-Phenylmaleimide (53) and diethyl azodicarboxylate (53) are among the electron-deficient unsaturated molecules which have been observed to give adducts with indoles involving reaction at C-3. Table X summarizes various types of reactions involving conjugate addition of indoles to unsaturated electron-deficient systems.

Base-catalyzed reactions of indoles with electron-deficient olefins result primarily in alkylation at the nitrogen. The reactions are regarded as Michael-type addition of the indolyl anion to the acceptor. The most nucleophilic site in the conjugate base is at nitrogen and 1-alkylation takes place. Cyanoethylations have been extensively examined. These reactions, along with examples involving other acceptors, are summarized in Table XI and do not seem to require additional comment.

Noland and Hammer (162) have observed a novel rearrangement which is apparently initiated by intramolecular reaction of the 3-position of an indole ring with fumaryl and maleyl residues. The rearrangement results in the conversion of maleyl and fumaryl diindoles to indole-3-succinic acids. The likely mechanism of the reaction is outlined below:

H B N N C—CH=CHCO$_2^-$ O

L71

N N $^-$O CO_2^-

L72

CO_2^- CHCH$_2$CO$_2^-$ N H + N H OH$^-$ N N O CO_2^-

L75 **L74** **L73**

The favorable steric environment for reaction at the 3-position of the indole nucleus is undoubtedly an important factor in permitting this rearrangement to proceed with facility. Studies with maleyl and fumaryl derivatives of mixed dimers of indoles have established that an N–H group must be present to permit rearrangement and that the succinyl residue is bonded to the indole portion of the diindole after the rearrangement.

TABLE XI ☐ BASE-CATALYZED ADDITIONS OF INDOLES TO ELECTRON-DEFICIENT OLEFINS AND ACETYLENES

Indole substituent	Acceptor	Position of reaction	Yield (%)	Ref.
	Acrylic acid derivatives			
None	Ethyl acrylate	1	30	(255)
2-Carbethoxy	Ethyl acrylate	1	48	(207)
	Acrylonitrile derivatives			
None	Acrylonitrile	1	95	(255)
None	Methacrylonitrile	1	74	(71)
2-Me	Acrylonitrile	1	77	(255)
2-Ph	Acrylonitrile	1	82	(253)
2-Carbethoxy	Acrylonitrile	1	90	(13)
2-Carbethoxy-5-methoxy	Acrylonitrile	1	83	(13)
5-Chloro	Acrylonitrile	1	—	(12)
	Acetylenes			
None	Acetylene	1	—	(209)
2-Me	Acetylene	1	—	(209)
	Vinylpyridines			
3-Me	4-Vinylpyridine	1	38	(86)
	Fluoroolefins			
None	Chlorotrifluoroethylene	1	88	(70)
None	Tetrafluoroethylene	1	74	(70)

M. MISCELLANEOUS ELECTROPHILIC SUBSTITUTIONS

Reaction of indolylmagnesium bromide with ethanesulfenyl chloride gives a complex mixture containing 3,3′-indolyl sulfide, 3-ethylindole, and 3-ethylthioindole as well as other products (112). 3-Methylindole gives 2-(2,4-dinitrophenylthio)-3-methylindole on reaction with 2,4-dinitrophenylsulfenyl chloride (73). 3-Phenylindole gives an 84% yield of 2-methylthio-3-phenylindole on reaction with methylsulfenyl chloride (282). A number of 3-substituted indoles have been found to give 2,2′-indolyl disulfides on reaction with disulfur dichloride (282).

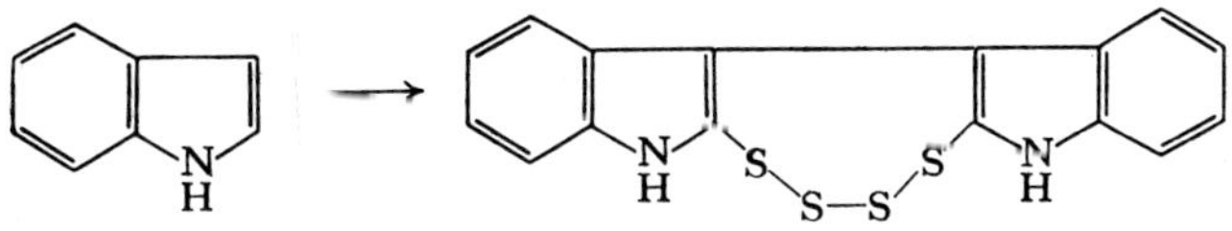

N H S S S S N H

M1 **M2**

Indole gives an excellent yield of 3-thiocyanoindole on reaction with thiocyanogen (84). Reaction of sulfur in dimethylformamide gives a compound containing two indole rings and four sulfur atoms (43). The structure **M2** has been proposed, but the evidence presented is not unequivocal (43).

Szmuszkovicz (250) has studied the reactions of indoles with thionyl chloride and sulfuryl chloride. Methyl 1-methylindole-2-carboxylate gives sulfinyl chloride **M4** on reaction with thionyl chloride, and subsequent reactions of **M4** have been described. Ethyl indole-2-carboxylate also gives

$$\text{M3} + SOCl_2 \longrightarrow \text{M4}$$

a sulfinyl chloride on reaction with thionyl chloride (250). Chlorination reactions occur with sulfuryl chloride. Indoles react with thiourea in the presence of iodine to give 3,3′-indolyl disulfides (286).

2-Methylindole is reported to react with phosphorus pentachloride to give 3-(dichlorophosphoryl)-2-methylindole (**M6**) (198).

$$\text{M5} \xrightarrow[(2)\ H_2O]{(1)\ PCl_5} \text{M6}$$

N. THE MECHANISM OF ELECTROPHILIC SUBSTITUTION IN 3-SUBSTITUTED INDOLES

Two general possibilities for 2-substitution in 3-substituted indoles must be considered. The electrophile can attack C-2 directly or it can attack the more electron-rich 3-position and subsequently rearrange intramolecularly to C-2. Reversible, nonproductive attack at C-3 could accompany either type of C-2 substitution. (See Chapter VI for a general discussion of rearrangements involving migration of substituents on the indole ring.)

The intermediates **N4** must be considered to be more stable than **N2** in the absence of overriding steric repulsions between R and E (for example, recall the protonation of 3-methylindole). But, for steric reasons, one can argue that **N2** might be the kinetically favored intermediate. Jackson and Smith (109) have recently reported a very interesting experiment which provides evidence for a substitution following the mechanism **N4** → **N5** → **N6**. The tritium-labeled indole **N7** was cyclized to tetrahydrocarbazole. The label was found to be distributed in the product as predicted by scheme **N8** → **N9** + **N10** but not by **N11** → **N12**.

N1 + E^+ → **N2** $\xrightarrow{-H^+}$ **N3**

N1 + E^+ → **N4** → **N5** $\xrightarrow{-H^+}$ **N6**

N7 ($C^*H_2CH_2CH_2OH$) $\xrightarrow{BF_3 \cdot OEt_2}$ **N8** → **N9** + **N10**

N7 $\xrightarrow{BF_3 \cdot OEt_2}$ **N11** → **N12**

Jackson and Smith (107) have suggested that 2,3-disubstituted indoles which arise from treatment of 3-substituted indolylmagnesium halides with various allylic halides are also formed by initial attack at C-3, followed by rearrangement. It has been demonstrated that the rearrangement of 3-allyl-3-methylindolenine occurs very readily in acidic solution. Jackson and co-workers have demonstrated the generality of such rearrangements and shown that the group which migrates is that which has the highest electron density (108).

Me
$CH_2CH{=}CH_2$ $\xrightarrow{\text{3N HCl}}$ $\xrightarrow[\text{PtO}_2]{\text{H}_2}$ Me, $CH_2CH_2CH_3$

N13 **N14**

Me, MgI + $(Me)_2C{=}CHCH_2Br$ $\longrightarrow$ $\xrightarrow[\text{PtO}_2]{\text{H}_2}$

N15 **N16**

Me, $CH_2CH_2CH(Me)_2$ + Me, $CH_2CH_2CH(Me)_2$

N17 **N18**

Similarly, 3-benzyl-2-*p*-methoxybenzylindole is formed when 3-*p*-methoxybenzylindole is benzylated with benzyl chloride as well as when 3-benzylindole is alkylated with *p*-methoxybenzyl chloride. This result indicates that 3-benzyl-3-*p*-methoxybenzyl-3*H*-indole is an intermediate in both reactions (21a).

Morrison and co-workers (152) have reported an interesting series of results which can be explained in terms of Jackson's views on the mechanism of substitution of 3-alkylindoles. The cyclization of the olefin **N19** in acidic media gives **N24**, whereas intramolecular alkylation involving the halide **N20** gives **N23**. Both processes would be expected to involve the carbonium ion **N21**. The formation of **N22** may be involved in the production of **N24** but it would be unstable to rearrangement, giving **N24** in acidic solution.

Robinson and Smith (213) have discussed the question of C-2 versus C-3 as the site of initial attack of electrophiles in connection with the structure of the product formed from 3-methylindole and mesityl oxide. The orientation of the mesityl oxide residue in the product suggests that in this

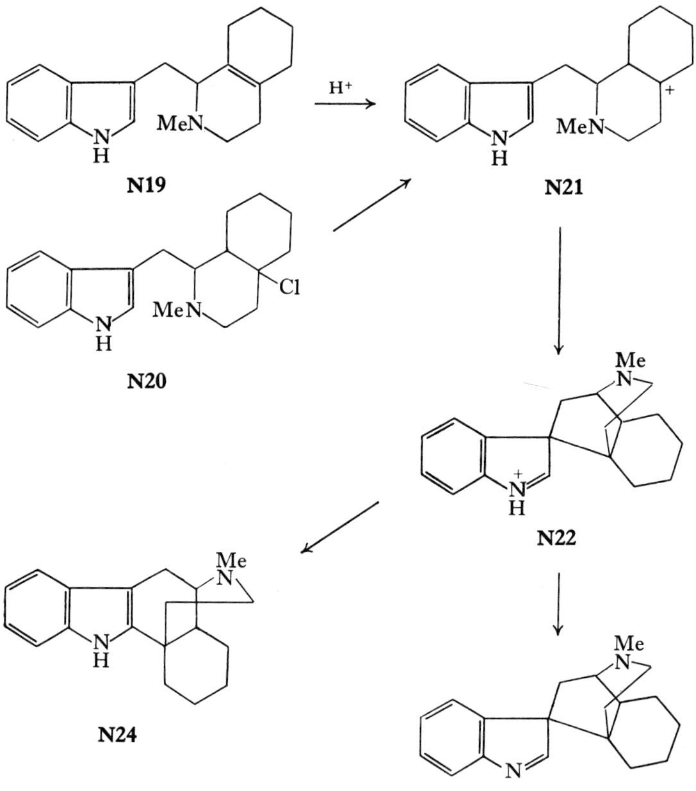

H+

MeN

N
H

N19

MeN

N
H

N21

MeN

Cl

N
H

N20

Me
N

+
N
H

N22

Me
N

N
H

N24

Me
N

N

N23

case the initial attack was at C-3. The view (172) that initial attack of the mesityl oxide would be at C-2 predicts the wrong structure for the adduct.

The internal cyclization **N27** → **N28** is interesting in that it proceeds in high yield (91%) in neutral ethanol at room temperature (262). Since such conditions are normally not expected to cause skeletal rearrangements, it was suggested (172) that the cyclization of **N27** should be viewed as a direct cyclization at C-2. Recent synthesis of the carbinolamine **N30**, a precursor of the indolenine **N31**, and investigation of its stability toward rearrange-

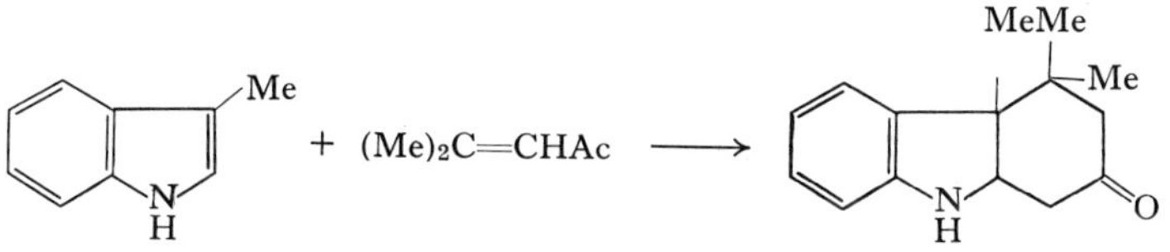

Me

N
H

N25

+ $(Me)_2C{=}CHAc$ ⟶

MeMe

Me

O

N
H

N26

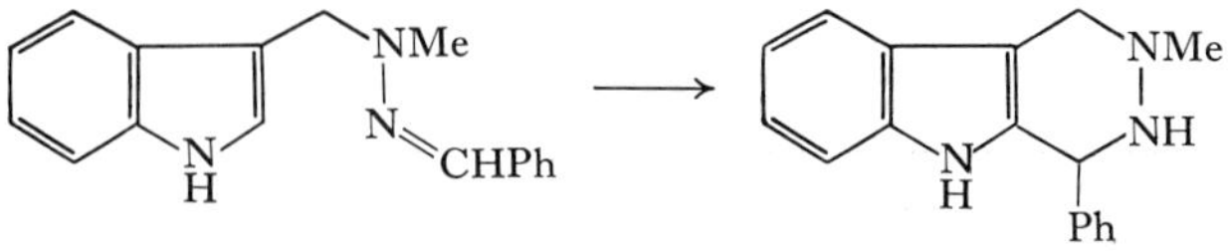

N27 → **N28**

ment has shown that even spectroscopic grade ethanol is sufficiently acidic to cause rearrangement of **N31** to **N32** (105). A rearrangement in the course of the conversion of **N27** to **N28** cannot, therefore, be ruled out.

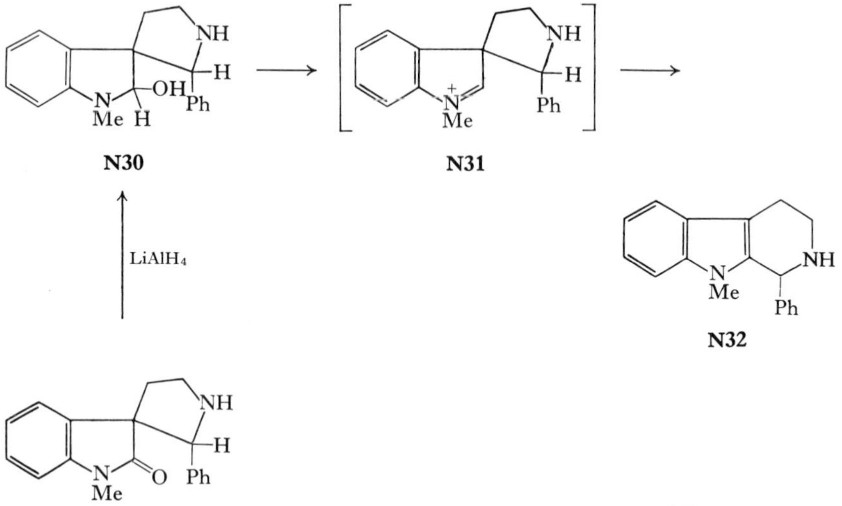

N29 —LiAlH₄→ **N30** → [**N31**] → **N32**

N33 + **N34** —H⁺→ **N35** → **N36** —several steps→ **N37**

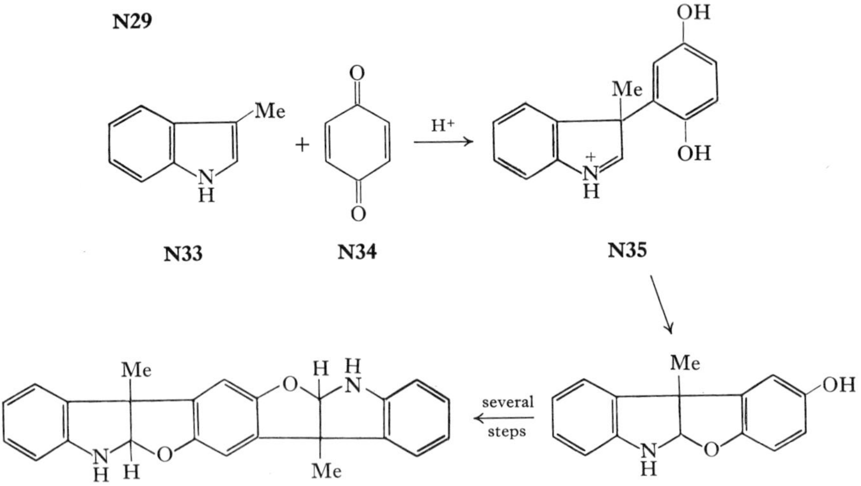

The adducts formed by quinones and 3-substituted indoles (160) can be considered to be the result of intramolecular trapping of the initial 3-substitution product. (For a more complete discussion see Section L.)

The bulk of the evidence available at this time strongly suggests that electrophilic substitution in 3-substituted indoles usually proceeds by initial attack at C-3 followed by rearrangement to the 2-substituted indole. The mechanism explains such results as the tritium scrambling observed by Jackson and Smith (109) and there seems to be no evidence which demands direct substitution at C-2.

O. THEORETICAL TREATMENT OF INDOLE REACTIVITY AND SUBSTITUENT EFFECTS

The results of experimental work on electrophilic aromatic substitution would seem to establish the general order of positional reactivity of the neutral indole molecule as $3 > 2 > 6 > 4 > 5 > 7$. The partial ordering $3 > 6 > 4 > 5$ is firmly established by studies on nitration (Section C). Data in Section I indicate the order $3 > 2 > 6 > 5$ for Friedel–Crafts type acylations. Substitution at C-7 is seldom observed. Combination of these data give the order above. Ridd (211) has presented a useful discussion of theoretical methods for predicting reactivities at various positions on heterocyclic rings. One of the approaches has involved the assumption that the pi electron distribution of the ground state heterocycle will govern orientation in substitution reactions. The predicted order of reactivity toward electrophilic substitution is assumed to correspond to the pi electron density of the molecule as obtained by molecular orbital calculations. The positions of high density, of course, are assumed to be the most reactive. In terms of reaction mechanism, this assumes that the transition states must greatly resemble the ground state molecule. In fact, since the electrophile is not taken into account in any way in the calculation, all polarization and partial bonding which occur as the electrophile approaches the ring are ignored. Brown and Coller have calculated pi electron densities for indole using the Hückel method (34). When appropriate parameters are introduced into the calculation to take account of the electronegativity of the nitrogen atom, the pi density order is $3 > 2 > 4 \sim 5 \sim 6 \sim 7$. This order is in reasonable agreement with the experimental order although it provides no meaningful prediction of the order of reactivity in the carbocyclic ring. The pi density shown in Fig. 3 gives the order $3 > 5 > 7 > 2 > 6 \sim 4$ and thus, with the exception of C-3, does not correlate with observed reactivity.

A second theoretical approach calculates the localization energy for electrophilic attack at each ring position. The localization energy is essentially the energy required to remove two electrons from the pi system for bonding to an attacking electrophile. Thus, to predict relative reactivity at C-2 and C-3, the relative energies of intermediate **O1** and **O2** are calculated. The mechanistic assumption implicit in this approach is that the transition

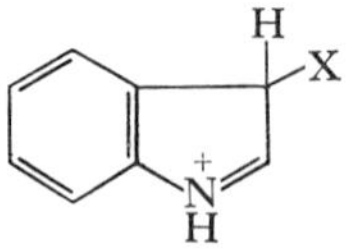

O1 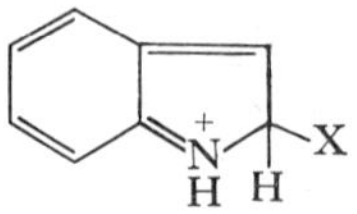 **O2**

states for substitution resemble σ-complexes such as **O1** and **O2**. Brown and Coller (34) used the Hückel method to calculate localization energies for the various positions on the indole ring. The order $3 > 2 > 4 > 7 > 6 > 5$ was obtained for electrophilic substitution when optimum parameter values were used. Thus, the order $3 > 2 >$ carbocyclic is again obtained but the carbocyclic positions are not correctly ordered. Recently, more sophisticated molecular orbital approaches have been applied (94a) to calculation of localization energies for C-2 and C-3 of the indole ring. Both the extended Hückel theory and the complete neglect of differential overlap method (CNDO/2) correctly predict $3 > 2$ but the various carbocyclic positions were not considered.

In the frontier electron density method (79) electrophilic attack is assumed to occur at the ring position which has the highest density of the two electrons in the highest occupied orbital of the ground state molecule. Specific effects of the electrophile are neglected, but attention is focused on only a single orbital as opposed to the calculation of total pi density. The distribution of the highest occupied level is considered to govern the site of attack. The ordering calculated is $3 \gg 4 > 7 > 2 \sim 6 > 5$. Again, the most reactive position is correctly recognized but the remaining order is not in agreement with the experimental order.

Each of these methods neglects possible steric effects which might operate to decrease the observed reactivity of positions 3, 4, and 7. They also, of course, neglect differences in the structure of attacking electrophiles. Highly reactive electrophiles may result in transition states which closely resemble starting material, whereas less reactive species may result in a transition state more closely resembling a σ-complex. Thus, while ground state properties such as dipole moment (23) and structural parameters (63) (see Section A) may now be calculated with some accuracy by molecular orbital methods, the problem of theoretical prediction of reactivity is incompletely solved. Each approach mentioned above successfully predicts high reactivity at C-3 but none is able to order the other positions satisfactorily. Since the underlying assumptions in each approach are quite different from one another, the fact that each method is partially successful makes it difficult to judge the relative merits of the methods and leads to little mechanistic insight.

The transmission of resonance and inductive effects within the indole ring has only recently begun to attract attention. Study of the dissociation constants of 2- and 3-indolecarboxylic acids (25, 110, 148) and the rates of esterification and hydrolysis reactions (25, 110) do not provide a clear picture of the relative effectiveness of conjugation with groups at the 5- and 6-positions with other positions on the ring. 6-Nitroindole-3-carboxylic acid is a very slightly stronger acid than the 5-isomer. The 5- and 6-methyl- and 5- and 6-methoxyindole-2-carboxylic acids also have similar dissociation constants (110). Dissociation constant data for six 5- and 6-substituted indole-3-carboxylic acids are correlated satisfactorily by use of σ_m for the substituents at C-5 and σ_p for substituents at C-6 (148). Other attempts to separate substituent effects into portions transmitted through the two connections to a reacting site (110) give empirical equations which correlate available data but fail to provide insight into transmission mechanisms, since the relative importance of the transmission paths seems to vary with the nature of the reaction.

REFERENCES

1. Acheson, R. M., and Hands, A. R., *J. Chem. Soc.* p. 744 (1961).
2. Akkerman, A. M., de Jongh, D. K., and Veldstra, H., *Rec. Trav. Chim.* **70**, 899 (1951).
3. Alessandri, L., *Atti. Accad. Lincei* **24**, II, 194 (1915); *Chem. Abstr.* **10**, 1350 (1916).
4. Ames, D. E., Bowman, R. E., Evans, D. D., and Jones, W. A., *J. Chem. Soc.* p. 1984 (1956).
5. Anthony, W. C., *J. Org. Chem.* **25**, 2049 (1960).
6. Arnall, F., and Lewis, T., *J. Soc. Chem. Ind.*, (*London*) **48T**, 159 (1929).
7. Arnold, R. D., Nutter, W. M., and Stepp, W. L., *J. Org. Chem.* **24**, 117 (1959).
8. Bader, H., and Oroshnik, W., *J. Am. Chem. Soc.* **81**, 163 (1959).
9. Bailey, A. S., and Merer, J. J., *J. Chem. Soc., C*, p. 1345 (1966).
10. Ballantine, J. A., Barrett, C. B., Beer, R. J. S., Boggiano, B. G., Eardley, S., Jennings, B. E., and Robertson, A., *J. Chem. Soc.* p. 2227 (1957).
11. Bartlett, M. F., Dickel, D. F., and Taylor, W. I., *J. Am. Chem. Soc.* **80**, 126 (1958).
12. Basangoudar, L. D., and Siddappa, S., *J. Chem. Soc., C*, p. 2599 (1967).
13. Bell, J. B., Jr., and Lindwall, H. G., *J. Org. Chem.* **13**, 547 (1948).
14. Benington, F., Morin, R. D., and Clark, L. C., Jr., *J. Org. Chem.* **23**, 1977 (1958).
15. Berti, G., Da Settimo, A., and Livi, O., *Tetrahedron* **20**, 1397 (1964).
16. Berti, G., Da Settimo, A., and Segnini, D., *Tetrahedron Letters* No. 26, 13 (1960).
17. Berti, G., and Da Settimo, A., *Gazz. Chim. Ital.* **91**, 728 (1961).
18. Berti, G., Da Settimo, A., and Segnini, D., *Gazz. Chim. Ital.* **91**, 571 (1961).
19. Binks, J. H., and Ridd, J. H., *J. Chem. Soc.* p. 2398 (1957).
20. Birkofer, L., and Frankus, E., *Chem. Ber.* **94**, 216 (1961).
21. Biswas, K. M., and Jackson, A. H., *Tetrahedron* **24**, 1145 (1968).

21a. Biswas, K. M., and Jackson, A. H., *Tetrahedron* **25**, 227 (1969).
22. Blicke, F. F., *Org. Reactions* **1**, 303 (1942).
23. Bloor, J. E., and Breen, D. L., *J. Am. Chem. Soc.* **89**, 6835 (1967); Bloor, J. E. (private communication).
24. Borsche, W., and Groth, H., *Ann. Chem.*, **549**, 238 (1941).
25. Bowden, K., and Parkin, D. C., *Can. J. Chem.* **44**, 1493 (1966).
26. Bradsher, C. K., and Litzinger, E. F., Jr., *J. Org. Chem.* **29**, 3584 (1964).
26a. Bramley, R. K., and Grigg, R., *Chem. Commun.*, p. 99 (1969).
27. Brehm, W. J., *J. Am. Chem. Soc.* **71**, 3541 (1949).
28. Brehm, W. J., and Lindwall, H. G., *J. Org. Chem.* **15**, 685 (1950).
29. Breslow, R., "Organic Reaction Mechanisms," p. 13. Benjamin, New York, 1966.
30. Brieskorn, C., and Reiners, W., *Arch. Pharm.* **295**, 544 (1962).
31. British Patent, 705,652 (1954); *Chem. Abstr.* **49**, 4722 (1955).
32. Brown, J. B., Henbest, H. B., and Jones, E. R. H., *J. Chem. Soc.* p. 3172 (1952).
33. Brown, K., and Katritzky, A. R., *Tetrahedron Letters* p. 803 (1964).
34. Brown, R. D., and Coller, B. A. W., *Australian J. Chem.* **12**, 152 (1959).
35. Bruce, J. M., *J. Chem. Soc.* p. 1514 (1962).
36. Bruce, J. M., *J. Chem. Soc.* p. 2366 (1959).
37. Buchmann, G., and Trautmann, P., *J. Prakt. Chem.* [4] **32**, 1 (1966).
38. Bu'Lock, J. D., and Harley-Mason, J., *J. Chem. Soc.* p. 703 (1951).
39. Burr, G. O., and Gortner, R. A., *J. Am. Chem. Soc.* **46**, 1224 (1924).
40. Buu-Hoi, N. P., Bisagni, E., Royer, R., and Routier, C., *J. Chem. Soc.* p. 625 (1957).
41. Cardani, C., Piozzi, F., and Casnati, G., *Gazz. Chim. Ital.* **85**, 263 (1955).; *Chem. Abstr.* **50**, 10709 (1956).
42. Cardillo, B., Casnati, G., Pochini, A., and Ricca, A., *Tetrahedron* **23**, 3771 (1967).
43. Carpenter, W., Grant, M. S., and Snyder, H. R., *J. Am. Chem. Soc.* **82**, 2739 (1960).
44. Challis, B. C., and Long, F. A., *J. Am. Chem. Soc.* **85**, 2524 (1963).
45. Chen, F. Y., and Leete, E., *Tetrahedron Letters* p. 2013 (1963).
46. Closs, G. L., and Schwartz, G. M., *J. Org. Chem.* **26**, 2609 (1961).
47. Closson, W. D., Roman, S. A., Kwiatkowski, G. T., and Corwin, D. A., *Tetrahedron Letters* p. 2271 (1966).
48. Cockerill, D. A., Robinson, R., and Saxton, J. E., *J. Chem. Soc.* p. 4369 (1955).
49. Colonna, M., and Bruni, P., *Boll. Sci. Fac. Chim. Ind. Bologna* **23**, 401 (1965); *Chem. Abstr.* **64**, 17536 (1966).
50. Colonna, M., Bruni, P., and Guerra, A. M., *Gazz. Chim. Ital.* **96**, 1410 (1966).
51. Colonna, M., Bruni, P., and Monti, A., *Gazz. Chim. Ital.* **95**, 868 (1965).
52. Colonna, M., and Marchetti, L., *Gazz. Chim. Ital.* **96**, 1175 (1966).
53. Colonna, M., and Monti, A., *Gazz. Chim. Ital.* **92**, 1401 (1962).
54. Cooper, A. R., Crowne, C. W. P., and Farrell, P. G., *Trans. Faraday Soc.* **62**, 18 (1966).
55. Cornforth, R. H., and Robinson, R., *J. Chem. Soc.* p. 680 (1942).
56. Curtin, D. Y., and Fraser, R. R., *J. Am. Chem. Soc.* **80**, 6016 (1958).
57. Dahlbom, R., and Misiorny, A., *Acta Chem. Scand.* **9**, 1074 (1955); *J. Am. Chem. Soc.*, **82**, 2397 (1960).
58. Daly, J. W., and Witkop, B., *J. Am. Chem. Soc.* **89**, 1032 (1967).
59. Da Settimo, A., *Gazz. Chim. Ital.* **92**, 150 (1962).
60. Da Settimo, A., and Saettone, M. F., *Tetrahedron* **21**, 823 (1965).
61. Da Settimo, A., and Saettone, M. F., *Tetrahedron* **21**, 1923 (1965).
62. De Graw, J. I., Kennedy, J. G., and Skinner, W. A., *J. Heterocyclic Chem.* **3**, 67 (1966).
63. Dewar, M. J. S., and Gleicher, G. J., *J. Chem. Phys.* **44**, 759 (1966).
64. Diels, O., and Dürst, W., *Chem. Ber.* **47**, 284 (1914).

65. Dobbs, H. E., *Chem. Commun.* p. 56 (1965); *J. Org. Chem.* **33**, 1093 (1968).
66. Doering, W. E., and Weil, R. A. N., *J. Am. Chem. Soc.* **69**, 2461 (1947).
67. Dostál, V., *Chem. Listy* **32**, 13 (1938).
68. Dyke, S. F., and Sainsbury, M., *Tetrahedron* **21**, 1907 (1965).
69. Eliel, E. L., and Murphy, N. J., *J. Am. Chem. Soc.* **75**, 3589 (1953).
70. England, D. C., Melby, L. R., Dietrich, M. A., and Lindsey, R. V., Jr., *J. Am. Chem. Soc.* **82**, 5116 (1960).
71. Erdtman, H., and Jönsson, A., *Acta Chem. Scand.* **8**, 119 (1954).
72. Fischer, E., *Ann. Chem.* **242**, 372 (1887).
73. Fontana, A., Marchiori, F., Moroder, L., and Scoffone, E., *Tetrahedron Letters* p. 2985 (1966).
74. Foster, R., and Fyfe, C. A., *J. Chem. Soc., B* p. 926 (1966).
75. Foster, R., and Hanson, P., *Tetrahedron* **21**, 255 (1965).
76. Freter, K. R., *Can. J. Chem.* **45**, 2628 (1967).
77. Fritz, H. E., *J. Org. Chem.* **28**, 1384 (1963).
78. Fuhlhage, D. W., and VanderWerf, C. A., *J. Am. Chem. Soc.* **80**, 6249 (1958).
79. Fukui, K., Yonezawa, T., Nagata, C., and Shingu, H., *J. Chem. Phys.* **22**, 1433 (1954).
80. Ganellin, C. R., Hollyman, D. R., and Ridley, H. F., *J. Chem. Soc., C* p. 2220 (1967).
81. Ganellin, C. R., and Ridley, H. F., *Chem. & Ind. (London)* p. 1388 (1964).
82. Gaudion, W. J., Hook, W. H., and Plant, S. G. P., *J. Chem. Soc.* p. 1631 (1947).
83. Gompper, R., and Kutter, E., *Chem. Ber.* **98**, 1365 (1965).
84. Grant, M. S., and Snyder, H. R., *J. Am. Chem. Soc.* **82**, 2742 (1960).
85. Gray, A. P., and Archer, W. L., *J. Am. Chem. Soc.* **79**, 3554 (1957).
86. Gray, A. P., Kraus, H., and Heitmeier, D. E., *J. Org. Chem.* **25**, 1939 (1960).
87. Hamana, M., and Kumadaki, J., *Chem. & Pharm. Bull. (Tokyo)* **15**, 363 (1967).
88. Hanson, A. W., *Acta Cryst.* **17**, 559 (1964).
89. Harley-Mason, J., *J. Chem. Soc.* p. 2433 (1952).
90. Heath-Brown, B., and Philpot, P. G., *J. Chem. Soc.* p. 7165 (1965).
91. Heinzelman, R. V., Anthony, W. C., Lyttle, D. A., and Szmuszkovicz, J., *J. Org. Chem.* **25**, 1548 (1960).
92. Hellmann, H., Hallmann, G., and Lingens, F., *Chem. Ber.* **86**, 1346 (1953), and references therein.
93. Hellmann, H., and Opitz, G., *Ann. Chem.* **604**, 214 (1957).
94. Hellmann, H., and Teichmann, K., *Chem. Ber.* **91**, 2432 (1958).
94a. Hermann, R. B., *Intern. J. Quantum Chem.* **2**, 165 (1968).
95. Hinman, R. L., and Bauman, C. P., *J. Org. Chem.* **29**, 2437 (1964).
95a. Hinman, R. L., and Bauman, C. P., *J. Org. Chem.* **29**, 1206 (1964).
96. Hinman, R. L., and Lang, J., *J. Am. Chem. Soc.* **86**, 3796 (1964).
97. Hinman, R. L., and Lang, J., *Tetrahedron Letters* No. 21, 12 (1960).
97a. Hinman, R. L., and Lang, J., *J. Org. Chem.* **29**, 1449 (1964).
98. Hinman, R. L., and Shull, E. R., *J. Org. Chem.* **26**, 2339 (1961).
99. Hinman, R. L., and Whipple, E. B., *J. Am. Chem. Soc.* **84**, 2534 (1962).
100. Hishida, S., *J. Chem. Soc. Japan, Pure Chem. Sect.* **72**, 312 (1951); *Chem. Abstr.* **46**, 5038 (1952).
101. Hobbs, C. F., McMillin, C. K., Papadopoulos, E. P., and VanderWerf, C. A., *J. Am. Chem. Soc.* **84**, 43 (1962).
102. Hodson, H. F., and Smith, G. F., *J. Chem. Soc.* p. 3544 (1957).
103. Horner, L., and Spietschka, W., *Ann. Chem.* **591**, 1 (1955).
104. Hoshino, T., *Chem. Ber.* **85**, 858 (1952).

105. Jackson, A. H., and Smith, A. E., *Tetrahedron* **24**, 403 (1968).
106. Jackson, A. H., and Smith, A. E., *J. Chem. Soc.* p. 5510 (1964).
107. Jackson, A. H., and Smith, A. E., *Tetrahedron* **21**, 989 (1965).
108. Jackson, A. H., and Smith, P., *Tetrahedron* **24**, 2227 (1968).
109. Jackson, A. H., and Smith, P., *Chem. Commun.* p. 264 (1967); Jackson, A. H., Naidoo, B., and Smith, P., *Tetrahedron* **24**, 6119 (1968).
110. Jaffé, H. H., and Jones, H. L., *Advan. Heterocyclic Chem.* **3**, 209 (1964).
111. Jardine, R. V., and Brown, R. K., *Can. J. Chem.* **41**, 2067 (1963).
112. Jardine, R. V., and Brown, R. K., *Can. J. Chem.* **43**, 1298 (1965).
113. Johnson, A. W., *J. Chem. Soc.* p. 1626 (1947).
114. Johnson, H. E., U.S. Patent 3,197,479 (1965); *Chem. Abstr.* **63**, 13217 (1965).
115. Johnson, H. E., and Crosby, D. G., *J. Org. Chem.* **28**, 1246 (1963).
116. Julia, M., Igolen, H., and Lenzi, J., *Bull. Soc. Chim. France* p. 2291 (1966).
117. Julia, M., and Manoury, P., *Bull. Soc. Chim. France* p. 1946 (1964).
118. Julia, M., Sliwa, H., and Caubére, P., *Bull. Soc. Chim. France* p. 3359 (1966).
119. Julian, P. L., Meyer, E. W., and Printy, H. C., *in* "Heterocyclic Compounds" (R. C. Elderfield, ed.), Vol. 3, p. 42. Wiley, New York, 1952.
120. Kamal, A., Ali Quershi, A., and Ahmad, I., *Tetrahedron* **19**, 681 (1963).
121. Kamlet, M. J., and Dacons, J. C., *J. Org. Chem.* **26**, 220 (1961).
122. Kašpárek, S., and Heacock, R. A., *Can. J. Chem.* **44**, 2805 (1966).
123. Katritzky, A. R., *J. Chem. Soc.* p. 2581 (1955).
124. Katritzky, A. R., and Robinson, R., *J. Chem. Soc.* p. 2481 (1955).
125. Keller, K., *Chem. Ber.* **46**, 726 (1913).
126. Kiang, A. K., and Mann, F. G., *J. Chem. Soc.* p. 594 (1953).
127. Kiang, A. K., Mann, F. G., Prior, A. F., and Topham, A., *J. Chem. Soc.* p. 1319 (1956).
128. Kissman, H. M., and Witkop, B., *J. Am. Chem. Soc.* **75**, 1967 (1953).
129. Knowlton, M., Dohan, F. C., and Sprince, H., *Anal. Chem.* **32**, 666 (1960).
130. Kobayashi, G., Furukawa, S., Matsuda, Y., and Washida, Y., *Chem. & Pharm. Bull.* (*Tokyo*) **15**, 1871 (1967).
131. Koizumi, M., *Bull. Chem. Soc. Japan* **14**, 453 and 491 (1939).
132. Kost, A. N., Yudin, L. G., Budylin, V. A., and Yaryshev, N. G., *Khim. Geterotsikl. Soedin., Akad. Nauk Latv. SSR* p. 632 (1965); *Chem. Abstr.* **64**, 3457 (1966).
133. Kucherova, N. F., Evdakov, V. P., and Kochetkov, N. K., *J. Gen. Chem. USSR* (*English Transl.*) **27**, 1131 (1957).
134. Kühn, H., and Stein, O., *Chem. Ber.* **70**, 567 (1937).
135. Kunori, M., *Nippon Kagaku Zasshi* **81**, 1431 (1960); *Chem. Abstr.* **56**, 3441 (1962).
136. Kunori, M., *Nippon Kagaku Zasshi* **78**, 1798 (1957); *Chem. Abstr.* **54**, 1487 (1960).
137. Leete, E., *J. Org. Chem.* **23**, 631 (1958).
138. Leete, E., *J. Am. Chem. Soc.* **81**, 6023 (1959).
139. Leete, E., and Marion, L., *Can. J. Chem.* **31**, 775 (1953).
140. Leggetter, B. E., and Brown, R. K., *Can. J. Chem.* **38**, 1467 (1960).
141. Levkoev, I. I., and Bashkirova, A. Ya., *Zh. Prikl. Khim.* **35**, 688 (1962); *Chem. Abstr.* **57**, 2176 (1962).
142. Lind, C. J., and Sogn, A. W., U.S. Patent 3,012,040 (1961); *Chem. Abstr.* **58**, 508 (1963).
143. Lippmann, E., Richter, K., and Mühlstädt, M., *Z. Chem.* **5**, 186 (1965).
144. Majima, R., and Kotake, M., *Chem. Ber.* **63**, 2237 (1930).
145. Mann, F. A., and Tetlow, A. J., *J. Chem. Soc.* p. 3352 (1957).
146. Marchant, R. H., and Harvey, D. G., *J. Chem. Soc.* p. 1808 (1951).
147. Marchetti, L., and Passalacqua, V., *Ann. Chim.* (*Rome*) **57**, 1275 (1967).

148. Melzer, M. S., *J. Org. Chem.* **27**, 496 (1962).
149. Mingoia, Q., *Gazz. Chim. Ital.* **56**, 772 (1926); *Chem. Abstr.* **21**, 1117 (1927).
150. Möhlau, R., and Redlich, A., *Chem. Ber.* **44**, 3605 (1911).
151. Morrison, G. C., Waite, R. O., Caro, A. N., and Shavel, J., Jr., *J. Org. Chem.* **32**, 3691 (1967).
152. Morrison, G. C., Waite, R. O., Serafin, F., and Shavel, J., Jr., *J. Org. Chem.* **32**, 2551 (1967).
153. Murphy, H. W., *J. Pharm. Sci.* **53**, 272 (1964).
154. Nakazaki, M., *Bull. Chem. Soc. Japan* **32**, 838 (1959).
155. Nakazaki, M., *Bull. Chem. Soc. Japan* **34**, 334 (1961).
156. Nakazaki, M., *Bull. Chem. Soc. Japan* **32**, 588 (1965).
157. Neklyudov, A. D., Shchukina, L. A., and Suvorov, N. N., *J. Gen. Chem. USSR (English Transl.)* **37**, 747 (1967).
158. Nesmeyanov, A. N., Sazonova, V. A., and Drozd, V. N., *Dokl. Akad. Nauk SSSR* **165**, 575 (1965); *Chem. Abstr.* **64**, 6687 (1966).
159. Netherlands Patent Appl., 6,505,983 (1965); *Chem. Abstr.* **64**, 12646 (1966).
160. Noland, W. E., and Baude, F. J., *J. Org. Chem.* **31**, 3321 (1966).
161. Noland, W. E., Christensen, G. M., Sauer, G. L., and Dutton, G. G. S., *J. Am. Chem. Soc.* **77**, 456 (1955).
162. Noland, W. E., and Hammer, C. F., *J. Org. Chem.* **23**, 320 (1958); **25**, 1536 (1960).
163. Noland, W. E., and Hammer, C. F., *J. Org. Chem.* **25**, 1525 (1960).
164. Noland, W. E. and Hartman, P. J. *J. Am. Chem. Soc.* **76**, 3227 (1954).
165. Noland, W. E., and Hovden, R. A., *J. Org. Chem.* **24**, 894 (1959).
166. Noland, W. E., and Johnson, J. E., *J. Am. Chem. Soc.* **82**, 5143 (1960).
167. Noland, W. E., and Johnson, J. E., *Tetrahedron Letters* p. 589 (1962).
168. Noland, W. E., and Kuryla, W. C., *J. Org. Chem.* **25**, 486 (1960).
169. Noland, W. E., Kuryla, W. C., and Lange, R. F., *J. Am. Chem. Soc.* **81**, 6010 (1959).
170. Noland, W. E., Richards, C. G., Desai, H. S., and Venkiteswaran, M. R., *J. Org. Chem.* **26**, 4254 (1961).
171. Noland, W. E., and Rieke, R. D., *J. Org. Chem.* **27**, 2250 (1962).
172. Noland, W. E., and Robinson, D. N., *Tetrahedron* **3**, 68 (1958).
173. Noland, W. E., and Robinson, D. N., *J. Org. Chem.* **22**, 1134 (1957).
174. Noland, W. E., and Rush, K. R., *J. Org. Chem.* **31**, 70 (1966).
175. Noland, W. E., Rush, K. R., and Smith, L. R., *J. Org. Chem.* **31**, 65 (1966).
176. Noland, W. E., Smith, L. R., and Johnson, D. C., *J. Org. Chem.* **28**, 2262 (1963).
177. Noland, W. E., Smith, L. R., and Rush, K. R., *J. Org. Chem.* **30**, 3457 (1965).
178. Noland, W. E., and Venkiteswaran, M. R., *J. Org. Chem.* **26**, 4263 (1961).
179. Noland, W. E., Venkiteswaran, M. R., and Lovald, R. A., *J. Org. Chem.* **26**, 4249 (1961).
180. Noland, W. E., Venkiteswaran, M. R., and Richards, C. G., *J. Org. Chem.* **26**, 4241 (1961).
181. Oddo, B., *Gazz. Chim. Ital.* **43**, 190 (1913); *Chem. Abstr.* **8**, 85 (1914).
182. Oddo, B., *Gazz. Chim. Ital.*, **43**, 385 (1913).
183. Oddo, B., and Alberti, C., *Gazz. Chim. Ital.* **63**, 236 (1933).
184. Oddo, B., and Perotti, L., *Gazz. Chim. Ital.* **60**, 13 (1930); *Chem. Abstr.* **24**, 3785 (1930).
185. Oddo, B., and Toffoli, C., *Gazz. Chim. Ital.* **64**, 359 (1934); *Chem. Abstr.* **28**, 6436 (1934).
186. Okuda, T., *Bull. Chem. Soc. Japan* **32**, 1165 (1959).
187. Pappalardo, G., and Vitali, T., *Gazz. Chim. Ital.* **88**, 1147 (1958).

187a. Parham, W. E., Davenport, R. W., and Biasotti, J. B., *Tetrahedron Letters* p. 557 (1969).
188. Perekalin, V. V., and Slavachevskaya, N. M., *Zh. Obshch. Khim.* **24**, 2164 (1954); *Chem. Abstr.* **50**, 300 (1956).
189. Piers, K., Meimaroglou, C., Jardine, R. V., and Brown, R. K., *Can. J. Chem.* **41**, 2399 (1963).
190. Plant, S. G. P., and Rogers, K. M., *J. Chem. Soc.* p. 40 (1936).
191. Plant, S. G. P., and Tomlinson, M. L., *J. Chem. Soc.* p. 955 (1933).
192. Plieninger, H., *Chem. Ber.* **87**, 127 (1954).
193. Pople, J. A., and Segal, G. A., *J. Chem. Phys.* **44**, 3289 (1966).
194. Potts, K. T., *J. Org. Chem.* **26**, 4719 (1961).
195. Potts, K. T., and Saxton, J. E., *J. Chem. Soc.* p. 2641 (1954).
196. Potts, K. T., and Saxton, J. E., *Org. Syn.* **40**, 68 (1960).
197. Potts, K. T., and Shin, H., *Chem. Commun.* p. 857 (1966).
198. Powers, J. C., *J. Org. Chem.* **31**, 2627 (1966).
199. Powers, J. C., *J. Org. Chem.* **30**, 2534 (1965).
200. Powers, J. C., Meyer, W. P., and Parsons, T. G., *J. Am. Chem. Soc.* **89**, 5812 (1967).
201. Pozharskii, A. F., Martsokha, B. K., and Siminov, A. M., *J. Gen. Chem. USSR (English Transl.)* **33**, 994 (1963).
202. Pratt, E. F., and Botimer, L. W., *J. Am. Chem. Soc.* **79**, 5248 (1957).
203. Rees, C. W., and Sabet, C. R., *J. Chem. Soc.* p. 680 (1965).
204. Rees, C. W., and Smithen, C. E., *J. Chem. Soc.* p. 938 (1964).
205. Rees, C. W., and Smithen, C. E., *J. Chem. Soc.* p. 928 (1964); *Advan. Heterocyclic Chem.* **3**, 57 (1964).
206. Reinecke, M. G., Johnson, H. W., and Sebastian, J. F., *Tetrahedron Letters* p. 1183 (1963).
207. Remers, W. A., and Weiss, M. J., *J. Med. Chem.* **8**, 700 (1965).
208. Reppe, W., *et al.*, *Ann. Chem.* **596**, 158 (1955).
209. Reppe, W., *et al.*, *Ann. Chem.* **601**, 128 (1956).
210. Reppe, W., and Uffer, H., German Patent 698,273 as quoted by Snyder and MacDonald (238).
211. Ridd, J. H., *Phys. Methods Heterocyclic Chem.* **2**, 109–160 (1963).
212. Robinson, B., *Tetrahedron Letters* p. 139 (1962).
213. Robinson, B., and Smith, G. F., *J. Chem. Soc.* p. 4574 (1960).
214. Robinson, R., and Saxton, J. E., *J. Chem. Soc.* p. 976 (1952).
215. Robinson, R., and Saxton, J. E., *J. Chem. Soc.* p. 3136 (1950).
216. Robinson, R., and Saxton, J. E., *J. Chem. Soc.* p. 2596 (1953).
217. Robinson, R. A., U.S. Patent 2,908,691 (1959); *Chem. Abstr.* **56**, 3455 (1962).
218. Runti, C., *Gazz. Chim. Ital.* **81**, 613 (1951); *Chem. Abstr.* **49**, 1700 (1955).
219. Sakakibara, H., and Kobayashi, T., *Tetrahedron* **22**, 2475 (1966).
220. Sakan, T., Matsubara, S., Takagi, H., Tokunaga, Y., and Miwa, T., *Tetrahedron Letters* p. 4925 (1968).
221. Salgar, S. S., and Marchant, J. R., *J. Prakt. Chem.* [4] **14**, 108 (1961).
222. Sausen, G. N., Engelhardt, V. A., and Middleton, W. J., *J. Am. Chem. Soc.* **80**, 2815 (1958).
223. Saxton, J. E., *J. Chem. Soc.* p. 3592 (1952).
224. Schach von Whittenau, M., and Els, H., *J. Am. Chem. Soc.* **85**, 3425 (1963).
225. Schellenberg, K. A., and McClean, G. W., *J. Am. Chem. Soc.* **88**, 1077 (1966).
226. Schofield, K., *Quart. Rev. (London)* **4**, 382 (1950).
227. Schofield, K., and Theobald, R. S., *J. Chem. Soc.* p. 796 (1949).
228. Schmitz-Du Mont, O., *Ann. Chem.* **514**, 267 (1934).

229. Schmitz-Du Mont, O., Hamann, K., and Geller, K. H., *Ann. Chem.* **504**, 1 (1933).
230. Seka, R., *Chem. Ber.* **56**, 2058 (1923).
231. Shabica, A. C., Howe, E. E., Ziegler, J. B., and Tishler, M., *J. Am. Chem. Soc.* **68**, 1156 (1946).
232. Shagalov, L. B., Sorokina, N. P., and Suvorov, N. N., *J. Gen. Chem. USSR* (*English Transl.*) **34**, 1602 (1964).
233. Shen, T., U.S. Patent, 3,242,185 (1966); *Chem. Abstr.* **64**, 17555 (1966).
234. Smith, G. F., *J. Chem. Soc.* p. 3842 (1954).
235. Smith, G. F., *Advan. Heterocyclic Chem.* **2**, 300–309 (1963).
236. Smith, G. F., *Chem. & Ind.* (*London*) p. 1451 (1954).
237. Smith, G. F., and Waters, A. E., *J. Chem. Soc.* p. 940 (1961).
238. Snyder, H. R., and MacDonald, J. A., *J. Am. Chem. Soc.* **77**, 1257 (1955).
239. Snyder, H. R., and Matteson, D. S., *J. Am. Chem. Soc.* **79**, 2217 (1957).
240. Speeter, M. E., U.S. Patent 2,703,325 (1955); *Chem. Abstr.* **50**, 1921 (1956).
241. Staab, H. A., *Chem. Ber.* **90**, 1320 (1957).
242. Staab, H. A., Otting, W., and Ueberle, A., *Z. Elektrochem.* **61**, 1000 (1957).
243. Stroh, R., Ebersberger, J., Haberland, H., and Hahn, W., *Angew. Chem.* **69**, 124 (1957).
244. Stroh, R., and Hahn, W., *Ann. Chem.* **623**, 176 (1959).
245. Suvorov, N. N., and Sorokina, N. P., *J. Gen. Chem. USSR* (*English Transl.*) **30**, 2036 (1960).
246. Swaminathan, S., and Narasimhan, K., *Chem. Ber.* **99**, 889 (1966).
247. Swaminathan, S., and Ranganathan, S., *J. Org. Chem.* **22**, 70 (1957).
248. Swaminathan, S., Ranganathan, S., and Sulochana, S., *J. Org. Chem.* **23**, 707 (1958).
249. Szmuszkovicz, J., *J. Am. Chem. Soc.* **79**, 2819 (1957).
250. Szmuszkovicz, J., *J. Org. Chem.* **29**, 178 (1964).
250a. Taborsky, R. G., Delvigs, P., Page, I. H., and Crawford, N., *J. Med. Chem.* **8**, 460 (1965).
251. Terent'ev, A. P., Belen'kiĭ, L. I., and Yanovskaya, L. A., *Zh. Obshch. Khim.* **24**, 1265 (1954); *Chem. Abstr.* **49**, 12327 (1955).
252. Terent'ev, A. P., Golubeva, S. K., and Tsymbal, L. V., *Zh. Obshch. Khim.* **19**, 781 (1949); *Chem. Abstr.* **44**, 1095 (1950).
253. Terent'ev, A. P., Kost, A. N., and Gurvich, S. M., *Zh. Obshch. Khim.* **22**, 1977 (1952); *Chem. Abstr.* **47**, 8663 (1953).
254. Terent'ev, A. P., Kost, A. N., and Smit, V. A., *Zh. Obshch. Khim.* **26**, 557 (1956); *Chem. Abstr.* **50**, 13871 (1956).
255. Terent'ev, A. P., Kost, A. N., and Smit, V. A., *Zh. Obshch. Khim.* **25**, 1959 (1955); *Chem. Abstr.* **50**, 4910 (1956).
256. Terent'ev, A. P., and Yanovskaya, L. A., *Zh. Obshch. Khim.* **21**, 1295 (1951); *Chem. Abstr.* **46**, 2048 (1952).
257. Thesing, J., *Chem. Ber.* **87**, 692 (1954).
258. Thesing, J., and Binger, P., *Chem. Ber.* **90**, 1419 (1957).
259. Thesing, J., Klussendorf, S., Ballach, P., and Mayer, H., *Chem. Ber.* **88**, 1295 (1955).
260. Thesing, J., Mayer, H., and Klüssendorf, S., *Chem. Ber.* **87**, 901 (1954).
261. Thesing, J., and Semler, G., *Ann. Chem.* **680**, 52 (1964).
262. Thesing, J., and Willersinn, C., *Chem. Ber.* **89**, 1195 (1956).
263. Thesing, J., Zeig, H., and Mayer, H., *Chem. Ber.* **88**, 1978 (1958).
264. Treibs, A., and Bhramaramba, A., *Tetrahedron* Suppl. **8**, 165 (1966).
265. Treibs, A., and Herrmann, E., *Z. Physiol. Chem.* **299**, 168 (1955); *Chem. Abstr.* **50**, 943 (1956).
266. Treibs, A., and Herrmann, E., *Ann. Chem.* **589**, 207 (1954).

267. Treibs, A., and Herrmann, E., *Ann. Chem.* **592**, 1 (1955).
268. Treibs, W., and Wahren, M., *Chem. Ber.* **94**, 2142 (1961).
269. Troxler, F., *Helv. Chim. Acta* **51**, 1214 (1968).
270. Troxler, F., Bormann, G., and Seemann, F., *Helv. Chim. Acta* **51**, 1203 (1968).
271. Tyson, F. T., and Shaw, J. T., *J. Am. Chem. Soc.* **74**, 2273 (1952).
272. van Tamelen, E. E., and Knapp, G. G., *J. Am. Chem. Soc.* **77**, 1860 (1955).
273. von Dobeneck, H., and Goltzsche, W., *Chem. Ber.* **95**, 1484 (1962).
274. von Dobeneck, H., and Maas, I., *Chem. Ber.* **87**, 455 (1954).
275. von Dobeneck, H., and Preietzel, H., *Z. Physiol. Chem.* **299**, 214 (1955); *Chem. Abstr.* **50**, 1765 (1956).
276. von Walther, P., and Clemen, J., *J. Prakt. Chem.* [2] **61**, 249 (1900).
277. Weissgerber, R., *Chem. Ber.* **46**, 651 (1913).
278. Weller, L. E., Rebstock, T. L., and Sell, H. M., *J. Am. Chem. Soc.* **74**, 2690 (1952).
279. Wenkert, E., *Accounts Chem. Res.* **1**, 78 (1968).
280. Wenkert, E., Dave, K. G., and Haglid, F., *J. Am. Chem. Soc.* **87**, 5461 (1965).
281. Wenkert, E., Dave, K. G., Haglid, F., Lewis, R. G., Oishi, T., Stevens, R. V., and Terashima, M., *J. Org. Chem.* **33**, 747 (1968).
282. Wieland, T., Weiberg, O., Fischer, E., and Hörlein, G., *Ann. Chem.* **587**, 146 (1954).
283. Williamson, W. R. N., *J. Chem. Soc.* p. 2834 (1962).
284. Winterfeldt, E., Radunz, H., and Strehlke, P., *Chem. Ber.* **99**, 3750 (1966).
285. Winterfeldt, E., and Strehlke, P., *Chem. Ber.* **98**, 2579 (1965).
286. Woodbridge, R. G., III, and Dougherty, G., *J. Am. Chem. Soc.* **72**, 4320 (1950).
286a. Yagil, G., *Tetrahedron* **23**, 2855 (1967); *J. Phys. Chem.* **71**, 1034 (1967).
287. Yamada, S., Shioiri, T., Itaya, T., Hara, T., and Matsueda, R., *Chem. & Pharm. Bull.* (*Tokyo*) **13**, 88 (1965).
288. Young, T. E., and Mizianty, M. E., *J. Med. Chem.* **9**, 635 (1966).
289. Youngdale, G. A., Anger, D. G., Anthony, W. C., Da Vanzo, J. P., Greig, M. E., Heinzelman, R. V., Keasling, H. H., and Szmuszkovicz, J., *J. Med. Chem.* **7**, 415 (1964).
289a. Zee, S. H., Ph.D. Thesis with W. E. Noland, University of Minnesota (1966); *Dissertation Abstr.* **27**, 123 (1967).
290. Zimmermann, H., and Geisenfelder, H., *Z. Elektrochem.* **65**, 368 (1961).
291. Zinner, G., Böhlke, H., and Kleigel, W., *Arch. Pharm.* **299**, 245 (1966).
292. Zinnes, H., Zuleski, F. R., and Shavel, J., Jr., *J. Org. Chem.* **33**, 3605 (1968).

II
GENERAL REACTIONS OF FUNCTIONALLY SUBSTITUTED INDOLES

A. INTRODUCTION

A number of reactions of functional groups attached to C-2 and C-3 of the indole ring are modified by the conjugation possible between these substituents and the indole nitrogen. The hydrogenolysis of carbonyl groups attached to C-3 and the facile displacement reactions which gramine and its derivatives undergo are important examples. This ease of displacement has been alluded to in connection with the formation of diindolylmethanes from indoles and carbonyl compounds. In general, it is true that any electronegative atom will be relatively easily eliminated from a 2-indolylcarbinyl carbon or, with even more facility, from a 3-indolylcarbinyl carbon. Two extreme transition states can be considered for such eliminations. Heterolysis of the C–X bond may be very complete in the transition state (E1 behavior) or removal of a hydrogen from nitrogen may be concerted with or precede heterolysis of the C–X bond (E2 and E1cb behavior, respectively). The ease of elimination of X^- must be attributed to the fact that conjugation involving the nitrogen atom greatly lowers the energy of the ions **A2** and **A5** relative to a benzyl carbonium ion. Generally, the elimination of a given X^- from a C-2 carbinyl system is less facile than elimination from the corresponding C-3 system. The structures **A2** and **A3**, having benzenoid aromaticity, are expected to be of lower energy than **A5** and **A6** and, if the transition states in the reactions reflect the relative stabilities of the intermediates, the transition states leading to **A2** and **A3** are expected to be of lower energy than those leading to **A5** and **A6**. Unfortunately, although displacements from indolylcarbinyl systems have found extensive synthetic use, there are very little kinetic or other data suitable for quantitative comparison of the ease of departure of leaving groups from indolylcarbinyl systems with similar reactions in other systems.

CH_2X A1 → A2 $+ X^-$

A1 $\xrightarrow{B^-}$ A3 $+ BH + X^-$

CH_2X A4 → A5 $+ X^-$

A4 $\xrightarrow{B^-}$ A6 $+ BH + X^-$

B. SUBSTITUTION AND ELIMINATION REACTIONS AT INDOLYLCARBINYL CARBON ATOMS

From a synthetic point of view, the most important reaction of indolylcarbinyl systems has been the displacement of dimethylamine from gramine (**B1**) or of trimethylamine from quaternary salts of gramine by a variety of carbon nucleophiles. Brewster and Eliel (16) reviewed alkylations by gramine in the early 1950's. The potential importance of the reaction was established by its application to the synthesis of tryptophan (131). Snyder and Eliel recognized the possibility of an elimination–addition mechanism for substitution reactions involving gramine and formulated the reaction between cyanide ion and the methiodide of 1-methylgramine as an elimination–addition sequence (129). The reaction of cyanide ion with 1-methylgramine is particularly interesting in that the by-product **B6** accompanies the principal product (129). Its formation from the intermediate **B3** formed by a prior elimination step can readily be rationalized. The structure of **B6** was proven by hydrolysis to **B7** which was then synthesized independently from **B8** (129).

Albright and Snyder (3) investigated the stereochemistry of displacement reactions using optically active 3-[1-(*N*-isopropylamino)ethyl]indole

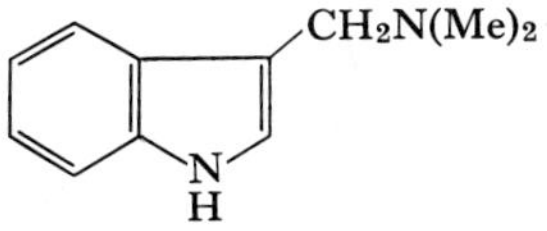

B1

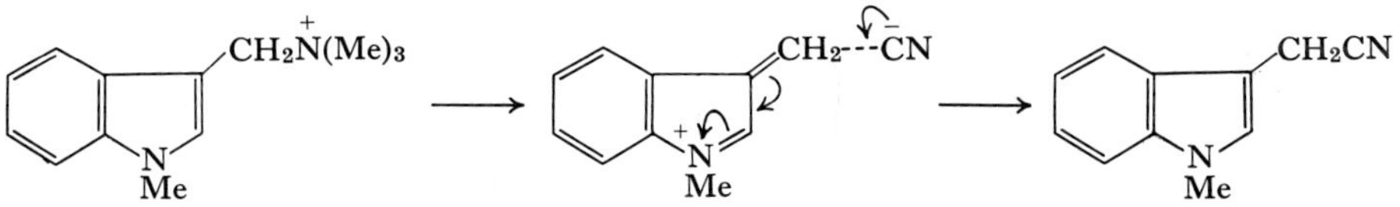

B2 B3 B4

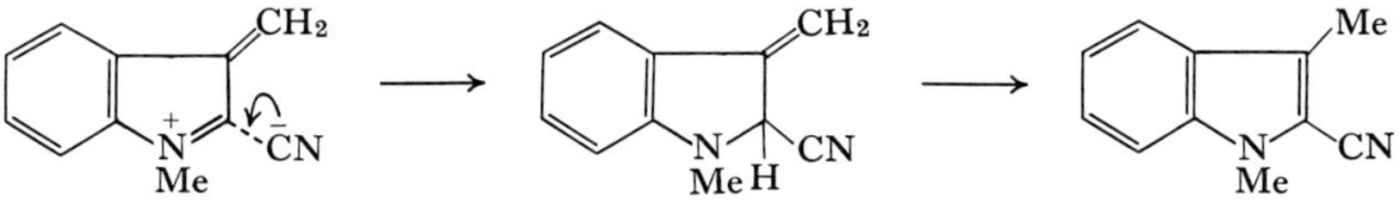

B3 B5 B6

B6 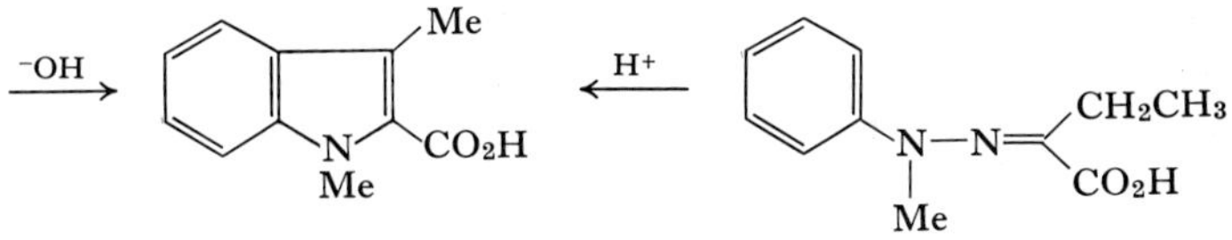

B7 B8

(**B9**). Alkylations of diethyl acetamidomalonate, diethyl malonate, and piperidine all gave racemic products, in accord with an elimination–addition mechanism. The feasibility of the addition step has been demonstrated (3) by addition of diethyl malonate to the benzylidene-3*H*-indole **B10**. The reactivity of the unstable alkylidene-3*H*-indole **B13** has been briefly examined (68) and it also shows an affinity for nucleophilic addition.

Some confusion as to reaction conditions arose in early work on alkylations with gramine because normal reaction procedures for quaternization of tertiary amines fail to yield gramine methiodide, but lead instead to a mixture of tetramethylammonium iodide and dimethyl-bis(3-indolylmethyl)ammonium iodide (45) by the mechanism shown below. Slow addition of gramine to a large excess of methyl iodide gives the authentic gramine methiodide (45).

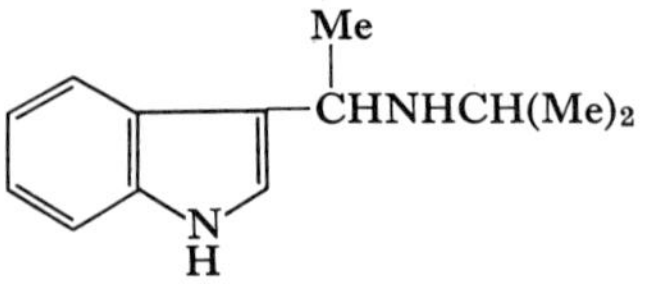

Me
CHNHCH(Me)$_2$
N
H

B9

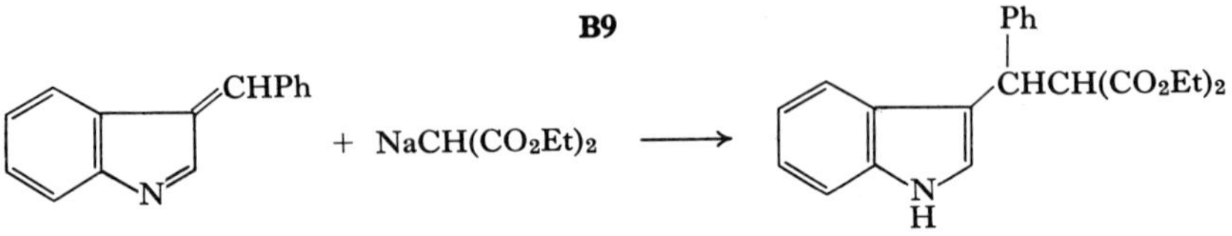

CHPh + NaCH(CO$_2$Et)$_2$ ⟶ Ph CHCH(CO$_2$Et)$_2$

B10 **B11** **B12**

Me Me Me + H$_2$O ⟶ OH C(Me)$_2$ Me

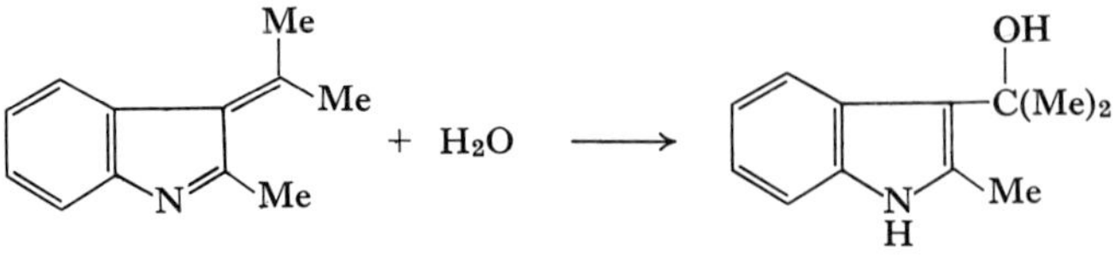

B13 **B14**

A variety of carbon nucleophiles have been successfully alkylated by gramine and its derivatives, and the reaction has general significance as a synthetic tool for elaboration of indole derivatives (99, 100, 107). Typical alkylations by gramine, its methiodide, methosulfate, and *N*-oxide are recorded in Table XII.

The parent primary amine, 3-aminomethylindole, has received much less attention than gramine. It has recently been prepared by Gower and Leete (48) by catalytic reduction of 3-cyanoindole. The amino group can be readily displaced by nucleophiles. Displacement by cyanide, thiophenoxide, and benzylamine has been reported (48). 1-Methyl-3-aminomethylindole is substantially less reactive.

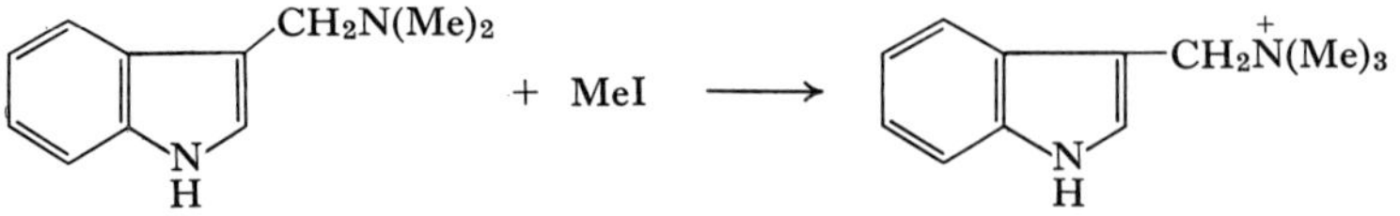

CH$_2$N(Me)$_2$ + MeI ⟶ CH$_2\overset{+}{N}$(Me)$_3$

B15 **B16**

B15 + **B16** ⟶ [indolyl-CH$_2$]$_2$$\overset{+}{N}(Me)_2$ + (Me)$_3$N

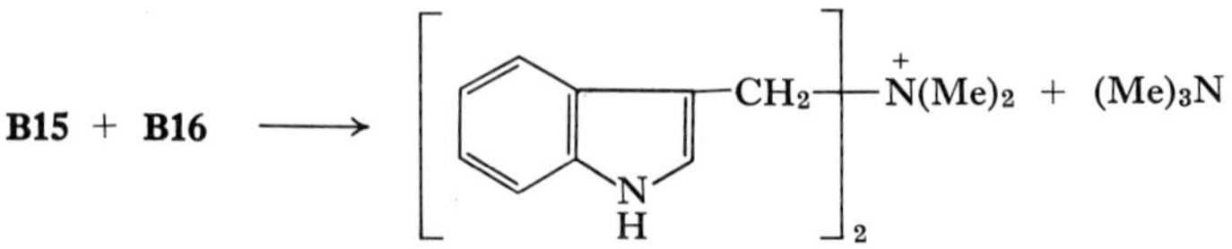

B17

TABLE XII □ ALKYLATIONS BY GRAMINE DERIVATIVES

Gramine substitution	Alkylated nucleophile	Yield (%)	Ref.
None	Acetate ion	88	(84)
None	1-Pyrrolidinocyclohexene	50	(124, 151)
None	Nitroacetonitrile	47[a]	(116)
None	2,5-Dimethylindole	39	(26)
None	Sulfide ion	—[b]	(85)
None	*n*-Amyl mercaptan	—	(85)
None	*p*-Tolyl mercaptan	85	(46)
None	Benzenesulfonamide	—	(85)
Methiodide	Hydroxide ion	66	(84)
Methiodide	Methoxide ion	79	(45)
Methiodide	Cyanide ion	76	(45)
Methiodide	Phenylmagnesium bromide	31	(130)
Methiodide	Sodio diethyl acetamidomalonate	63	(131)
Methiodide	Triethyl phosphite	72	(149)
Ethiodide	Sodio diethyl acetamidomalonate	73	(2)
Methosulfate	Sodium diethylphosphite	75	(149)
N-Oxide	Methoxide ion	63	(54)
N-Oxide	Cyanide ion	83	(54)
N-Oxide	Nitromethane	31	(54)
1-Methyl methiodide	Cyanide ion	83	(129)
1-Methyl methiodide	Phenylmagnesium bromide	72	(130)

[a] Product is bis(3-indolylmethyl)nitroacetonitrile.
[b] Product is di-(3-indolylmethyl)sulfide.

Oxygen-containing leaving groups are also easily displaced from the 3-indolylcarbinyl system. 1-Acetyl-3-acetoxymethylindole gives a 38% yield of 3-indoleacetonitrile when heated with potassium cyanide in ethanolic solution (25) and 3-indolylmethanol undergoes displacement reactions with cyanide ion and secondary amines (119). Elimination–addition sequences provide the most satisfactory explanation for these transformations.

The indole ring also interacts strongly with a developing center of electron deficiency at a C-2 substituent. The enhanced stability of the intermediate ion should favor ionization of electronegative substituents from the 2-carbinyl carbon atom, relative to ionization from a simple benzyl carbon atom. The "*o*-quinoid" nature of the residue, however, would lead to the prediction that a substituent at C-2 should be less reactive than the corresponding substituent at a C-3 carbinyl atom. 2-Dimethylaminomethylindole (isogramine) and its quaternary derivatives have been successfully subjected to displacement reactions comparable to those carried out on gramine. The sodium enolate of diethyl acetamidomalonate displaces trimethylamine from isogramine methiodide in refluxing ethanol (76, 128), although the reaction is incomplete even after 16 hours. The displacement is markedly slower than comparable displacements from derivatives of gramine (2). The difference in reactivity between gramine methiodide and isogramine methiodide is further illustrated by the relative ease of preparation of the two substances. As has been noted, special precautions must be taken in the preparation of gramine methiodide because it readily quaternizes unreacted gramine (45, 128) while preparation of isogramine methiodide proceeds normally (128). No quantitative comparison of the reactivities of gramine and isogramine systems appears to have been made, however.

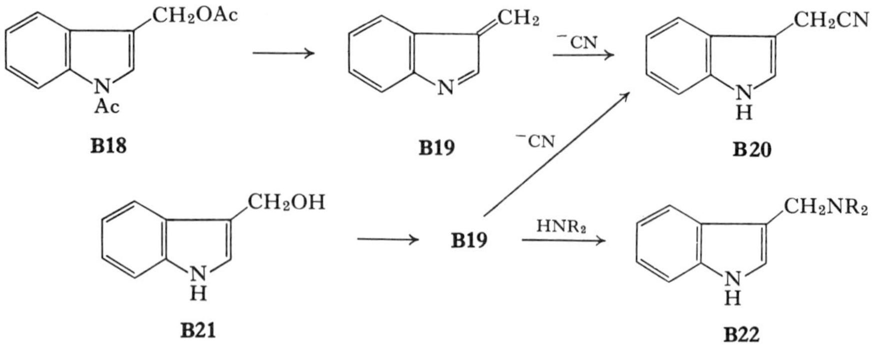

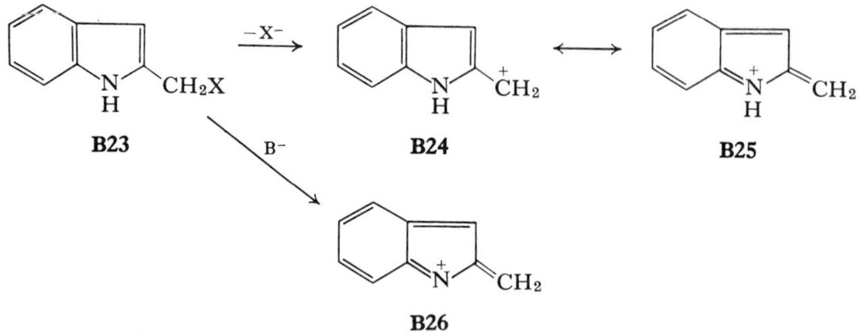

Several successful alkylations with isogramine methiodide are collected in Table XIII. Alkylation reactions in which pyridine is displaced from 1-(2-indolylmethyl)pyridinium salts have also been recorded (120).

A gramine type displacement with ring opening occurs when the quaternary ion **B27** reacts with cyanide ion (51). This reaction has been shown to be very useful for the synthesis of certain alkaloid skeletons containing large C rings. The reaction may not be general, however, since reaction of the desethyl analog of **B27** with cyanide ion takes a different course (37).

TABLE XIII □ NUCLEOPHILIC SUBSTITUTION REACTIONS OF ISOGRAMINE METHIODIDES

Substituent	Nucleophile	Yield (%)	Ref.
None	Cyanide ion	51	(122)
None	Sodio diethyl acetamidomalonate	74	(128)
5-Benzyloxy	Cyanide ion	50	(123)
5-Benzyloxy	Sodio diethyl formamidomalonate	83	(123)

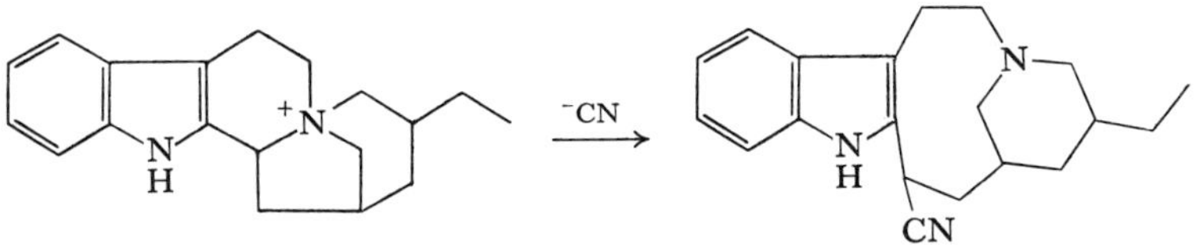

N
H
+N
$^{-}$CN
N
N
H
CN
B27
B28

The reactivity of indolylcarbinyl systems is further illustrated by a number of reactions of indole alkaloids which incorporate this structural feature. The alkaloid uleine (**B29**) is a 3-indolylcarbinylamine (21). Uleine methiodide reacts with sodium methoxide at room temperature in 1 hour to give the displacement product **B31** (67). Cyanide effects a similar dis-

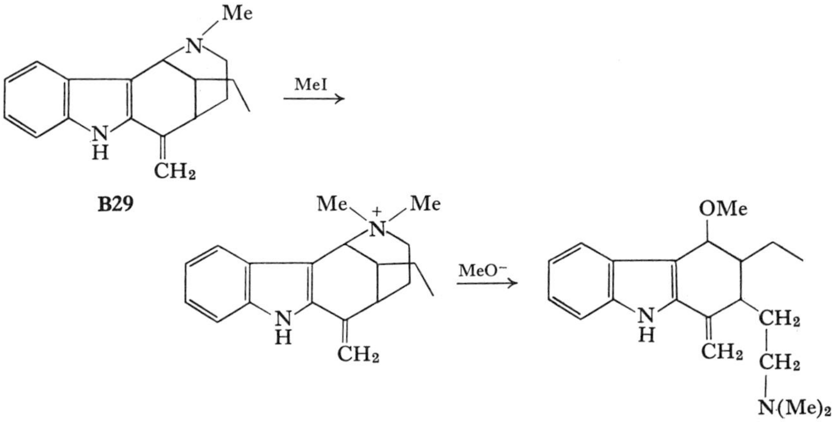

Me
N
MeI
N
H
CH_2
B29

Me
+
Me
N
MeO^{-}
N
H
CH_2
OMe
N
H
CH_2
CH_2
CH_2
$N(Me)_2$
B30
B31

placement reaction. Treatment of uleine with acetic anhydride and pyridine, followed by heating in methanol, gives **B34** (67). Two gramine-type displacements rationalize this result. Analytical data support the intermediacy of the pyridinium salt **B33**.

B29 $\xrightarrow[\text{pyridine}]{Ac_2O}$ Ac, ^+N—Me, N H, CH_2 ⟶

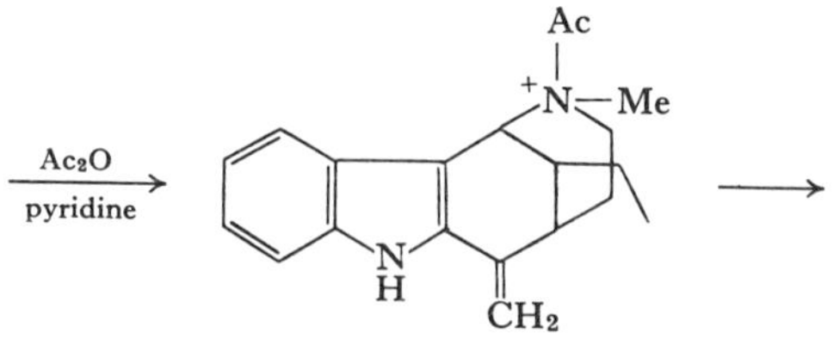

B32

N_+, N H, CH_2, CH_2, CH_2, NMe Ac $\xrightarrow{MeOH}$ OMe, N H, CH_2, CH_2, CH_2, NMe Ac

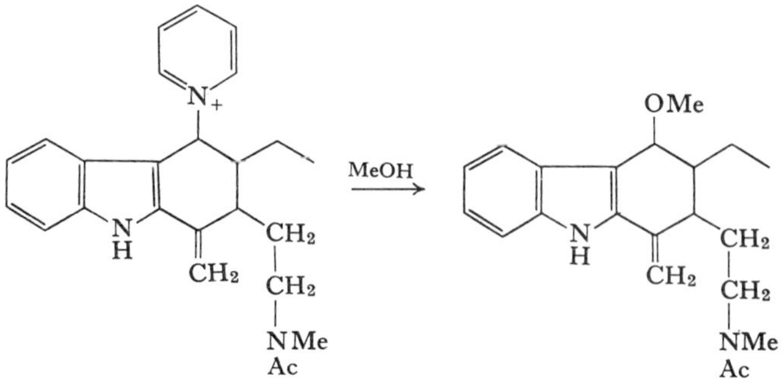

B33 B34

A very large number of indole alkaloids incorporate a tetrahydro-β-carboline system (**B35**) which is, of course, a substituted 2-indolylcarbinylamine. Foster, Harley-Mason, and Waterfield (38) have recently discovered a facile cleavage of the C–N bond in 2-indolylcarbinyl amines which promises

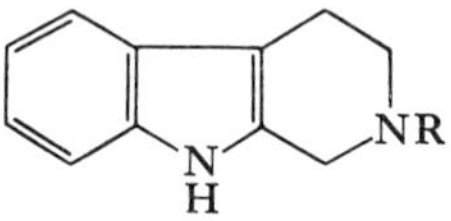

NR, N H

B35

to be of significance in synthesis and interrelation of indole alkaloids. The tetrahydrocarboline **B36** is cleaved by cold acetic anhydride, giving **B39**. 2-(2-Quinuclidinyl)indole is cleaved under similarly mild conditions (134). A reasonable mechanism for this reaction would involve acetylation of the basic nitrogen followed by displacement by acetate, perhaps via an elimination sequence. This transformation is related to a C–N cleavage of the tetrahydrocarboline system developed by Dolby and Sakai and illustrated by the transformation of **B42** into **B43** (33). In this instance, the acylation is intramolecular, perhaps via a mixed anhydride, but otherwise parallels the cleavage observed by Foster, Harley-Mason, and Waterfield (38). The

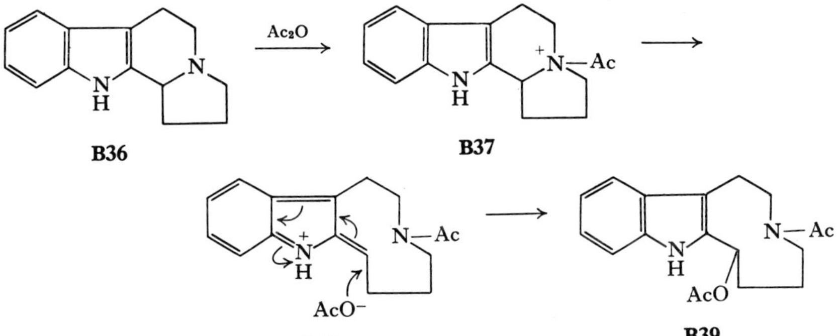

B36 $\xrightarrow{Ac_2O}$ **B37** ⟶ **B38** ⟶ **B39**

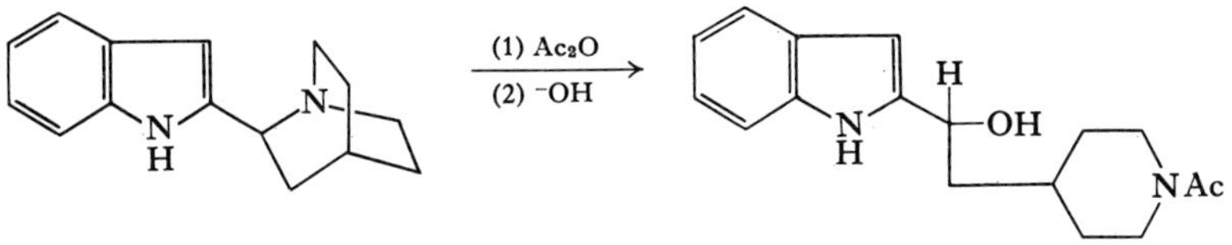

B40 $\xrightarrow{(1)\ Ac_2O;\ (2)\ ^{-}OH}$ **B41**

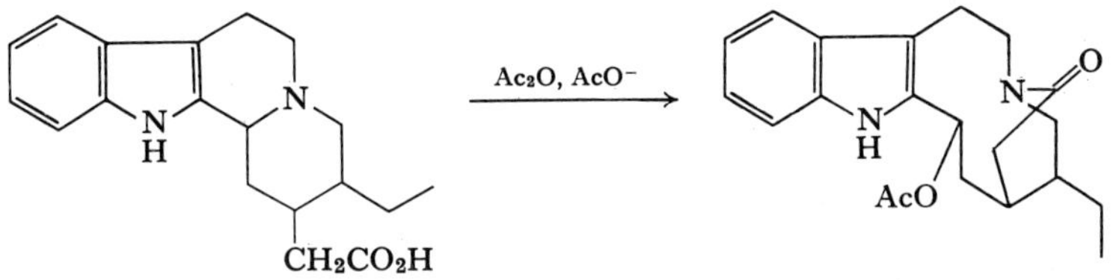

B42 $\xrightarrow{Ac_2O,\ AcO^{-}}$ **B43**

cleavage of cinchonamine by hot acetic anhydride (**B44** → **B45**) is another example of the process (47). Recent investigation (14) of the latter cleavage has shown that the reaction occurs relatively rapidly even below room temperature. Cleavage of simple benzylamines by acid anhydrides is known (148), but there are no data to permit quantitative comparison of the facility of this reaction with the corresponding cleavages of indolylmethylamines.

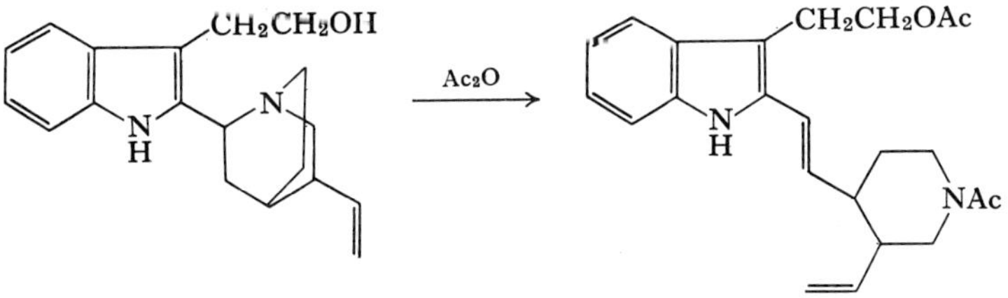

B44 $\xrightarrow{Ac_2O}$ **B45**

The conversion of gramine to 1-acetyl-3-acetoxymethylindole (45, 84) and the cleavage of 1-phenyl-2-methyl-1,2,3,4-tetrahydro-β-carboline (39) are other examples of this reaction. The cleavage of tetrahydro-β-carbolines by cyanogen bromide in hydroxylic solvents (2a) is a related process.

The transformation **B46** → **B47** is effected by refluxing **B46** in xylene for 8 hours (19). This cyclization which is, formally, the displacement of a hydroxyl group by an amine in basic solution, is paralleled by the trans-

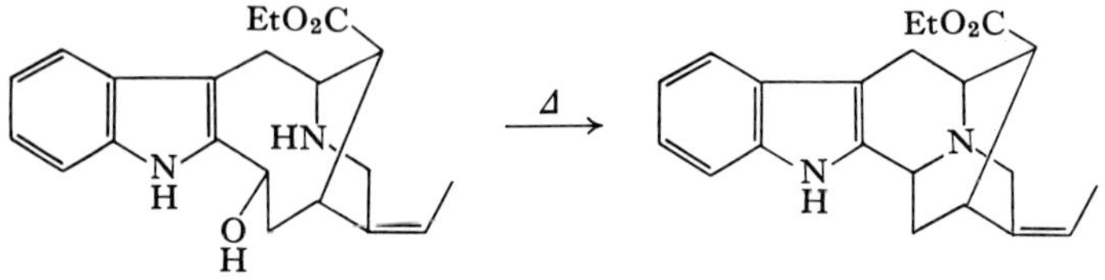

B46 **B47**

formation of **B48** to 2-quinuclidinylindole (134). Both of these displacement reactions are best formulated as elimination–addition processes. Intramolecular catalysis by the indole NH may be important in the elimination step.

Other examples of facile displacements from 2-indolylcarbinyl carbons can be drawn from the chemistry of the iboga alkaloids. The ether **B54** and alcohol **B53** are readily interconverted (20). These reactions, if formulated

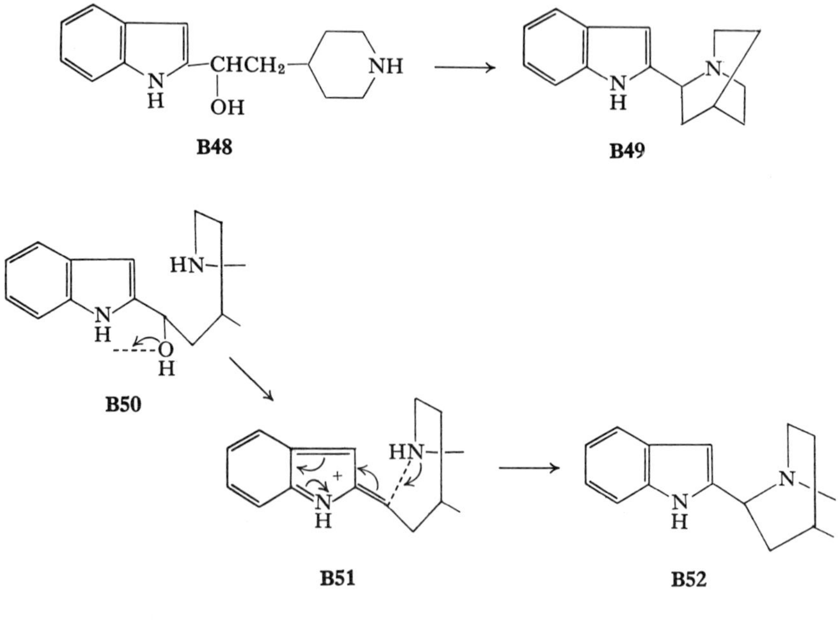

B48 **B49**

B50

B51 **B52**

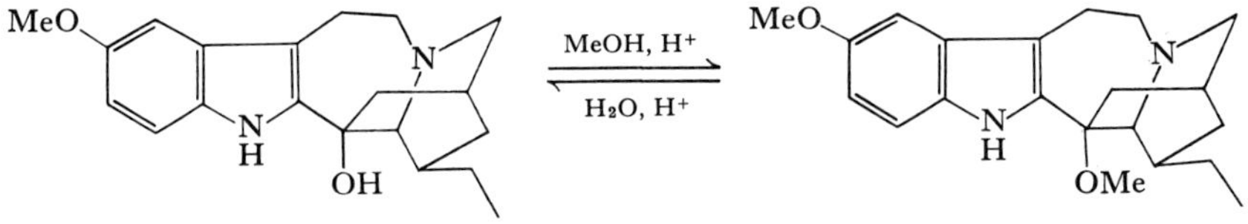

MeOH, H+ / H₂O, H+

B53 **B54**

as elimination–addition sequences, proceed through the ion **B55**. It has been suggested (20) that participation by the basic nitrogen may result in an alternative intermediate **B56**. The conversion of **B57** to **B59** is a step in the

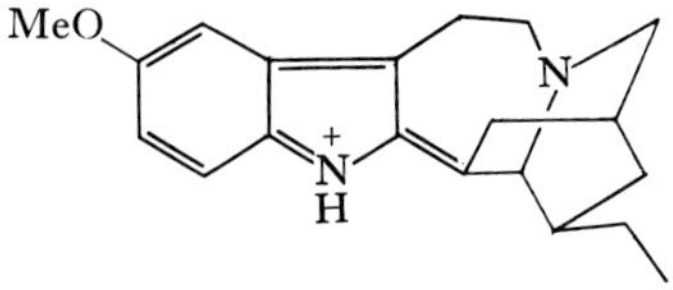

B55 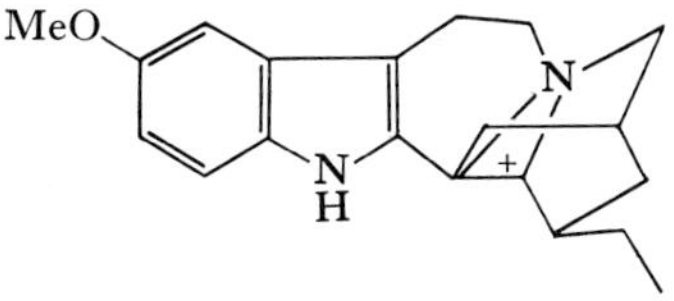 **B56**

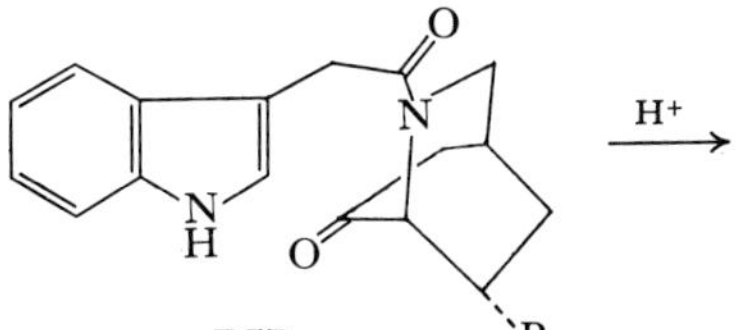

B57 $\xrightarrow{H^+}$

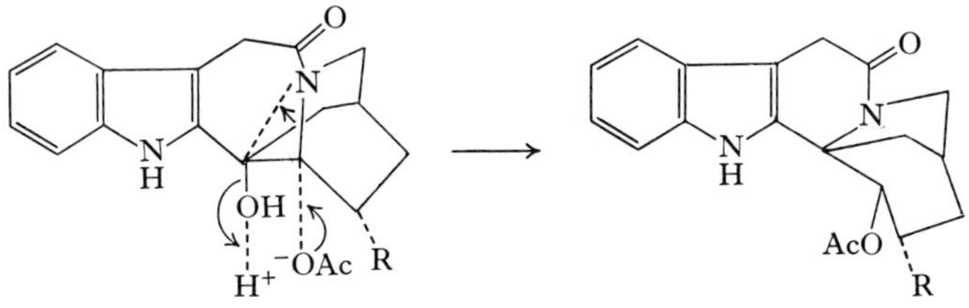

B58 → **B59**

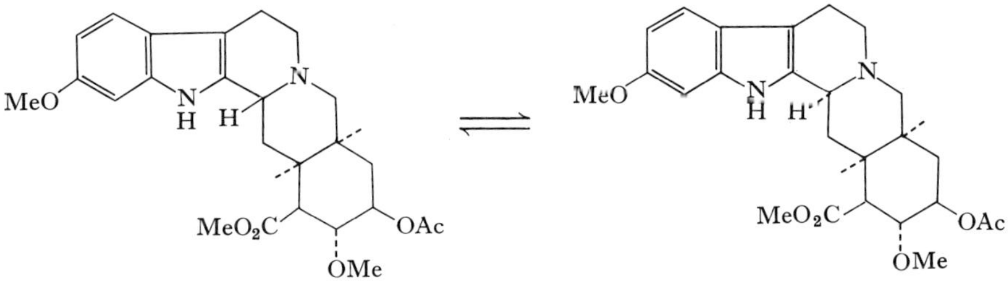

B60 ⇌ **B61**

Mechanism A

MeO N H H N **B60** ⇌(H+) MeO N+ H H N **B62**

⇅

MeO N+ H H N **B64** ⇌ MeO N H N+ **B63**

⇅

MeO N H H N **B61**

total synthesis of ibogaine and ibogamine (18). The transformation, as formulated below, is seen to involve displacement of a carbinyl substituent accompanied by 1,2-shift of the nitrogen substituent. The observation of such a rearrangement is consistent with the hypothesis that the quinuclidine nitrogen atom can participate in the departure of leaving groups from C-18 of the iboga skeleton. The intermediate **B56** has been referred to as a "non-classical carbonium ion." Actually, effective participation by the nitrogen atom would generate an aziridinium ion with most of the positive charge on nitrogen.

Reversible ring opening involving cleavage of a 2-indolylcarbinyl C–N bond has been considered as a possible mechanism for the acid-catalyzed epimerization of certain tetrahydro-β-carboline systems. Gaskell and Joule

Mechanism B

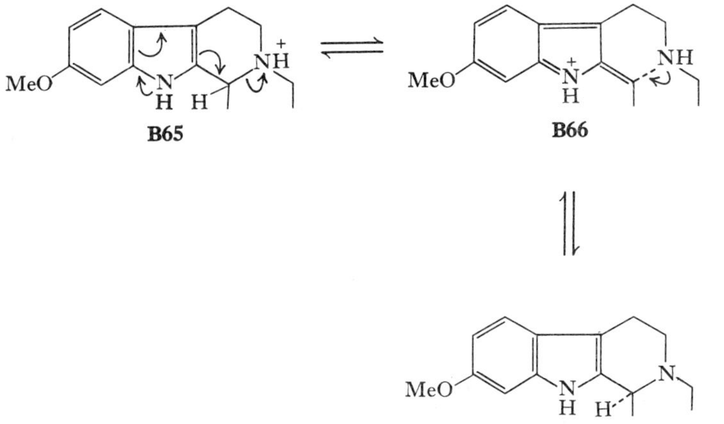

B65

B66

B61

have discussed the acid-catalyzed epimerizations of reserpine and deserpidine (43). Reserpine (**B60**), for example, is equilibrated with isoreserpine (**B61**) in refluxing acetic acid. Gaskell and Joule have presented evidence which favors mechanism A over mechanism B. Mechanism B is rejected by Gaskell and Joule on the basis of the failure of quaternary reserpine and isoreserpine salts to equilibrate. A quaternary ion would be expected to ring

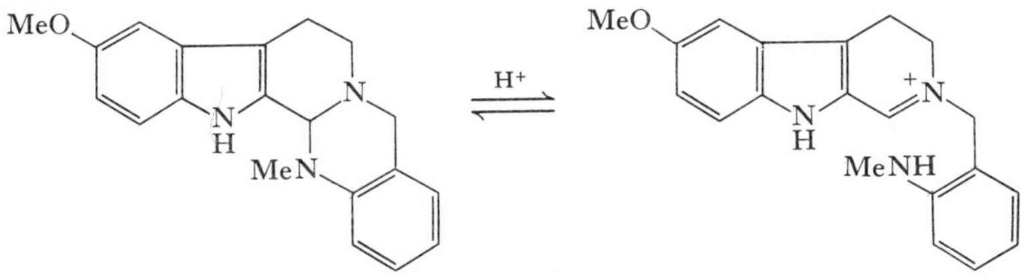

B67

B68

open with facility comparable to the protonated amine **B65**. This conclusion is supported by other evidence and mechanism A seems to be the one operating despite the fact that the C-2 protonated intermediate **B62** must certainly be present in very small concentration relative to the nitrogen-protonated ion **B65**. Another interesting example of facile ionization of C–N bonds in tetrahydro-β-carbolines is found in the chemistry of the compound **B67**. Ring opening occurs in acid solution, but the pentacyclic molecule is regenerated by base (98).

A series of alkylations of indoles by ring-opened 1-aryltetrahydro-β-carbolines has been reported (39) and examples are shown below:

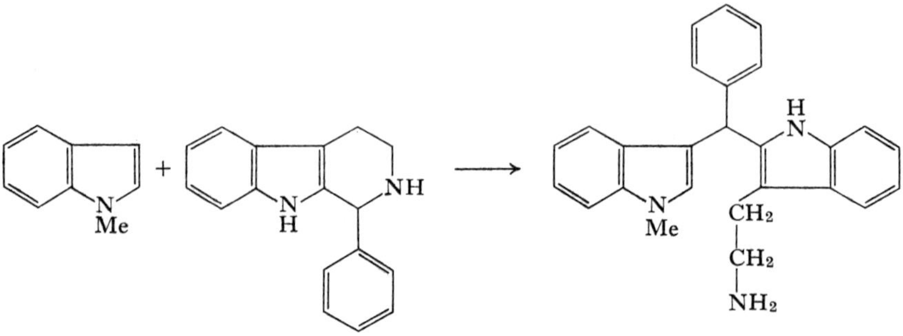

B69 **B70** **B71**

B69 + **B72** → **B73**

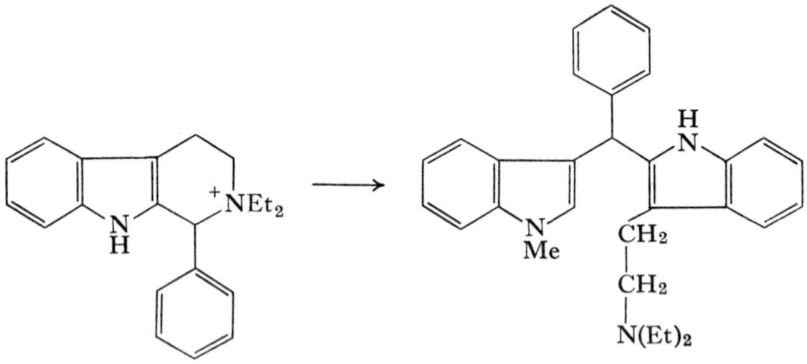

Acid-catalyzed cleavage of the C–N bond generates the ion **B74** which can then alkylate the unsubstituted indole ring. 2,3-Disubstituted indoles are

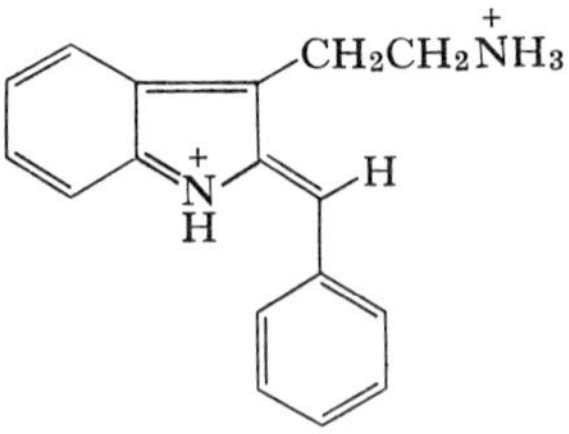

B74

alkylated in the carbocyclic ring. The presence of an aromatic ring at the reaction site is apparently essential for the reaction, as 1-alkytetrahydro-β-carbolines are reported not to react under similar conditions.

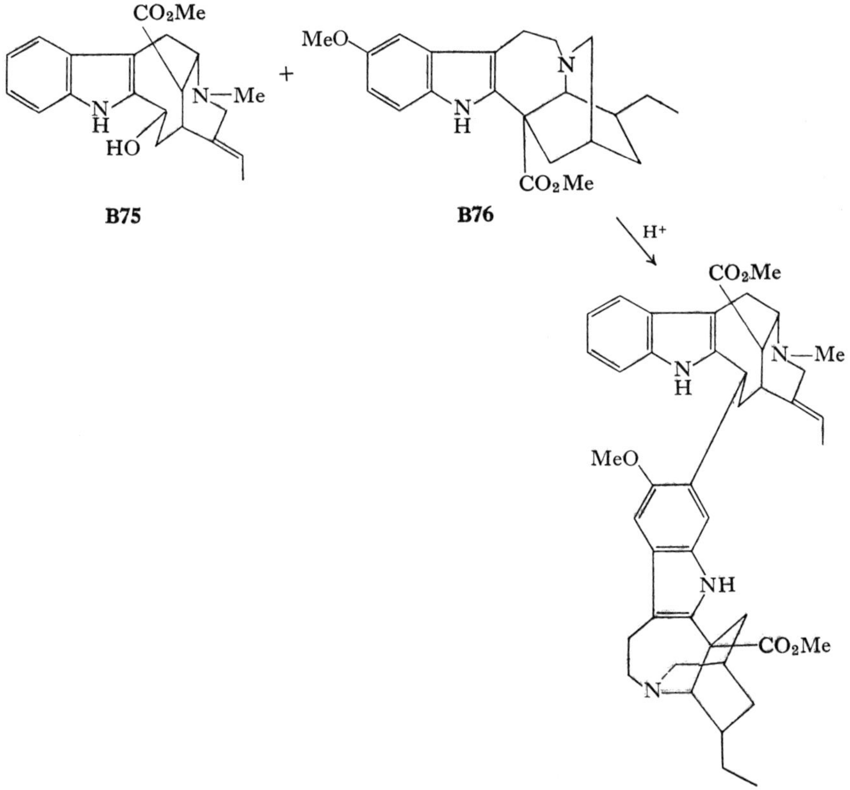

CO_2Me N—Me N H HO + MeO N N H CO_2Me

B75 B76

H^+

CO_2Me N—Me N H MeO NH CO_2Me N

B77

The generation of the so-called dimeric indole alkaloids follows the same mechanistic pattern. Vobasinol (**B75**) and voacangine (**B76**) give voacamine (**B77**) when heated in methanolic hydrochloric acid (19, 114). Acid-cata-

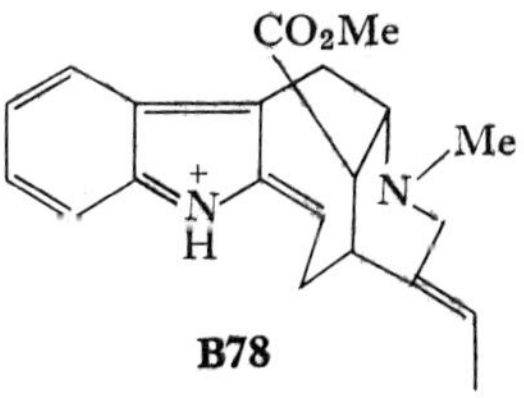

CO_2Me Me N $\overset{+}{N}$ H

B78

lyzed elimination of water from **B75** generates the ion **B78** which alkylates the carbocyclic ring of **B76**. Similar alkylations probably are involved in the biogenesis of "dimeric" indole alkaloids such as voacamine.

C. HYDROGENOLYSIS OF INDOLYLCARBINYL SUBSTITUENTS

A number of substituents attached to C-2 and C-3 of the indole ring behave somewhat abnormally toward reduction conditions. The abnormalities can be traced to easy displacement of indolylcarbinyl substituents. In 1953 two groups reported hydrogenolysis of oxygen-containing functional groups at C-3 of the indole ring (84, 118). Lithium aluminum hydride in refluxing ether or tetrahydrofuran was reported to reduce both 3-formylindole and 3-carbethoxylindole to 3-methylindole, and 3-acetylindole was similarly reduced to 3-ethylindole (84, 118). Leete and Marion showed that 3-hydroxymethylindole, the expected product of normal hydride reduction, was reduced by lithium aluminum hydride to 3-methylindole and they suggested that the reaction proceeds via metal ion-catalyzed elimination of the hydroxyl group (84). In view of the acidity of the indole N–H, base catalysis of the elimination step must also be included (29, 80). The importance of the ionization of the N–H bond in facilitating hydrogenolysis has

CH_2OM ; B^- ; NH — C1 → CH_2 ; N — C2 —($^-AlH_4$)→ CH_3 ; $\bar{N}$ — C3

been demonstrated (29, 80). Unlike indole-3-carboxaldehyde, 1-methylindole-3-carboxaldehyde is converted to 3-hydroxymethyl-1-methylindole by lithium aluminum hydride (80). 5-Bromo-1-methylindole-3-carboxaldehyde is likewise reduced by lithium aluminum hydride to the alcohol stage (92). Similarly, *N*-methyl-2-hydroxytetrahydrocarbazole (**C5**) is resistant to hydrogenolysis by lithium aluminum hydride, whereas the unmethylated ketone **C6** is extensively hydrogenolyzed under similar conditions (29).

N Me, O — **C4** —($LiAlH_4$, ether)→ N Me, OH — **C5** (97%)

N H, O — **C6** —($LiAlH_4$, ether)→ N H, OH — **C7** (44%) + N H — **C8** (49%)

2-Hydroxymethylindole is not rapidly hydrogenolyzed by lithium aluminum hydride in ether (80), indicating that 2-indolylcarbinyl systems are less active toward displacement than the corresponding 3-derivatives.

Diborane effects hydrogenolysis of 3-acylindoles and indole-3-carboxaldehydes (15). Indole aldehydes give rise to dimeric and polymeric products in certain cases (15). Extensive hydrogenolysis of 2-acyl substituents is also observed. Substitution on nitrogen does not prevent hydrogenolysis, in contrast to the situation with lithium aluminum hydride. The electrophilic diborane presumably assists in ionization of the oxygen substituent. It has been suggested that a neutral boron-containing group may be more easily eliminated than a negatively charged aluminate group (15). Diborane leaves ester substituents on the indole ring unreduced. Although the reaction has not been extensively investigated, lithium aluminum hydride–aluminum trichloride mixtures appear to be able to effect hydrogenolysis of *N*-alkyl-3-acyl indoles. 1-Methyl-3-acetylindole gives a 74% yield of 1-methyl-3-ethylindole with this reagent (102).

A convenient preparation of tryptophol is based upon reduction of ethyl indolyl-3-gloxylate and involves hydrogenolysis at the conjugated carbonyl group (90). The presence of electronegative substituents on the β-carbon of

N H **C9** —ClCOCOCl→ COCOCl N H **C10** —EtOH→

COCO$_2$Et N H **C11** —LiAlH$_4$→ CH$_2$CH$_2$OH N H **C12**

the acyl side chain seems to have a retarding effect on hydrogenolysis in certain instances. Compound **C13** is incompletely hydrogenolyzed by lithium aluminum hydride in ether (135).

3-Aminomethylindole is reduced to 3-methylindole by lithium aluminum hydride in refluxing dioxane (48). Amides of 1-benzyl-5-methoxyindole-3-carboxylic acid, however, can be successfully reduced to the corresponding amines with lithium aluminum hydride in ether or tetrahydrofuran (34). The diminished tendency of *N*-substituted systems to suffer hydrogenolysis is again evident.

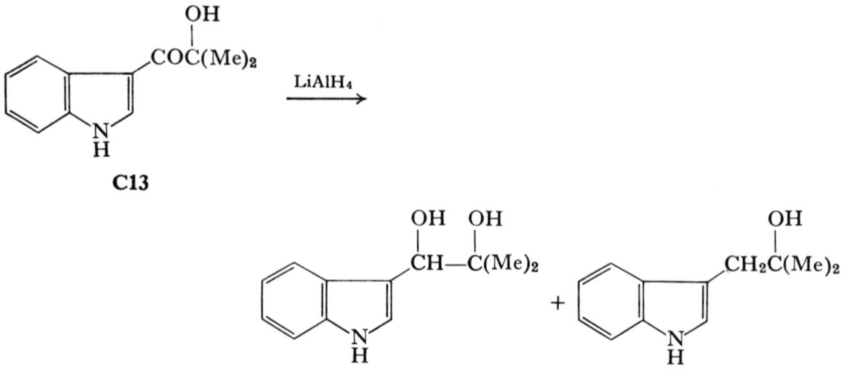

C13 → ($LiAlH_4$) C14 + C15

Cyanide is displaced from the indole **C16** whereas the *N*-methyl derivative **C18** undergoes normal reduction to the amine **C19** (111). Base-catalyzed elimination of cyanide followed by reduction of the resulting alkylidene indolenine accounts for the abnormal behavior of **C16**.

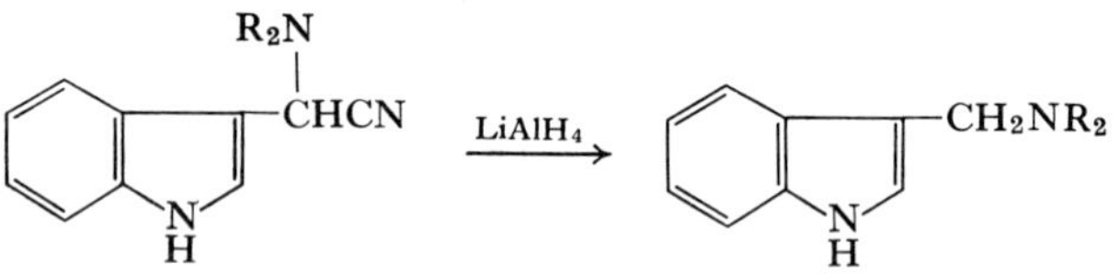

C16 → ($LiAlH_4$) C17

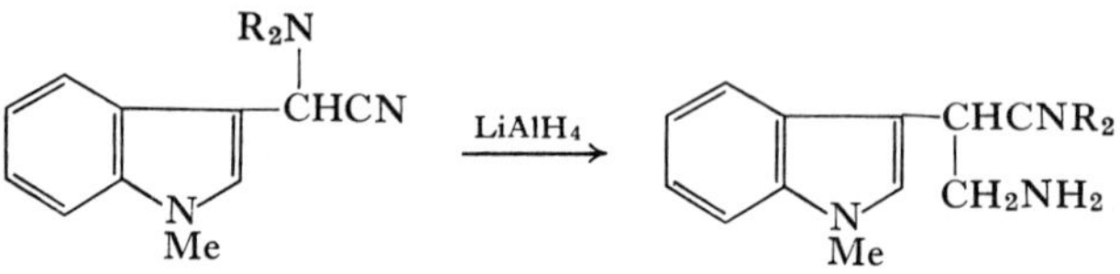

C18 → ($LiAlH_4$) C19

Hydrogenolysis of 3-substituted systems can be largely avoided by using sodium borohydride or lithium borohydride as the reducing agent. The decreased basicity of the borohydride ion is responsible for the diminished extent of hydrogenolysis. 2-Methylindole-3-carboxaldehyde is reduced to 3-hydroxymethyl-2-methylindole in 88% yield by sodium borohydride (80), lithium borohydride reduction of indole-3-carboxaldehyde gives 3-hydroxymethylindole in 89% yield (5). The transformation is effected equally well with sodium borohydride. In contrast, 3-acetylindole undergoes extensive hydrogenolysis to 3-ethylindole with lithium borohydride in refluxing tetrahydrofuran (5) and with sodium borohydride in propanol (153a).

The reductive ring opening of quaternary tetrahydro-β-carbolines is an important degradative and synthetic technique. Various reducing agents have been employed. Herbst and co-workers (55) effected the cleavage of **C20** to **C21** with lithium in ammonia and the reaction was found useful for the preparation of several large ring systems. Dolby and Booth (29) used lithium aluminum hydride in refluxing morpholine to effect a similar ring opening.

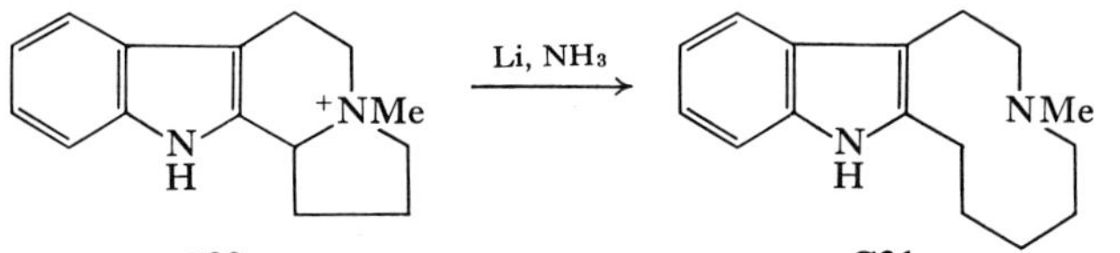

Li, NH_3

C20 **C21**

The methiodide **C22** is reported to show contrasting reactivity toward reduction with lithium aluminum hydride in *N*-methylmorpholine. Demethylation occurs in preference to ring opening. Sodium in liquid ammonia effects the ring opening satisfactorily (31).

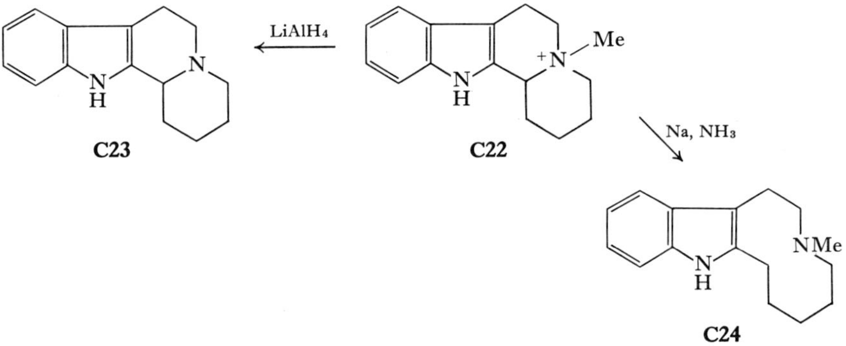

$LiAlH_4$

Na, NH_3

C23 **C22** **C24**

Leete and co-workers (82, 83) have used the reductive cleavage (Emde reduction) to advantage in degradation of radioactively labeled indole alkaloids. A key step in the synthesis of the quebrachamine skeleton also makes use of this cleavage (78, 154).

Cleavage of tetrahydro-β-carbolines has also been accomplished by zinc in acetic acid (44). The simplest example of several reported is the conversion of **C25** to **C26**. The mechanism postulated for the conversion is outlined below. The mechanism of reduction by alkali metals in ammonia is probably basically similar, with the indole ring being the initial site of reduction. Protonation of the indole ring, of course, would not be involved in the latter case.

Zn, AcOH

N—Ac

N H

C25

C26

C25 $\overset{2H^+}{\rightleftharpoons}$ C27 $\xrightarrow{Zn}$ C28

$^+$NH

Zn

C29

C26 ← C31 ← C30

NH

C32 $\xrightarrow[H_2,\ MeOH]{PtO_2}$

OH

OMe

HN

C33 + C34

CH2

+ C35

A number of examples of catalytic hydrogenolysis of 3-indolylcarbinyl C–O and C–N bonds are recorded although little data exist which would permit comparison of these hydrogenolyses with typical benzylic hydrogenolysis (53). The hydrogenolysis of the dimethyamino group in gramine and 5-benzyloxygramine has been observed (86). A British patent (17) reports that the ketone **C32** gives a mixture of the alcohol **C33**, the ether **C34** (presumably formed from **C33** by elimination–addition) and the hydrogenolysis product **C35**. 3-Indolyl-2-pyridylcarbinol suffers extensive

TABLE XIV ☐ TYPICAL REDUCTIONS OF INDOLES WITH REDUCIBLE SUBSTITUENTS AT C-2 AND C-3

Indole substituents	Reducing agent	Yield major product (%)	Ref.
Hydrogenolysis minimized			
1-Me-2-carboxy	$LiAlH_4$	86	(36)
1,2-diMe-3-formyl	$LiAlH_4$	55	(80)
N-Me-1-oxotetrahydro-carbazole	$LiAlH_4$	98	(29)
1-Me-3-formyl	$NaBH_4$	88	(80)
N-Me-amide of 1-methyl-indole glyoxylic acid	$LiAlH_4$	72	(133)
2-Carbethoxy	$LiAlH_4$	—	(138)
2-Carboxamido	$LiAlH_4$	78	(89)
2-Me-3-formyl	$NaBH_4$	95	(80)
3-Formyl	$NaBH_4$	86	(126)
3-Formyl	$LiBH_4$	90	(5)
3-Dimethylaminoacetyl	$NaBH_4$	80	(6)
3-(3-Dimethylaminopropionyl)	$NaBH_4$	95	(136)
3-(Phenylacetamidoacetyl)	$LiBH_4$	80	(5)
3-Formyl-4-cyano	$LiAlH_4$	64	(150)
Extensive hydrogenolysis			
3-Formyl	$LiAlH_4$	84	(84)
3-Acetyl	$LiAlH_4$, $NaBH_4$	72, 61	(84, 153a)
3-Carbethoxy	$LiAlH_4$	92	(84)
3-Hydroxymethyl	$LiAlH_4$	87	(84)
3-Methoxymethyl	$LiAlH_4$	71	(84)
3-Benzoyl	Diborane	70–95	(15)
3-Dimethylaminomethyl	Pd-C, H_2	84	(86)
3-(3-Dimethylaminopropionyl)	$LiAlH_4$	97	(136)
1-Me-3-acetyl	$LiAlH_4$-$AlCl_3$	74	(102)
1-Me-3-benzoyl	Diborane	85	(15)
1-Oxotetrahydrocarbazole	$LiAlH_4$	100	(29)

hydrogenolysis as well as reduction of the pyridine ring when reduced over platinum in acidic solution (7). 3-Hydroxymethylindole apparently resists catalytic hydrogenolysis over platinum oxide (84).

Typical 3-alkylidene-3*H*-indolenium salts, which are presumed to be intermediates generated by eliminations from 3-indolylcarbinyl systems in acidic solution, have been examined as models for biological hydrogen acceptors (57, 121). 3-Benzylidene-2-methyl-3*H*-indolium ions accept

SR, CHPh, Me, N H — $\xrightarrow{H^+,\ Na_2S_2O_4}$ — CH_2Ph, Me, N H

C36 **C37**

hydrogen from dihydropyridines (57, 121). Interestingly, 2-methyl-3-indolyl phenyl carbinyl thioethers are cleaved to benzylindoles by acidic dithionite solutions (**C36** → **C37**) (121). An elimination–reduction mechanism seems the most probable explanation.

Typical reductions of 2- and 3-substituted indoles are tabulated in Table XIV.

D. CLEAVAGE OF SUBSTITUENTS FROM THE INDOLE RING. DECARBOXYLATIONS AND RELATED REACTIONS

Ethyl indole-3-carboxylate (**D1**) and substituted derivatives undergo loss of the ethoxycarbonyl groups in high yield when heated with ethanolic solutions of sulfuric acid or sodium ethoxide (4, 11, 103). These reactions are examples of a number of processes in which a C-2 or C-3 substituent is lost from the indole ring. It has been proposed that such cleavages occur via the 3*H*-indole or indolenine tautomers (103).

Heating indole-3-carboxaldehyde or 3-acetylindole in strongly basic solutions for extended periods also results in cleavage of the substituent (103). Loss of the acetyl groups results when either 4- or 6-nitro-3-acetylindole is refluxed with concentrated hydrochloric acid (93). 3-Acetyl-2-methyl-6-nitroindole is deacetylated under similar conditions (94). Loss of the 3-cyano group occurs when 4- or 6-nitroindole-3-carbonitrile is refluxed for extended periods with acidic ethanol (93). Ethanolysis of the nitrile function may occur prior to cleavage of the cyano group. Acid-catalyzed hydrolysis of a 2-acyl group probably accounts for the formation of 5-nitro-3-phenylindole along with the 2-acylindole **D10** during the hydrolysis of **D9** (42).

Displacement of formyl and acetyl groups during nitration has been observed (13, 94) and these are probably related mechanistically to acid-catalyzed cleavages with the nitronium ion acting as the electrophile at C-3.

Tertiary butyl groups are cleaved from 2- and 3-positions on the indole ring by acid. 3-*t*-Butylindole is reported to give an 80% yield of *t*-butyl bromide after 15 minutes' reflux with 48% hydrobromic acid (28). Similarly, 3-*t*-butyl-2,3-dimethyl-3*H*-indole gives 2,3-dimethylindole on reaction with hydrochloric acid (88).

$$\textbf{D12} \xrightarrow[H_2O]{HBr} (Me)_3CBr \ \textbf{D13}$$

$$\textbf{D14} \xrightarrow{HCl} \textbf{D15}$$

Each of the cleavage reactions discussed above can be accommodated by a generalized mechanism in which an indole ring acts as a leaving group from an intermediate generated by protonation (or nitration) at C-3. The cleavage will occur with facility when the substituents can accommodate the electron

$$\textbf{D16} \longrightarrow \textbf{D17} + S^+$$

$$S = -C(OH)_2(OEt),\ -C(OH)_2Me,\ -CMe_3,\ \text{etc.}$$

deficiency which develops as the C-3–"S" bond is cleaved. In the case of cleavage of carbonyl substituents, the cleavage presumably occurs via tetrahedral intermediates formed by addition of solvent to the carbonyl group.

This mode of substituent cleavage can be recognized in some acid catalyzed transformations in the alkaloid area. Tabersonine (**D18**) is converted by hot acetic acid to allocatharanthine (**D21**) (17a, 110). Similar cleavage reactions may be involved as crucial steps in the biogenesis of the indole alkaloids.

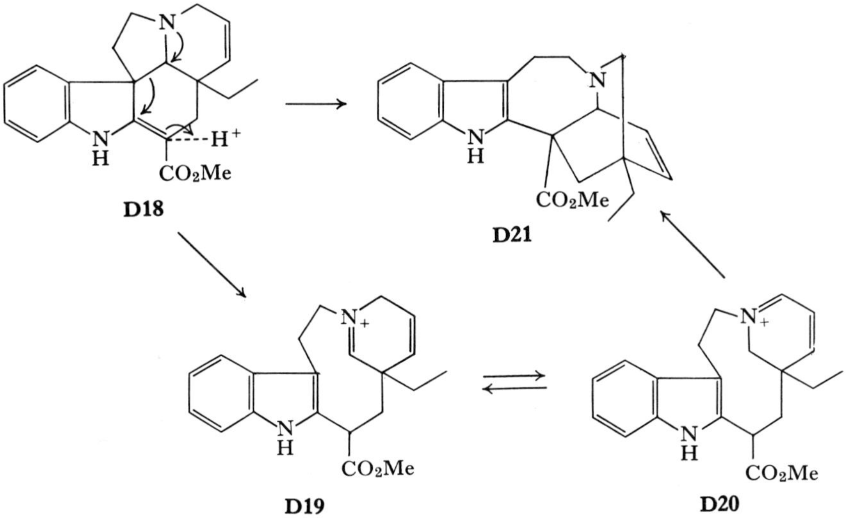

D18

D21

D19

D20

Joule and Smith have hypothesized that a similar cleavage is involved in the oxidation of isostrychnic acid (68).

The decomposition of akuammicine (**D22**) in methanol at 80°C is another example of this type of cleavage. The product is the inner salt **D26** formed by the steps shown below (35):

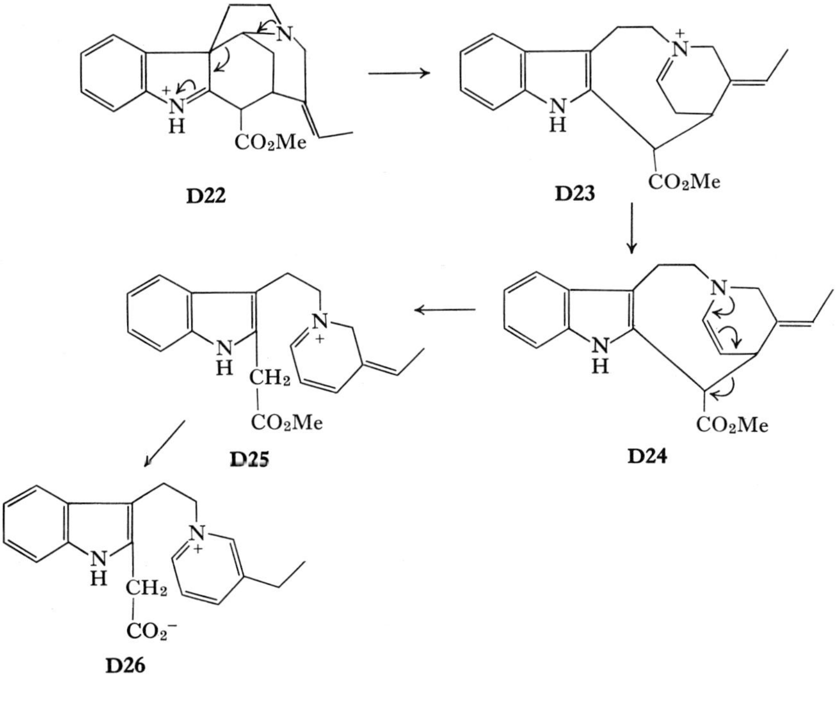

D22

D23

D24

D25

D26

The acid-catalyzed conversion of **D27** to **D30** involves cleavage of the substituent group from C-2. The intermediate recloses to a pentacyclic skeleton via electrophilic attack at C-3 followed by cyclization involving electrophilic substitution on the reactive dihydroxybenzene ring (52). The same cleavage pattern has been discussed in Section B in connection with epimerization of reserpine (43).

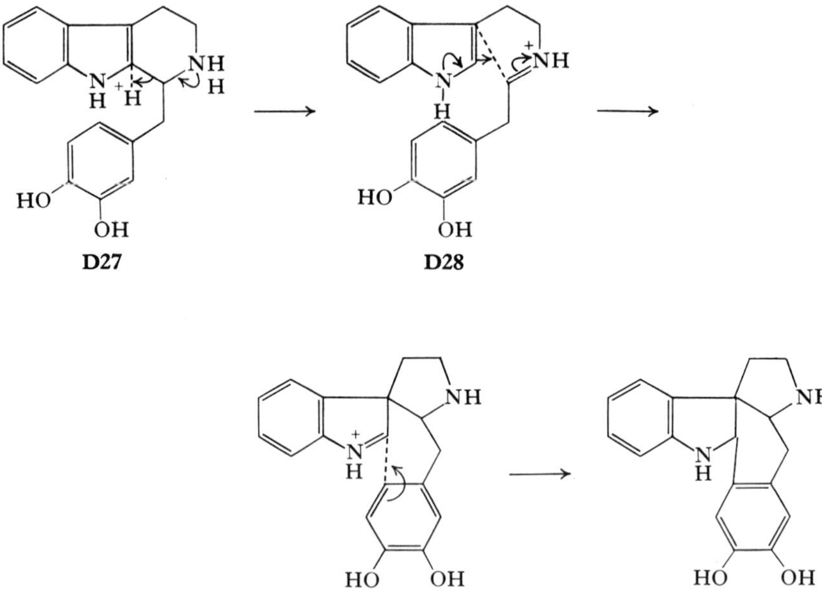

NH H N+ H H HO OH **D27** → +NH N H HO OH **D28** →

NH +N H HO OH **D29** → NH N H HO OH **D30**

The decarboxylation of indole-2-carboxylic acids and indole-3-carboxylic acids in refluxing aqueous acid has been a valuable synthetic procedure for some time. The mechanism of decarboxylation is related to the general substituent cleavage reaction discussed above. A more complete discussion of such decarboxylations is found in Chapter IX, C, 1.

It has been noted that indole alkaloids having the iboga skeleton and carboxy functions at C-18 decarboxylate readily (46a). It has been suggested that decarboxylations in such systems are initiated by protonation at C-3 on the indole ring (9), generating a β-iminoacid. However, decarboxylation via **D32** generates the highly strained intermediate **D33** (and violates Bredt's rule). Gorman and co-workers have put forward a modified mechanism (46a). They suggest that a cleavage of the C-5—C-18 bond is involved in decarboxylation. Amines, particularly hydrazine, also effect decarbomethoxylation of such systems (115). Decarboxylation of such compounds

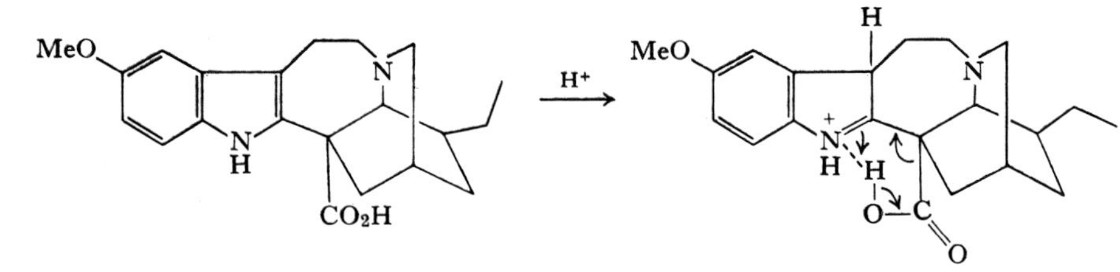

MeO N H+ CO2H H
D31

MeO H N + N H H O C O
D32

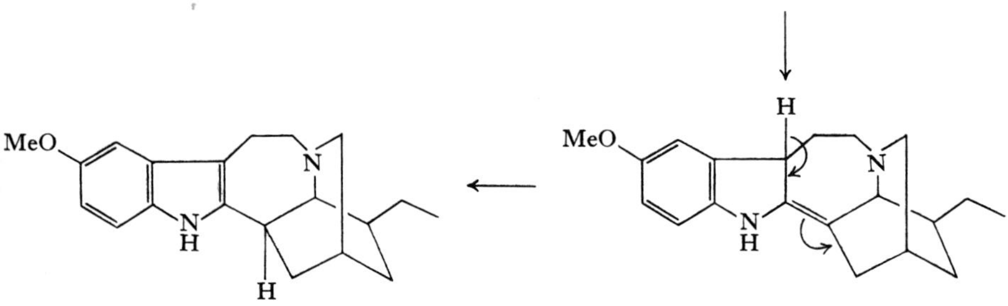

MeO N H N
D34

MeO H N N H
D33

D32 → MeO H + N + N H H—O—C O →
D35

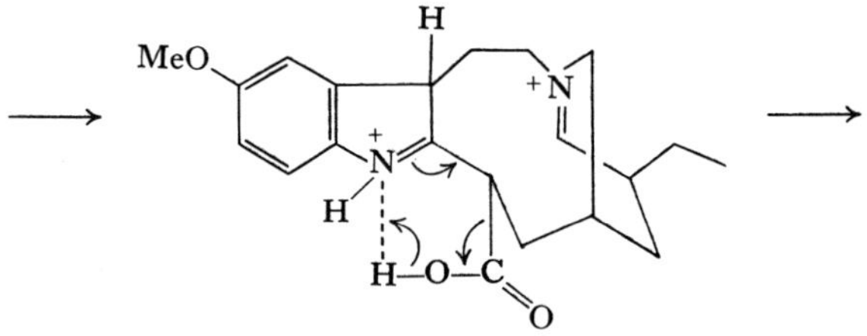

MeO H + N + N H H → D34
D36

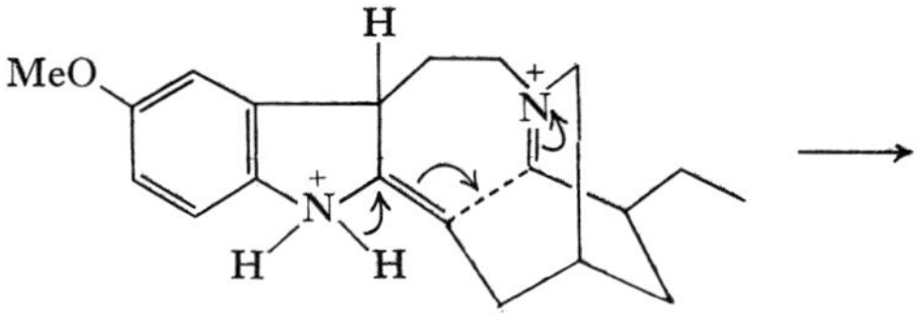

O N—CH2 CH2 CH2—N O CH2 N H CHCH2 CO2H N H
D37

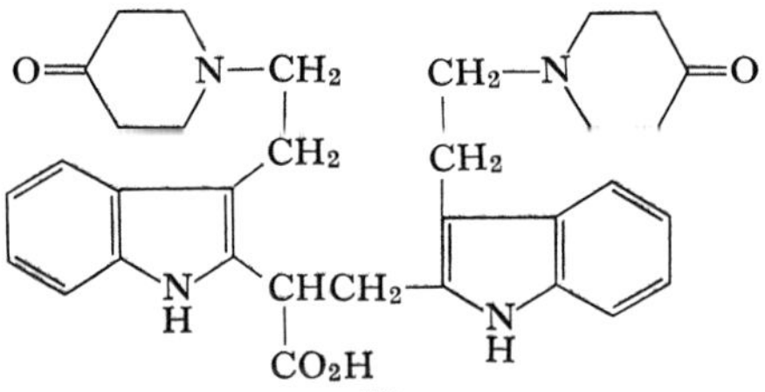

N H NH CO2H CH2Ph
D38

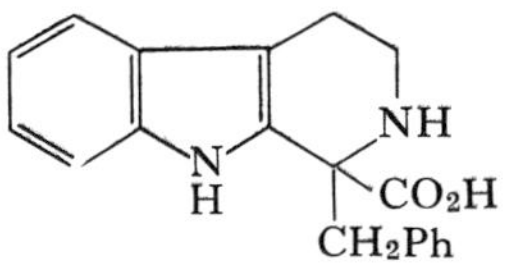

as **D37** and **D38** has also been observed to be facile (50a, 155). Again, the lability of the acid is best explained in terms of decarboxylation via the conjugate acid formed by protonation at C-3.

E. REACTIONS OF 3*H*-INDOLE DERIVATIVES

The conjugate acids which result from 3-protonation of indoles are derivatives of 3*H*-indole. The polarized carbon-nitrogen double bond which

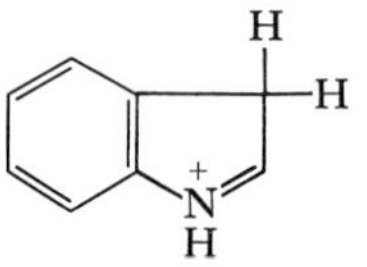

E1 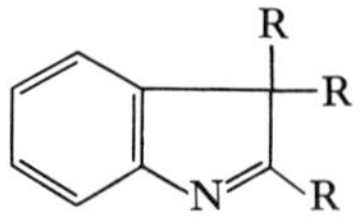 **E2**

is present in these systems presents a site for nucleophilic attack. Several examples of nucleophilic addition to C-2 of 3*H*-indoles are noteworthy. Indole reacts with sodium bisulfite in aqueous ethanolic solution to give the adduct **E4** (147). The mechanism for this reaction is presumed to involve

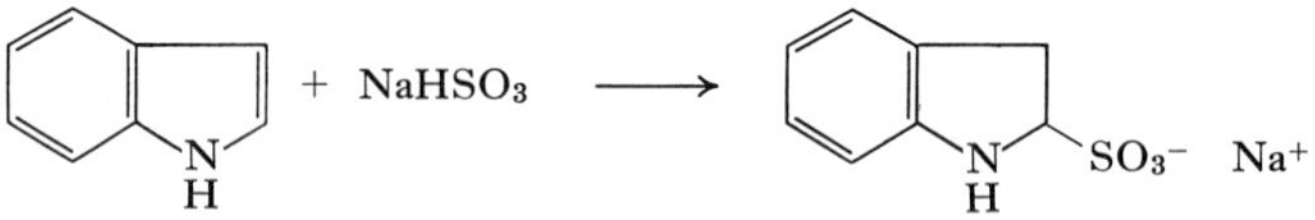

E3 + $NaHSO_3$ ⟶ **E4** (SO_3^- Na^+)

addition of the strongly nucleophilic bisulfite anion to the conjugate acid of indole. The adduct **E4** is of some synthetic interest, since, as an aniline derivative, it undergoes electrophilic substitution readily in the 5-position. 5-Bromoindole and 5-nitroindole have been prepared via **E4** in this manner (see Section H for further discussion).

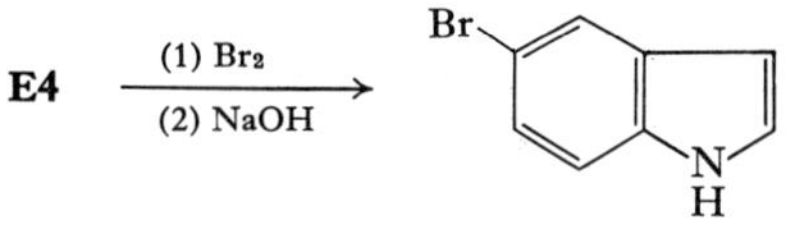

E4 —(1) Br_2 (2) NaOH⟶ **E5**

E6 (SO_3Na, Ac) ⟶ **E7** (O_2N)

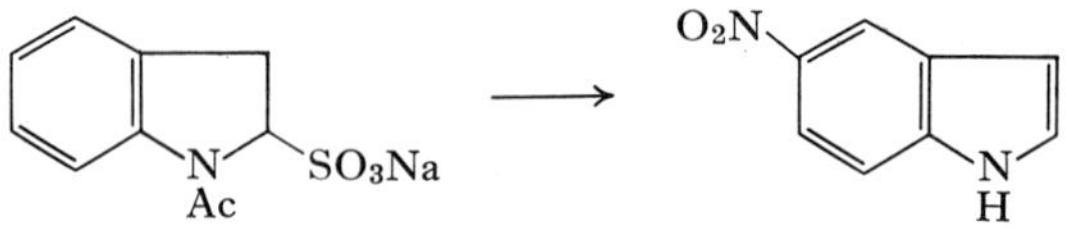

Simple 3,3-disubstituted 3*H*-indoles are usually prepared by Fischer cyclization of appropriate phenylhydrazones (Chapter III,A,3) or by alkylation of indoles (Chapter I,G). Recent physical measurements, particularly nmr studies (1a, 41, 61, 62) have established that equilibrium exists in many cases between 3,3-dialkyl-3*H*-indoles and the corresponding trimer. Dissociation is favored by elevated temperatures (41), acidic media,

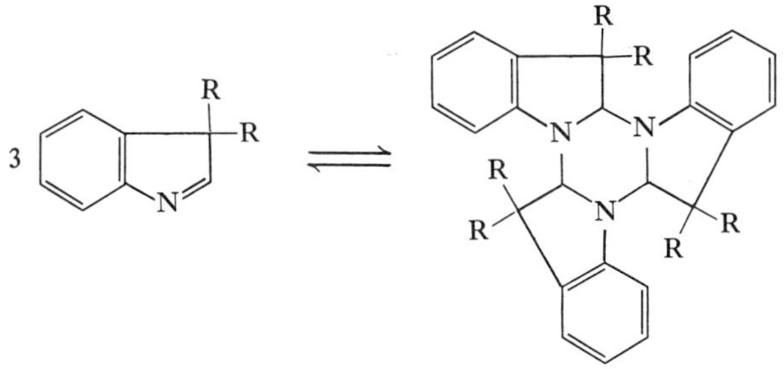

E8 **E9**

(62), and bulky 3-substituents (62). 3*H*-Indoles with 2-alkyl substituents apparently have no tendency to trimerize.

Carson and Mann (23) have discussed the properties of the carbinolamine **E10**. Ultraviolet spectral data indicate that the equilibrium **E10** → **E11** is established in solution. The equilibrium is shifted to the right in polar solvents and to the left in nonpolar solvents or in the presence of excess hydroxide ion. 3,3-Dialkyl-3*H*-indoles with hydrogen on the C-2 side chain

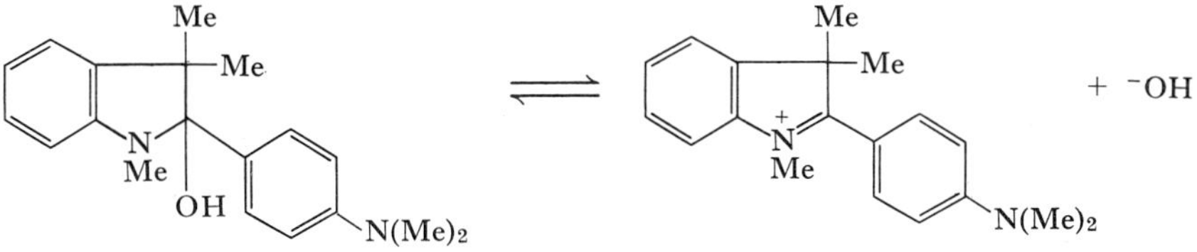

E10 **E11**

show an alternative mode of reaction as illustrated by the behavior of the 1,2,3,3-tetramethyl-3*H*-indolium ion **E12**. On treatment with base, the 2-methylenindoline **E13** is formed (117). The transformation **E14** → **E15**

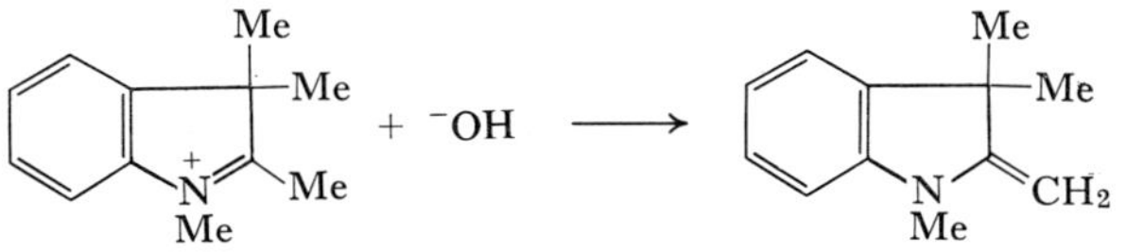

E12 **E13**

presumably proceeds via a nucleophilic addition–elimination mechanism (152).

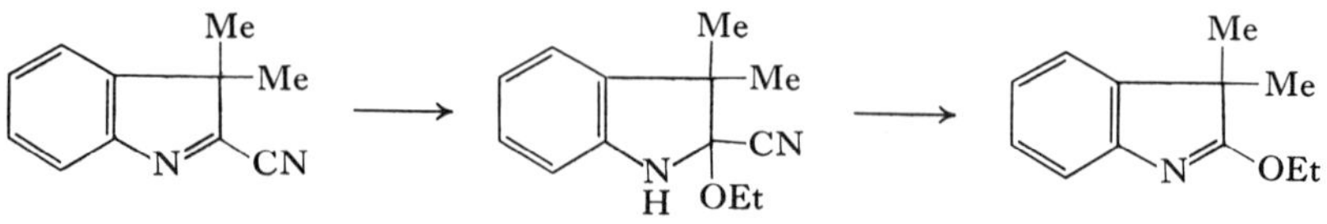

E14 E15

Intramolecular additions to the imine bonds in 3*H*-indole systems are perhaps the most important examples of the reaction type under discussion. The amine **E16** has been synthesized. The cyclic form **E17** is favored in solution. In acidic solution ring opening occurs and **E18** is formed (40).

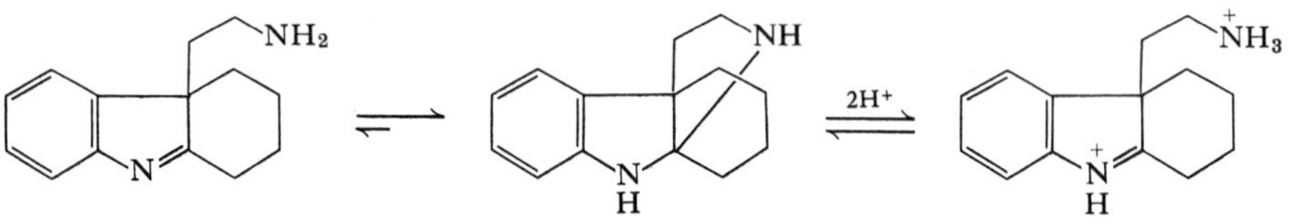

E16 E17 E18

The transformations **E19** → **E21** and **E22** → **E24** are further examples of intramolecular nucleophilic additions, as illustrated by the mechanisms shown (87).

Other examples of both intramolecular and intermolecular nucleophilic additions to 3*H*-indole systems can be found in the chemistry of indole alkaloids (9a, 14a, 14b, 123a).

The reaction of tryptophan with 2-hydroxy-5-nitrobenzyl bromide, **E26**, (Koshland's reagent) is an important reaction for modification and assay of tryptophan in proteins (8). The reactions of this reagent with certain 3-substituted indoles have recently been shown (132) to involve formation of products explicable in terms of 3*H*-indole intermediates. The product of the reaction of **E26** with 3-methylindole is **E28**. The mechanism shown below can account for formation of **E28** and is seen to involve an intramolecular nucleophilic addition to a 3*H*-indole. A similar reaction product is formed from **E26** and 2,3-dimethyindole. The ethyl ester of tryptophan is also initially alkylated at C-3 but the side-chain amino group then adds to the 3*H*-indole ring (85a).

Certain 3*H*-indoles with potential leaving groups at C-3 react to give products of nucleophilic substitution at the C-2 substituent, as represented by the generalized scheme below. Taylor (137) has pointed out that this sequence of reactions is a general feature of indole chemistry. The pattern

Me CH₂CN N Me **E19** $\xrightarrow[\text{EtOH}]{H^+}$ Me NH CH₂C—OEt N H Me **E20** →

Me N H N OEt Me **E21**

CH₂CH₂OH N H Me **E22** $\xrightarrow[\text{(2) MeI}]{\text{(1) MeMgI}}$ [Me CH₂CH₂O⁻ N Me] **E23** →

Me N H O Me **E24**

Me N H **E25** + BrH₂C NO₂ HO **E26** →

Me NO₂ + N H O H **E27** → Me NO₂ N H H O **E28**

X R N CH₂R′ **E29** ⇌ X R R′ N H H Y **E30** → R N H CHR′ Y **E31**

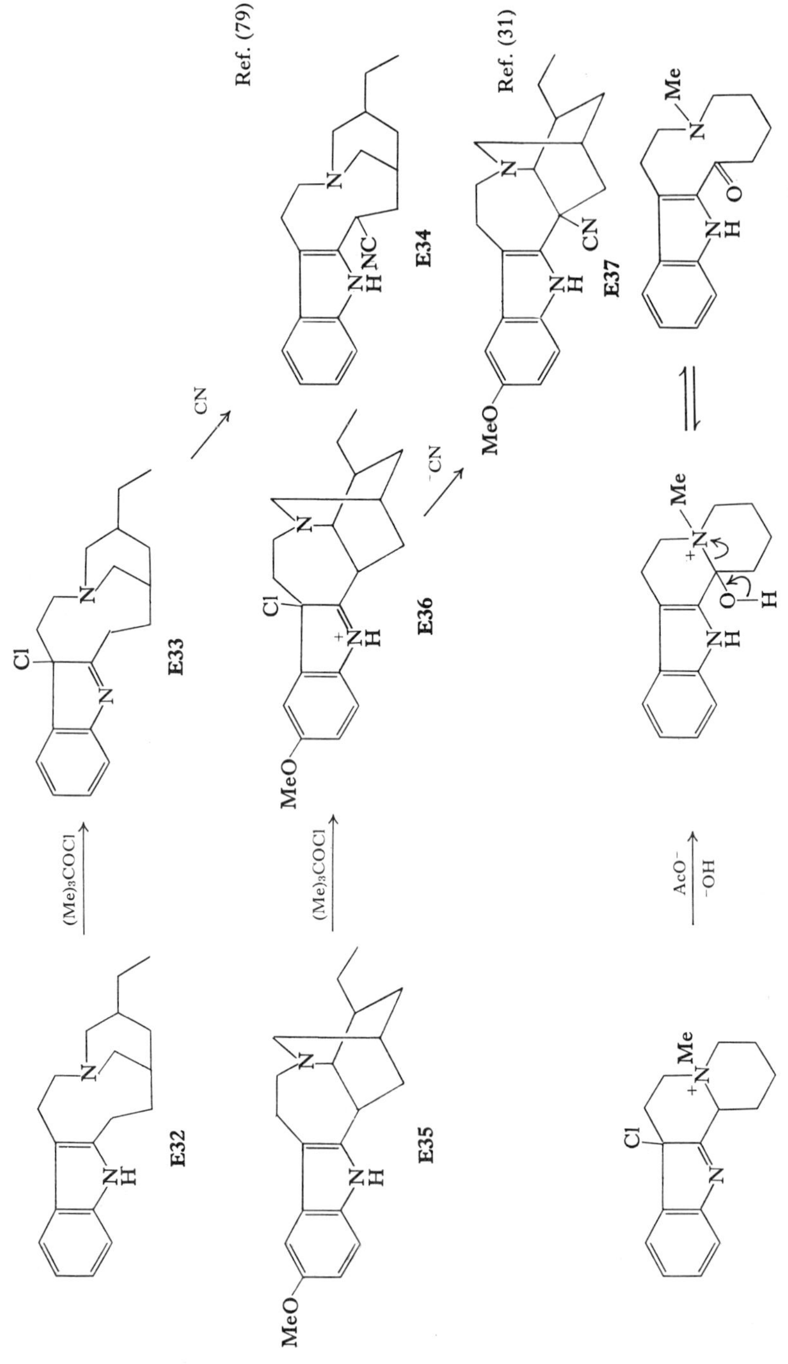

E32 $\xrightarrow{(Me)_3COCl}$ E33 $\xrightarrow{CN}$ E34 Ref. (79)

E35 $\xrightarrow{(Me)_3COCl}$ E36 $\xrightarrow{^{-}CN}$ E37 Ref. (31)

E38 $\xrightarrow[^{-}OH]{AcO^{-}}$ E39 $\rightleftharpoons$ E40 is frequently encountered in the synthetic modification of alkaloid skeletons. As Dolby (31) has pointed out, at least three schemes, elimination–addition, S_N2' displacement (as in **E29** → **E31**) or S_N1' displacement, can account for the net bonding changes in the above reactions. It is entirely possible that each of these mechanisms operates in appropriate systems.

Intramolecular reactions of nucleophilic groups with 3-halo-3*H*-indoles result in elaboration of a new ring. The cyclization of tryptamine and tryptophan derivatives to pyrrolo[2,3-b] indoles fits into this reaction pattern (97).

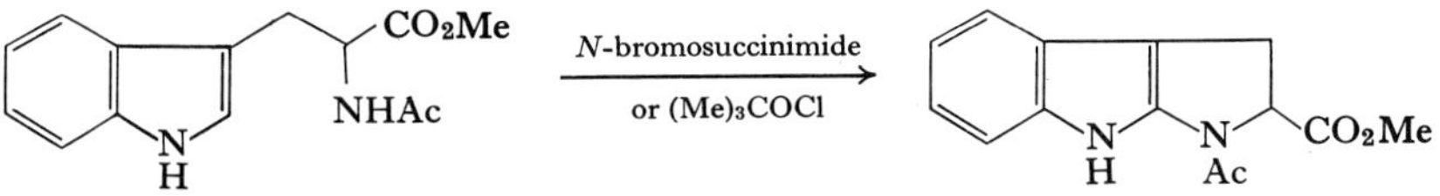

CO₂Me NHAc N H *N*-bromosuccinimide or $(Me)_3COCl$ N H N Ac CO₂Me

E41 **E42**

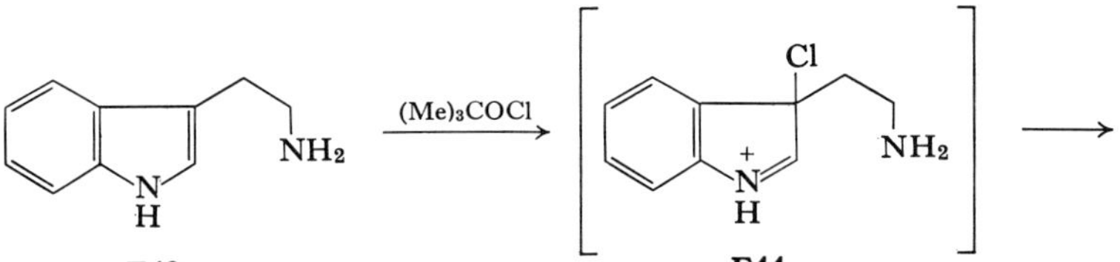

NH_2 $(Me)_3COCl$ Cl NH_2 N H

E43 **E44**

Cl N H H N H

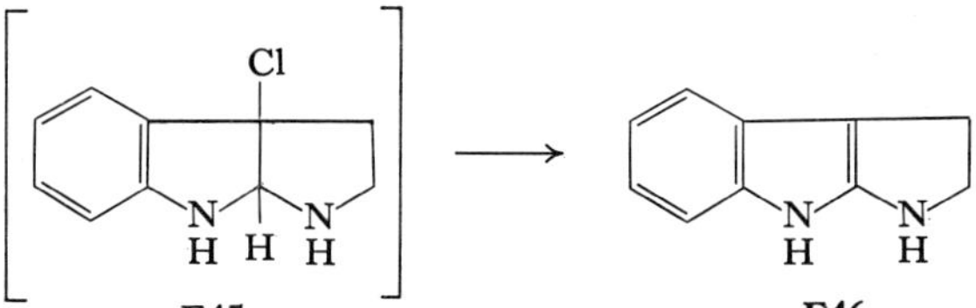

E45 **E46**

F. REACTIONS OF VINYLINDOLES

Derivatives of 3-vinylindole and, to a lesser extent, those of 2-vinylindole are expected to be relatively electron-rich olefins because of conjugation

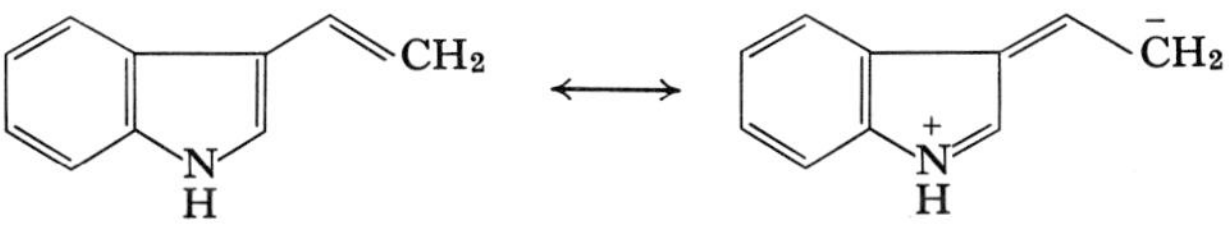

CH_2 ⟷ $\bar{C}H_2$

F1 **F2**

with the indole nucleus. This nucleophilicity manifests itself in the reaction of vinylindoles with electron-deficient olefins. 3-Vinylindole has been

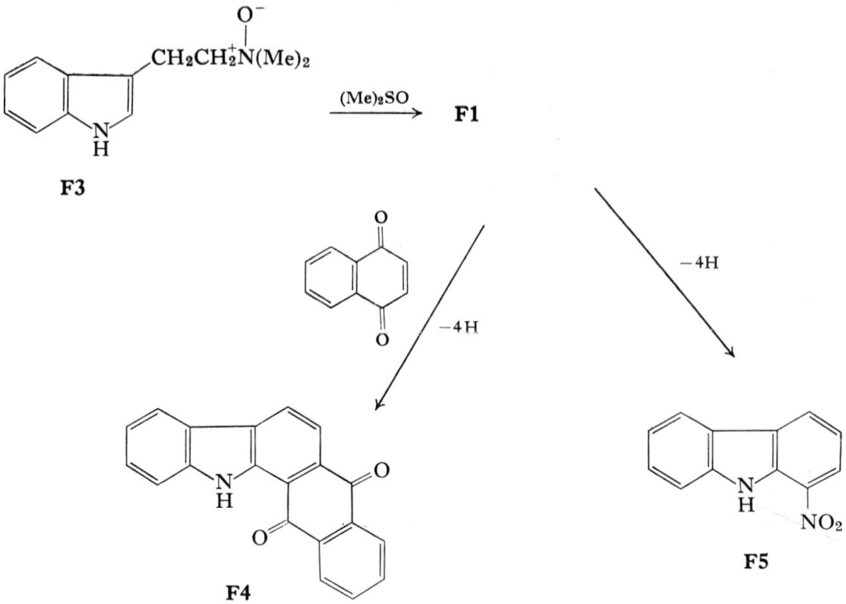

F3 $\xrightarrow{(Me)_2SO}$ F1

F4 ← $-4H$ — F1 — $-4H$ → F5

prepared (95) and found to react with naphthoquinone and, in very low yield, with nitroethylene. Diels–Alder type adducts presumably are initially formed. The isolated products are in the fully aromatic oxidation state. Tricyanovinylindoles, which are obtained from indoles and tetracyanoethylene, react with dimethyl acetylendicarboxylate to give substituted carbazoles (91). A number of other examples of Diels–Alder type cyclizations of vinylindoles have been studied by Noland and co-workers. The examples shown below are among the successful cases.

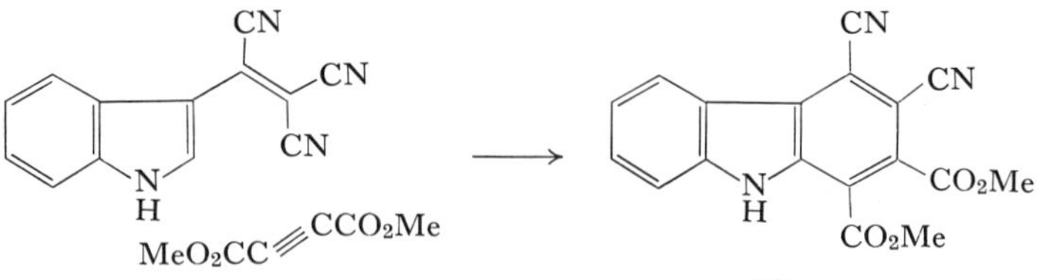

F6 + $MeO_2CC{\equiv}CCO_2Me$ → F7

A reaction between the vinylindole **F14** and dimethyl acetylenedicarboxylate is initiated by cyclization but takes a novel course, resulting in ring opening (66).

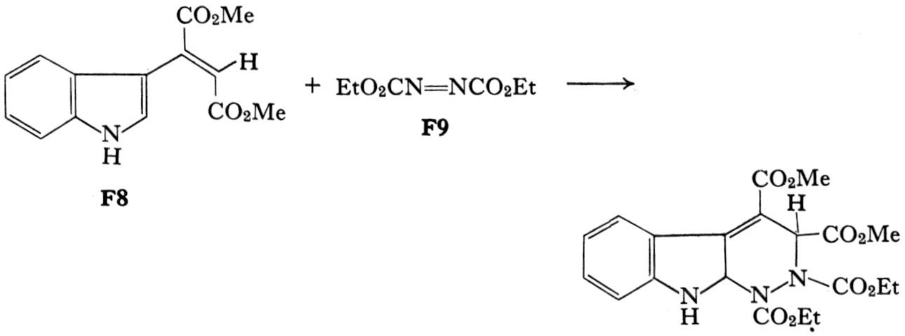

CO_2Me H CO_2Me N H **F8** + $EtO_2CN{=}NCO_2Et$ **F9** →

CO_2Me H CO_2Me N CO_2Et N H N CO_2Et **F10** Ref. (153a)

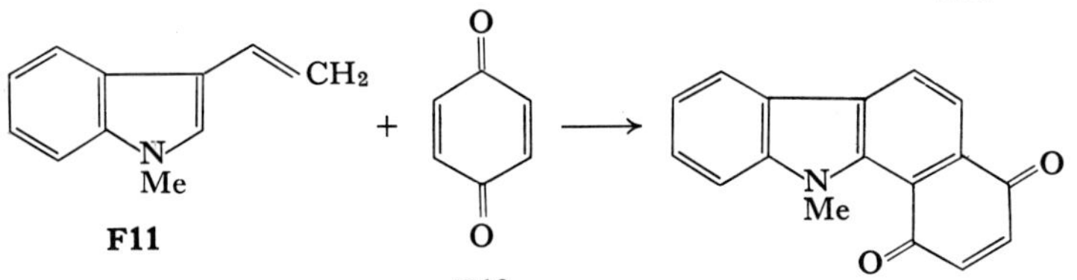

CH_2 N Me **F11** + O O **F12** → N Me O O **F13** Ref. (153a)

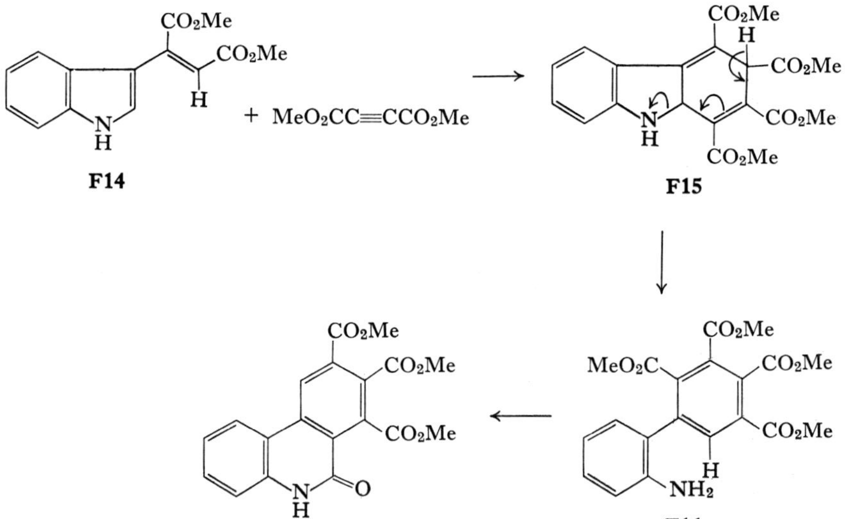

CO_2Me CO_2Me H N H **F14** + $MeO_2CC{\equiv}CCO_2Me$ → CO_2Me H CO_2Me N H CO_2Me CO_2Me **F15**

↓

CO_2Me CO_2Me CO_2Me N H O **F17** ← MeO_2C CO_2Me CO_2Me CO_2Me H NH_2 **F16**

Beck and Schenker (10) have observed several successful cyclizations with 2- and 3-(1,2,3,6-tetrahydro-4-pyridyl)indoles with *N*-phenylmaleimide and with acrylonitrile. A limiting factor in the success of many of these reactions may be the efficiency of the oxidation to stable products with carbazole-type conjugation.

Me
Me
$(Me)_3CO^-$
H
CH_2
O
$(CH_2)_4N(Me)_2$
F18
F19

$(Me)_3CO^-$

Me
$(CH_2)_3N(Me)_2$
CHMe
O
$^-CH(CH_2)_3N(Me)_2$
F21
F20

H
CH_2
+ ^-CN
H CN Me
F22
F23

OH
HO
Me
H
CH_2
^-OH
Me
F24
F25
F26

H
$(CH_2)_4NMe$
Me
Me
F28
F27

3-Vinylindoles have been found to add nucleophiles at the carbon atom adjacent to the indole ring. The most complete study in this area is that of Dolby and Gribble (32). The reaction shown can be reasonably explained in terms of intramolecular nucleophilic addition to the alkylidene indolenine. 3-Vinylindole reacts with cyanide ion to give 3-(1-cyanoethyl)indole in good yield (32). Foster and Harley-Mason have also reported an apparent example of nucleophilic addition by cyanide ion to a 3-vinylindole (37). It has been observed that the 3-vinyl group in **F24** can be removed by alkaline hydrolysis. The mechanism is believed to involve hydration of the vinyl group, followed by base-catalyzed cleavage of the resulting indolylcarbinol. Subsequent transformations lead to **F28** (32).

Certain other vinylindoles have been prepared (27, 81, 136) but little is known about their chemistry.

G. REDUCTION OF THE INDOLE RING AND INTERCONVERSION OF INDOLES AND DIHYDROINDOLES (INDOLINES)

Synthetic interconversion of substituted indoles and the corresponding dihydro derivatives (indolines) is an important process. The conversion of the indole ring to an indoline ring can be accomplished by catalytic hydrogenation (1). Moderately vigorous conditions are usually required and many easily reduced functional groups can be selectively reduced without reduction of the indole ring. The sodium salt of indole-3-propionic acid is reduced to the dihydro derivative at room temperature over Raney nickel under a hydrogen pressure of 3000–4000 psi (77). A temperature of 100° was found to be required for reduction of potassium indole-3-acetate over Raney nickel (65). Under somewhat more vigorous conditions reduction can proceed to the octahydro stage. Hydrogenation of indole in ethanol at 100–110°C and 90–100 atmospheres of hydrogen pressure gives indoline whereas at 150–160°C reduction of the benzene ring also occurs and is accompanied by *N*-alkylation (72). The pyridine ring is among the systems which can often be reduced under conditions which do not affect the indole ring (49).

G1 —(Ni, H_2, EtOH, 100°)→ G2

G1 —(Ni, H_2, EtOH, 160°)→ G3 (N-Et)

G4 →(Pt, H_2, H^+) G5 Ref. (49)

Hydrogenation of 2-(3-pyridyl)indole, however, proved troublesome and reduction of the pyridine ring was accomplished indirectly (56).

Certain indoles having 3-nitroethyl substituents show a peculiar tendency to undergo hydrogenation of the indole ring in preference to reduction of the nitro group (24, 158). A satisfactory explanation has not yet been put forward.

A significant recent finding is the observation that the indole ring undergoes facile catalytic reduction in strongly acidic solution (127). It is presumably the 3*H*-indolium ion which is reduced under these conditions (127).

G6 →(HBF_4, EtOH; H_2, Pt) G7

The indole ring can also be reduced to the indoline system by reduction with zinc or tin in acid solution. The transformations below are specific examples.

G8 →(Zn, HCl) G9 Ref. (101)

G10 →(Zn, H_3PO_4) G11 (64%) Ref. (30)

G12 →(Sn, HCl) G13 Ref. (50)

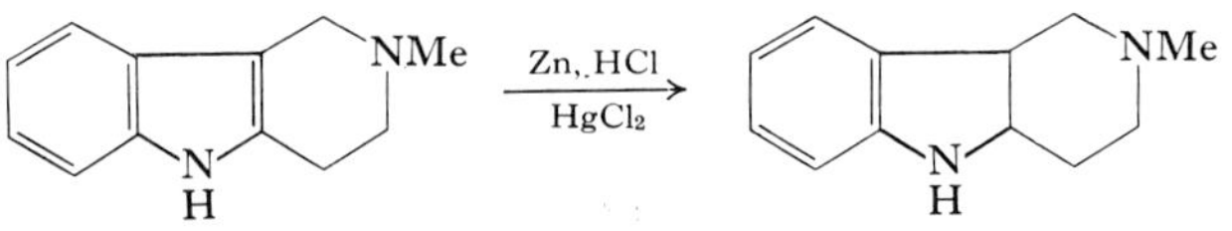

G14 —(Zn, HCl / $HgCl_2$)→ G15   Ref. (75)

The indole ring is reduced by lithium metal in liquid ammonia but this reaction has attracted little attention until recently. O'Brien and Smith (96) observed the reduction of indole by lithium metal in liquid ammonia followed by addition of methanol. The product was considered to be a mixture of 4,7-dihydroindole and 4,5,6,7-tetrahydroindole. 1-Methylindole is more easily reduced, giving 1-methylindoline, which is subsequently further reduced. Recent work has established the importance of the mode of addition of the proton source, methanol (113). Reduction of 5-methoxy-1-methylindole by lithium in ammonia in the absence of methanol gives the corresponding indoline but if methanol is present in the reducing media **G17** is formed. 5-Methoxyindole is reduced to the 4,7-dihydro derivative in 80% yield by lithium in ammonia containing methanol. This behavior is

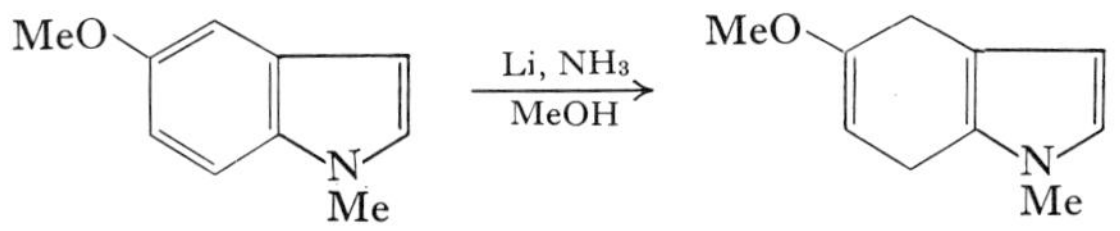

G16 —(Li, NH_3 / MeOH)→ G17

suggested to be the result of different protonation sites for the radical anion **G18** and the dianion **G19** formed by transfer of one and two electrons, respectively, to the indole ring. It is suggested that in the presence of the

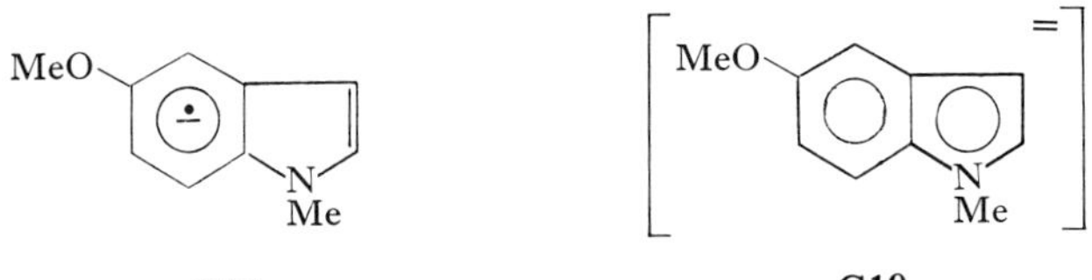

G18   G19

relatively acidic OH group in methanol, **G18** is protonated preferentially on the carbocyclic ring. In the absence of hydroxylic groups **G19** may be formed by a second electron transfer. Protonation of **G19** then occurs on the heterocyclic ring.

Indole is reduced in part to indoline by diborane, but the scope and usefulness of this reaction has not yet been fully explored (15). A photochemical reduction of tryptophan to 4,7-dihydrotryptophan in the presence of sodium borohydride has also been observed (157).

The dehydrogenation of indolines is a process of considerable synthetic significance. Chemical oxidation and catalytic dehydrogenations have been used to effect the transformation. Quinones have been the most widely used oxidizing reagents for dehydrogenation of indolines to indoles. Chloranil (tetrachloro-*p*-benzoquinone) and 2,3-dichloro-5,6-dicyano-*p*-benzoquinone have been successfully applied as dehydrogenating reagents. Table XV records a number of examples.

TABLE XV □ DEHYDROGENATION OF INDOLINES TO INDOLES WITH BENZOQUINONES

Indoline substituents	Dehydrogenation agent	Yield (%)	Ref.
1-(2,3,4,6-tetra-*O*-acetyl-β-D-glucopyranosyl)	2,3-Dichloro-5,6-dicyano-1,4-benzoquinone	91	(153)
1-(2,3,4,6-tetra-*O*-benzyl-β-glucopyransosyl)	2,3-Dichloro-5,6-dicyano-1,4-benzoquinone	100	(108)
1-Me-5-bromo	Chloranil	53	(144)
1-Me-5-phthalimido	Chloranil	49	(144)
1-Me-5-thiocyanato	Chloranil	51	(141)
5-Ac	Chloranil	60	(145)
5-Br	Chloranil	42	(60)
5-Cyano	Chloranil	45	(60)
5-Formyl	Chloranil	64	(140)
5-Amidosulfonyl	Chloranil	55	(143)
5-Thiocyanato	Chloranil	68	(141)

O_2N … CH_2CO_2H, N–Ac (**G20**) $\xrightarrow[\text{NaOH}]{\text{Raney Ni}}$ H_2N … CH_2CO_2H, N–H (**G21**, 93%) Ref. (64)

O_2N … N–H (**G22**) $\xrightarrow[\text{NaOH}]{\text{Raney Ni}}$ H_2N … N–H (**G23**, 43%) Ref. (64)

Aromatization of indolines over palladium on carbon catalyst or with metallic oxides constitutes an alternative approach to dehydrogenation. Aromatization of 5-, 6-, and 7-methoxyindoline to the respective indoles occurs over palladium in refluxing mesitylene (58). In some procedures a hydrogen acceptor such as maleic anhydride is employed with the palladium

$$\xrightarrow{MnO_2}$$

G24 G25

catalyst, as in the dehydrogenation of 6-nitroindoline (73). Johnson (64) has reported a series of reactions involving the conversion of nitroindolines to aminoindoles over large amounts of Raney nickel. Transformations **G20** → **G21** and **G22** → **G23** are typical. Here the nitro group apparently acts as the hydrogen acceptor.

Indoline has been oxidized to indole by manganese dioxide (105). The same reagent was found to be the reagent of choice for the conversion of **G24** to **G25** (63).

Dehydrogenations of indoline and 1-methylindoline with cupric chloride-pyridine complex have been reported (139). Recently, a novel method

TABLE XVI □ DEHYDROGENATIONS OF INDOLINES OVER METALLIC CATALYSTS AND BY OXIDES AND SALTS

Indoline substituent	Dehydrogenating agent	Yield (%)	Ref.
None	Pd/C	60	(139)
None	MnO_2	59	(104)
None	$CuCl_2$-pyridine	55–60	(139)
1-Me-4-methoxy	$CuCl_2$-pyridine	52	(69)
1-Me-4-methoxy	Pd/C-cinnamic acid	64	(69)
1-Me-6-methoxy	$CuCl_2$-pyridine	27	(69)
2-Me	Pd/C	65	(104)
3-(2-Carbethoxyethyl)	MnO_2	86	(63)
5-Carboxy	Pd/C	80	(60)
5-(3-Indolylmethyl)	Pd/C	70	(146)
5-Trimethylsilyl	Pd/C	75	(12)
6-Cyano	Pd/C	—	(60)
6 Methoxy	Pd/C	83	(58)
6-Methoxy	Na, NH_3-toluene	78	(156)
6-Nitro	Pd/C-maleic anhydride	40	(73)
7-Carboxy	Pd/C	81	(60)
7-Methoxy	Pd/C	95	(58)
1,11α-Imino-3-methoxy-estra-1,3,5(10)-trien-17-one	Pd/C	95	(22)

for the dehydrogenation of indoline and 6-methoxyindoline in good yield has been discovered (156). The procedure involves forming the sodium salt of the indoline in liquid ammonia, allowing the ammonia to evaporate, and heating the salt in toluene. The mechanism has been only briefly discussed.

Table XVI records typical dehydrogenations of indoles involving metal catalysts and metallic oxides.

H. THE INDOLINE–INDOLE SYNTHETIC METHOD

The dihydroindole or indoline ring system is a substituted aniline system. As such, *N*-acylation and *N*-alkylation reactions can be accomplished by standard methods appropriate for secondary amines. Furthermore, the strong directing effect of the substituted amino function results in the carbocyclic ring being readily substituted by electrophilic reagents with the 5-position being the primary site of attack and the 7-position a secondary site in the neutral amine and its *N*-acyl derivatives. These methods for synthesis of indolines substituted on nitrogen or the carbocyclic ring, when coupled with efficient methods for subsequent dehydrogenation of the

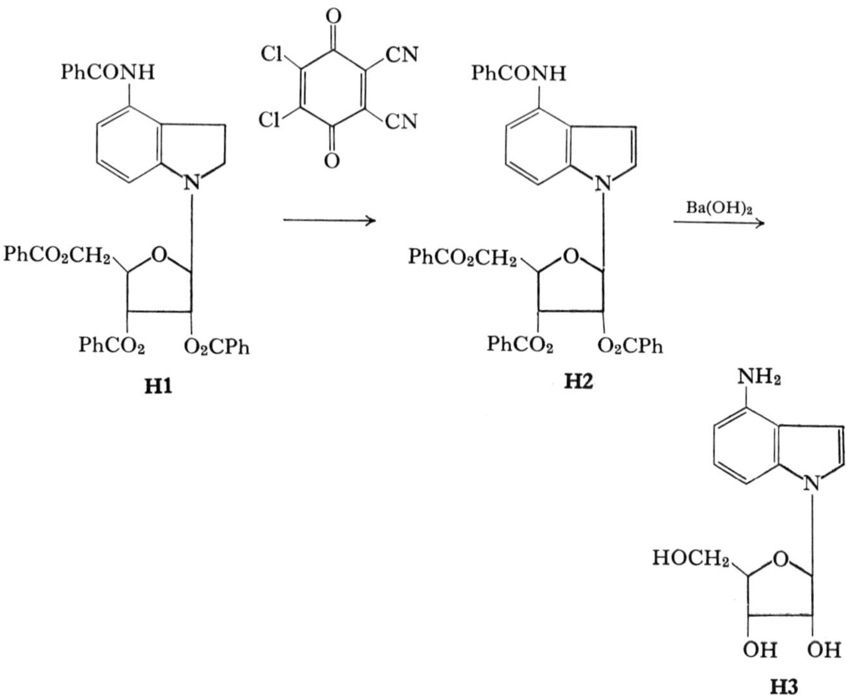

indoline to the corresponding indole, constitute a useful synthetic approach to substituted indoles. The method has largely been developed by Russian workers and a recent review admirably summarizes the work in this area (105).

The method has recently, for example, been applied to synthesis of 1-glycosylindoles (109, 153).

The method has also been applied to the synthesis of 5- and 6-haloindoles (59, 60, 106), 5-formyl- and 5-acylindoles (140, 145), cyanoindoles (60), nitroindoles (142), trimethylsilylindoles (12), and protected aminoindoles (143, 144).

Similar application of the indoline derivative **H5**, which is easily obtained from indole, is noteworthy (147).

H4 —($NaHSO_3$)→ H5 (N-H, SO_3Na) —(Ac_2O)→ H6 (N-Ac, SO_3Na) —(Br_2)→ H7 (Br, N-Ac, SO_3Na) —(H_2O, NaOH)→ H8 (Br, N-H)

H4 **H5** **H6** **H7** **H8**

I. METALATION OF INDOLES

As discussed in Chapter I, G, indoles unsubstituted at nitrogen react with organometallics to give *N*-metalloindoles. 1-Substituted indoles, however, can be metalated at C-2 by alkyllithium reagents. Shirley and Roussel prepared 2-lithio-1-methylindole by reaction of 1-methylindole with butyllithium, and carried out the reactions shown below as well as others (125). This potentially versatile method for preparation of 1,2-disubstituted indoles has received little subsequent attention although Kebrle and Hoffmann have recorded several applications (70, 71), typified by the syntheses of **I9** and **I11**.

Indole undergoes mercuration with mercuric acetate to give a diacetoxymercury derivative. Contradictory evidence (74, 112) has been presented

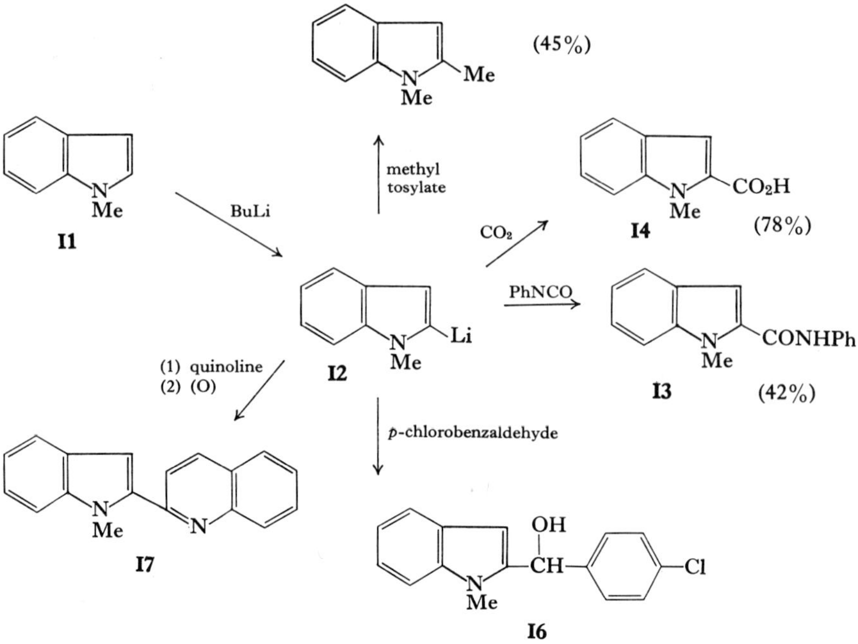

(45%)

Me

N
Me

N
Me

I1

BuLi

methyl tosylate

CO_2

N
Me

I4

CO_2H

(78%)

PhNCO

N
Me

Li

I2

(1) quinoline
(2) (O)

N
Me

CONHPh

I3

(42%)

p-chlorobenzaldehyde

N
Me

N

I7

OH

CH

Cl

N
Me

I6

concerning the structure of the product. The most recent results (74) suggest that the most likely structure is the 1,3-dimercurated derivative. Neither this substance nor related mono- and triacetoxymercury derivative (112) have yet found significant application.

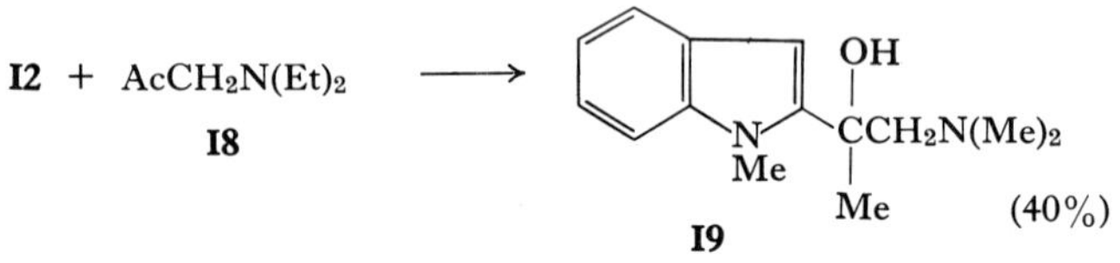

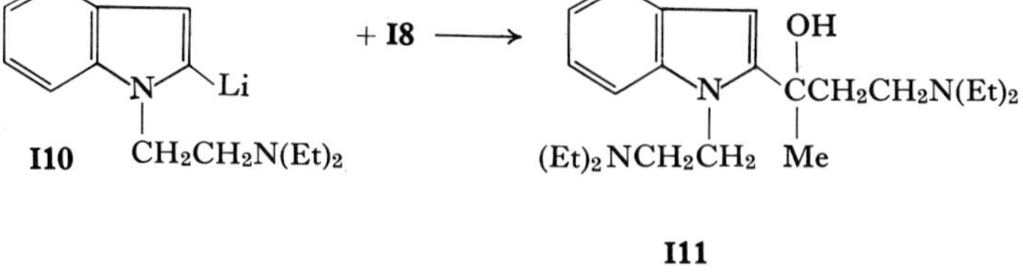

I2 + $AcCH_2N(Et)_2$ ⟶
I8

OH
$CCH_2N(Me)_2$
N
Me
Me
I9
(40%)

N
Li
$CH_2CH_2N(Et)_2$
I10
+ I8 ⟶

OH
$CCH_2CH_2N(Et)_2$
N
$(Et)_2NCH_2CH_2$ Me
I11

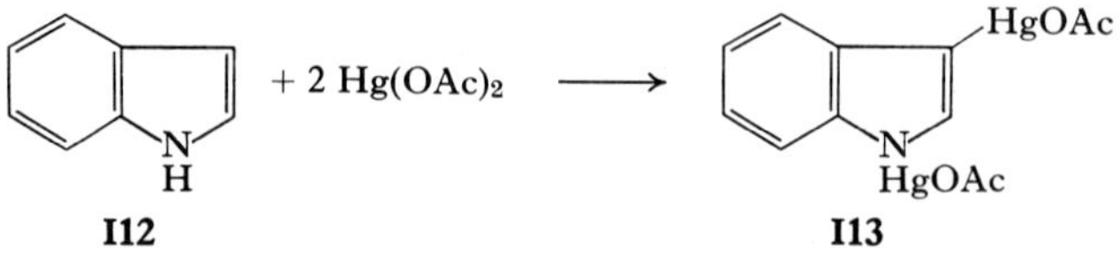

N
H
I112
+ 2 $Hg(OAc)_2$ ⟶

HgOAc
N
HgOAc
I113

REFERENCES

1. Adkins, H., and Burks, R. E., Jr., *J. Am. Chem. Soc.* **70**, 4174 (1948).
1a. Ahmed, M., and Robinson, B., *J. Chem. Soc., B* p. 411 (1967).
2. Alberton, N. F., Archer, S., and Suter, C. M., *J. Am. Chem. Soc.* **66**, 500 (1944).
2a. Albright, J. D., and Goodman, L., *J. Am. Chem. Soc.* **91**, 4317 (1969).
3. Albright, J. D., and Snyder, H. R., *J. Am. Chem. Soc.* **81**, 2239 (1959).
4. Allen, G. R., Jr., Pidacks, C., and Weiss, M. J., *J. Am. Chem. Soc.* **88**, 2536 (1966).
5. Ames, D. E., Bowman, R. E., Evans, D. D., and Jones, W. A., *J. Chem. Soc.* p. 1984 (1956).
6. Anthony, W. C., and Szmuszkovicz, J., U.S. Patent 2,821,532 (1958); *Chem. Abstr.* **52**, 10203 (1958).
7. Bader, H., and Oroshnik, W., *J. Am. Chem. Soc.* **79**, 5686 (1957).
8. Barman, T. E., and Koshland, D. E., Jr., *J. Biol. Chem.* **242**, 5771 (1967).
9. Bartlett, M. F., Dickel, D. F., and Taylor, W. I., *J. Am. Chem. Soc.* **80**, 126 (1958).
9a. Bartlett, M. F., Lambert, B. F., Werblood, H. M., and Taylor, W. I., *J. Am. Chem. Soc.* **85**, 475 (1963).
10. Beck, D., and Schenker, K., *Helv. Chim. Acta* **51**, 264 (1968).
11. Beer, R. J. S., Clarke, K., Davenport, H. F., and Robertson, A., *J. Chem. Soc.* p. 2029 (1951).
12. Belsky, I., Gertner, D., and Zilkha, A., *J. Org. Chem.* **33**, 1348 (1968).
13. Berti, G., Da Settimo, A., and Livi, O., *Tetrahedron* **20**, 1397 (1964).
14. Beugelmans, R., Potier, P., LeMen, J., and Janot, M.-M., *Bull. Soc. Chim. France* p. 2207 (1966).
14a. Bickel, H., Schmid, H., and Karrer, P., *Helv. Chim. Acta* **38**, 649 (1955).
14b. Birch, A. J., Hodson, H., Moore, B., and Smith, G. F., *Proc. Chem. Soc.* p. 62 (1961).
15. Biswas, K. M., and Jackson, A. H., *Tetrahedron* **24**, 1145 (1968).
16. Brewster, J. H., and Eliel, E. L., *Org. Reactions* **7**, 99 (1953).
17. British Patent, 851,780 (1960); *Chem. Abstr.* **55**, 11442 (1961).
17a. Brown, R. T., Hill, J. S., Smith, G. F., and Stapleford, K. S., *Chem. Commun.* p. 1475 (1969).
18. Büchi, G., Coffen, D. L., Kocsis, K., Sonnet, P. E., and Ziegler, F. E., *J. Am. Chem. Soc.* **88**, 3099 (1966).
19. Büchi, G., Manning, R. E., and Monti, S. A., *J. Am. Chem. Soc.* **86**, 4631 (1964).
20. Büchi, G., and Manning, R. E., *J. Am. Chem. Soc.* **88**, 2532 (1966).
21. Büchi, G., and Warnhoff, E. W., *J. Am. Chem. Soc.* **91**, 4433 (1959).
22. Cantrall, E. W., Conrow, R. B., and Bernstein, S., *J. Org. Chem.* **32**, 3445 (1967).
23. Carson, D. F., and Mann, F. G., *J. Chem. Soc.* p. 5819 (1965).
24. Cohen, A., and Heath-Brown, B., *J. Chem. Soc.* p. 7179 (1965).
25. Coker, J. N., Mathre, O. B., and Todd, W. H., *J. Org. Chem.* **28**, 589 (1963).
26. Dahlbom, R., and Misiorny, A., *Acta Chem. Scand.* **9**, 1074 (1955).
27. Daly, J. W., and Witkop, B., *J. Org. Chem.* **27**, 4104 (1962).
28. David, S., and Régent, P., *Bull. Soc. Chim. France* p. 101 (1964).
29. Dolby, L. J., and Booth, D. L., *J. Org. Chem.* **30**, 1550 (1965).
30. Dolby, L. J., and Gribble, G. W., *J. Heterocyclic Chem.* **3**, 124 (1966).
31. Dolby, L. J., and Gribble, G. W., *J. Org. Chem.* **32**, 1391 (1967).
32. Dolby, L. J., and Gribble, G. W., *Tetrahedron* **24**, 6377 (1968).
33. Dolby, L. J., and Sakai, S., *J. Am. Chem. Soc.* **86**, 1890 (1964); *Tetrahedron* **23**, 1 (1967).
34. Domschke, G., and Fürst, H., *Chem. Ber.* **93**, 2097 (1960).
35. Edwards, P. N., and Smith, G. F., *J. Chem. Soc.* p. 1458 (1961).
36. Eiter, K., and Sveirak, O., *Monatsh. Chem.* **83**, 1453 (1952).

37. Foster, G. H., and Harley-Mason, J., *Chem. Commun.* p. 1440 (1968).
38. Foster, G. H., Harley-Mason, J., and Waterfield, W. R., *Chem. Commun.* p. 21 (1967).
39. Freter, K., Hübner, H. H., Merz, H., Schroeder, H. D., and Zeile, K., *Ann. Chem.* **684**, 159 (1965).
40. Fritz, H., and Fischer, O., *Tetrahedron* **20**, 1737 and 2047 (1964).
41. Fritz, H., and Pfaender, P., *Chem. Ber.* **98**, 989 (1965).
42. Fryer, R. I., Earley, J. V., and Sternbach, L. H., *J. Org. Chem.* **32**, 3798 (1967).
43. Gaskell, A. J., and Joule, J. A., *Tetrahedron* **23**, 4053 (1967).
44. Gaskell, A. J., and Joule, J. A., *Tetrahedron* **24**, 5115 (1968).
45. Geissman, T. A., and Armen, A., *J. Am. Chem. Soc.* **74**, 3916 (1952).
46. Gill, N. S., James, K. B., Lions, F., and Potts, K. T., *J. Am. Chem. Soc.* **74**, 4923 (1952).
46a. Gorman, M., Neuss, N., and Cone, N. J., *J. Am. Chem. Soc.* **87**, 93 (1965).
47. Goutarel, R., Janot, M.-M., Prelog, V., and Taylor, W. I., *Helv. Chim. Acta* **33**, 150 (1950).
48. Gower, B. G., and Leete, E., *J. Am. Chem. Soc.* **85**, 3683 (1963).
49. Gray, A. P., and Kraus, H., *J. Org. Chem.* **26**, 3368 (1961).
50. Gurney, J., Perkin, W. H., Jr., and Plant, S. G. P., *J. Chem. Soc.* p. 2676 (1927).
50a. Hahn, G., Bärwald, L., Schales, O., and Werner, H., *Ann. Chem.* **520**, 107 (1935).
51. Harley-Mason, J., and Atta-ur-Rahman, *Chem. Commun.* p. 208 (1967).
52. Harley-Mason, J., and Waterfield, W. R., *Tetrahedron* **19**, 65 (1963).
53. Hartung, W. H., and Simonoff, R., *Org. Reactions* **7**, 263 (1953).
54. Henry, D. W., and Leete, E., *J. Am. Chem. Soc.* **79**, 5254 (1957).
55. Herbst, D., Rees, R., Hughes, G. A., and Smith, H., *J. Med. Chem.* **9**, 864 (1966).
56. Huffman, J. W., *J. Org. Chem.* **27**, 503 (1962).
57. Huffman, R. W., and Bruice, T. C., *J. Am. Chem. Soc.* **89**, 6243 (1967).
58. Hunt, R. R., and Pickard, R. L., *J. Chem. Soc., C* p. 344 (1966).
59. Ikan, R., Hoffmann, E., Bergmann, E. D., and Galun, A., *Israel J. Chem.* **2**, 37 (1964); *Chem. Abstr.*, **61**, 5596 (1964).
60. Ikan, R., and Rapaport, E., *Tetrahedron* **23**, 3823 (1967).
61. Jackson, A. H., and Smith, A. E., *Tetrahedron* **21**, 989 (1965).
62. Jackson, A. H., and Smith, P., *Tetrahedron* **24**, 2227 (1968).
63. Jansen, A. B. A., Johnson, J. M., and Surtee, J. R., *J. Chem. Soc.* p. 5573 (1964).
64. Johnson, H. E., U.S. Patent 3,226,396 (1965); *Chem. Abstr.* **64**, 11179 (1966).
65. Johnson, H. E., and Crosby, D. G., *J. Org. Chem.* **28**, 2794 (1963).
66. Johnson, R. A., Ph.D. Thesis with W. E. Noland, University of Minnesota (1965); *Dissertation Abstr.* **26**, 5719 (1966).
67. Joule, J. A., and Djerassi, C., *J. Chem. Soc.* p. 2777 (1964).
68. Joule, J. A., and Smith, G. F., *Proc. Chem. Soc.* p. 322 (1959).
69. Julia, M., and Gaston-Breton, H., *Bull. Soc. Chim. France* p. 1335 (1966).
70. Kebrle, J., and Hoffmann, K., *Gazz. Chim. Ital.* **93**, 238 (1963).
71. Kebrle, J., Rossi, A., and Hoffmann, K., *Helv. Chim. Acta* **42**, 907 (1959).
72. King, F. E., Barltrop, J. A., and Walley, R. J., *J. Chem. Soc.* p. 277 (1945).
73. Kinoshita, T., Inoue, H., and Imato, E., *Nippon Kagaku Zasshi* **78**, 1372 (1957); *Chem. Abstr.* **54**, 491 (1960).
74. Kirby, G. W., and Shah, S. W., *Chem. Commun.* p. 381 (1965).
75. Kochetkov, N. K., Kucherova, N. F., and Zhukova, I. G., *Zh. Obshch. Khim.* **31**, 924 (1961); *Chem. Abstr.* **55**, 23523 (1961).
76. Kornfeld, E. C., *J. Org. Chem.* **16**, 806 (1951).
77. Kornfeld, E. C., Fornefeld, E. J., Kline, G. B., Mann, M. J., Morrison, D. E., Jones, R. G., and Woodward, R. B., *J. Am. Chem. Soc.* **78**, 3087 (1956).

78. Kutney, J. P., Abdurahman, N., Le Quesne, P., Piers, E., and Vlattas, I., *J. Am. Chem. Soc.* **88**, 3656 (1966).
79. Kutney, J. P., Cretney, W. J., Le Quesne, P., McKague, B., and Piers, E., *J. Am. Chem. Soc.* **88**, 4756 (1966).
80. Leete, E., *J. Am. Chem. Soc.* **81**, 6023 (1959).
81. Leete, E., *Tetrahedron* **14**, 35 (1961).
82. Leete, E., *Chem. & Ind.* (*London*) p. 692 (1960).
83. Leete, E., Ahmad, A., and Kompis, I., *J. Am. Chem. Soc.* **87**, 4168 (1965).
84. Leete, E., and Marion, L., *Can. J. Chem.* **31**, 775 (1953).
85. Licari, J. J., and Dougherty, G., *J. Am. Chem. Soc.* **76**, 4039 (1954).
85a. Loudon, G. M., Portsmouth, D., Lukton, A., and Koshland, D. E., Jr., *J. Am. Chem. Soc.* **91**, 2793 (1969).
86. Marchand, B., *Chem. Ber.* **95**, 577 (1962).
87. Nakazaki, M., *Bull. Chem. Soc. Japan* **32**, 588 (1959).
88. Nakazaki, M., Isoe, S., and Tanno, K., *Nippon Kagaku Zasshi* **76**, 1262 (1955); *Chem. Abstr.* **51**, 17878 (1957).
89. Nógrádi, T., *Monatsh. Chem.* **88**, 1087 (1958).
90. Nogrady, T., and Doyle, T. W., *Can. J. Chem.* **42**, 485 (1964).
91. Noland, W. E., Kuryla, W. C., and Lange, R. F., *J. Am. Chem. Soc.* **81**, 6010 (1959).
92. Noland, W. E., and Reich, C., *J. Org. Chem.* **32**, 828 (1967).
93. Noland, W. E., and Rush, K. R., *J. Org. Chem.* **31**, 70 (1966).
94. Noland, W. E., Smith, L. R., and Rush, K. R., *J. Org. Chem.* **30**, 3457 (1965).
95. Noland, W. E., and Sundberg, R. J., *J. Org. Chem.* **28**, 884 (1963).
96. O'Brien, S., and Smith, D. C. C., *J. Chem. Soc.* p. 4609 (1960).
97. Ohno, M., Spande, T. F., and Witkop, B., *J. Am. Chem. Soc.* **90**, 6521 (1968).
98. Pachter, I. J., Mohrbacker, R. J., and Zacharias, D. E., *J. Am. Chem. Soc.* **83**, 635 (1961).
99. Plieninger, H., and Müller, W., *Chem. Ber.* **93**, 2024 (1960).
100. Plieninger, H., and Suehiro, T., *Chem. Ber.* **88**, 550 (1955).
101. Pope, W. J., and Clarke, G., Jr., *J. Chem. Soc.* **85**, 1330 (1904).
102. Potts, K. T., and Liljegren, D. R., *J. Org. Chem.* **28**, 3202 (1963).
103. Powers, J. C., *Tetrahedron Letters* p. 655 (1965).
104. Pratt, E. F., and McGovern, T. P., *J. Org. Chem.* **29**, 1540 (1964).
105. Preobrazhenskaya, M. N., *Russian Chem. Rev.* (*English Transl.*) **36**, 753 (1967).
106. Preobrazhenskaya, M. N., Fedotova, M. V., Sorokina, N. P., Orgareva, O. B., Uvarova, N. V., and Suvorov, N. N., *J. Gen. Chem. USSR* (*English Transl.*) **34**, 1310 (1964).
107. Preobrazhenskaya, M. N., Orlova, L. M., and Suvorov, N. N., *J. Gen. Chem. USSR* (*English Transl.*) **33**, 1347 (1963).
108. Preobrazhenskaya, M. N., and Suvorov, N. N., *J. Gen. Chem. USSR* (*English Transl.*) **35**, 896 (1965).
109. Preobrazhenskaya, M. N., Vigdorchik, M. M., and Suvorov, N. N., *Tetrahedron* **23**, 4653 (1967).
110. Qureshi, A. A., and Scott, A. I., *Chem. Commun.* p. 945 (1968).
111. Rajagopolan, P., and Advani, B. G., *Tetrahedron Letters* p. 2197 (1965).
112. Ramachandran, L. K., and Witkop, B., *Biochemistry* **3**, 1603 (1964).
113. Remers, W. A., Gibs, G. J., Pidacks, C., and Weiss, M. J., *J. Am. Chem. Soc.* **89**, 5513 (1967).
114. Renner, U., and Fritz, H., *Tetrahedron Letters* p. 283 (1964).
115. Renner, U., Prins, D. A., and Stoll, W. G., *Helv. Chim. Acta* **42**, 1572 (1959).
116. Ried, W., Köhler, E., and Königstein, F. J., *Ann. Chem.* **598**, 145 (1956).

117. Robinson, B., *J. Chem. Soc.* p. 586 (1963).
118. Rossiter, E. D., and Saxton, J. E., *J. Chem. Soc.* p. 3654 (1953).
119. Runti, C., and Orlando, G., *Ann. Chim.* (*Rome*) **43**, 308 (1953); *Chem. Abstr.* **49**, 3940 (1955).
120. Sakakibara, H., and Kobayashi, T., *Tetrahedron* **22**, 2475 (1966).
121. Schellenberg, K. A., McLean, G. W., Lipton, H. L., and Lietman, P. S., *J. Am. Chem. Soc.* **89**, 1948 (1967).
122. Schindler, W., *Helv. Chim. Acta* **40**, 2156 (1957).
123. Schindler, W., *Helv. Chim. Acta* **40**, 1130 (1957).
123a. Schnoes, H. K., and Biemann, K., *J. Am. Chem. Soc.* **86**, 5693 (1964).
124. Semenov, A. A., and Terenteva, I. V., *Khim. Geterotsikl. Soedin., Akad. Nauk Latv. SSR* p. 235 (1965); *Chem. Abstr.* **63**, 11478 (1965).
125. Shirley, D. A., and Roussel, P. A., *J. Am. Chem. Soc.* **75**, 375 (1953).
126. Silverstein, R. M., Ryskiewicz, E. E., and Chaikin, S. W., *J. Am. Chem. Soc.* **76**, 4485 (1954).
127. Smith, A., and Utley, J. H. P., *Chem. Commun.* p. 427 (1965).
128. Snyder, H. R., and Cook, P. L., *J. Am. Chem. Soc.* **78**, 969 (1956).
129. Snyder, H. R., and Eliel, E. L., *J. Am. Chem. Soc.* **70**, 1703 and 1857 (1948).
130. Snyder, H. R., Eliel, E. L., and Carnahan, R. E., *J. Am. Chem. Soc.* **73**, 970 (1951).
131. Snyder, H. R., and Smith, C. W., *J. Am. Chem. Soc.* **66**, 350 (1944).
132. Spande, T. F., Wilchek, M., and Witkop, B., *J. Am. Chem. Soc.*, **90**, 3257 (1968).
133. Speeter, M. E., U.S. Patent 2,825,734 (1958); *Chem. Abstr.* **52**, 12923 (1958).
134. Sundberg, R. J., *J. Org. Chem.* **33**, 487 (1968).
135. Szmuszkovicz, J., *J. Org. Chem.* **27**, 515 (1962).
136. Szmuszkovicz, J., *J. Am. Chem. Soc.* **82**, 1180 (1960).
137. Taylor, W. I., *Proc. Chem. Soc.* p. 247 (1962).
138. Taylor, W. I., *Helv. Chim. Acta* **33**, 164 (1950).
139. Terent'ev, A. P., Ban-Lun, G., and Preobrazhenskaya, M. N., *J. Gen. Chem. USSR* (*English Transl.*) **32**, 173 (1962).
140. Terent'ev, A. P., Ban-Lun, G., and Preobrazhenskaya, M. N., *J. Gen. Chem. USSR* (*English Transl.*) **32**, 1311 (1962).
141. Terent'ev, A. P., and Preobrazhenskaya, *Dokl. Akad. Nauk USSR* **121**, 481 (1958); *Chem. Abstr.* **53**, 1303 (1959).
142. Terent'ev, A. P., and Preobrazhenskaya, M. N., *Dokl. Akad. Nauk USSR* **118**, 302 (1958).
143. Terent'ev, A. P., and Preobrazhenskaya, M. N., *J. Gen. Chem. USSR* (*English Transl.*) **30**, 1238 (1960).
144. Terent'ev, A. P., and Preobrazhenskaya, M. N., *J. Gen. Chem. USSR* (*English Transl.*) **29**, 322 (1959).
145. Terent'ev, A. P., Preobrazhenskaya, M. N., and Sorokina, G. M., *J. Gen. Chem. USSR* (*English Transl.*) **29**, 2835 (1959).
146. Thesing, J., Klüssendorf, S., Ballach, P., and Mayer, H., *Chem. Ber.* **88**, 1295 (1955).
147. Thesing, J., Semler, G., and Mohr, G., *Chem. Ber.* **95**, 2205 (1962).
148. Tiffeneau, M., *Bull. Soc. Chim. France* **9**, 825 (1911).
149. Torralba, A. F., and Myers, T. C., *J. Org. Chem.* **22**, 972 (1957).
150. Uhle, F. C., and Harris, L. S., *J. Am. Chem. Soc.* **79**, 102 (1957).
151. von Strandtmann, M., Cohen, M. P., and Shavel, J., Jr., *J. Org. Chem.* **30**, 3240 (1965).
152. Walls, F., *Bol. Inst. Quim. Univ. Nacl. Auton. Mex.* **10**, 3 (1958); *Chem. Abstr.* **53**, 4252 (1959).

153. Walton, E., Holly, F. W., and Jenkins, S. R., *J. Org. Chem.* **33**, 192 (1968).
153a. Weinman, J. M., Ph.D. Thesis with W. E. Noland, University of Minnesota (1964); *Dissertation Abstr.* **25**, 1588 (1964).
154. Wenkert, E., Garratt, S., and Dave, K. G., *Can. J. Chem.* **42**, 489 (1964).
155. Winterfeldt, E., and Strehlke, P., *Chem. Ber.* **98**, 2579 (1965).
156. Yakhontov, L. N., Uritskaya, M. Ya., and Rubstov, M. V., *J. Gen. Chem. USSR (English Transl.)* **34**, 1460 (1964).
157. Yonemitsu, O., Cerutti, P., and Witkop, B., *J. Am. Chem. Soc.* **88**, 3941 (1966).
158. Young, D. V., and Snyder, H. R., *J. Am. Chem. Soc.* **83**, 3160 (1961).

III
SYNTHESIS OF THE INDOLE RING

In this chapter emphasis is placed on discussion of the scope and mechanism of reactions which result in the formation of an indole ring from nonindolic starting materials. In Chapter IV synthetic transformations which elaborate preformed indoles are discussed.

A. THE FISCHER INDOLE SYNTHESIS

1. Mechanism of the Reaction

The cyclization of arylhydrazones to indoles, widely known as the Fischer indole synthesis, remains the most versatile and widely applied reaction for the formation of the indole ring. The conversion of the phenylhydrazone of ethyl methyl ketone to 2,3-dimethylindole and small amounts of 2-ethylindole (76) can serve as an example for discussion. The mechanism

$$\text{A1 (PhNHN=C(Me)Et)} \xrightarrow{ZnCl_2} \text{A2 (2,3-dimethylindole)} + \text{A3 (2-ethylindole)}$$

of this cyclization has been the subject of considerable study and the mechanism originally proposed by Robinson and Robinson (219) as reformulated by Carlin and Fisher (55) has obtained wide acceptance.

Two recent review articles (214) have summarized the extensive experimental evidence supporting this mechanism. Another review (228) emphasizes the relationship of the Fischer reaction to other aromatic re-

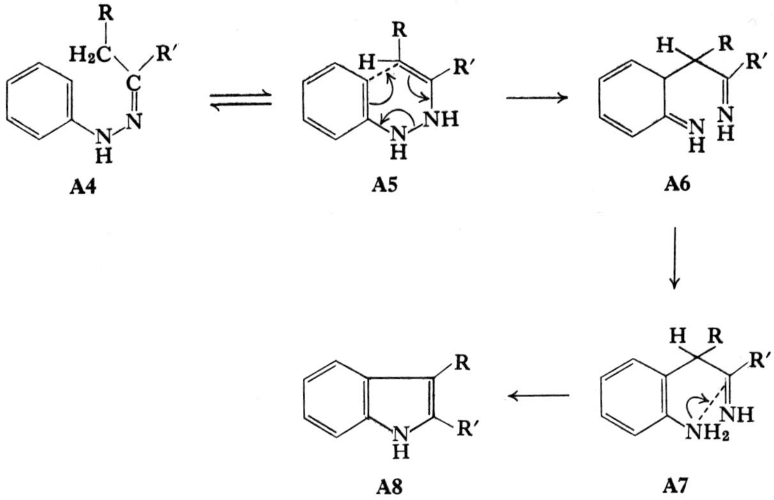

A4 A5 A6 A7 A8

arrangements. One line of effort has involved attempts to detect or trap the intermediates proposed for the various steps in the reaction. The hydrazone–vinylhydrazine equilibrium in the first step of the reaction is analogous to ketone–enol and imine–enamine equilibria, but no firm physical evidence for the existence of measurable concentrations of the vinylhydrazine has been found to date (25a, 137a, 182, 183).

Refluxing the phenylhydrazone of methyl ethyl ketone in acetic anhydride gives the diacylated vinylhydrazine **A10**. This can be considered to constitute "trapping" of the first intermediate in the Fischer cyclization. Heating

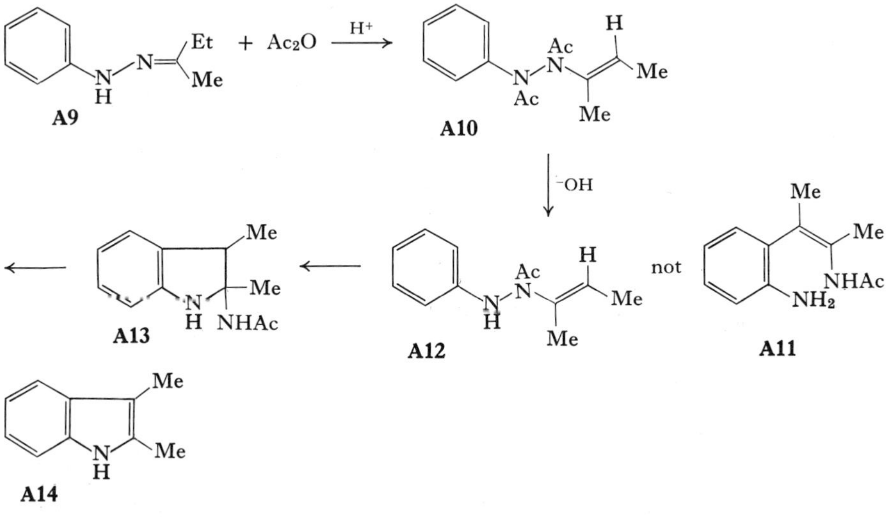

A9 A10 A11 A12 A13 A14

A10 with dilute mineral acids gives 2,3-dimethylindole (259). These transformations appear to constitute the closest analogy observed to date for the first step in the Robinson mechanism. Alkaline hydrolysis of **A10** gave a substance formulated as **A11** (259). This material gave 2,3-dimethylindole and acetamide on heating. Reexamination (72a) of the structure of this material has shown that it is actually **A12**. The formation of **A11** can no longer be cited as an analogy for the second stage of the Robinson mechanism. There are, however, products formed during Fischer reactions in certain cyclic systems which correspond to interruption of the reaction after the ortho substitution step.

Plieninger (203, 206) has reported the isolation of **A16** by treatment of **A15** with hydrochloric acid in acetic acid. Subsequent nmr spectral evidence (185) has shown that the substance actually exists as the tautomer **A17**. It has been suggested that **A17** is isolated because the enamine is very weakly basic and therefore the iminium species **A18** is not generated in sufficient concentration to lead to indolization (185).

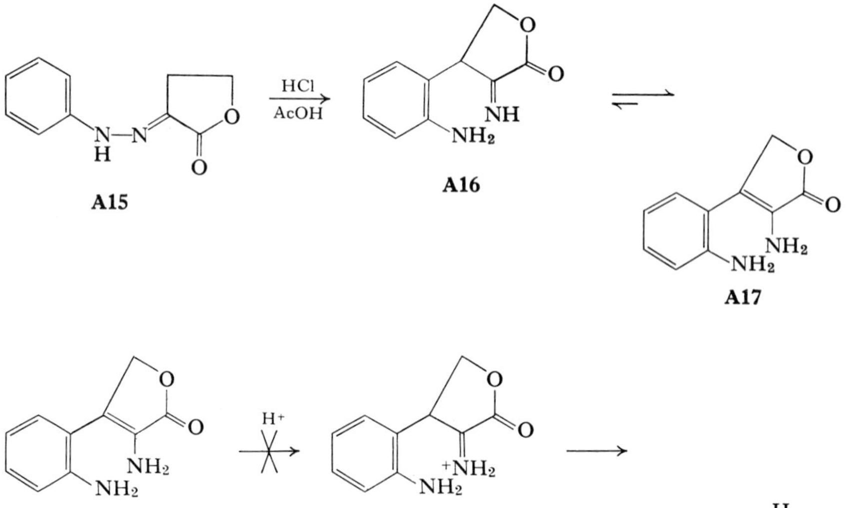

A15 —(HCl, AcOH)→ A16 ⇌ A17

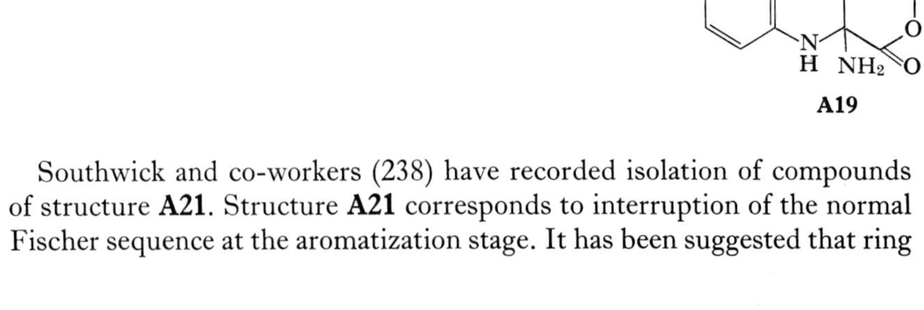

A17 —(H+)✗→ A18 → A19

Southwick and co-workers (238) have recorded isolation of compounds of structure **A21**. Structure **A21** corresponds to interruption of the normal Fischer sequence at the aromatization stage. It has been suggested that ring

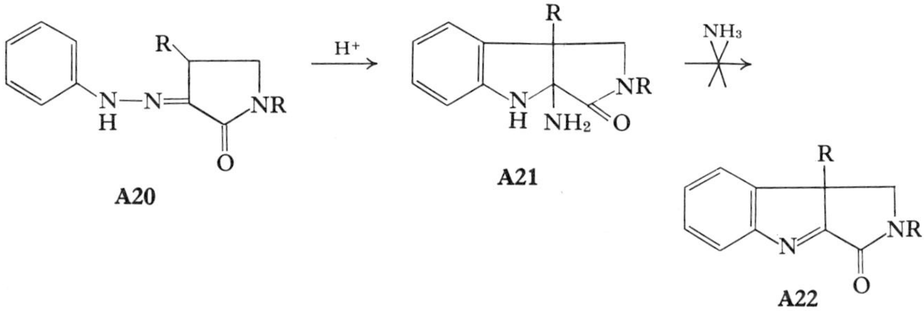

A20 A21 A22

strain effects are responsible for the failure of **A21** to lose ammonia. Ring C of structure **A22** contains three sp^2 atoms, representing considerable angle strain.

Carlin and co-workers have examined the behavior of 2,6-disubstituted arylhydrazones under conditions of the Fischer synthesis (51, 54, 57, 59). Diortho substitution prevents the tautomerization **A6** → **A7** in the Robinson mechanism. Heating the 2,6-dimethylphenylhydrazone of acetophenone in the presence of zinc chloride gives as the main product **A24**, as well as 4,7-dimethyl-2-phenylindole and cleavage products (54). The intermediate **A26** can account for the formation of **A24** and **A25**.

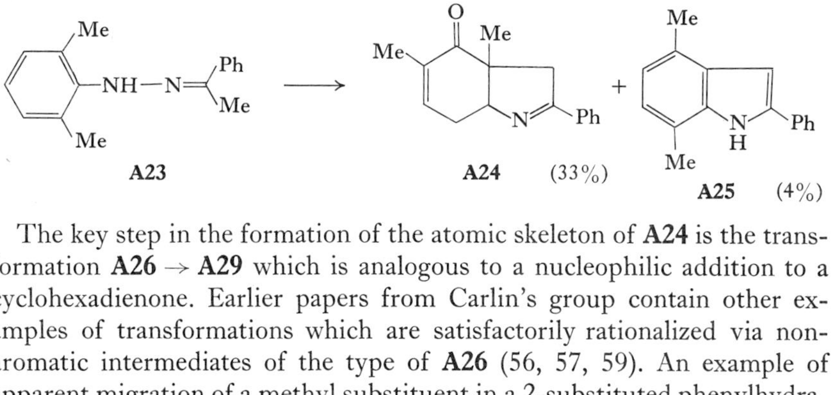

A23 A24 (33%) A25 (4%)

The key step in the formation of the atomic skeleton of **A24** is the transformation **A26** → **A29** which is analogous to a nucleophilic addition to a cyclohexadienone. Earlier papers from Carlin's group contain other examples of transformations which are satisfactorily rationalized via nonaromatic intermediates of the type of **A26** (56, 57, 59). An example of apparent migration of a methyl substituent in a 2-substituted phenylhydrazone has been recorded (101).

Robinson and Brown (217) have examined the reaction of *N*-methyl-2,6-dichlorophenylhydrazine with cyclohexanone and observed the formation of **A31** and **A32**. The formation of **A32** can be readily rationalized in terms of a dienone imine intermediate similar to **A26**.

Further support for the Robinson mechanism has been reported by Bajwa and Brown who obtained the substance **A34** when **A33** and isobutyraldehyde were heated in benzene (20). The elimination of ammonia and aromatization are blocked by the presence of the several methyl substituents.

A26 ≡ A26

A29 → (H_2O,) A24

A27 → A28 → A25

The ortho substitution step in the Fischer reaction is believed to be electrophilic in character. An experimental indication of the electrophilic nature of the attack on the ring is found in the work of Ockenden and Schofield (181). Meta-substituted phenylhydrazones were cyclized and the ratio of 4-substituted to 6-substituted indole was studied. Ockenden and Schofield contend that, if the substitution step is an electrophilic attack on the ring, the ratio of **A39** to **A37** should increase as X becomes electron-withdrawing in character. They consider substitution at the para position to be retarded less than at the ortho position by electron-withdrawing substituents. The data obtained, mainly for $X = NO_2$ or CH_3, fit this pattern. The problem of estimating competing steric effects arising from the steric compression which will result from cyclization to **A37** were ignored. This work represents at best only a qualitative indication of the polarity in the

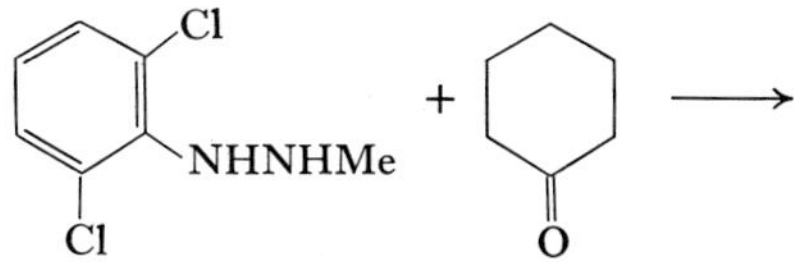

A30

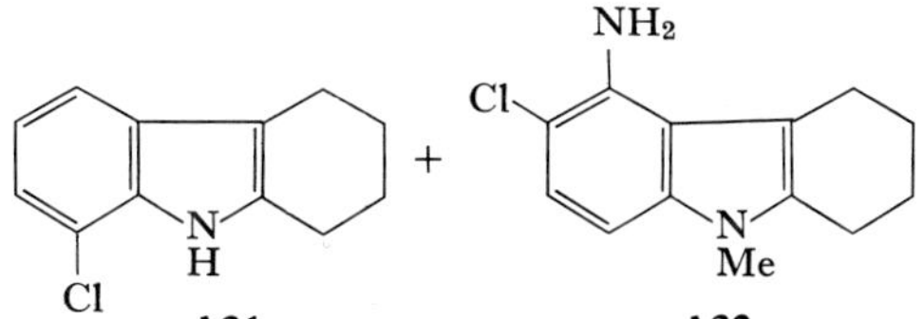

A31

A32

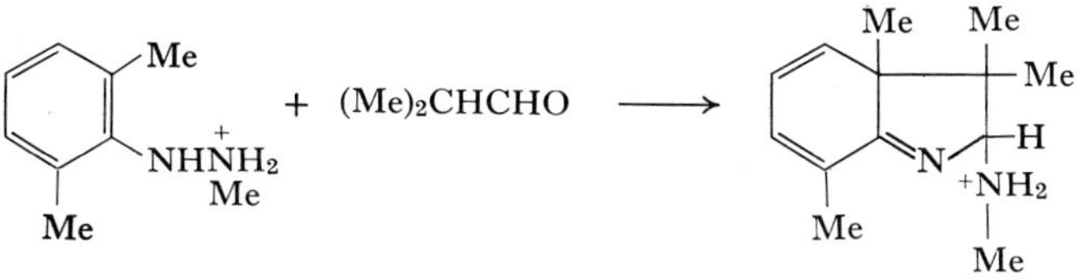

A33

A34

transition state. The limited kinetic data (165, 190) available on the Fischer reaction are of little help in discussion of the reaction mechanism since the rate-determining step has not been identified (53a). Electron-withdrawing groups on the aromatic ring are known to retard cyclization.

The intramolecularity of the Fischer cyclization has been convincingly demonstrated in the case of thermal, as opposed to acid-catalyzed, cyclizations (138) by "crossover" experiments. It was shown that simultaneous cyclization of closely related phenylhydrazones gave only the indoles expected from intramolecular cyclizations. The intramolecularity of acid-catalyzed Fischer cyclizations is more difficult to demonstrate since phenylhydrazones undergo exchange reactions by hydrolytic mechanisms at rates

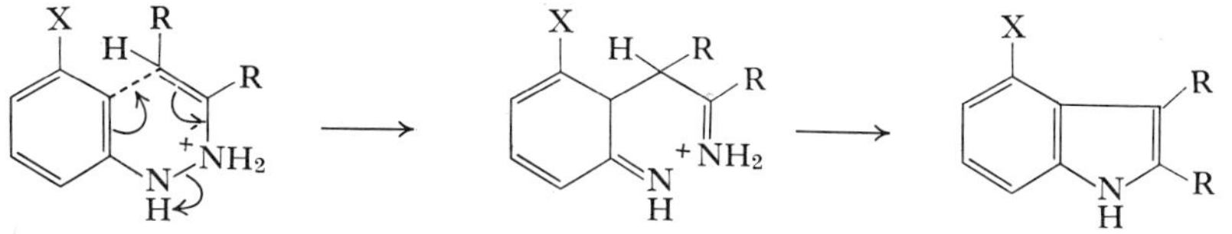

A35

A36

A37

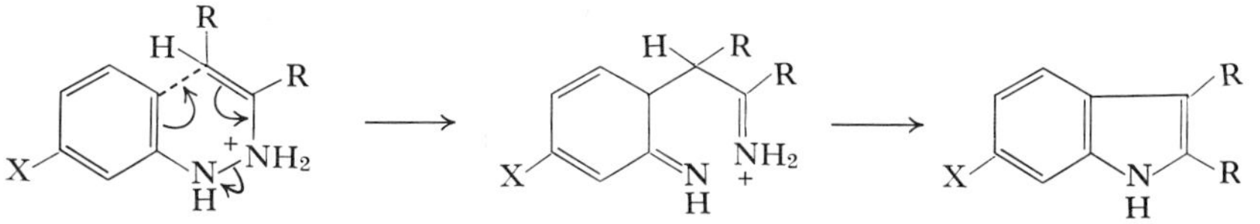

A35

A38

A39

competitive with cyclization (83) and thus give rise to spurious crossover products (191).

The Robinson mechanism predicts that N-2 of the phenylhydrazone will be eliminated as ammonia in the cyclization. Labeling experiments have confirmed this prediction. Acetophenone-1-phenylhydrazone-^{15}N gives ^{15}N-labeled 2-phenylindole (13) whereas acetone 2-phenylhydrazone-^{15}N gives unlabeled 2-methylindole (66).

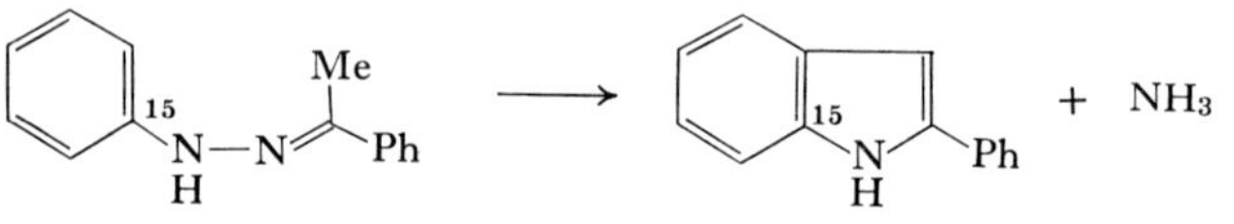

A40 → A41 + NH_3

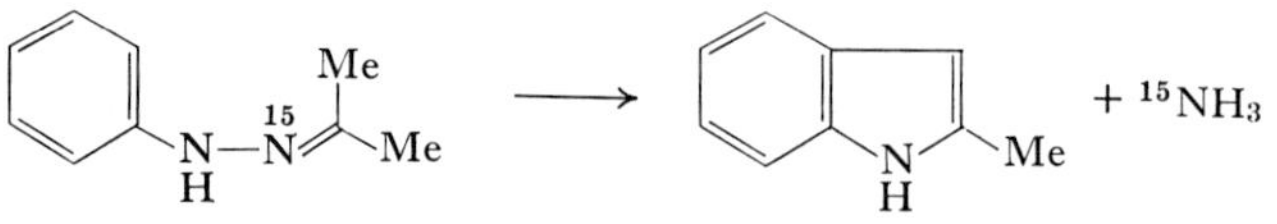

A42 → A43 + $^{15}NH_3$

Carlin and co-workers (58) have examined the question of the origin of the nitrogen atoms in products **A45** and **A46**. As predicted by the mechanism on page 146, **A45** contains N-2 from the phenylhydrazone while **A46** contains the nitrogen atom which was N-1 in the hydrazone.

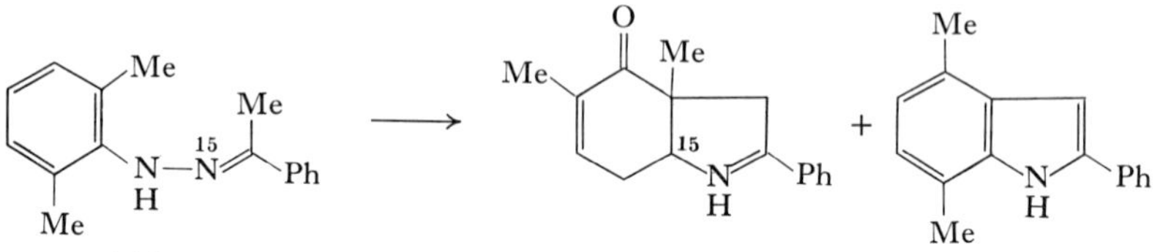

A44 → A45 + A46

2. The Direction of Cyclization of Unsymmetrical Phenylhydrazones

The cyclization of a phenylhydrazone of general structure **A47** could proceed in either of two directions. A good deal of data has been accumulated which demonstrate that in the case R = H, R′ = alkyl or aryl, the cyclization

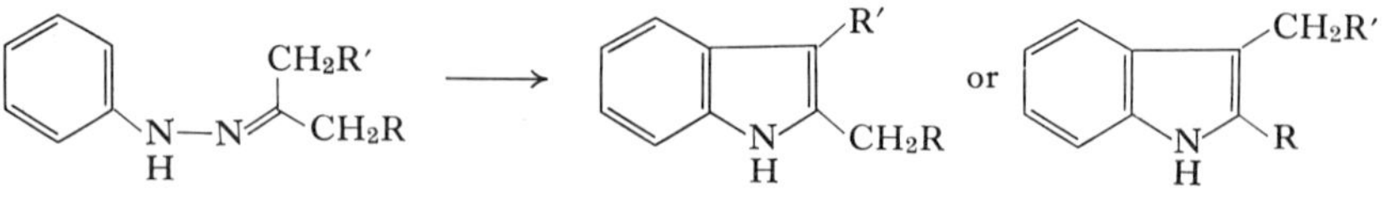

A47 → A48 or A49

proceeds to give **A48**, R = H, i.e., cyclization tends to occur into a substituted chain in preference to a methyl group. The phenylhydrazone of methyl ethyl ketone gives predominately 2,3-dimethylindole (76). Similarly, the nitrophenylhydrazones of methyl propyl ketones give 3-ethyl-2-methylindoles (222). In the case of ketones with one branched and one straight alkyl chain, cyclization of the phenylhydrazone can give a 2,3-disubstituted indole or a 2,3,3-trisubstituted-3*H*-indole.

Correlating and understanding the effects dictating the direction of cyclization is important from both synthetic and mechanistic viewpoints.

Several examples in which alternative modes of cyclization are possible are shown in Table XVII. The following general conclusions have been

TABLE XVII ☐ DIRECTION OF CYCLIZATION OF UNSYMMETRICAL PHENYLHYDRAZONES[a]

R_1	R_2	Catalyst	Selectivity[b]	Ref.
Ethyl	Methyl	Zinc chloride	4:1	(76, 145)
Isopropyl	Methyl	10% Sulfuric acid (5 moles)	44:1[c]	(114)
Isopropyl	Methyl	50% Sulfuric acid (5 moles)	19:1[c]	(114)
Isopropyl	Methyl	78% Sulfuric acid (5 moles)	3.4:1	(114)
Isopropyl	Methyl	Zinc chloride	Exclusive	(123, 200)
n-Hexyl	Methyl	Zinc chloride	?	(50)
n-Heptyl	Methyl	Hydrochloric acid, acetic acid	Exclusive	(48)
Cyclohexyl	Methyl	Acetic acid or zinc chloride	Exclusive	(159)
Cyclohexyl	Methyl	Polyphosphoric acid	16:1	(159)
Benzyl	Methyl	Hydrochloric acid, acetic acid	Exclusive	(49)
Benzyl	Methyl	Polyphosphoric acid	5:1	(49)
Benzyl	Ethyl	Cupric chloride	?	(121)
$—CH(CH_3)CH_2—$	$—CH_2(CH_2)_2—$[d]	Thermal cyclization	1:1	(138)
$—CH(CH_3)CH_2—$	$—CH_2(CH_2)_2—$[d]	Dilute sulfuric acid	2:1	(189)
$—CH(CH_3)CH_2—$	$—CH_2(CH_2)_2—$[d]	Acetic acid	10:1	(189)
$—CH(C_6H_5)CH_2—$	$—CH_2(CH_2)_2—$[e]	Acetic acid	?	(86)

[a] Direction of preferred cyclization is underscored.
[b] Ratio of major to minor product.
[c] Very complete data on this system are available in Ref. (114).
[d] Phenylhydrazone of 2-methylcyclohexanone.
[e] Phenylhydrazone of 2-phenylcyclohexanone.

drawn from these and related data. (1) In the phenylhydrazone of a methyl alkyl ketone, cyclization will occur into the alkyl branch. (2) In a methyl

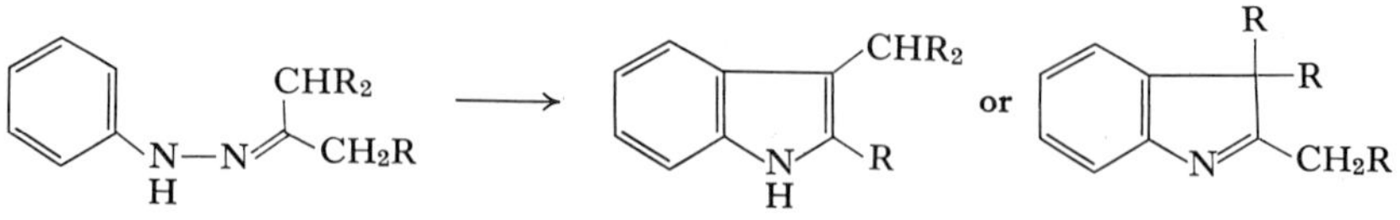

A50 → A51 or A52

alkyl ketone in which C-3 is disubstituted, cyclization will occur into the branched alkyl group, giving a 3,3-disubstituted-2-methyl-3*H*-indole. (3) In 2-substituted cyclohexanones, products derived from cyclization into both branches of the ring will be formed (201). (Systems of the type

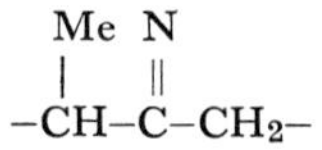

$$-CH(Me)-C(=N)-CH_2-$$

seem to have been studied only in cyclic cases to date.) Recent work has brought to light several exceptions to these venerable generalizations (49, 114, 138, 159, 189). It now is clear that the direction of cyclization in a Fischer reaction may be highly dependent on the identity of the catalyst used and on the concentration of the catalyst.

Consideration of the Robinson mechanism (page 143) shows that the direction of cyclization of an unsymmetrical hydrazone could be determined by the relative activation energies for formation of the two possible vinylhydrazines (if subsequent cyclization is fast) or by the relative energies of the transition states for the two alternate modes of cyclization. The latter possibility seems the more likely. Lyle and Skarlos (159) have put forward an explanation for the effect of the catalyst on the direction of cyclization based on this assumption and on consideration of steric effects in the alternative transition states.

In the case of methyl cyclohexyl phenylhydrazone, the two vinylhydrazines must cyclize through transition states **A53** or **A54**. Lyle and Skarlos suggest that steric repulsion between the phenyl group and axial hydrogen of the cyclohexane ring in **A54** will make **A53** the more easily attained

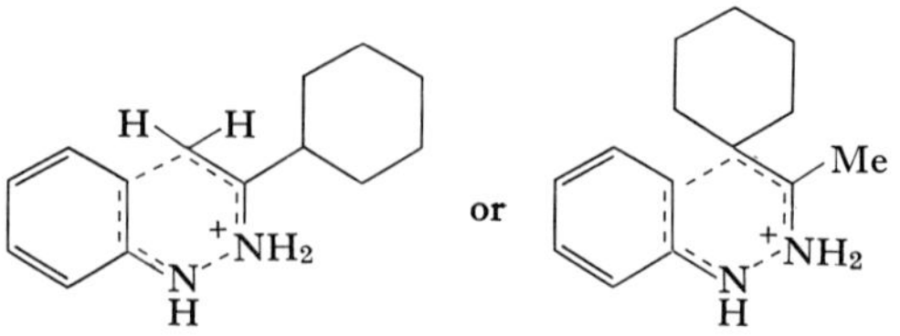

A53 or A54

transition state. They suggest that catalysts sterically larger than the proton (for example, X = coordinated metal atoms or acyl groups) will raise the energy of transition state **A55** and lead to cyclization via **A56**. Recent

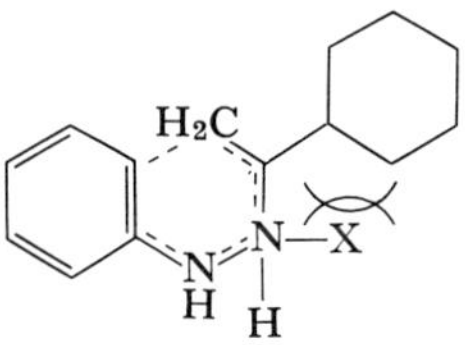

A55 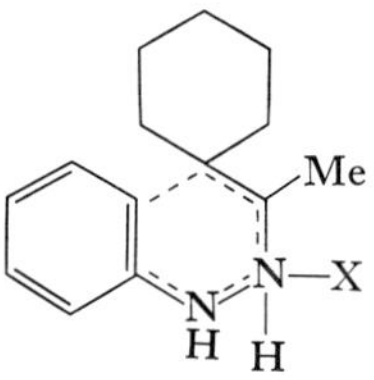 **A56**

data (114) indicate that the acidity of the reaction medium may be very important in determining the direction of cyclization. Illy and Funderbunk have convincingly demonstrated that concentration of the acid catalyst is important in determining the direction of cyclization of the phenylhydrazone of isopropyl methyl ketone (114). Acetic acid, bisulfate ion, and aqueous phosphoric acid all give predominately 2,3,3-trimethyl-3*H*-indole, the product of cyclization into the more branched chain of the ketone. Polyphosphoric acid and concentrated sulfuric acid in molar ratios of 5:1 over the hydrazone favor formation of 2-isopropylindole, the product of cyclization into the methyl group. A maximum yield of 84% for the indole was observed with 78% sulfuric acid present in a 6:1 ratio. These data point to a change in mechanism involving a diprotonated intermediate which cyclizes preferentially into the least-branched alkyl group. Palmer and McIntyre (186a) have reported data which show similar trends. The direction of cyclization of four unsymmetrical phenylhydrazones as a function of acid concentration was examined. At relatively low acid concentrations cyclization occurred into the most-branched chain but as acid concentration was increased, cyclization into the least-branched chain became dominant. A mechanistic interpretation suggesting involvement of a diprotonated species at high acid concentrations was suggested. Full description of the mechanistic basis for the acid concentration effect is not yet possible. The possibility that the diprotonated species is converted to a vinyl hydrazinium ion very rapidly, and with a kinetic preference for removal of one of the three primary hydrogens, can explain the data presently available.

An interesting ring-strain effect was noted by Stork and Dolfini when applying the Fischer cyclization to the synthesis of aspidospermine (247). The *o*-methoxyphenylhydrazone **A61** apparently cyclized to the indole **A62** whereas the amine **A63** gives the 3*H*-indole **A64**. It is proposed that ring strain (three sp^2 atoms) in the transition state prevents cyclization of **A61** in the direction observed for **A63**.

A57 $\xrightarrow{H^+}$ A59 $\xrightarrow{\text{fast}}$ A60 ⟶ product

A57 ⟶ A58 ⟶ product

A61 $\xrightarrow{AcOH}$ A62

A63 ⟶ A64

3. The Scope of the Reaction

In Tables XVIII–XXIII representative examples of synthesis of various indoles by the Fischer reaction are tabulated. In Table XVIII synthesis of 2-substituted indoles from various methyl ketones are recorded.

Table XIX lists syntheses of 3-substituted indoles from arylhydrazones of representative aldehydes.

Table XX lists selected examples of the synthesis of 2,3-disubstituted indoles by various Fischer cyclization procedures. Table XVII can be consulted for further examples.

TABLE XVIII □ SYNTHESIS OF 2-SUBSTITUTED INDOLES BY THE FISCHER REACTION

Arylhydrazone			Cyclization catalyst	Yield (%)	Ref.
Aryl substituent	*N*-Substituent	Ketone			
None	None	Ethyl pyruvate	Acetic acid, sulfuric acid	58	(73)
None	None	Diethyl oxaloacetate	Ethanol, hydrochloric acid	65	(160)
None	None	3-Acetylpyridine	Polyphosphoric acid	68	(107)
None	None	2-Acetyl-5-ethylthiophene	Zinc chloride	43	(47)
None	None	Acetylcyclohexane	Polyphosphoric acid	80[a]	(159)
None	None	4-Acetyl-1-methylpiperidine	Polyphosphoric acid	64[b]	(159)
None	Carbethoxymethyl	Ethyl pyruvate	Ethanol, hydrochloric acid	?	(231)
2-Me	None	Ethyl pyruvate	Polyphosphoric acid	45	(101)
2-Nitro	None	Ethyl pyruvate	Polyphosphoric acid	13, 66	(188, 230)
2,4-diCl	None	Ethyl pyruvate	Polyphosphoric acid	50	(101)
2-Nitro-4-methyl	None	Ethyl pyruvate	Polyphosphoric acid	85	(102)
2-Nitro-4-methoxy	None	Ethyl pyruvate	Polyphosphoric acid	74	(102)
2,4-Diisoamyl	None	3,3-Dimethyl-2-pentanone	Polyphosphoric acid	?	(60)
2-Methyl-4-benzyloxy	None	Ethyl pyruvate	Polyphosphoric acid	29	(101)
4-Carboxy	None	Ethyl pyruvate	Zinc chloride	48	(156)
4-Nitro	None	Ethyl pyruvate	Polyphosphoric acid	57	(188)

[a] Product is 2-cyclohexylindole.
[b] Product is 2-[4-(1-methyl)piperidyl]indole.

TABLE XIX □ SYNTHESIS OF 3-SUBSTITUTED INDOLES BY THE FISCHER REACTION

Arylhydrazone			Cyclization catalyst	Yield (%)	Ref.
Aryl substituent	*N*-Substituent	Aldehyde			
None	None	2-Ethylthioacetaldehyde	Boron trifluoride	67	(122)
None	None	2-Phenylthioacetaldehyde	Ethanol, hydrochloric acid	?	(277)
2-Me	None	3-Ethoxycarbonylpropionaldehyde	Acetic acid	35	(245)
2-Methoxy	None	4-Acetamido-4,4-dicarbethoxy-butyraldehyde	Amberlite IR-120 or sulfuric acid	10	(29)
2-Phenyl	None	Propionaldehyde	Zinc chloride	59	(146)
2-Phenyl	None	Phenylacetaldehyde	Oxalic acid	60	(146)
3-Nitro	None	Butyraldehyde	Benzene, hydrochloric acid	8[a]	(164)
3-Nitro	None	4-Chlorobutyraldehyde	Benzene, hydrochloric acid	15	(164)
4-F	None	4-Aminobutyraldehyde diethyl acetal	Zinc chloride	?	(10)
4-Nitro	None	4-Chlorobutyraldehyde	Benzene, hydrochloric acid	15	(227)
4-Benzyloxy	None	3-Carboxypropionaldehyde	Sulfosalicyclic acid	55	(257)
4-Benzyloxy	None	4-Nitrovaleraldehyde	Benzene, hydrochloric acid	~50	(155)
4-Benzyloxy	None	4-Carbethoxy-4-nitrovaleraldehyde	Sulfosalicyclic acid	47	(256)

[a] Product is a mixture of the 4- and 6-nitroindole.

TABLE XX □ SYNTHESIS OF 2,3-DISUBSTITUTED INDOLES BY THE FISCHER REACTION

None

Arylhydrazone			Cyclization catalyst	Yield (%)	Ref.
Aryl substituent	*N*-Substituent	Ketone			
None	None	2,2-Dimethyl-3-pentanone	Thermal cyclization	37	(216)
None	None	Propiophenone	Polyphosphoric acid	58	(141)
None	None	5-Hydroxy-2-pentanone	Cupric chloride	69	(85)
None	None	Phenylthioacetone	Ethanol, hydrochloric acid	80–90	(277)
None	None	5-Cyano-2-pentanone	Zinc chloride	66	(147)
None	None	Diethyl 2-oxo-5-carbethoxamido-adipate	Hydrochloric acid	50	(202)
None	None	Cyclohexane-1,3-dione (mono-phenylhydrazone)	Sulfuric acid	60[a]	(64)
None	Methyl	Ethyl 3-phenylacetoacetate	Acetic acid	80[b]	(262)
2-Ph	None	2-Butanone	Zinc chloride	49	(146)
2-Ph	None	Phenylacetone	Polyphosphoric acid	86	(146)
2-Carboxy	None	2-Butanone	Hydrochloric acid, acetic acid	?	(41)
2,5-diMe	None	Ethyl levulinate	Ethanol, sulfuric acid	39	(245)
3,5-diMe	None	Ethyl levulinate	Ethanol, sulfuric acid	67	(245)
4-Methoxy	None	5-Phthalimido-2-pentanone	Ethanol, hydrochloric acid	87	(37)
4-Me	None	2-Butanone	Boron trifluoride	80	(180)
4-Methylthio	Benzyl	5-Phthalimido-2-pentanone	Ethanol, hydrochloric acid	65	(38)
4-Benzyloxy	None	Ethyl 2-oxo-5-phthalimidovalerate	Ethanol, hydrochloric acid	51	(36)
4-Ac	None	Ethyl 2-oxo-4-dimethylamino-butyrate	Polyphosphoric acid	43	(225)
4-Carboxy	None	2-Butanone	?	81	(264)
4-Nitro	None	Cyclooctanone	Acetic acid, hydrochloric acid	64	(213a)

[a] Product is 4-oxo-1,2,3,4-tetrahydrocarbazole.
[b] Product is ethyl 1-methyl-3-phenylindole-2-acetate.

TABLE XXI □ SYNTHESIS OF 3,3-DISUBSTITUTED 3*H*-INDOLES BY THE FISCHER REACTION[a]

Arylhydrazone			Cyclization catalyst	Yield (%)	Ref.
Aryl substituent	*N*-Substituent	Ketone			
None	None	3-Methyl-2-butanone	Thermal cyclization	69	(215)
None	None	5-Cyano-3-methyl-2-pentanone	Sulfuric acid	59	(147)
None	None	2-Methylcyclohexanone	Thermal cyclization	?	(138)
None	None	Acetylcyclohexane	Acetic acid	73	(159)
None	None	2-Ethyl-1-indanone	Zinc chloride	?	(172)
None	None	4-Methyl-1,2-benzocyclooct-1-en-3-one	Ethanol, hydrochloric acid	?	(172)
2-Nitro-5-chloro	None	3-Methyl-2-butanone	Hydrochloric acid, acetic acid	?	(260)
4-Nitro	None	3-Methyl-2-butanone	Hydrochloric acid	48	(227)

[a] See Refs. (114) and (186a) for additional examples.

TABLE XXII □ SYNTHESIS OF TRYPTAMINES BY THE ABRAMOVITCH PROCEDURE

Arylhydrazone				
Aryl substituent	3-Carboxy-2-piperidone substituent	Cyclization catalyst	Yield (%)[a]	Ref.
None	None	Polyphosphoric acid	80	(2, 7)
None	1,6-diMe	Ethanol, hydrochloric acid	85	(100)
None	5-Me	Formic acid	80	(5)
None	5,6-diMe	Formic acid	62	(167)
None	6-Me	Sulfosalicyclic acid	90	(258)
2-Benzoyl	None	Formic acid	52	(271)
2-Me-5-F	None	Formic acid	73	(192)
2-Cl-4-Ac	None	Formic acid	69	(272)
3-Ac	None	Formic acid	78	(271)
4-F	None	Formic acid	68	(10)
4-Methoxy	None	Formic acid	30	(2, 7)
4-Nitro	None	Polyphosphoric acid	85	(2, 7)
4-Ac	None	Formic acid	?	(226)
4-Carbomethoxy	5,6-diMe	Formic acid	82	(263)
N-Oxide of 4-pyridylhydrazone of 3-carboxy-2-piperidone		Zinc chloride	30	(198)

[a] Yields tabulated are for the cyclization step.

TABLE XXIII □ USE OF ARYLHYDRAZONES OF HETEROCYCLIC KETONES IN THE FISCHER REACTION

Arylhydrazone			Cyclization catalyst	Yield (%)	Ref.
Aryl substituent	*N*-Substituent	Ketone			
None	None	1-Propyl-2,3-dioxopyrrolidine	Hydrochloric acid	68	(239)
None	None	1-(2-Phenylethyl)-2,3-dioxopyrrolidine	Hydrochloric acid	85	(239)
None	None	2,2,6,6-Tetramethyl-4-piperidone	Hydrochloric acid	42	(144)
None	None	1,3-Dimethyl-4-piperidone	Hydrochloric acid	78	(149)
None	None	3-Methyl-4-thiochromanone *S,S*-dioxide	Hydrochloric acid	81	(135)
None	None	3,4-Dihydrobenzothiepin-5(2*H*)one	Hydrochloric acid	55	(11)
None	None	3,4-Dihydrobenzothiepin-5(2*H*)one 1,1-dioxide	Hydrochloric acid	84	(11)
None	None	Tetrahydro-4-thiopyrone	Hydrochloric acid	77	(150)
None	None	4-Phenyl-4-phosphacyclohexanone	Hydrochloric acid	37	(81)
None	None	1,2,3,4-Tetrahydro-1-oxopyridocolium bromide	Hydrochloric acid	?	(208)
None	Methyl	1,2,3,4-Tetrahydro-1-oxopyridocolium bromide	Hydrochloric acid	?	(208)
4-Carbethoxy	None	1-Methyl-4-piperidone	Hydrochloric acid	86	(143)
4-Carbethoxy	None	2,2,6,6-Tetramethyl-4-piperidone	Hydrochloric acid	36	(144)
4-Ethoxy	None	Tetrahydro-4-thiopyrone *S,S*-dioxide	Hydrochloric acid	71	(148)
4-Me	None	3-(2-Carbomethoxyethyl)-4-thiopyrone	Hydrochloric acid	67	(135)
4-Nitro	None	3-(2-Carbomethoxyethyl)-4-thiopyrone	Hydrochloric acid	12	(135)

As discussed in Section A,2, cyclization of the phenylhydrazones of an α-substituted ketone will lead to 3*H*-indoles. Table XXI records typical examples of such cyclizations.

Among the specialized adaptions of the Fischer reaction which have been developed, one of the most widely used is the procedure of Abramovitch and Shapiro (2, 7) which provides a three-step approach to substituted tryptamines (3-(2-aminoethyl)indoles). Aryl diazonium ions react with 3-carboxy-2-piperidones, giving arylhydrazones of 3-oxo-2-piperidone. Fischer cyclization gives indoles **A68** which can be hydrolyzed and decarboxylated to tryptamines. Substitution in the piperidone ring gives trypta-

A65 + **A66** → **A67**

H^+

A68 $\xrightarrow{H_2O, H^+}$ **A69**

mines with branched chains, although introduction of a 4-methyl substituent in the piperidone ring has been reported to prevent cyclization. (1) Substituents can be introduced on the indole nitrogen prior to hydrolysis of the lactam. Table XXII records a number of cases of application of the Abramovitch tryptamine synthesis.

The Japp–Klingemann reaction permits important extension of the scope of the Fischer indole synthesis. The Japp–Klingemann reaction is the electrophilic substitution of electron-rich carbon–carbon multiple bonds by aryl diazonium ions (196). The reaction makes available many arylhydrazones which are not easily available directly from the ketone. It has been most widely used with various enolates but the recent extension of the reaction to enamines (229) also has broad potential and has been used to advantage by Jackson, Joule, and co-workers in the synthesis of the alkaloid dasycarpidone via the intermediate **A73** (118, 119).

The Japp–Klingemann reaction has been most widely applied to enolates of β-keto esters and the corresponding carboxylate salts. In the former case

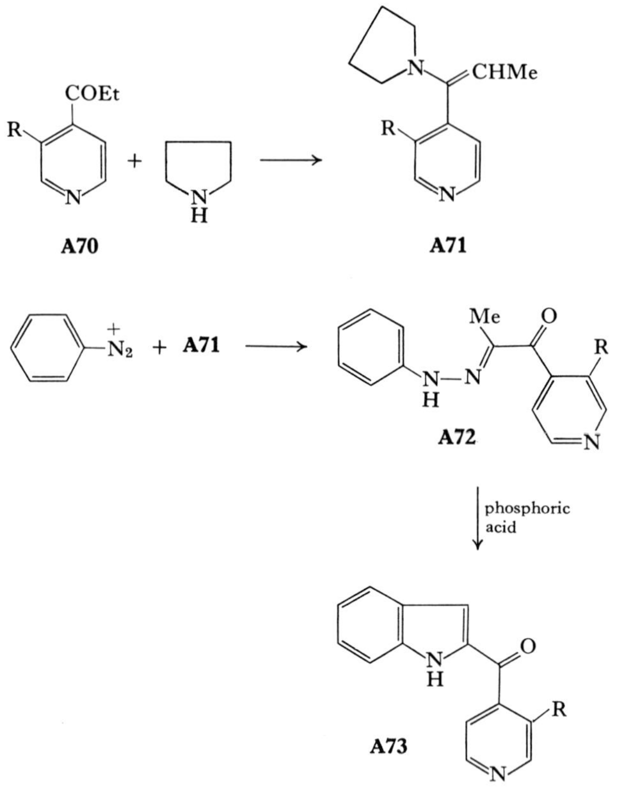

the acyl group is lost and the arylhydrazone of an α-keto ester is formed, whereas with the β-keto acids decarboxylation accompanies the reaction and the monophenylhydrazone of an α-diketone is obtained.

The first stage of the Abramovitch tryptamine synthesis will be recognized as a Japp–Klingemann reaction of the latter type. The review of Phillips (196) records many examples of Japp–Klingemann coupling reactions.

$$\underset{\textbf{A73}}{\mathrm{AcCH(Me)CO_2Et}} + \underset{\textbf{A74}}{\mathrm{Ph\overset{+}{N}_2}} \longrightarrow \underset{\textbf{A75}}{\mathrm{AcC(Me)(CO_2Et){-}N{=}N{-}Ph}} \longrightarrow \underset{\textbf{A76}}{\mathrm{MeC({=}NNHPh)CO_2Et}} + \mathrm{AcOH}$$

$$\underset{\textbf{A77}}{\mathrm{AcCH(Me)CO_2^-}} + \underset{\textbf{A74}}{\mathrm{Ph\overset{+}{N}_2}} \longrightarrow \underset{\textbf{A78}}{\mathrm{AcC(Me)(CO_2^-){-}N{=}N{-}Ph}} \longrightarrow \underset{\textbf{A79}}{\mathrm{AcC(Me){=}NNHPh}} + \mathrm{CO_2}$$

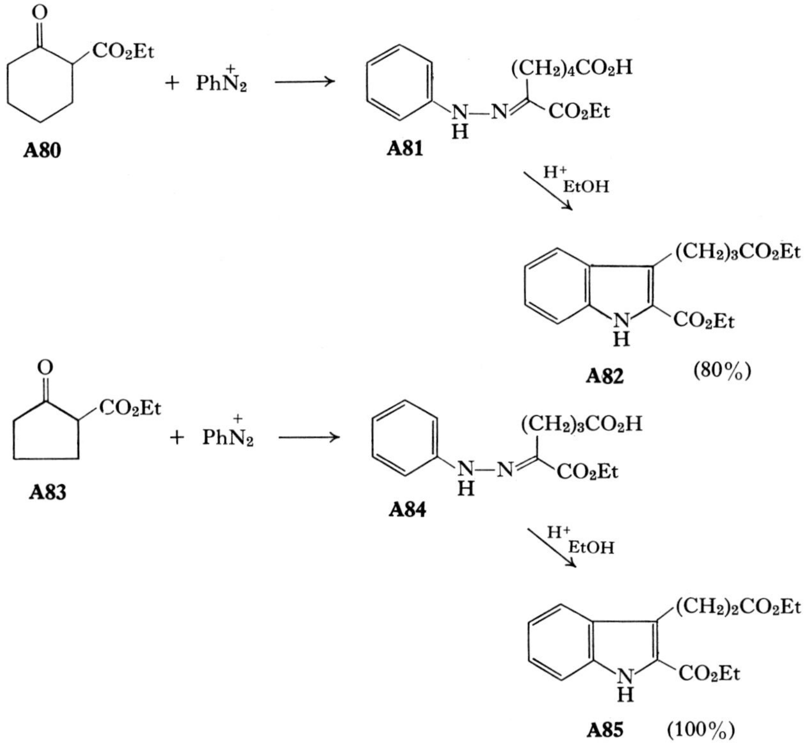

Cyclic β-keto esters undergo ring opening under the conditions of the Japp–Klingemann reaction. Fischer cyclization of the resulting hydrazones then gives indole-2-carboxylic acid derivatives having an alkanoic acid side chain at the 3-position (75, 120, 186).

Use of arylhydrazones of heterocyclic ketones in the Fischer reaction results in the formation of indoles having a heterocyclic system fused to the 2,3-side of the indole ring. A great variety of such reactions have been reported, especially in the Russian literature. The transformations below are illustrative. Further examples are recorded in Table XXIII.

Reference to the preceding tables shows that a variety of catalysts, most of which are either proton sources or Lewis acids, have been employed in effecting the Fischer cyclization in addition to the classic catalysts, ethanolic hydrogen chloride or zinc chloride. Polyphosphoric acid has probably been the most useful of the more recently developed catalysts for the reaction (141). Among the other catalyst systems that have been found to be of occasional advantage are boron trifluoride etherate (180, 237), polyphosphate ester (136) (a chloroform-soluble esterified polyphosphate material), and acidic ion-exchange resins such as Amberlite IR-120 (280).

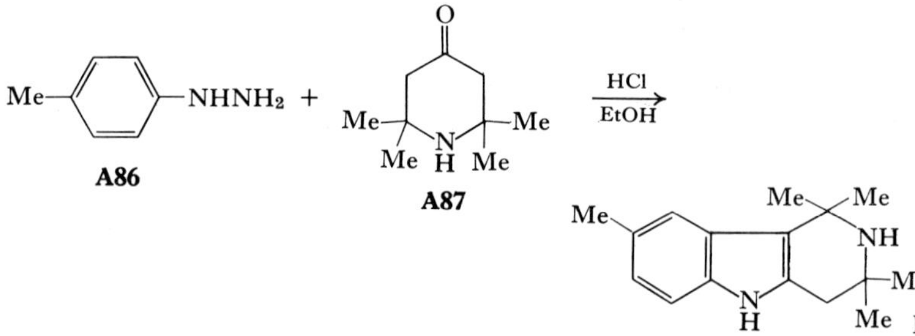

A86 + **A87** $\xrightarrow[\text{EtOH}]{\text{HCl}}$ **A88** (78%) Ref. (144)

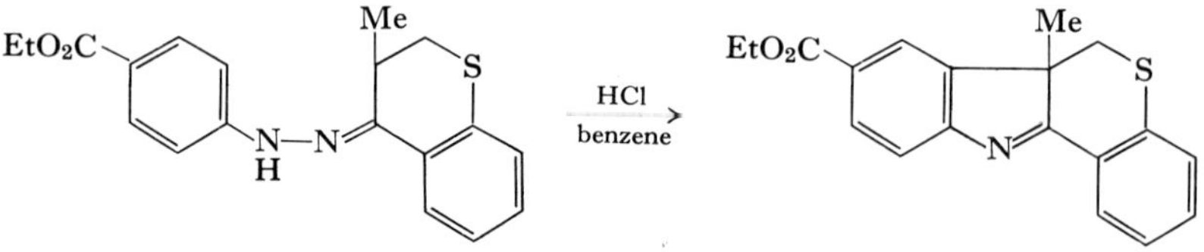

A89 $\xrightarrow[\text{benzene}]{\text{HCl}}$ **A90** (100%) Ref. (135)

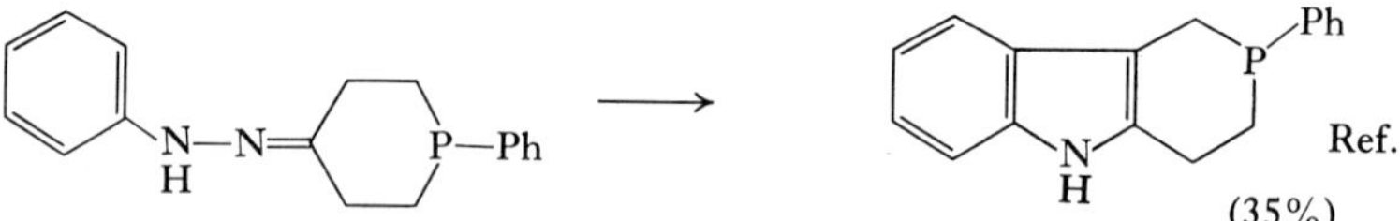

A91 $\longrightarrow$ **A92** (35%) Ref. (81)

Fischer cyclizations have also been effected at elevated temperature in the absence of any catalyst (77, 138, 215). The presence of a substituent on N-2 of the phenylhydrazone seems to result in Fischer-type cyclization proceeding under exceptionally mild conditions (217). Successful cyclizations in the absence of catalysts have somewhat clouded the mechanistic interpretation of the role of the acid in catalyzed reaction, but the view that the acid-catalyst influences the hydrazone–vinylhydrazine tautomerization more strongly than the other steps on the Robinson mechanism has found some acceptance (217, 259).

Relatively few Fischer cyclizations have been reported to follow abnormal courses. The phenylhydrazone of camphor gives **A94** instead of **A95**, which would be the expected product (23). A satisfactory explanation for the formation of **A94** either from **A93** or from **A95** is not apparent.

Phenylhydrazones derived from aldehydes containing phosphonic acid functional groups are reported (211) to give 2-substituted indoles instead of the normally expected 3-substituted indole. This result is attributed to the rearrangement of the initially expected product under the reaction conditions. The rearrangement has been demonstrated (211) to occur but under conditions quite different from those of the synthesis.

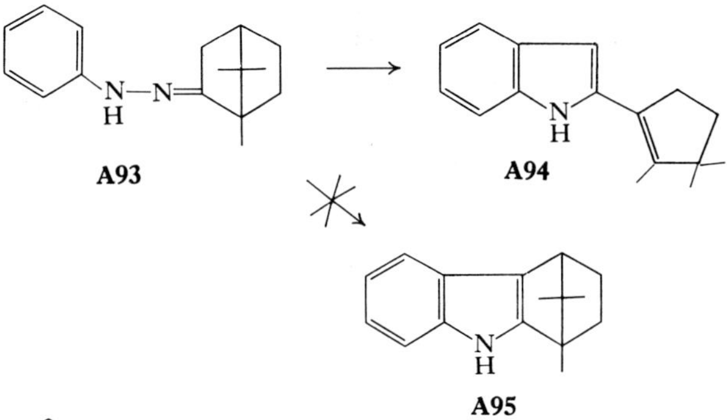

A93

A94

A95

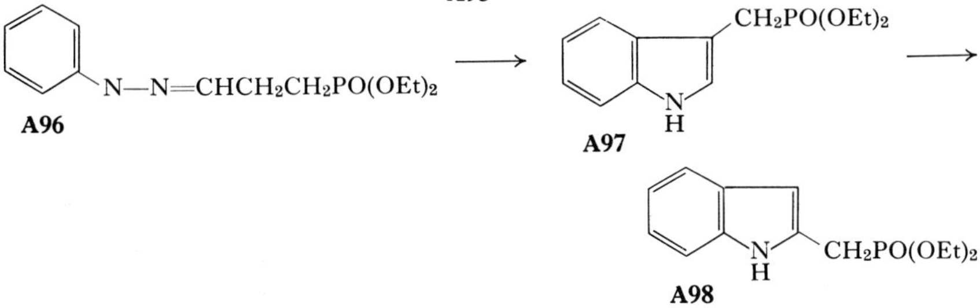

$CH_2PO(OEt)_2$

N—N=$CHCH_2CH_2PO(OEt)_2$

A96

A97

$CH_2PO(OEt)_2$

A98

An investigation of Fischer cyclization of some 1,3-disubstituted 4-piperidones with ethanolic hydrogen chloride has shown that 1,2,3,4-tetrahydropyrimido[3,4-a]-indoles are formed (71a). A reasonable mechanism involving formation and subsequent rearrangement of the expected 3*H*-indoles has been proposed. Robinson's most recent review discusses other classes of substrates which do not give normal Fischer cyclizations (214).

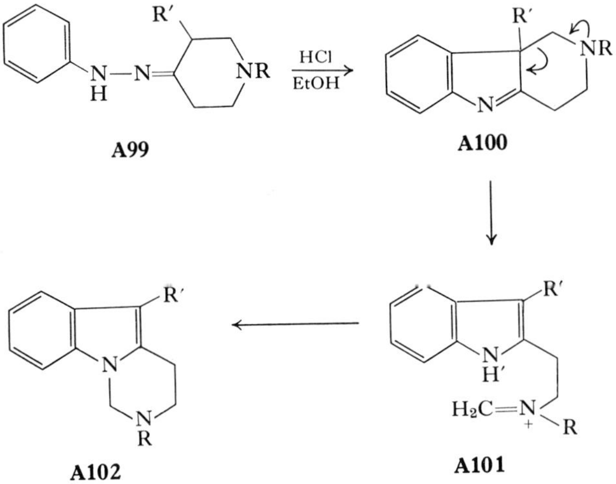

HCl / EtOH

A99

A100

A101

A102

B. SYNTHESIS OF INDOLES FROM DERIVATIVES OF α-ANILINOKETONES: THE BISCHLER SYNTHESIS

A number of indole syntheses fit into the general mechanistic outline **B1** → **B4**. The electrophilic character of the carbonyl or imine group of

B1 B2 B3 B4

an α-anilinoketone derivative is enhanced by coordination with a proton or other Lewis acid. Cyclization by electrophilic attack at the ortho position of the aniline ring ensues and the transformation is completed by aromatization of **B3** by the loss of AXH. Among the examples which can serve to typify the synthesis of indoles from anilinoketones are the syntheses **B5** → **B7** and **B8** → **B9**. Other typical examples are recorded in Table XXIV. Reactions **B10** → **B12**, **B13** → **B15**, **B16** → **B17** + **B18** and **B19** →

B5 + B6 (BrCH(Me)COMe) —(1) heat, (2) HCl→ B7 (61%) Ref. (31)

B8 —$MgCl_2$, $PhNH_2$→ B9 (88%) Ref. (52)

B21 serve to illustrate the fact that the products derived from Bischler-type reactions cannot always be directly predicted from the structure of the ketonic starting material.

Julia and co-workers have demonstrated that secondary aromatic amines react quite generally with ethyl 4-bromoacetoacetate and 4-bromoaceto-

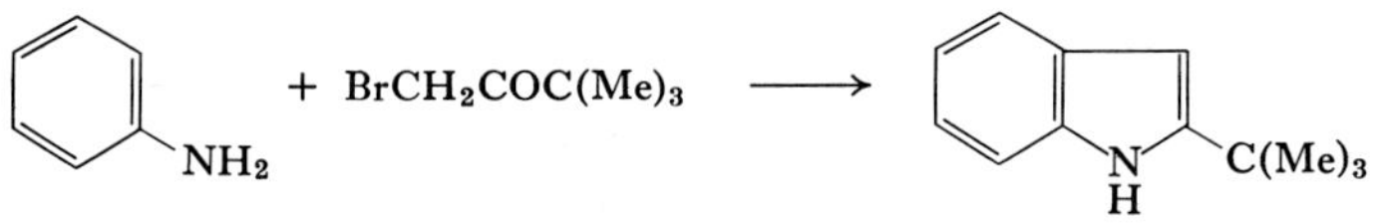

$+ BrCH_2COC(Me)_3 \longrightarrow$

B10 **B11** **B12** Ref. (124)

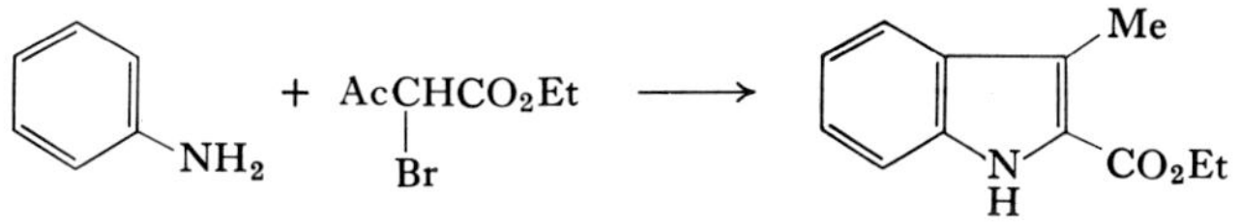

$+ AcCH(Br)CO_2Et \longrightarrow$

B13 **B14** **B15** Ref. (169)

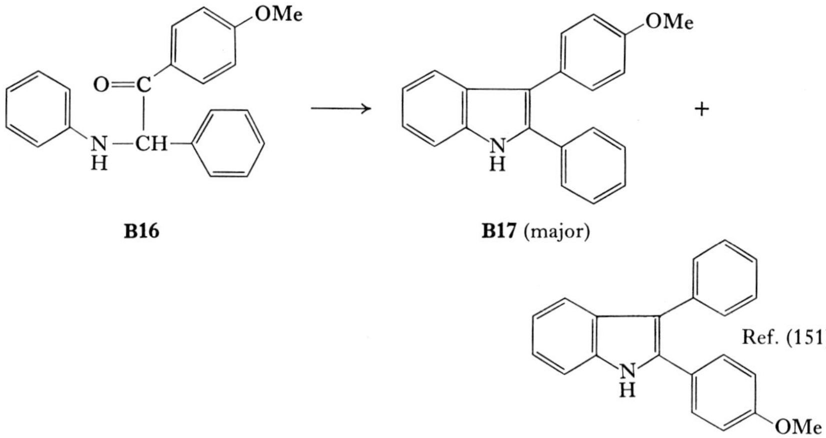

B16 $\longrightarrow$ **B17** (major) + **B18** (minor) Ref. (151)

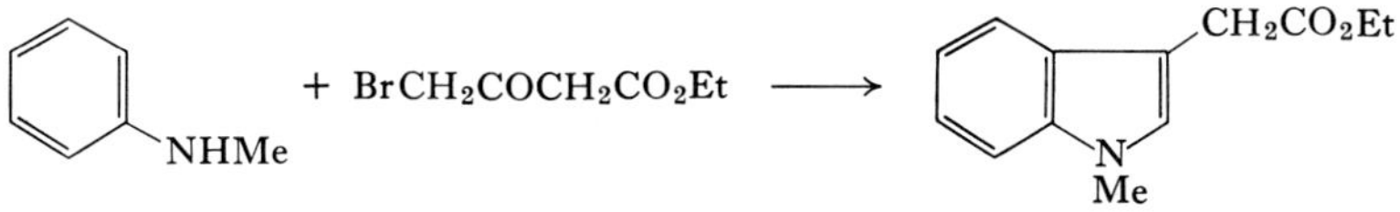

$+ BrCH_2COCH_2CO_2Et \longrightarrow$

B19 **B20** **B21** Ref. (133)

acetamides to give derivatives of indole-3-acetic acid. In the case of secondary amines the cyclization occurs in each case in such a fashion as to make the carbonyl carbon of the bromoketone the 3-carbon of the indole ring, i.e. cyclization occurs without any apparent rearrangement. With primary amines the reaction takes another course. Ethyl 4-bromoacetoacetate and aniline give a pyrrole derivative (169). Aniline and diethyl 2-bromoacetylsuccinate are reported to give diethyl indole-2-succinic acid (125). Julia's group has also extended the reaction with *N*-substituted anilines to 4-bromoacetoacetates carrying various 2-substituents (125, 126).

The results of Julia provide additional verification of the early conclusion

TABLE XXIV □ SYNTHESIS OF INDOLES BY THE BISCHLER REACTION

Aniline substituent	Ketone or aldehyde	Product	Yield (%)	Ref.
None	2-Chlorocyclohexanone	1,2,3,4-Tetrahydrocarbazole	89[a]	(52)
None	4,4′-Dimethoxybenzoin	2,3-di-(4-Methoxyphenyl)indole	58	(261)
2-Me	2-Hydroxycyclohexaone	8-Methyl-1,2,3,4-tetrahydrocarbazole	34	(142)
2-Ac	2-Chlorocyclohexanone	8-Acetyl-1,2,3,4-tetrahydrocarbazole	33	(139)
2,4-diMe	Phenacyl bromide	5,7-Dimethyl-2-phenylindole	95	(103)
2,5-Dimethoxy	3-Bromo-2-butanone	4,7-Dimethoxy-2,3-dimethylindole	61	(31)
N-Methyl	4-Bromoacetoacetanilide	1-Methylindole-3-acetanilide	42	(128)
N-benzyl	Ethyl 4-bromoacetoacetate	Ethyl 1-benzylacetoacetate	53	(132)
N-Cyanoethyl	Chloroacetone	1-(2-Cyanoethyl)-3-methylindole	23	(131)
4-Methoxy-*N*-methyl	Ethyl 4-bromoacetoacetate	Ethyl 5-methoxyl-1-methylindole-3-acetate	26	(129, 131)

[a] Yield is for cyclization of isolated 2-anilinocyclohexanone using anilinium hydrochloride as catalyst.

NH_2 + O, CH_2Br ⟶ N H

B22 B23 B24

that cyclization of *N*-alkyl-α-anilinoketones usually proceeds without apparent rearrangement whereas unsubstituted α-anilinoketones often give rise to products resulting from apparent rearrangement (40, 68, 134).

Historically, the first example of the formation of an indole in a Bischler-type synthesis was the conversion of phenacyl bromide and aniline to 2-phenylindole. The fact that 2-phenylindole and not 3-phenylindole is the product of this cyclization has led to a number of mechanistic investigations

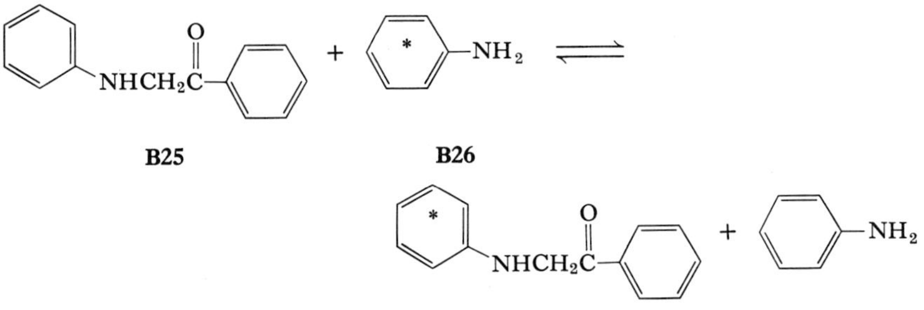

B25 B26

B27

which have been summarized in the significant paper of Weygand and Richter (276). It has been established that the cyclization of phenacylaniline is catalyzed by anilinium bromide and that complete exchange of the anilinium bromide and the anilino groups from phenacylaniline occurs during the cyclization. The Bischler synthesis of 2-phenylindole can be formulated as is shown below:

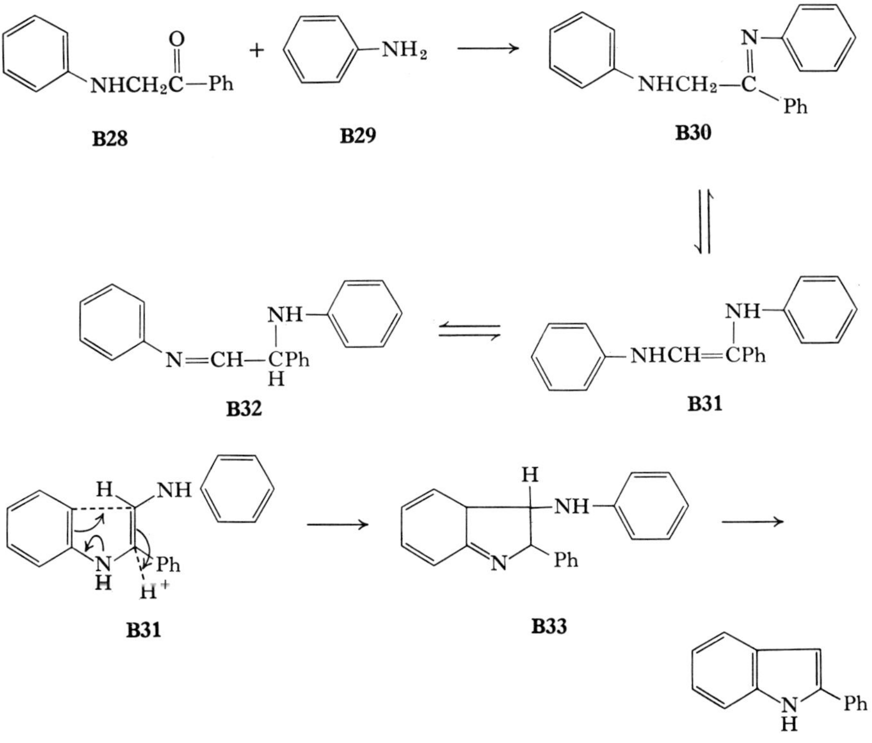

B28 B29 B30

B32 B31

B31 B33

B34

2-Phenylindole is formed in preference to 3-phenylindole because **B31** cyclizes to **B33** more readily than to a precursor of 3-phenylindole. The reason for this preference may be the greater steric crowding which results in the alternative transition state. In general, the structure of an indole generated in other Bischler-type cyclizations will depend upon similar factors. With primary anilines, especially in the presence of anilinium salts, equilibrations analogous to that of **B30** and **B32** are rapid relative to indolization. The relative rates of the competing modes indolization then determine the structure of the product. With secondary amines such equilibrations are retarded and the carbonyl carbon of the haloketone usually becomes C-3 of the indole ring.

The condensation of anilines with benzoin followed by cyclization constitutes another variation of the Bischler synthesis.

B35 + PhCH(OH)COPh (**B36**) ⟶ **B37** Ref. (184)

B38 + **B36** ⟶ **B39** (NHCH(Ph)COPh) ⟶ **B40** Ref. (65)

The anilino substituent is presumably introduced via imine formation and enolization.

Dialkylanilines have been successfully subjected to the conditions of the Bischler reaction (67). *N*-Dealkylation accompanies cyclization. Dealkylation occasionally accompanies cyclization of secondary anilines (40).

$PhNH_2$ + **B36** ⟶ PhCHOH—C(=NPh)—Ph ⇌ PhC(HO)=C(NHPh)—Ph ⇌ PhC(=O)—CHPh(NHPh)

B41 **B42** **B43** **B44**

$N(Me)_2$ + $BrCH_2COMe$ ⟶ Me N Me (54%)

B45 **B46**

B45 + $ClCH_2COMe$ ⟶ **B46** (45%) + N Me Me (5%)

B47

Reactions which parallel the Bischler cyclization have been observed between aryl diazomethyl ketones and anilinium salts (32). In this procedure 1-alkyl-3-arylindoles are the major products when *N*-alkylanilinium salts are used. The major product is accompanied by lesser amounts of 1-alkyl-2-arylindoles and, again, dealkylated 2-arylindoles were encountered.

Chastrette (63) has prepared a number of diethyl acetals of 2-arylamino-propionaldehyde. These can be cyclized to 2-methylindoles in 29–53% yield. No structural ambiguity is expected in this procedure.

X— $CH(OEt)_2$ CHMe N H $\xrightarrow{BF_3}$ X— N H Me

B48 **B49**

The earlier review of Julian, Meyer and Printy (134) contains a compilation of the Bischler-type cyclizations reported in the literature prior to 1950.

The generalized Bischler mechanism can be recognized in several other reactions used for indole synthesis. Martynov and co-workers have obtained α-anilinocarbinols by reactions between aromatic amines and epoxides (161). The decomposition of the α-hydroxy ester in sulfuric acid generates a carbonyl group and cyclization occurs to give the intermediate **B55** which gives rise to the observed indole by migration of a butyl group (161). A general preference for the migration of the larger substituent is apparent

NH_2 + Me–(epoxide, O)–CO_2Et, $(CH_2)_3Me$ → $NHC(Me)((CH_2)_3Me)CHOHCO_2Et$

B50 B51 B52

H_2SO_4 ↓

$(CH_2)_3Me$, Me, N, H

B53

B52 → $\overset{+}{O}H$, CH, $\overset{+}{N}$, $(CH_2)_3Me$, Me, H, H → ^{+}H, $\overset{+}{N}$, $(CH_2)_3Me$, H, H, Me

B54 B55

↓

B53 ← H, $(CH_2)_3Me$, $\overset{+}{N}$, H, Me

B56

(162) but it is difficult to judge the extent of selectivity from published data. Martynov and co-workers have provided a number of examples of such processes. Use of identical β-substituents in the starting epoxy ester removes the possible structural ambiguity introduced by the possibility of competitive migration of unlike groups. Cyanoepoxides can also be employed (163).

NH_2 + Ph–(epoxide, O)–CN $\xrightarrow[\text{heat}]{\text{HCl}}$ N, H, Ph

B57 B58 B59 (82%)

In this case an aldehyde is presumably generated by acid-catalyzed loss of hydrogen cyanide from the intermediate α-anilinocarbinol.

C. INDOLES FROM THE REACTIONS OF ENAMINES WITH QUINONES: THE NENITZESCU SYNTHESIS

The Nenitzescu synthesis (174) of indoles involves the condensation of an enamine and a quinone to generate a hydroxyindole. Extensive interest has developed recently in this synthesis because it can provide directly 5-hydroxyindoles. These are of considerable interest because of their various physiological activities. The mechanism of the reaction has been discussed recently by Allen, Pidacks, and Weiss (15). The essential steps appear to include addition of the enamine to the electrophilic quinone, aromatization, oxidation, cyclization, and reduction. The reduction (**C7** → **C8**) and oxidation (**C5** → **C6**) steps are coupled with one another.

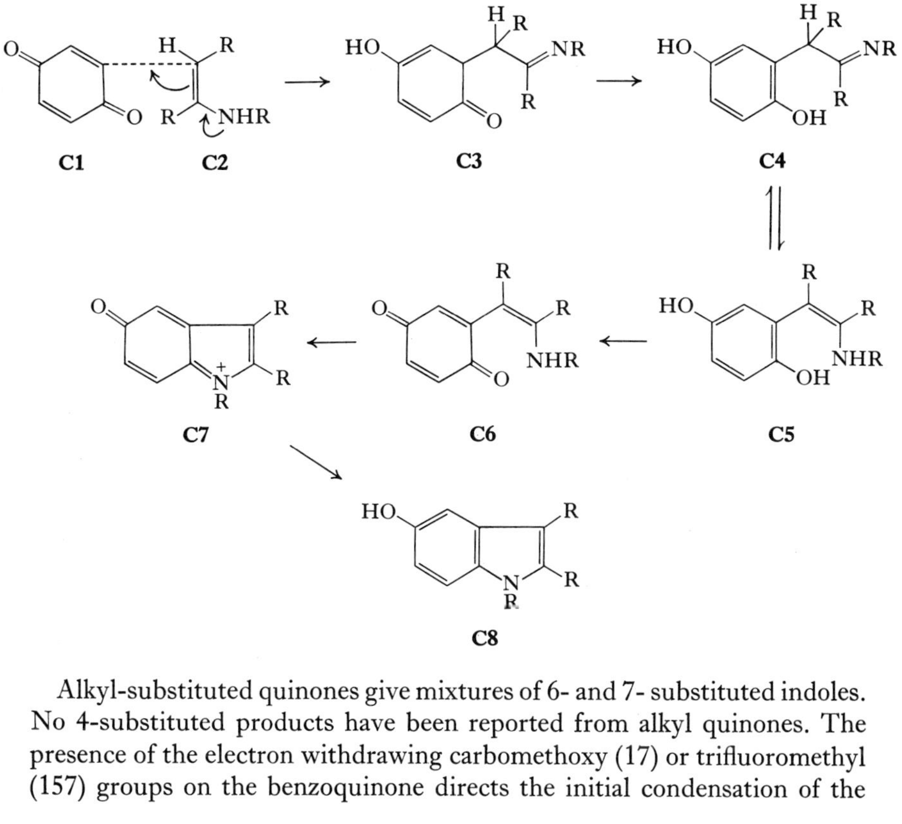

C1 **C2** → **C3** → **C4** ⇌ **C5** → **C6** → **C7** → **C8**

Alkyl-substituted quinones give mixtures of 6- and 7- substituted indoles. No 4-substituted products have been reported from alkyl quinones. The presence of the electron withdrawing carbomethoxy (17) or trifluoromethyl (157) groups on the benzoquinone directs the initial condensation of the

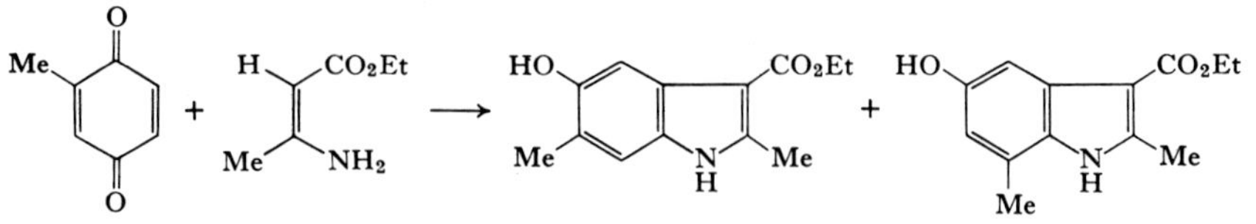

C9 C10 C11 C12

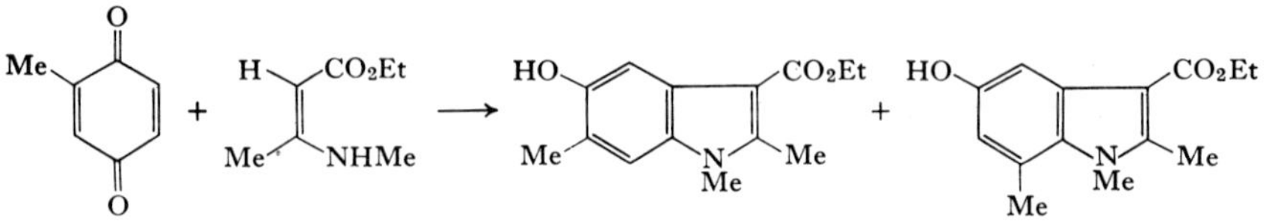

C13 C14 C15 C16

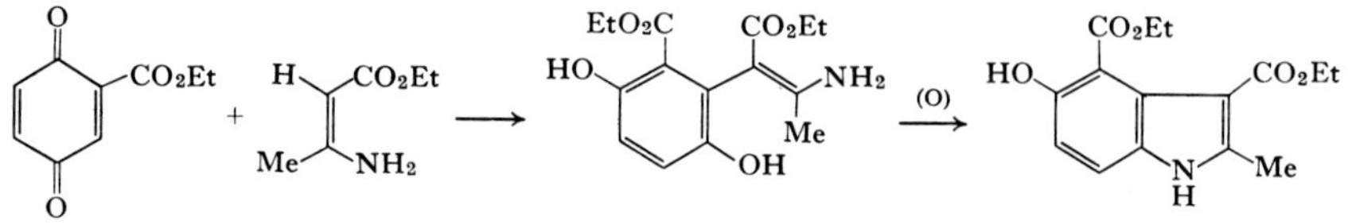

C17 C18 C19 C20

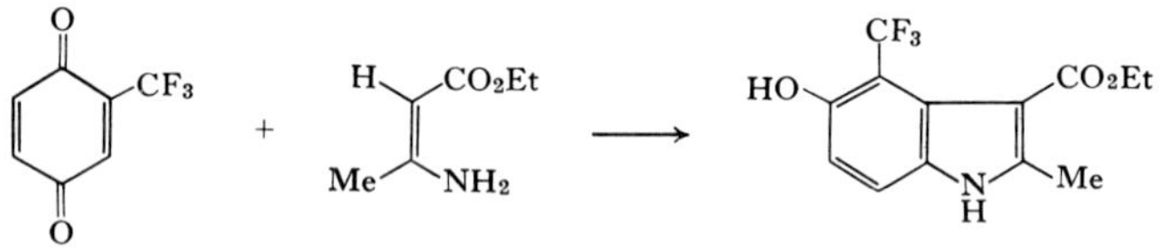

C21 C22 C23

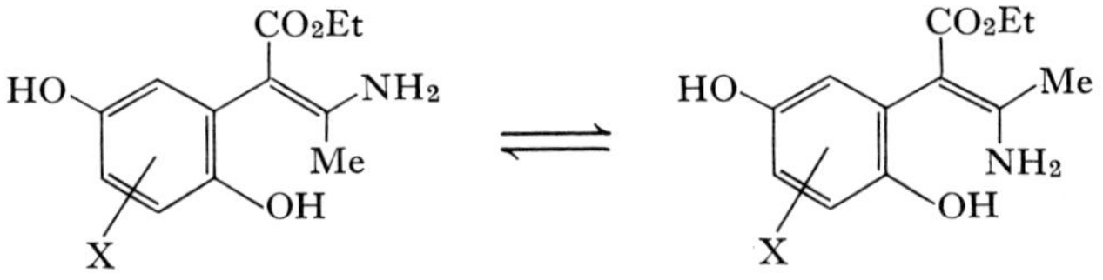

C24 C25

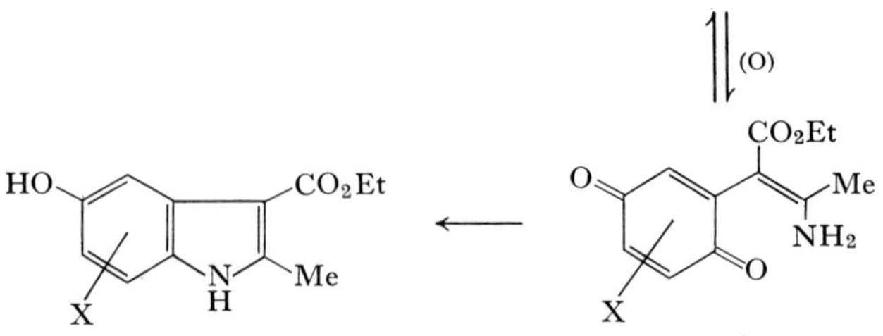

C27 C26

enamine to the 3-position. 4-Substituted indoles are then obtained. When, as in the case of **C24**, the initial adduct has stereochemistry inappropriate for direct cyclization, indolization can be accomplished by an oxidant in the presence of an acid (15, 17, 157). The acid effects configurational equilibration of the enamine via its carbon conjugate acid. The oxidant is necessary to convert the hydroquinone **C25** to the quinone oxidation state.

The 3-carbethoxy substituent can be removed efficiently by hydrolysis (15, 24). The Nenitzescu reaction thus provides a route to indoles unsubstituted in the 3-position and these can then be elaborated by appropriate synthetic procedures.

Table XXV records typical preparations of indoles by the Nenitzescu synthesis.

TABLE XXV □ TYPICAL PREPARATIONS OF INDOLES BY THE NENITZESCU SYNTHESIS

Indole substituents	Yield (%)	Ref.
3-Acetyl-5-hydroxy-2-methyl	38	(92)
3-Acetyl-5-hydroxy-2-methyl-1-phenyl	53	(88, 90)
3-Acetyl-5-hydroxy-1,2,6,7-tetramethyl	28	(88)
1-Benzyl-3-carbethoxy-5-hydroxy-2-methyl	58	(71)
1-Benzyl-3-carbethoxy-5-hydroxy-6-methyl-2-phenyl	38	(30)
1-Benzyl-3-carbethoxy-5-hydroxy-2-phenyl	42	(30)
3-Carbethoxy-6,7-dichloro-1-ethyl-5-hydroxy-2-methyl	54	(92)
3-Carbethoxy-1,2-dimethyl-5-hydroxy	48	(89)
3-Carbethoxy-5,6-dihydroxy-2-methyl	35	(24)
3-Carbethoxy-1-ethyl-5-hydroxy-2-methyl	75	(89a)
3-Carbethoxy-6-ethyl-5-hydroxy-2-methyl	30	(15)
3-Carbethoxy-5-hydroxy-6-methoxy-2-methyl	35	(24)
3-Carbethoxy-5-hydroxy-2-methyl	34	(240)
3-Carbethoxy-5-hydroxy-2-methyl-1-phenyl	60	(87)
3-Carbethoxy-5-hydroxy-2-methyl-1-*o*-tolyl	53	(87)
3-Carbethoxy-5-hydroxy-2 phenyl	46	(210)

Although the Nenitzescu synthesis has been most widely applied to 3-carbethoxyindoles, use of β-aminovinyl ketones permits the synthesis of 3-acylindoles. In certain cases the 3-acylindoles are accompanied by benzofuran derivatives (92). The extent of benzofuran formation apparently depends on the nature of the nitrogen substituent in the β-aminovinyl ketone (89a) as well as on the structure of the quinone (86, 91).

Closely related to the Nenitzescu synthesis is the reaction between quinone imines and β-dicarbonyl compounds to give 5-aminoindole derivatives (9). The reaction is illustrated by the transformations which follow.

C28 + $MeC{=}CAc$ (NH_2) C29 ⟶ C30 + C31

HO, Ac, N H, O

C32 + $(Ac)_2CH_2$ C33 ⟶ C34

PhCON, NCOPh, $PhCONH$, $CH(Ac)_2$, NHCOPh

C34 —H^+→ C35

H_2N, Me, N H

C36 —H^+→ C37 (94%) Ref. (9)

$PhSO_2NH$, $CH(COPh)_2$, $NHSO_2Ph$, COPh, Ph, N H

C38 C39 ⟶ C40 ⟶ C41 —$-H_2O$→ C42

XN, Y, CH_2COR, NX, XNH, H, COR, NHX, R, OH, N X

The adduct formation step is analogous to the first step of the Nenitzescu reaction. Cyclization then occurs followed by aromatization via loss of water.

It is interesting to note that 3-acetyl, 3-carbethoxy, and 3-benzoyl substituents are cleaved under the usual conditions of the cyclization reaction involving refluxing aqueous hydrochloric acid (8, 9). (See Chapter II, D).

$(Me)_2NSO_2NH$ … $CH(Ac)_2$ … $NHSO_2N(Me)_2$ $\xrightarrow[H_2O]{H^+}$ H_2N … Me (N–H) Ref. (8)

C43 **C44** (86%)

$PhSO_2N$ … NSO_2Ph + (O, O) ⟶ $PhSO_2NH$ … (O, O) … $NHSO_2Ph$

C45 **C46** **C47**

AcOH

$PhSO_2NH$ … (O) … N, SO_2Ph

C48 (98%)

$PhSO_2NH$ … Me, O … N, SO_2Ph ⟵ pyridine ⟵ $PhSO_2NH$ … Me, O, O … NH, SO_2Ph

C51 **C49**

AcOH

$PhSO_2NH$ … Me … $(CH_2)_3CO_2H$ … N, SO_2Ph

C50

Use of cold concentrated sulfuric acid as the cyclizing reagent generally leads to cyclization without loss of the 3-substituent. Use of cyclic diketones in this procedure generates a tetrahydrocarbazole derivative (105).

Use of 2-substituted cyclohexane-1,3-diones gives products of the type **C50** or **C51**, depending on the conditions of the reaction. The indole **C50** is presumably formed from **C51** via a fragmentation reaction. The ring opening is another manifestation of substituent cleavage from a 3*H*-indole system (Chapter II, D).

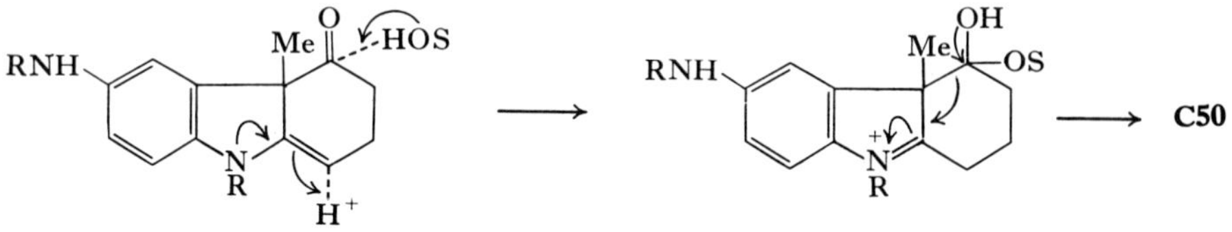

RNH Me O HOS N R H+ → RNH Me OH OS + N R → **C50**

C51 **C52**

D. REDUCTIVE CYCLIZATIONS

1. Reductive Cyclization of o-Nitrobenzyl Ketones and Cyanides

The chemical or catalytic reduction of an *o*-nitrobenzyl ketone generates an *o*-aminobenzyl ketone which can subsequently cyclize and aromatize by dehydration to an indole. If an *o*-nitrobenzyl cyanide is reduced in such a

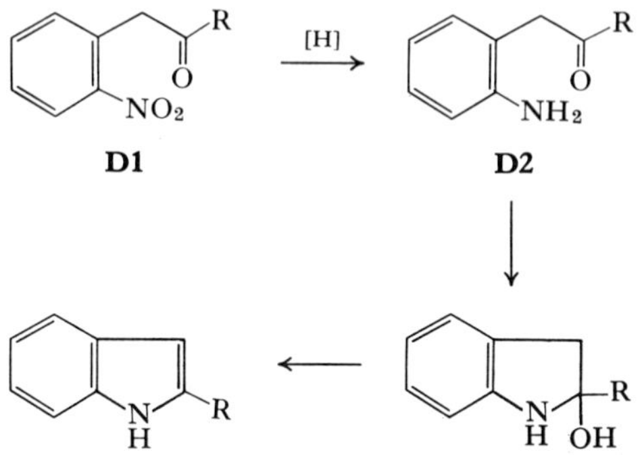

R O NO_2 —[H]→ R O NH_2

D1 **D2**

N H R ← N H R OH

D4 **D3**

manner that partial reduction of the nitrile function accompanies reduction of the nitro group, a similar cyclization and aromatization is feasible.

These generalized reaction patterns are realized in a number of indole syntheses. Perhaps most widely applied is the Reissert procedure (212) involving base-catalyzed condensation of an *o*-nitrotoluene derivative with an oxalate ester followed by reductive cyclization to an indole-2-carboxylic acid derivative.

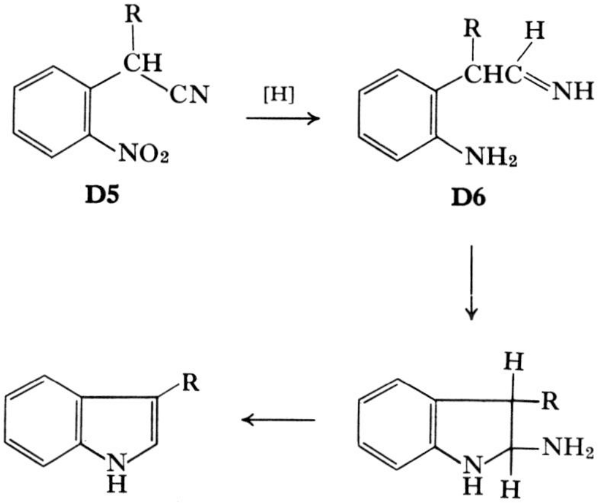

D5 → [H] → D6

D6 → D7 → D8

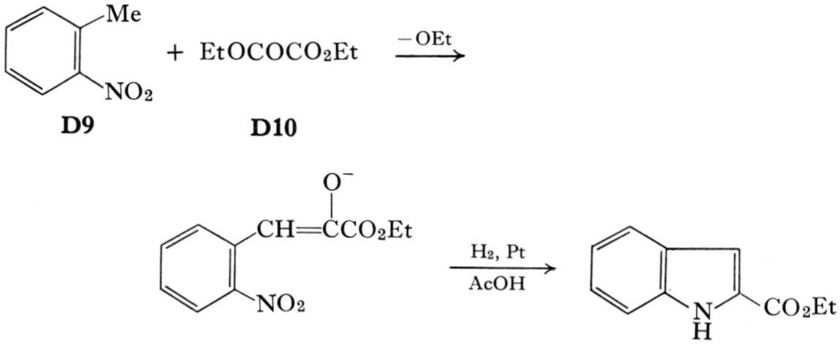

D9 + $EtOCOCO_2Et$ (D10) $\xrightarrow{-OEt}$

D11 $\xrightarrow[AcOH]{H_2, Pt}$ D12 Ref. (175) (47–51%)

An alternative route to the *o*-nitrophenylpyruvate intermediates involves condensation of an *o*-nitrobenzaldehyde with an oxazolone followed by hydrolysis (25). The Reissert procedure has found wide use for the synthesis

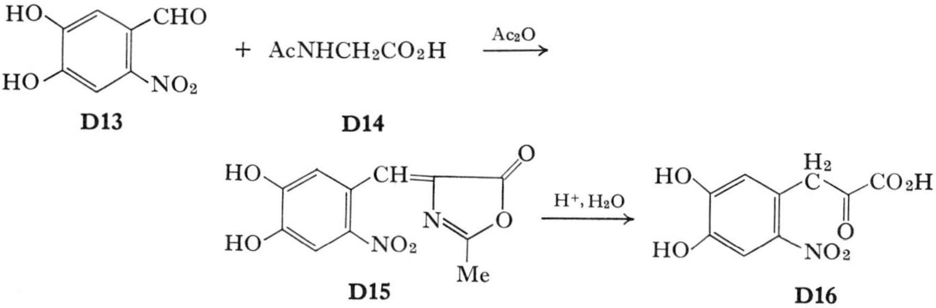

D13 + $AcNHCH_2CO_2H$ (D14) $\xrightarrow{Ac_2O}$

D15 $\xrightarrow{H^+, H_2O}$ D16

of indole-2-carboxylic acids and Table XXVI records typical examples of reductive cyclizations of *o*-nitrophenylpyruvates.

TABLE XXVI □ PREPARATION OF INDOLE-2-CARBOXYLIC ACIDS BY THE REISSERT PROCEDURE

Indole-2-carboxylic acid substituent	Condensing agent	Reducing agent	Yield[a] (%)	Ref.
None	Potassium ethoxide	Hydrogen, platinum	66	(175)
4-Cl	Potassium ethoxide	Ferrous sulfate, ammonia	90	(221, 268)
4-Methoxy	Potassium ethoxide	Ferrous sulfate, ammonia	73	(33, 84)
4-Benzyloxy	Potassium ethoxide	Sodium bisulfite	64	(246)
5-Benzyloxy	Potassium ethoxide	Sodium bisulfite	48	(246)
5-Methoxy-6-methyl	Potassium *t*-butoxide	Ferrous sulfate, ammonia	56	(16)
5,6-Diacetoxy	[b]	Iron, acetic acid	44	(25)
6-F	Potassium etoxide	Ferrous sulfate, ammonia	22	(14)
6-Cl	Potassium ethoxide	Ferrous sulfate, ammonia	85	(221)
6-Dimethylamino	Potassium ethoxide	Iron, hydrochloric acid	60	(269)
6-(1-Pentenyl)	Potassium ethoxide	Ferrous sulfate, ammonia	65	(235)
6-Benzylthio	Sodium ethoxide	Ferrous sulfate, ammonia	62	(197)
7-Methoxy	Sodium ethoxide	Ferrous sulfate, ammonia	63	(33)
5-Methoxy-6-aza	Potassium ethoxide	Hydrogen, palladium	85	(79)

[a] Yields recorded are for the cyclization step.
[b] The pyruvic acid was prepared from the aldehyde via the oxazolone.

Rosemund and Hasse have examined the reaction of *o*-nitrophenylacetyl chlorides with enamines (220). Following a general pattern of enamine reactivity (108), good yields of β-diketones are formed. The reduction of the resulting nitrobenzyl ketones gives good yields of indoles accompanied by lesser amounts of azepinones such as **D21** formed by condensation at the ring carbonyl group (220). The preparation of 2-methyl-7-methoxyindole has been effected by the sequence **D22** $\rightarrow$ **D26** (34). A number of halogen-substituted indoles have also been prepared in this manner (199), with zinc and acetic acid being used as the reducing agent to prevent hydrogenolysis. Young has employed the reductive cyclization of *o*-nitrobenzyl ketones **D27** and **D28** in the preparation of 2,3′-biindolyls (282). *o*-Nitrophenyl-

CH_2COCl NO_2 + (morpholino enamine) → CH_2CO NO_2

D17 **D18** **D19**

$SnCl_2$, HCl

(10%) (72%)

D21 **D20**

CH_2COCl NO_2 OCH_3 + $EtOMgCH(CO_2Et)_2$ → $CH_2COCH(CO_2Et)_2$ NO_2 OMe

D22 **D23** **D24**

H_2O, H^+

H_2, Ni ← CH_2COMe NO_2 OMe

MeO Me **D26** **D25**

MeO MeO NO_2 R → $SnCl_2$ or $NaHSO_3$ → MeO MeO R

D27 R = H
D28 R = Me

D29 R = H
D30 R = Me

acetaldehydes can also be efficiently converted to indoles by reductive cyclization. The major limitation on this synthetic route is the relative inaccessibility of the required aldehydes (176).

Interest in the reductive cyclization (209, 241) of *o*-nitrophenylacetonitriles was reawakened by Walker (273) who prepared 5,6-dimethoxyindole from 2-nitro-4,5-dimethoxyphenylacetonitrile by catalytic reduction over palladium on carbon at 80°C. Walker prepared several 3-substituted

MeO, MeO, CH_2CN, NO_2 $\xrightarrow[H_2]{Pd/C}$ MeO, MeO, N, H + NH_3

D31 **D32**

indoles by analogous reactions and observed that in some instances *o*-aminobenzyl cyanides could be isolated and they were assumed to be intermediates in indole formation. He further noted two examples of an alternate mode of reduction and cyclization, represented by the transformation **D33** → **D34** in which a "2-aminoindole" (see Chapter VIII for a discussion of other 2-aminoindoles) was isolated. Chemical reduction of the nitrile **D35** takes a similar course (236).

$CH_2CH_2N(Me)_2$, CHCN, MeO, MeO, NO_2 $\xrightarrow[H_2]{Pd/C}$ MeO, MeO, $CH_2CH_2N(Me)_2$, N, NH_2

D33 **D34**

CO_2Et, CHCN, EtO_2C, NO_2 $\xrightarrow{SnCl_2HCl}$ EtO_2C, CO_2Et, N, H, NH_2

D35 **D36**

Walker suggested that "2-aminoindoles" were intermediates in the formation of indole, but Snyder and co-workers (236) have rejected this mechanism on the basis of the observation that the 2-aminoindole **D36** could not be converted to the indole **D37** under conditions sufficient to form the latter from starting nitro compound **D35**.

Snyder and co-workers suggest that indole formation takes place via imines formed by partial reduction of the cyano group. The overall course of *o*-nitrophenylacetonitrile reductions will in general depend upon the relative rates of the cyclization and reduction steps. Indole formation is believed to result when reduction to **D40** is faster than cyclization leading to **D39**. Acidic reaction conditions may favor formation of aminoindoles by

CO_2Et
CHCN
EtO_2C
NO_2
$\xrightarrow[\text{Pd/C}]{H_2}$

D35

CO_2Et
EtO_2C
N
H

D37

CO_2Et
EtO_2C
N
H
NH_2

D36

R
CHCN
NH_2

D38

R
CHCH=NH
NH_2

D40

H
R
NH_2
N
H H

D41

H
R
N
H
NH

D39

R
N
H
H

D42

MeO
MeO
CH_2CN
NH_2

D43

+ $EtOCH{=}C(CO_2Et)_2$

D44

MeO
MeO
$CH_2C{\equiv}N$
N—CH=C$(CO_2Et)_2$
H

D45

MeO
MeO
N
NH
O
CO_2Et

D47

MeO
MeO
N
N
O
CO_2Et

D46

catalyzing the cyclization to **D39**. The amidine group in **D39** is resistant to reduction under mild conditions. Other examples of successful syntheses of indoles from *o*-nitrophenylacetonitriles have been reported by Plieninger and Nogradi (205) and by Hoffman and co-workers (104).

The *o*-aminobenzyl cyanide **D43** condenses to a pyrimido[1,2-a]-indole on heating in ethanol containing sodium ethoxide (187). This cyclization is effectively the intramolecular trapping of the "aminoindole" intermediate.

2. Reductive Cyclization of o,β-Dinitrostyrenes

The conversion of *o*,β-dinitrostyrenes to indoles is perhaps the most widely applied indole synthesis of the reductive cyclization type. The mechanism of this reaction presumably involves reduction of the aryl nitro groups to the amino stage, partial reduction of the α,β-unsaturated nitro group, cyclization, and aromatization.

CH=CHNO$_2$, NO$_2$ **D48** —[H]→ CH$_2$CH=NOH, NH$_2$ **D49** → NHOH, N H, H **D50** → N H **D51**

The *o*,β-dinitrostyrenes are usually obtained by base-catalyzed condensation of an *o*-nitrobenzaldehyde with nitromethane but nitration of β-nitrostyrenes has been used occasionally (94, 99, 117). Reduction has most frequently been effected with iron and acetic acid, but catalytic reduçtion over palladium on carbon in a solvent containing some acid is an alternative.

Baxter and Swan have investigated the efficiency of numerous types of catalysts in the reduction of 4,5-dialkoxynitrostyrenes (22). Palladium on carbon catalyst was satisfactory for synthesis of 5,6-dimethoxyindole. The use of a rhodium catalyst gave the nitrophenylacetaldehyde oxime in one case. Raney nickel and hydrazine, lithium aluminum hydride and sodium

MeO, PhCH$_2$O, CH=CHNO$_2$, NO$_2$ **D52** —rhodium / alumina→ MeO, PhCH$_2$O, CH$_2$CH=NOH, NO$_2$ **D53**

borohydride were also examined. Cinnolines were formed using lithium aluminum hydride, and selective reduction of the conjugated olefinic bond was observed with sodium borohydride. Dehalogenation has been observed during the catalytic cyclizations (26, 99). Heacock and co-workers (99) have reported an intriguing iodine migration during the catalytic cyclization of 2,β-dinitro-4,5-dimethoxy-3-iodo-β-methylstyrene but aside from this instance, the reaction seems to have led to no cases of structural ambiguity. The reaction has so far been applied mainly to synthesis of indoles lacking 2- or 3-substitution but the use of nitroethane in place of nitromethane has led to *o*,β-dinitro-β-methylstyrenes which were successfully cyclized (99). The method would therefore seem to be readily adaptable to synthesis of various 2-alkylindoles.

Table XXVII records typical examples of cyclizations of *o*,β-dinitrostyrenes.

TABLE XXVII □ SYNTHESIS OF INDOLES BY REDUCTIVE CYCLIZATION OF *o*,β-DINITROSTYRENES

Indole substituents	Reducing agent	Yield (%)	Ref.
4-F	Hydrogen, palladium	20–25	(28)
4-Cl	Iron, acetic acid	84	(224)
5-Benzyloxy	Iron, acetic acid	61	(72)
5,6-Dimethoxy	Hydrogen, palladium	60	(106)
5-Methoxy-6-benzyloxy	Iron, acetic acid	44	(132a)
5,6,7-Trimethyl	Hydrogen, palladium	43	(26)
5,6,7-Trimethoxy	Iron, acetic acid	50	(170)
5,6-Dimethoxy-7-iodo-2-methyl	Iron, acetic acid	47	(99)
6-F	Hydrogen, palladium	62	(27)
6-Cl	Iron, acetic acid	72	(255)
6-Benzyloxy	Iron, acetic acid	72	(255)
6,7-Dimethoxy	Hydrogen, palladium	23	(26)
7-Benzyloxy	Iron, acetic acid	75	(72)

E. ELECTROPHILIC CYCLIZATION OF STYRENE DERIVATIVES

In this section are considered cyclizations which proceed by the general paths **E1** → **E3** or **E4** → **E6**.

The deoxygenation of *o*-nitrostyrenes by trivalent derivatives of phosphorus is an example of the first type of reaction. The transformations which follow serve to illustrate this reaction.

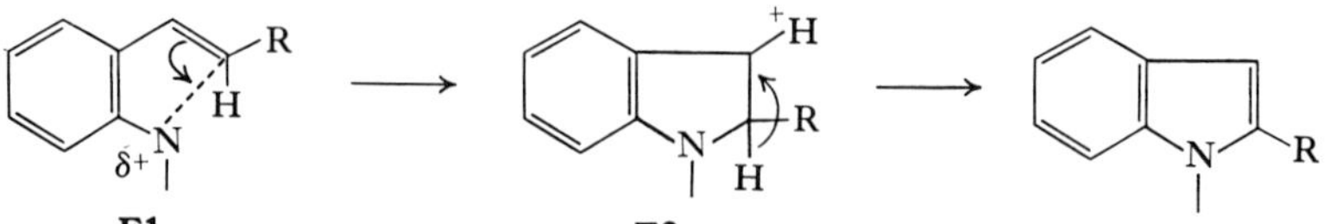

E1 → E2 → E3

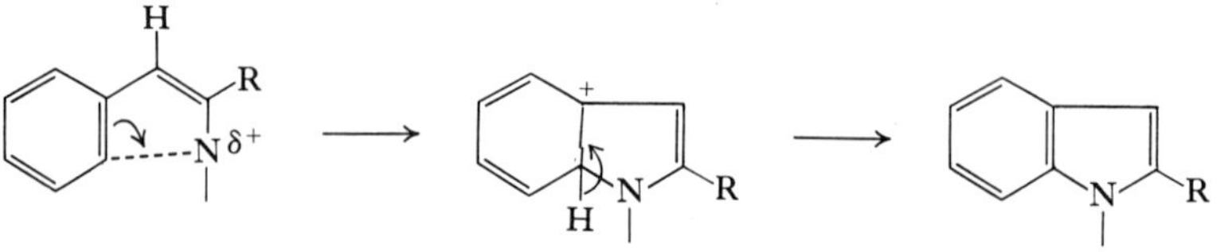

E4 → E5 → E6

The detailed mechanism of these reactions has not been firmly established but the gross outline seems firm. The thermal deoxygenation of the nitro compound probably produces the corresponding nitroso derivative (51, 252). Nitrosoaromatics are also deoxygenated by triethyl phosphite, and presumably produce arylnitrenes. Thus, the pathway shown below rationalizes the formation of 2-phenylindole. However, it has been shown that

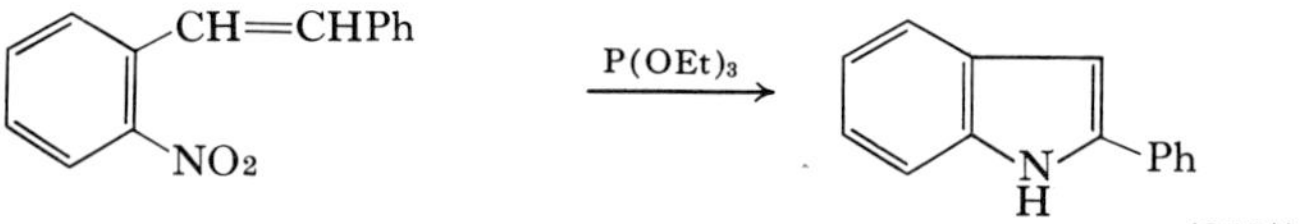

E7 $\xrightarrow{P(OEt)_3}$ E8 (85%) Ref. (51)

E9 $\xrightarrow{P(OEt)_3}$ E10 (60%) Ref. (250)

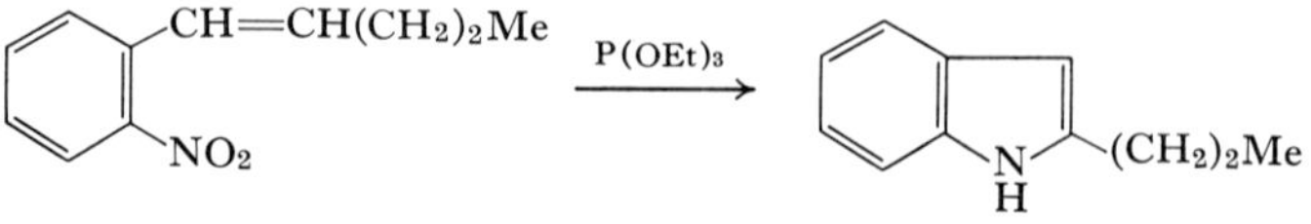

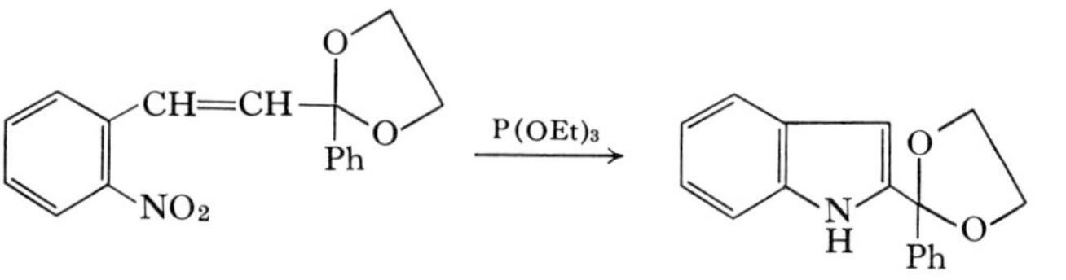

E11 $\xrightarrow{P(OEt)_3}$ E12 Ref. (251)

some, and possibly all, of the 2-phenylindole is formed via 1-hydroxy-2-phenylindole (250) and therefore other paths must also be considered.

The isolation of small amounts of 1-ethoxyindoles, along with indoles, in several reactions (250, 251) suggests that cyclization to *N*-hydroxyindoles is at least competitive with formation and cyclization of the nitrene. Alkylation of *N*-hydroxyindole intermediates by triethyl phosphate or other phosphorus derivatives is the most likely source of the 1-ethoxyindoles.

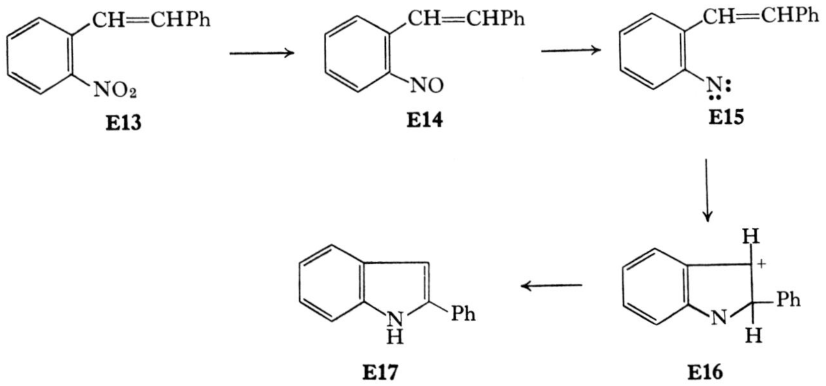

CH=CHPh
NO_2
E13

CH=CHPh
NO
E14

CH=CHPh
N:
E15

H
+
Ph
N
H
E16

N
H
Ph
E17

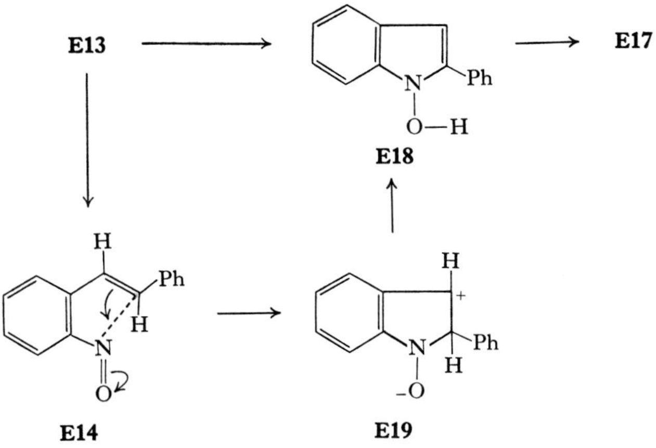

E13

Ph
N
O—H
E18

E17

H
Ph
H
N
O
E14

H
+
N
Ph
H
-O
E19

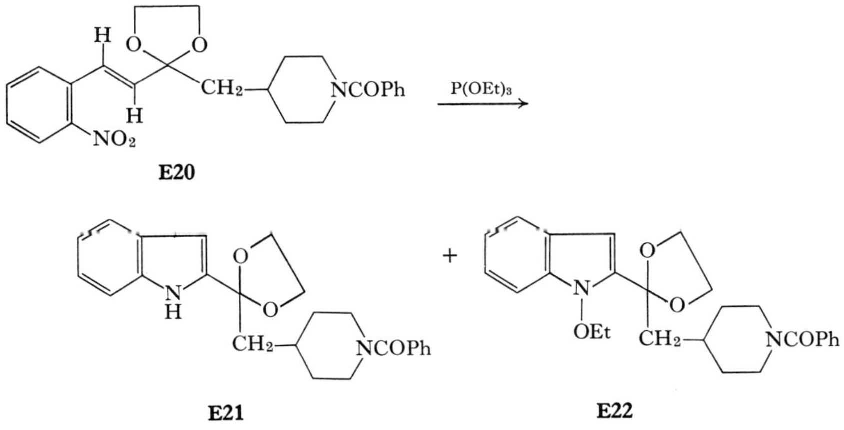

H
O
O
CH_2
NCOPh
H
NO_2
$P(OEt)_3$
E20

O
N
H
O
CH_2
NCOPh
E21

\+

O
N
O
OEt
CH_2
NCOPh
E22

The deoxygenations of nitrostyrenes **E20** and **E23** give indole as a by-product. A fragmentation process (93) initiated by electrophilic attack on the olefin is considered to be responsible for the formation of indole (251).

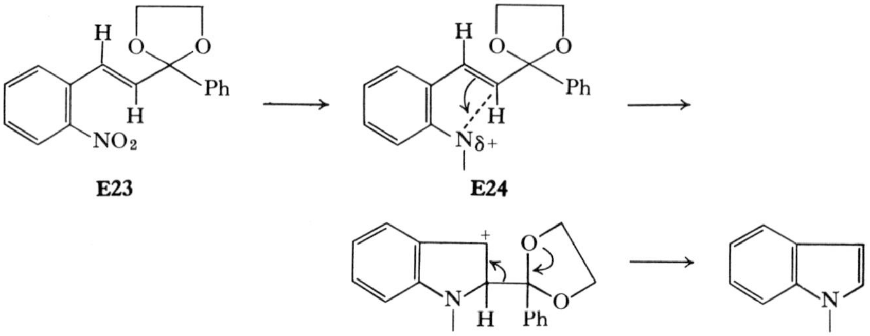

E23 → **E24** → **E25** → **E26**

Indoles are also formed, along with considerable amounts of other products, in the deoxygenation of β,β-disubstituted *o*-nitrostyrenes. Here a migration of one of the substituents is necessary and a structural ambiguity arises in the case of unlike β-substituents (254).

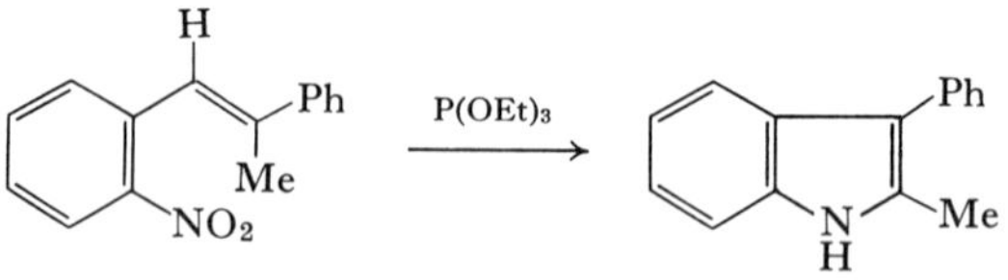

E27 $\xrightarrow{P(OEt)_3}$ **E28**

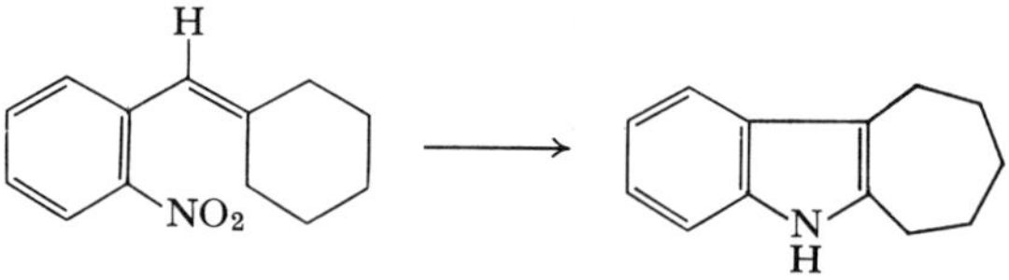

E29 → **E30**

The pyrolytic deoxygenation of aromatic nitro compounds also generates electrophilic nitrogen species (3, 6). One instance of indole formation in an oxalate pyrolysis has been reported (4). Photolytic or pyrolytic decom-

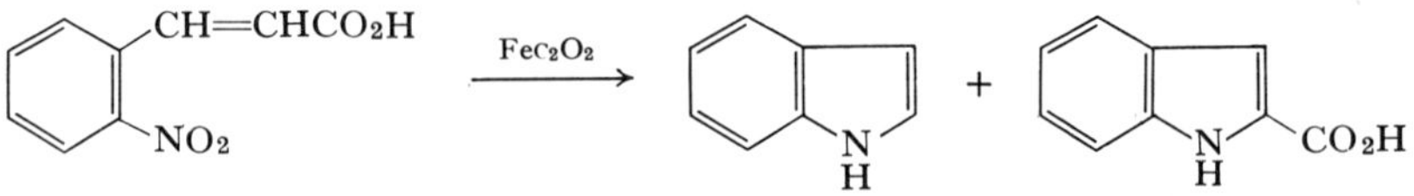

E31 $\xrightarrow{FeC_2O_2}$ **E32** + **E33**

position of 1-(*o*-azidophenyl)1-pentene gives 3-propylindole, presumably via a nitrene intermediate, and several other *o*-azidostyrenes have also been converted thermally to indoles (253).

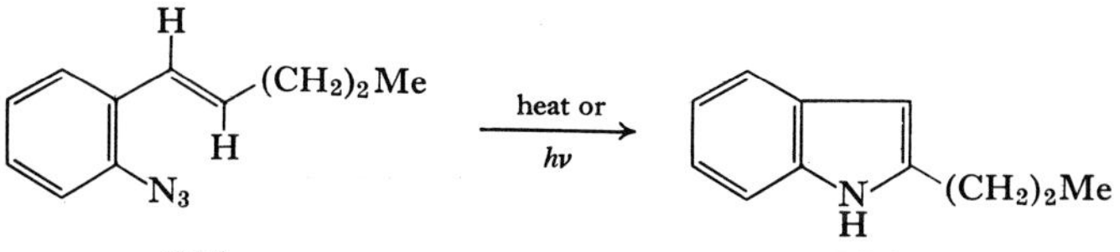

H
(CH$_2$)$_2$Me
H
N$_3$
heat or *hv*
N H (CH$_2$)$_2$Me

E34 **E35**

Cadogan and co-workers observed the formation of 2-phenylindole in 16% yield when α-nitrostilbene was deoxygenated by triethyl phosphite (51). It is assumed that the cyclization is effected by a nitrene or other electrophilic nitrogen species. This cyclization is seen to be an example of the

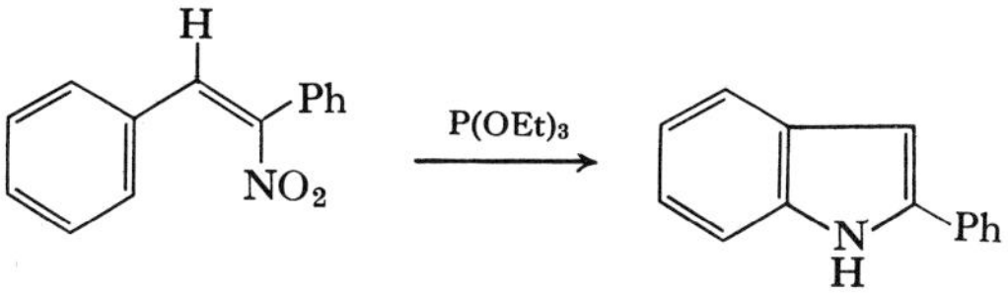

H
Ph
NO$_2$
P(OEt)$_3$
N H Ph

E36 **E37**

mechanistic pattern **E4** → **E6**. The transformation is related to a number of decompositions of styryl azides. Although Boyer and co-workers reported that no indole is formed on pyrolysis or photolysis of β-azidostyrene (35), (phenylacetonitrile was the principal recognized product) (35) it has subsequently been reported (117) that both *cis*- and *trans*-β-azidostyrene give indole in about 45% yield when pyrolyzed in high-boiling hydrocarbon

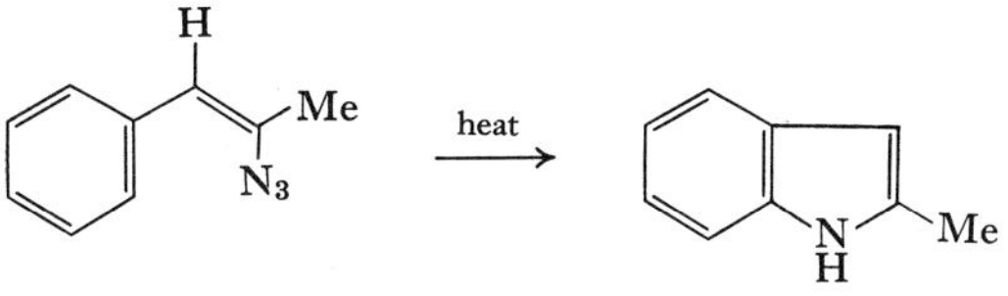

H
Me
N$_3$
heat
N H Me

E38 **E39**

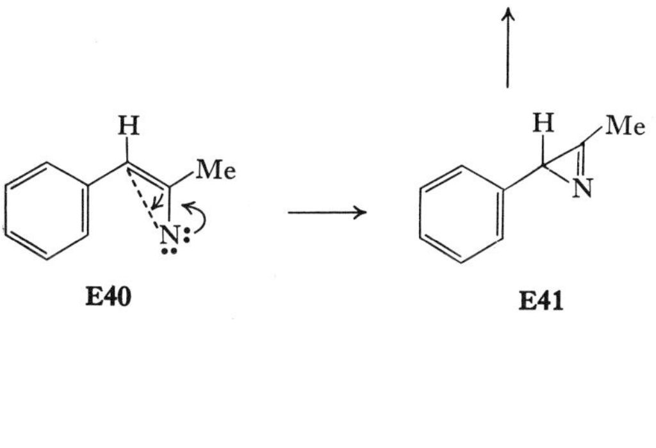

H
Me
N:
H Me
N

E40 **E41**

solution. Similarly, *cis*- and *trans*-β-azido-β-methylstyrene are reported to give 2-methylindole. Isomuro and co-workers (117) proposed azirines as intermediates in formation of the indoles. Smolinsky and Pryde have also found examples of formation of indoles from azidostyrenes (234). They found the ratio of indole to other products to be solvent dependent. Hydroxylic solvents tended to reduce the indole yields drastically.

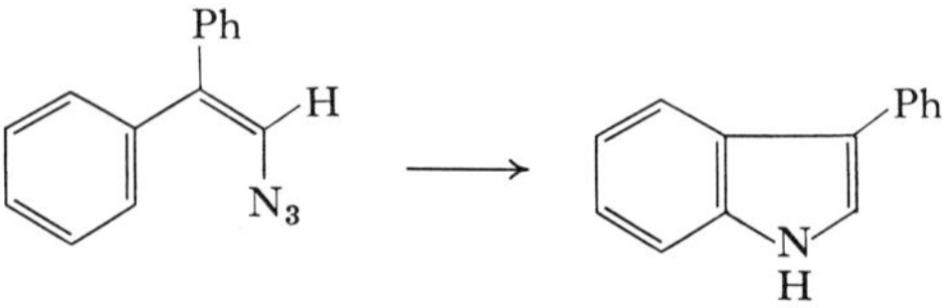

Ph H N_3 → Ph N H

E42 **E43**

The pyrolysis of *o*-alkylarylazides or the deoxygenation of *o*-alkylnitrobenzenes generates aromatic nitrenes which give C-H insertion reactions generating indolines and tetrahydroquinolines (232, 233, 252). Indoles can then be obtained by dehydrogenation. The application of this sequence to an aryl azide derived from a steroid nucleus has been reported by Cantrall and co-workers (53).

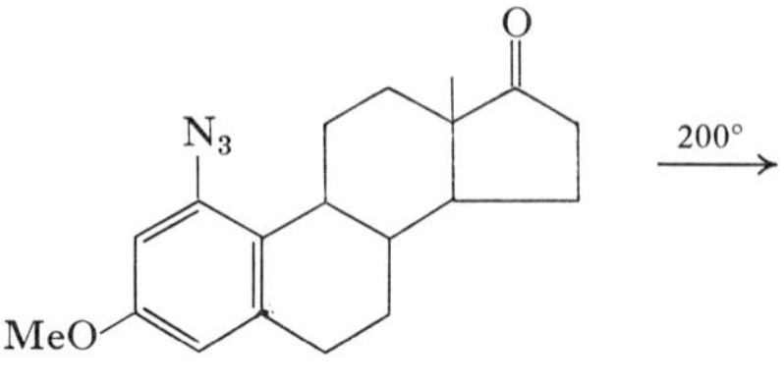

N_3 O MeO $\xrightarrow{200°}$

E44

H N O MeO $\xrightarrow{Pd/C}$ H N O MeO

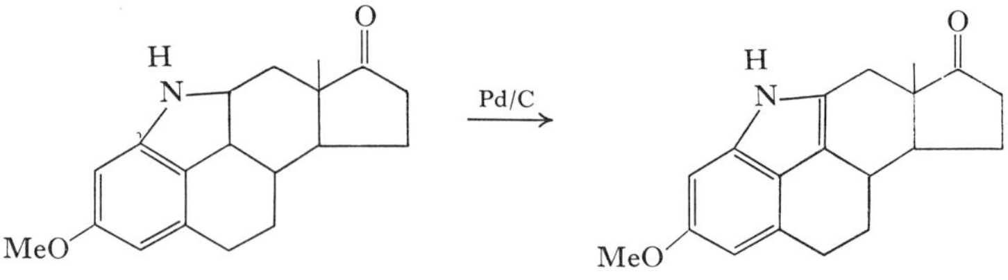

E45 (73%) **E46** (90%)

The pyrolytic conversion of α-phenylcycloalkanone oximes to indoles (82, 275) may be a related process but neither the mechanism or synthetic value of the reaction have received much attention.

F. THE MADELUNG SYNTHESIS

Base-catalyzed condensation of *N*-acyl-*o*-alkylanilines gives indoles. The synthesis of 2-methylindole can be readily effected by this reaction (12). The mechanism presumably involves proton abstraction from the weakly acidic *o*-alkyl group, followed by intramolecular addition to the amide carbonyl although little work on the mechanism of the reaction has been reported. Prior ionization of the amide NH is expected, greatly reducing the electrophilicity of the carbonyl group. The reaction is initiated only by strong bases at elevated temperatures and the useful scope of the synthesis is therefore

Me NHAc $\xrightarrow{^{-}NH_2}$ CH_2^- NAc $\longrightarrow$ N Me O^-

F1 **F2** **F3**

N Me H $\longleftarrow$ N Me

F5 (80–83%) **F4**

limited to molecules which can survive very strongly basic conditions. The reaction has usually been applied to indoles bearing only alkyl substituents. Table XXVIII records typical recent applications of the Madelung method.

TABLE XXVIII □ SYNTHESIS OF INDOLES BY THE MADELUNG REACTION

Indole substitution	Base	Solvent	Yield (%)	Ref.
None	Sodium *o*-toluidide	—	68	(267)
None	Potassium *t*-butoxide	—	31	(152)
2,3-diMe	Sodium amide	*N,N*-Diethylaniline	41	(198a)
2-Dimethylaminomethyl	Sodium amide	—	67	(281)
2-(1,1-Dimethylpropyl)-5,7-bis(4-methylbutyl)	Potassium *t*-butoxide	—	45	(60)
2-Ethyl-5-methyl	Sodium amide	*N,N*-Diethylaniline	67	(274)
2-Me	Sodium amide	—	80	(12)
3-Me	Sodium alkoxides	—	11–28	(154)
5-Methyl-2-propyl	Sodium amide	*N,N*-Diethylaniline	28	(274)
7-(2-Ethyl-4,5-dimethoxyphenyl)	Potassium *t*-butoxide	—	50	(166)

In the case of *o*-dialkylanilides, condensation takes place at a methyl group in preference to a more highly substituted group (198a).

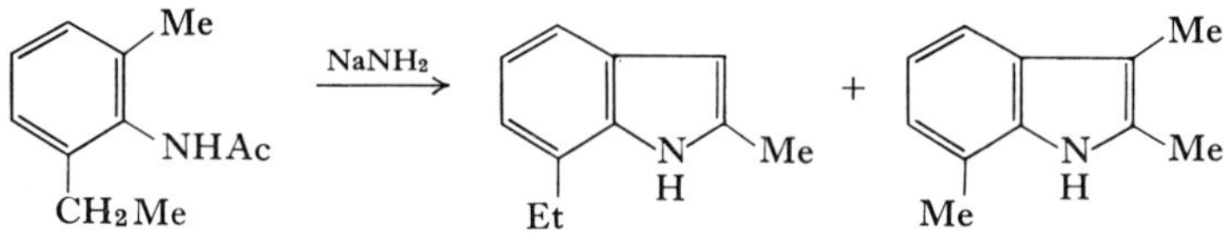

F6 → ($NaNH_2$) F7 + F8

Ratio **F7**:**F8** = 35:1

It has been reported that amides of *o*-aminophenylacetic acid give 2-substituted indoles on heating with soda lime (171). This conversion is clearly of the Madelung pattern but the sequence of cyclization and decarboxylation is not established.

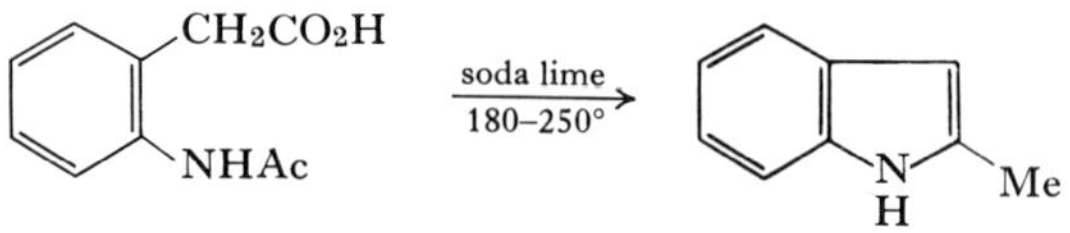

F9 → (soda lime, 180–250°) F10

A recent development involves the use of amidines in place of amides in a Madelung-type condensation (158).

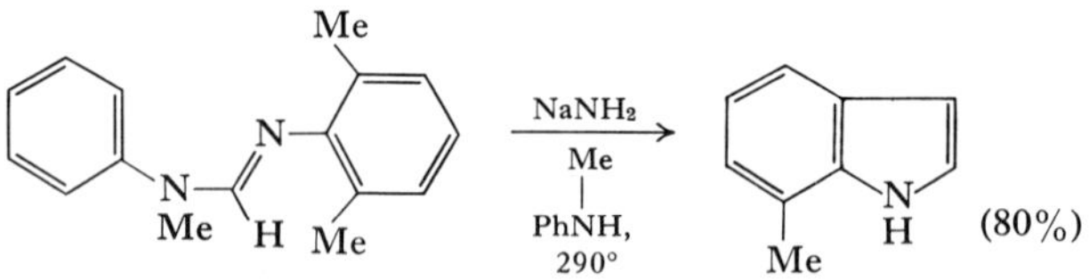

F11 → ($NaNH_2$, PhN(Me)H, 290°) F12 (80%)

The mechanism suggested for the standard Madelung cyclization can be expanded to accommodate this reaction. In contrast to primary acetanilides, the secondary amidine linkage presents no acidic proton and this factor should make the amidine version of the condensation somewhat more facile.

The transformation **F13** → **F15**, which is formally similar to the Madelung cyclization, has been shown to proceed by the mechanism outlined below. Both the enol and keto forms of **F14** have been isolated from the

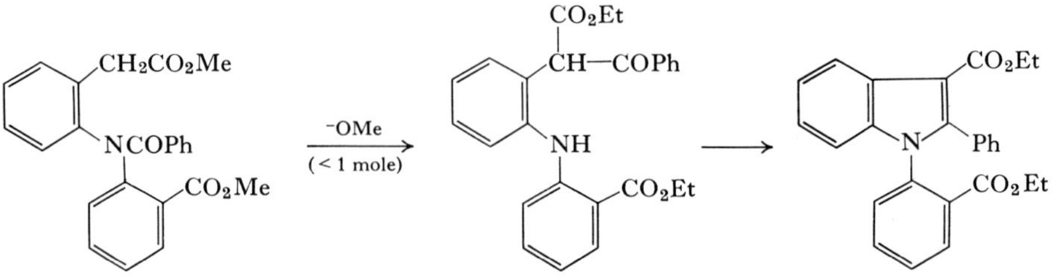

F13 → (^{-}OMe, (< 1 mole)) F14 → F15

reaction, demonstrating that transfer of the benzoyl group from nitrogen to carbon precedes cyclization. In the presence of excess base, **F14** and analogous compounds follow an alternative reaction path giving oxindoles (223). The isolation of **F14** as an intermediate in this cyclization suggests the possibility that N to C acyl transfers might be involved in the classic Madelung cyclization.

G. OXIDATIVE CYCLIZATIONS

Various oxidizing agents convert 2-phenylethylamines with aromatic hydroxyl substituents to indoles. The conversion shown below is an example of such a cyclization. The hydroxyindole **G2** is rapidly oxidized to **G3**

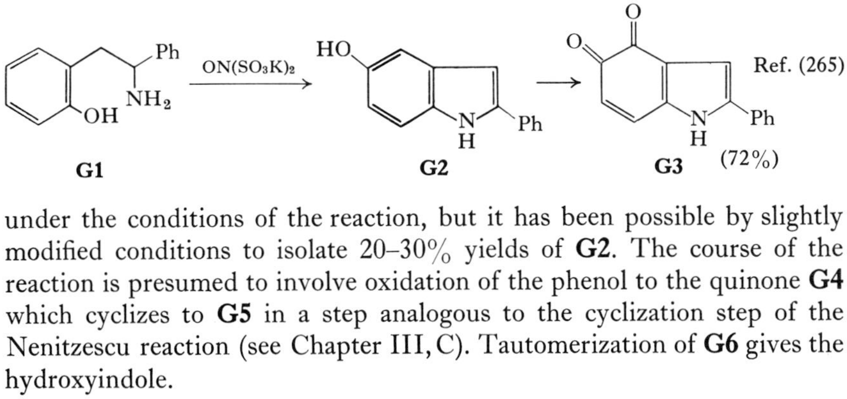

$ON(SO_3K)_2$ — Ref. (265)

G1 **G2** **G3** (72%)

under the conditions of the reaction, but it has been possible by slightly modified conditions to isolate 20–30% yields of **G2**. The course of the reaction is presumed to involve oxidation of the phenol to the quinone **G4** which cyclizes to **G5** in a step analogous to the cyclization step of the Nenitzescu reaction (see Chapter III, C). Tautomerization of **G6** gives the hydroxyindole.

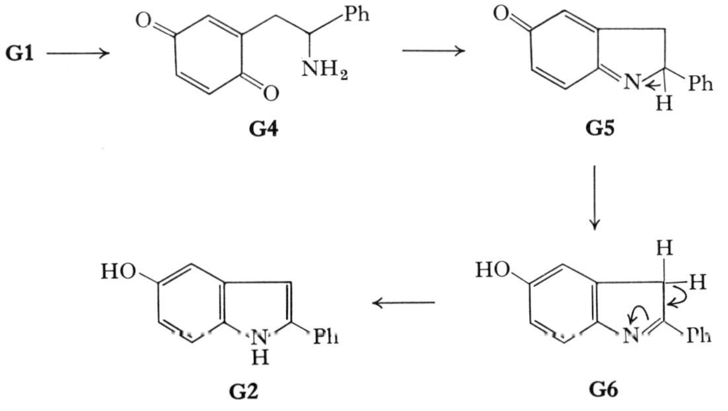

G1 → **G4** → **G5** → **G6** → **G2**

The understanding and application of oxidative cyclization of hydroxyphenylethylamines has its foundation in the chemical investigation of the adrenochromes (97) and particularly in the studies of Harley-Mason and co-workers in this area. Oxidation of 3,4-dihydroxyphenylalanine gives 5,6-dihydroxyindole (44). The expected indole is formed in 85% yield, by oxidation of 2-(2,5-dihydroxyphenyl)ethylamine (69). The methyl ester of 2,3-dihydroxyphenylalanine gives methyl 7-hydroxylindole-2-carboxylate (69). In this case an ortho quinone must be involved as the intermediate.

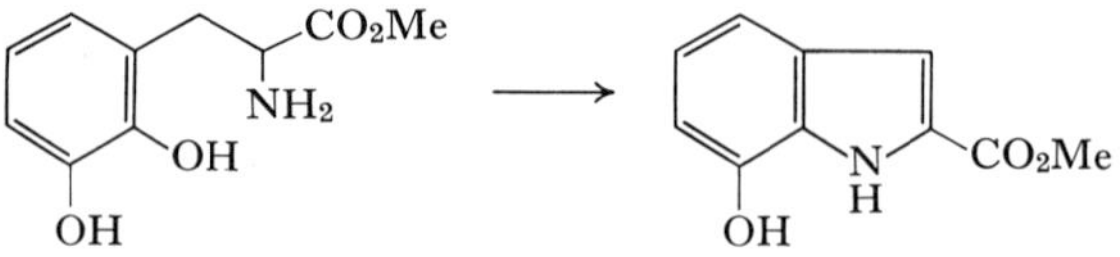

CO_2Me NH_2 OH OH → CO_2Me N H OH

G7 G8

An interesting variation of the cyclization is the conversion of 2-(2-amino-4,5-dihydroxyphenyl)ethylamine to 5,6-dihydroxyindole. Autoxidation occurs in aqueous solution and the indole is formed in 30–50% yield (94). Oxidation of 2-(2-hydroxy-4-aminophenyl) ethylamine gives 5-aminoindole (25%) (95).

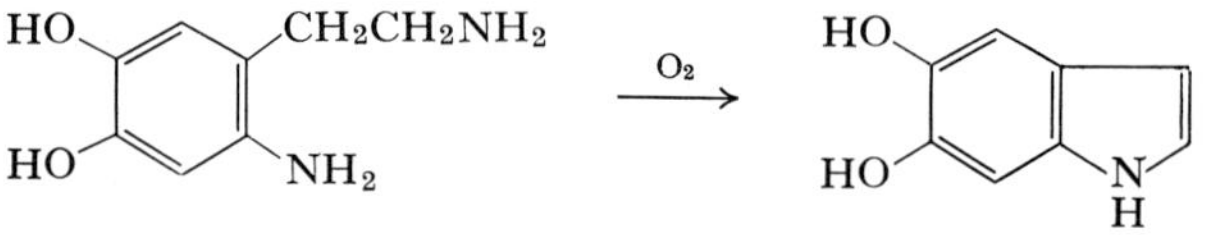

HO HO $CH_2CH_2NH_2$ NH_2 $\xrightarrow{O_2}$ HO HO N H

G9 G10

An example of application of oxidative cyclization to relatively more complex substrates is provided by the conversion **G11** to **G12** which was effected by Moore and Capaldi (168). *N*-Substituted indoles can be formed under similar conditions (78).

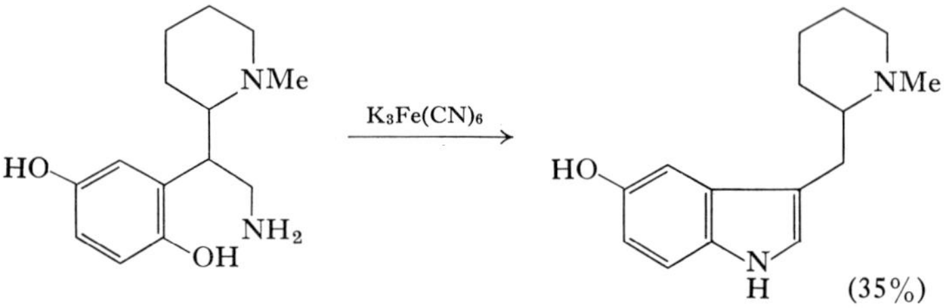

NMe HO NH_2 OH $\xrightarrow{K_3Fe(CN)_6}$ NMe HO N H (35%)

G11 G12

HO HO NH OH OH $\xrightarrow{K_3Fe(CN)_6}$ HO HO N OH OH

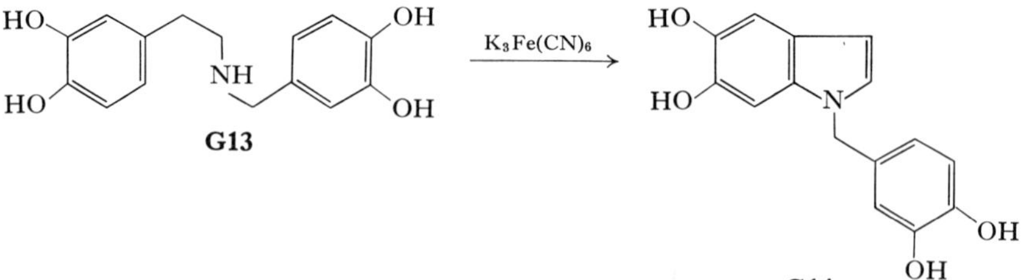

G13 G14

Oxidation of **G15** by silver oxide or ferricyanide gives **G16**, which can be reduced (and dehydrated) to **G17** by ascorbic acid. The pattern of the cyclization parallels that of **G13** → **G14** but the product **G16**, presumably

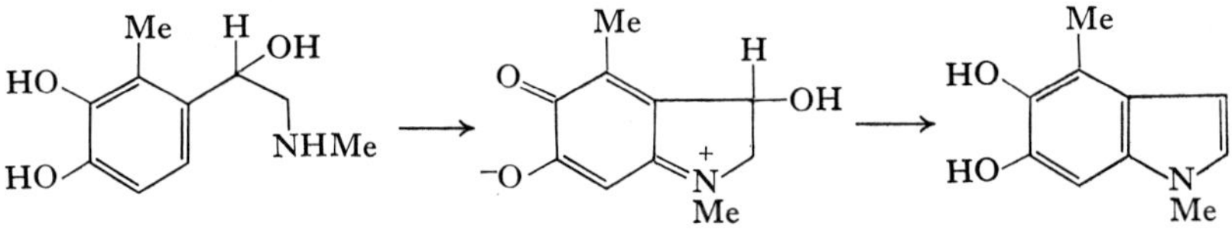

G15 G16 G17

because of the resonance stabilization present in the vinylogous amide group, does not tautomerize directly to an indole (98).

Treatment of the ethyl ester of tyrosine with *N*-bromosuccinimide gives ethyl 5,7-dibromo-6-hydroxyindole-2-carboxylate (278). The reaction proceeds by bromination of the phenolic ring to a tribromocyclohexadienone which subsequently undergoes cyclization. The cyclization step is formulated as an intramolecular Michael addition.

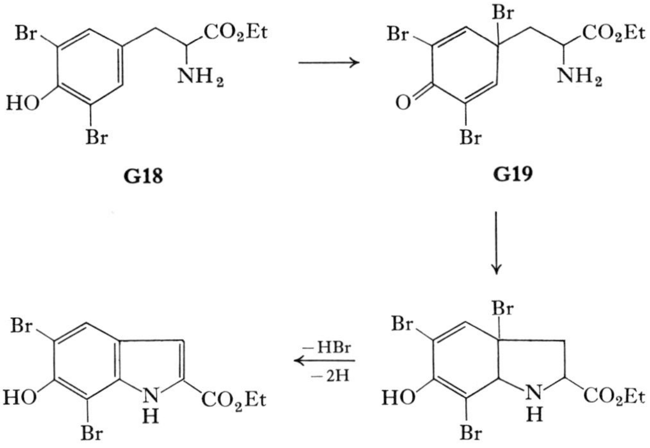

G18 G19

G21 G20

H. MISCELLANEOUS REACTIONS LEADING TO INDOLE RING FORMATION

In this section are considered a number of reactions which generate indole rings, but which have not as yet been very much applied in synthesis. Some are limited in scope while others have been little investigated, although not obviously limited as to scope.

1. Acid-Catalyzed Cyclization of *N*-Allylanilines

Heating *N*-crotylaniline (**H1**) with polyphosphoric acid gives 2,3-dimethylindole and 2,3-dimethylindoline in approximately equal amounts (113). The mechanism below, involving the nitrogen analog of a Claissen rearrangement, has been suggested.

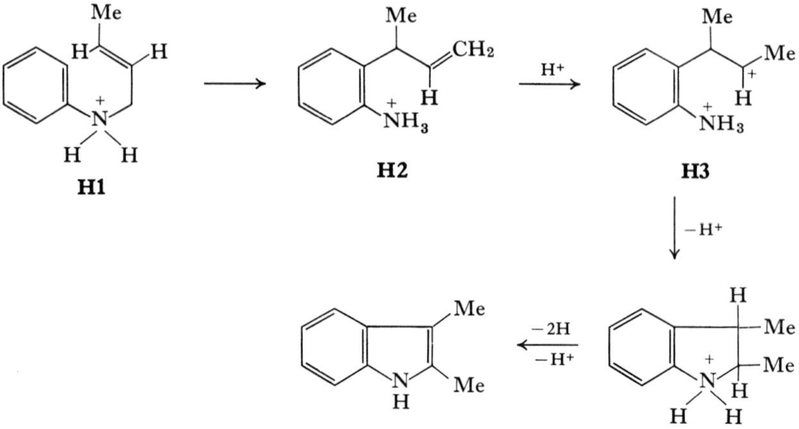

H1 → H2 —H^+→ H3 —$-H^+$→ H4 —$-2H$, $-H^+$→ H5

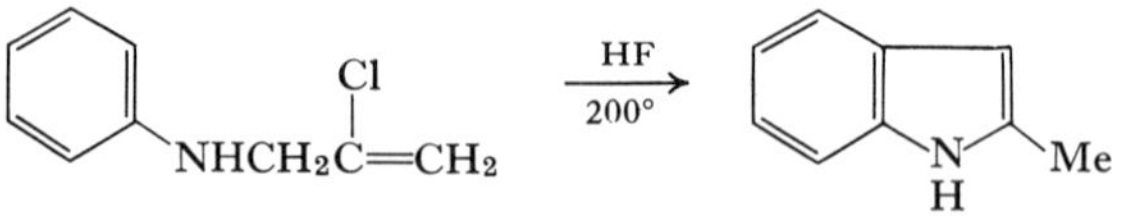

H6 —HF, 200°→ H7 Ref. (218, 266)

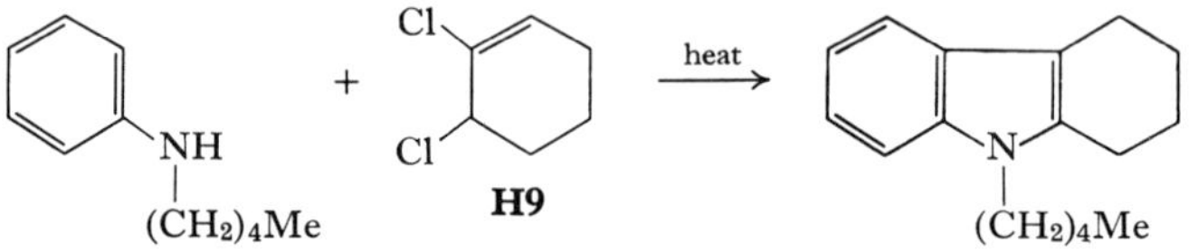

H8 + H9 —heat→ H10 Ref. (218)

Information on the mode of aromatization is lacking. A similar cyclization of allylanilines to 2-methylindoles can be effected by hydrochloric acid in hydrocarbon solvents (19). The same pattern is evident in the cyclization of a series of *N*-(2-chlorallyl)anilines reported by Towne and Hill (266). Aromatization can be effected by loss of hydrogen chloride. The most severe limit to synthetic utility of the acid-catalyzed cyclization of *N*-allylanilines lies in the fact that the strongly acidic conditions of the reaction would doubtless cause further reactions of many of the more sensitive indoles.

2. Generation and Cyclization of o-Aminophenylacetaldehydes

The generation of an *o*-aminophenylacetaldehyde is followed by cyclization and aromatization by dehydration. A few such syntheses have been effected by reductive approaches to aminophenylacetaldehydes (see Section D, 1). Several examples of oxidative approaches to derivatives of *o*-amino-

CH_2CHO / NH_2 ⇌ N H OH —($-H_2O$)→ N H

H11 **H12** **H13**

phenylacetaldehydes have led to indoles. The approach of Plieninger involves generation of 1-amino-5,8-dihydronaphthalenes, followed by ozonolysis with reductive workup. 4-Substituted indoles are generated by this approach. Typical sequences include **H14** → **H17** and **H18** → **H20**. [The report (204) that the compound **H20** exists as the hydroxymethylene tautomer and curious spectral data engender some reservations about the

NH_2 —(1) Na, (2) $(Ac)_2O$→ NHAc —(1) O_3, (2) Pd/C, H_2→ CH_2CHO, N Ac Ref. (207)

H14 **H15** **H16**

↓ (1) NH_2OH (2) $(Ac)_2O$

CH_2CN, N Ac

H17

O O, NHAc —(1) O_3, (2) Zn→ H, O O, HO, OH, NAc —H^+→ H, O O, HO, NAc Ref. (204)

H18 **H19** **H20**

correctness of structure **H20**.] Pennington and co-workers have generated *o*-aminophenylacetaldehydes by periodate oxidation of 3-hydroxy-1,2,3,4-tetrahydroquinolines. The aldehydes cyclize *in situ* to indoles as illustrated by **H21** → **H24**, **H25** → **H26**, and **H27** → **H28** (193–195).

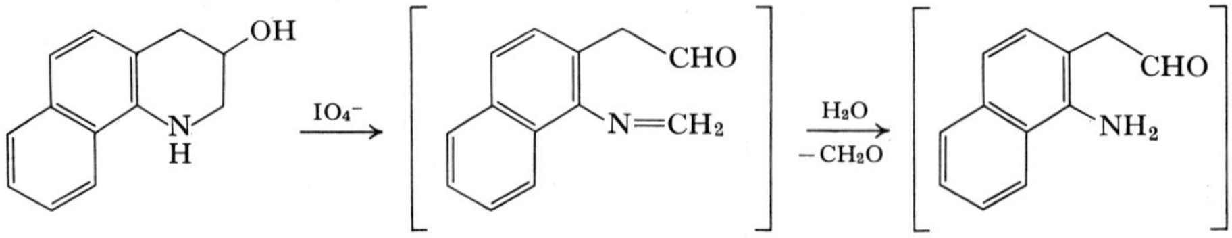

H21 $\xrightarrow{IO_4^-}$ [**H22**] $\xrightarrow[-CH_2O]{H_2O}$ [**H23**] $\xrightarrow{-H_2O}$ **H24**

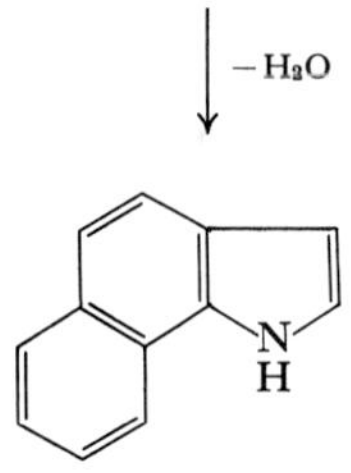

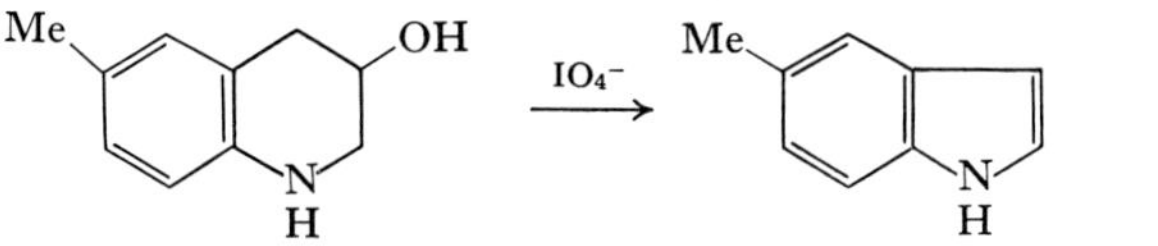

H25 $\xrightarrow{IO_4^-}$ **H26** (27%)

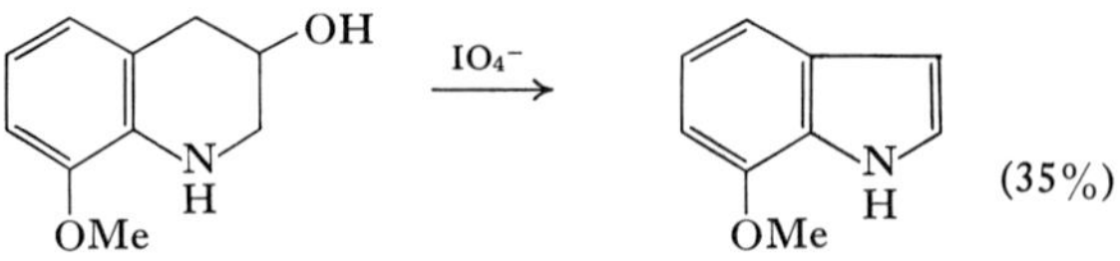

H27 $\xrightarrow{IO_4^-}$ **H28** (35%)

The requisite tetrahydroquinolines are available from anilines and 2-(1-haloalkyl)oxiranes.

3. Indoles by Nucleophilic Aromatic Displacement

The reaction pattern **H29** → **H30** has occasionally given indoles. The reactions in some instances probably proceed via elimination–addition (aryne mechanism). The first example of indole formation by reactions of

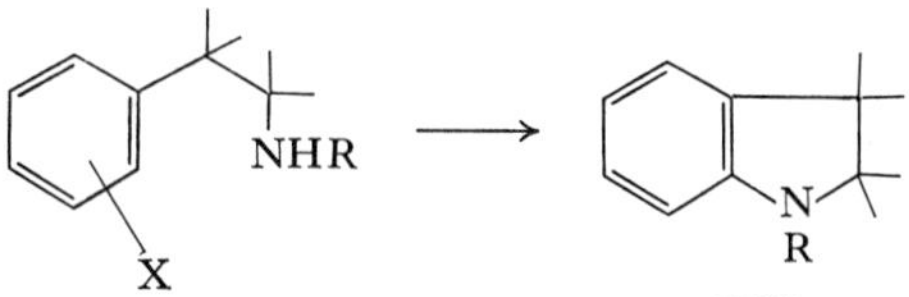

H29 → **H30**

this type was reported by Bunnett and Hrutfiord (45). Thus, *o*-chlorophenylacetone gives 2-methylindole (25% yield) on reaction with sodamide in liquid ammonia.

CH_2Ac $\xrightarrow{^-NH_2}$ Cl N H Me

H31 **H32**

N-Methyl-2,3-dihydroindole has been obtained from both *N*-methyl-2-(*o*-chlorophenyl)ethylamine and the *m*-chloro isomer (109). This syn-

CH_2CH_2NHMe $\xrightarrow{PhLi}$ Cl N Me (52%)

H33 **H34**

CH_2CH_2NHMe $\xrightarrow{PhLi}$ N Me (85%) Cl

H35 **H36**

OMe CH_2CH_2NHMe $\xrightarrow{PhLi}$ OMe N Me (80%) Ref. (127) Cl

H37 **H38**

thetic pattern has also been applied to synthesis of the azaindole **H41** from the pyridine **H39** (279). Nucleophilic displacement is particularly facile in this case because of the ease of nucleophilic attack on the pyridine ring.

Me CH_2CH_2Cl Cl N Cl $\xrightarrow[EtOH]{NH_3}$ Me Cl N N H $\xrightarrow{chloranil}$ Me Cl N N H

H39 **H40** **H41**

The base-catalyzed cyclization of the enamine **H42** probably involves displacement of fluoride by an addition-elimination mechanism (39).

NaH, dimethylformamide

H42 → H43

4. Generation of Indoles by Synthesis or Reactions of Pyrrole Rings

Treatment of 1,4-dicarbonyl compounds with amines is a general method of synthesis of the pyrrole ring. If a 2-acylmethylcyclohexanone is subjected to reaction with ammonia or amines, a 4,5,6,7-tetrahydroindolc is obtained. A number of reactions of this type have been successful. Stetter and co-

H44 + RNH_2 → H45

workers have reported the transformations which are illustrated below (243, 244).

H46 —NH_3→ H47 (82%)

H46 —$PhNH_2$→ H48 (76%)

H49 —$MeNH_2$→ H50 (95%)

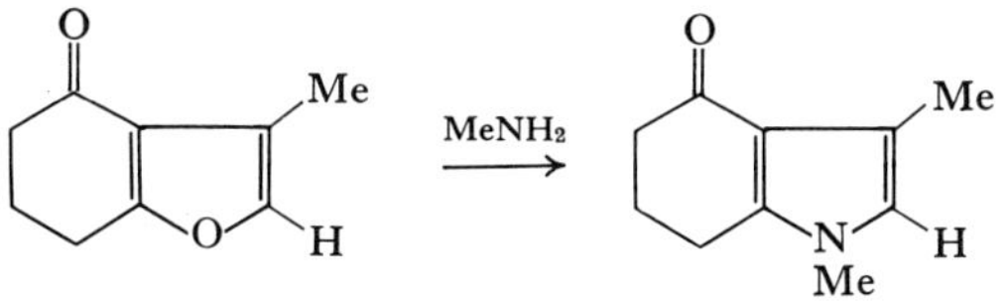

O Me MeNH2 O Me
O H N H
Me

H51 H52

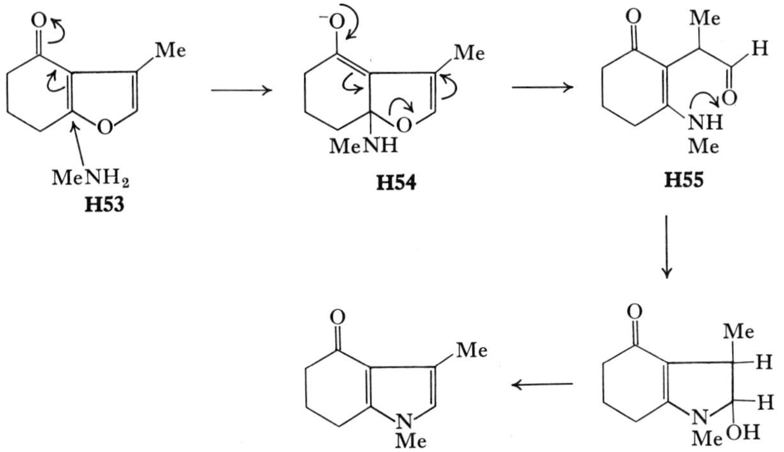

O Me O Me Me
O MeNH2 -O Me O H
O O NH
MeNH Me
H53 H54 H55

O O Me
Me H
H
N N
Me Me OH
H57 H56

H O O OH
HOC Pd/C
N Me N Me N Me
Et Et Et
H59 H58 H62
CF3CO2NHCOCF3 CH2=P(Ph)3
(1) MeI
(2) MeO⁻
O Me
NC
N Me N Me
Et Et
H60 E63

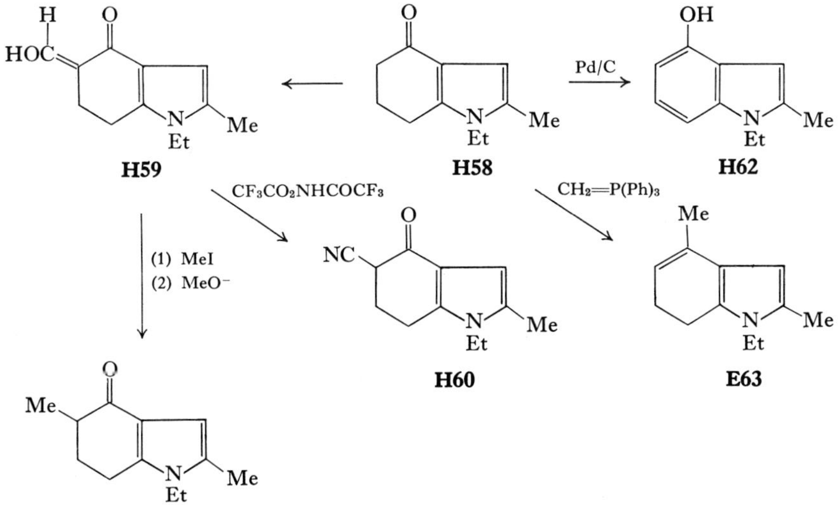

O
Me
N Me
Et
H61

Stetter and Lauterbach (243) also demonstrated that 3-acylfurans were readily converted to indoles by amines. This transformation must proceed via a ring-opening reaction and has an analogy in early work with 3-acylfurans (80).

The 4-oxo-4,5,6,7-tetrahydroindoles obtained in such reactions are versatile intermediates for further elaboration to various substituted indoles. A brief report (213) by Remers and Weiss includes the reactions in the scheme above in addition to a number of others.

An alternative route to 4-oxo-4,5,6,7-tetrahydroindoles has recently been briefly described (96). It constitutes a variation of the classic Knorr pyrrole synthesis. Good yields were reported for most systems which were described, including several 4-oxo-3-carbethoxy-4,5,6,7-tetrahydroindoles.

O O R R N HO Zn O R R N H

H64 H65 H66

O O + O CO_2Et HON CH_2Ph Zn O CO_2Et N H CH_2Ph (66%)

H67 H68 H69

5. Indoles from Enamines and α-β Unsaturated Carbonyl Compounds

Two brief reports appeared in 1958 and 1959 describing cyclizative condensation of dihydropyrroles with acrolein and methyl vinyl ketone. Aromatization of the adducts gave indoles (62, 115). The reactions seem not to have found subsequent application.

Me N Me + $AcCH{=}CH_2$ ⟶ Me N Me Pd/C Me N Me (62%)

H70 H71 H72 H73

Me N Me + $PhCOCH{=}CH_2$ ⟶ Ph N Me Pd/C Ph N Me (46%)

H74 H75 H76 H77

H78 → H79 $\xrightarrow[Al_2O_3]{Pd}$ H80

6. The Reaction of Hydrazobenzenes and Phenylhydroxylamine with Dimethyl Acetylenedicarboxylate

The reactions **H81** → **H83**, **H84** → **H85**, and **H86** → **H87** were reported some years ago by Huntress and co-workers (110–112).

H81 + $MeO_2CC{\equiv}CCO_2Me$ (H82) $\longrightarrow \xrightarrow{heat}$ H83 (74%) Ref. (110)

H84 + H82 ⟶ [?] $\xrightarrow{H^+}$ H85 (46%) Ref. (112)

PhNHNHPh (H86) + H82 ⟶ [?] ⟶ H87 (63%) Ref. (70, 111)

The mechanistic interpretation of these reactions is clouded by some uncertainty about the structures of some of the initial adducts which are formed in the reactions. In cases where the structures of the adducts have been investigated (10a), it is not clear that they are indole precursors. The adducts generated by addition of hydrazobenzene to dimethyl acetylenedicarboxylate may be the vinylhydrazine **H88**. Indole formation could proceed from **H88** by a mechanism closely related to the final stages of the Fischer indole synthesis.

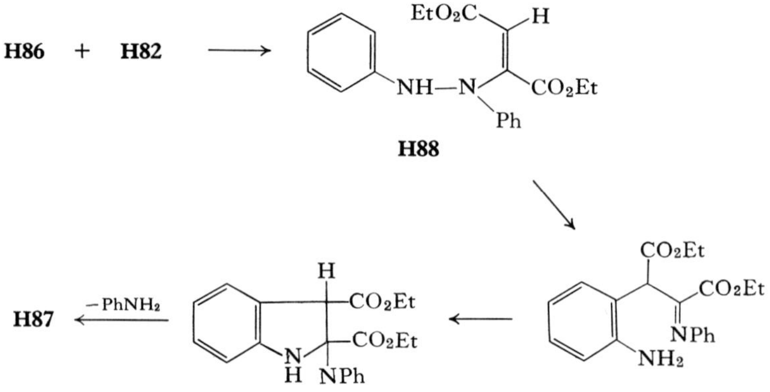

H86 + **H82** ⟶ **H88**

H88 ⟶ **H89** ⟶ **H90** $\xrightarrow{-PhNH_2}$ **H87**

7. Indoles from Other Heterocyclic Rings

Cinnolines (benzo[c]pyridazines) can be converted to indoles under reductive conditions (18, 21, 29a, 173). The reaction presumably proceeds

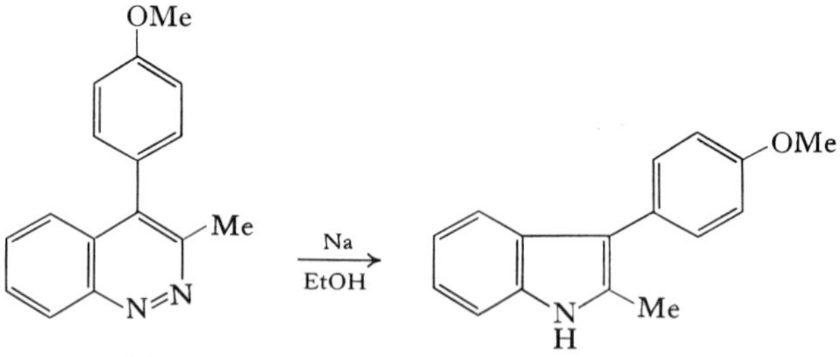

H91 $\xrightarrow[EtOH]{Na}$ **H92** Ref. (18)

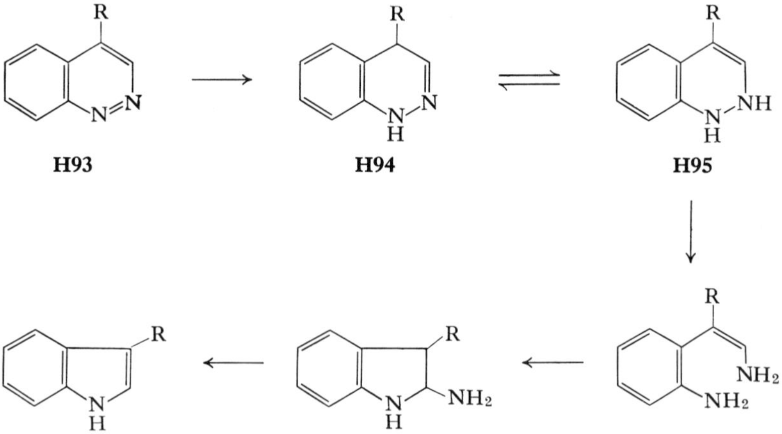

H93 ⟶ **H94** ⇌ **H95** ⟶ **H96** ⟶ **H97** ⟶ **H98**

via the reductive cleavage of the N–N bond followed by cyclization and loss of ammonia.

1-Aryltriazoles give indoles as well as other products on photolytically induced loss of nitrogen (46).

Ph
Ph
N
H
H101

Ph
N
N=N
Ph
H99

Ph
N
Ph
H100

Ph
C=C=N
Ph
H102

Indole formation is rationalized by the bonding process **H100** → **H101**.

Recent reports (42, 43, 116, 137, 248) that indole derivatives can be obtained from quinoline-*N*-oxides by a photochemical ring contraction

H100 ⟷

Ph
N
Ph
H103

H
Ph
N
Ph
H104

→ **H101**

Me
$\overset{+}{N}$
CN
^{-}O
H105

$h\nu$

Me
O
N
CN
H106

$h\nu$
MeOH

Me
N
H
CN
H108

Ref. (137)
(100%)

MeOH,
heat

Me
N
H
(50%)
H107

$\overset{+}{N}$
Ph
^{-}O
H109

$h\nu$

O
N
Ph
H110

EtOH
H_2O

N
OH
H
PhCO
H111

Ref. (42)
(65%)

have generated much interest. Typical examples are shown in the accompanying equations. The indoles can arise by hydrolysis of the benzoxazepines **H106** and **H110**.

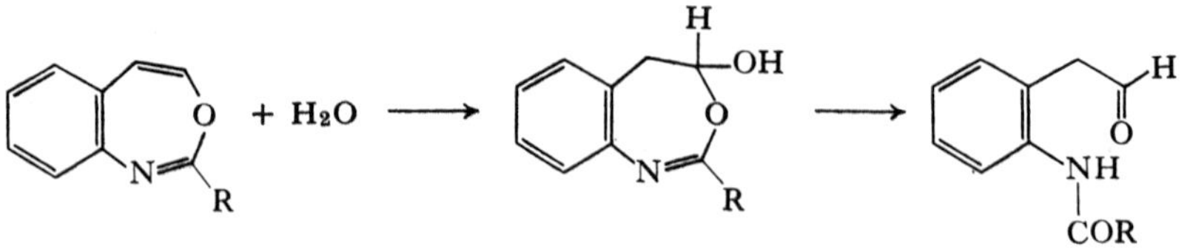

+ H₂O
OH
R
NH
COR

H112 **H113** **H114**

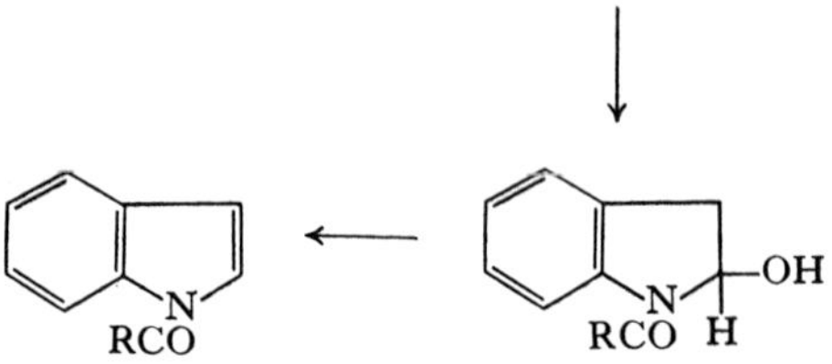

RCO
OH
RCO H

H116 **H115**

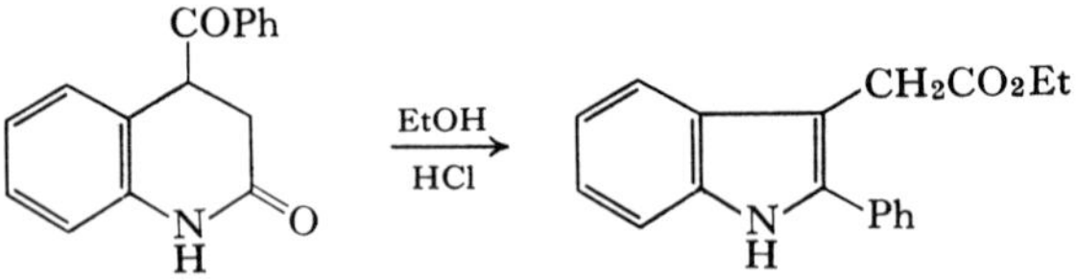

COPh
EtOH
HCl
CH₂CO₂Et
Ph

H117 **H118**

Ref. (179)

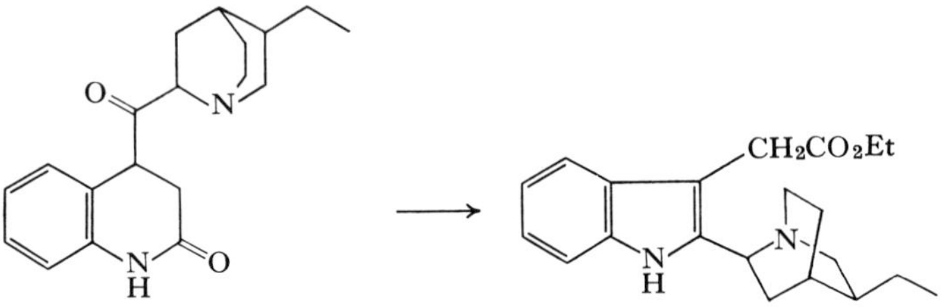

CH₂CO₂Et

Ref. (177)

H119 **H120**

Hydrolytic conditions effect ring contractions of dihydro-4-acyl-2-quinolones to indole-3-acetic acid derivatives (177–179). The indole ring arises by cyclization of the *o*-aminobenzyl ketones formed by hydrolysis of the lactam linkage.

8. Indoles from the Reaction of o-Iodoanilines with Acetylenes

Treatment of *o*-aminophenylacetylenes with cuprous iodide in dimethylformamide generates 2-substituted indoles (61, 242). The *o*-aminophenylacetylenes are prepared by coupling *o*-iodoaniline with cuprous acetylides in pyridine. Generally good yields have been reported for 2-substituted indoles and 1,2-disubstituted indoles.

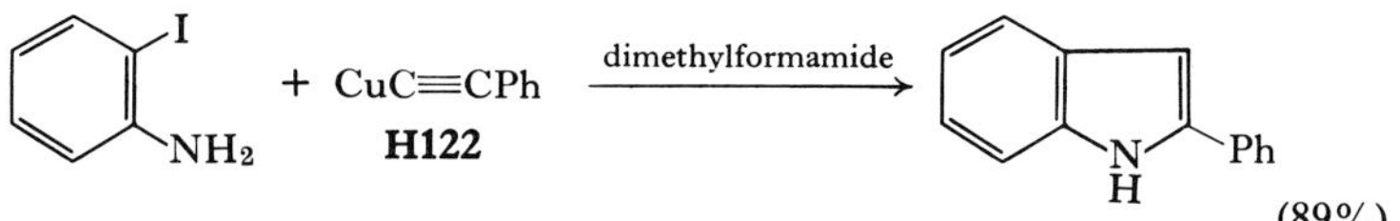

H121 + $CuC{\equiv}CPh$ (H122) —dimethylformamide→ H123 (89%)

9. Indoles by Intramolecular Michael Additions

Suh and Puma have obtained a number of 2-aryl-3-cyanoindoles by reduction of 2-nitro-α-cyanostilbenes with iron in acetic acid (249). The cyclization step is formulated as a Michael addition at the hydroxylamine oxidation stage. The cyano function is therefore required to activate the olefinic bond toward nucleophilic attack. The corresponding anilines do not cyclize under the reaction conditions.

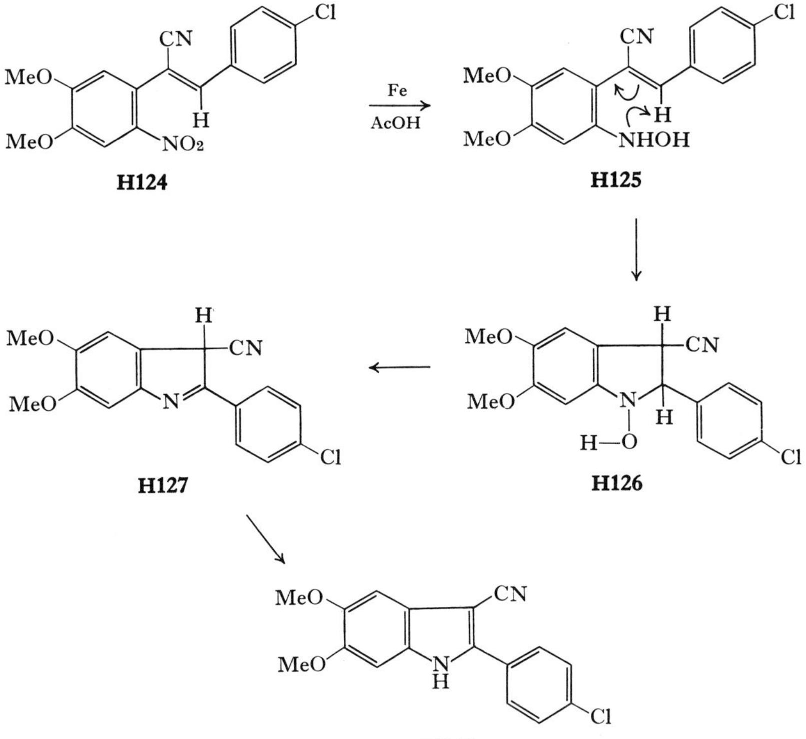

H124 —Fe, AcOH→ H125 → H126 → H127 → H128

10. High-Temperature Catalytic Cyclization of *o*-Ethylaniline

A number of catalyst systems have been found to catalyze the conversion of *o*-ethylaniline to indole at 400°–800°C (74, 140, 153, 270).

REFERENCES

1. Abramovitch, R. A., *Can. J. Chem.* **36**, 354 (1958).
2. Abramovitch, R. A., *J. Chem. Soc.* p. 4593 (1956).
3. Abramovitch, R. A., and Adams, K. A. H., *Can. J. Chem.* **39**, 2516 (1961).
4. Abramovitch, R. A., Ahmad, Y., and Newman, D., *Tetrahedron Letters* p. 752 (1961).
5. Abramovitch, R. A., and Muchowski, J. M., *Can. J. Chem.* **38**, 554 (1960).
6. Abramovitch, R. A., Newman, D., and Tertzakian, G., *Can. J. Chem.* **41**, 2390 (1963).
7. Abramovitch, R. A., and Shapiro, D., *J. Chem. Soc.* p. 4589 (1956).
8. Adams, R., and Samuels, W. P., Jr., *J. Am. Chem. Soc.* **77**, 5375 (1955).
9. Adams, R., Werbel, L. M., and Nair, M. D., *J. Am. Chem. Soc.* **80**, 3291 (1958).
10. Adlerova, E., Ernest, I., Hněvsová, V., Jílek, J. O., Novák, L., Pomykáček, J., Rajšner, M., Sova, J., Vejdêlek, Z. J., and Protiva, M., *Collection Czech. Chem. Commun.* **25**, 784 (1960).

10a. Agosta, W. C., *J. Org. Chem.* **26**, 1724 (1961).

11. Aksanova, L. A., Kucherova, N. F., and Zagorevskii, V. A., *J. Gen. Chem. USSR (English Transl.)* **34**, 3417 (1964).
12. Allen, C. F. H., and Van Allan, J., *in* "Organic Syntheses" (E. C. Horning, ed.), Collective Vol. 3, p. 597. Wiley, New York, 1955.
13. Allen, C. F. H., and Wilson, C. V., *J. Am. Chem. Soc.* **65**, 611 (1943).
14. Allen, F. L., Brunton, J. C., and Suschitzky, H., *J. Chem. Soc.* p. 1283 (1955).
15. Allen, G. R., Jr., Pidacks, C., and Weiss, M. J., *J. Am. Chem. Soc.* **88**, 2536 (1966).
16. Allen, G. R., Jr., Poletto, J. F., and Weiss, M. J., *J. Org. Chem.* **30**, 2897 (1965).
17. Allen, G. R., Jr., and Weiss, M. J., *J. Org. Chem.* **33**, 198 (1968).
18. Atkinson, C. M., and Simpson, J. C. E., *J. Chem. Soc.* p. 1649 (1947).
19. Bader, A. R., Bridgwater, R. J., and Freeman, P. R., *J. Am. Chem. Soc.* **83**, 3319 (1961).
20. Bajwa, G. S., and Brown, R. K., *Can. J. Chem.* **46**, 3105 (1968).
21. Baumgarten, H. E., and Furnas, J. L., *J. Org. Chem.* **26**, 1536 (1961).
22. Baxter, I., and Swan, G. A., *J. Chem. Soc., C* p. 468 (1968).
23. Beck, D., Schenker, K., Stuber, F., and Zürcher, R., *Tetrahedron Letters* p. 2285 (1965).
24. Beer, R. J. S., Clarke, K., Davenport, H. F., and Robertson, A., *J. Chem. Soc.* p. 2029 (1951).
25. Beer, R. J. S., McGrath, L., Robertson, A., and Woodier, A. B., *J. Chem. Soc.* p. 2061 (1949).

25a. Bellamy, A. J., and Guthrie, R. D., *J. Chem. Soc.* pp. 2788 and 3528 (1965).

26. Benington, F., Morin, R. D., and Clark, L. C., Jr., *J. Org. Chem.* **24**, 917 (1959); **25**, 1542 (1960).
27. Bentov, M., Kaluszyner, A., and Pelchowicz, Z., *J. Chem. Soc.* p. 2825 (1962).
28. Bentov, M., Pelchowicz, Z., and Levy, A., *Israel J. Chem.* **2**, 25 (1964); *Chem. Abstr.* **60**, 15814 (1964).
29. Bergmann, E. D., and Hoffmann, E., *J. Chem. Soc.* p. 2827 (1962).

29a. Besford, L. S., and Bruce, J. M., *J. Chem. Soc.* p. 4037 (1964).

30. Betkerur, S. N., and Siddappa, S., *J. Chem. Soc. C* p. 296 (1967).
31. Blackhall, A., and Thomson, R. H., *J. Chem. Soc.* p. 3916 (1954).

32. Blades, C. E., and Wilds, A. L., *J. Org. Chem.* **21**, 1013 (1956).
33. Blaikie, K. G., and Perkin, W. H., Jr., *J. Chem. Soc.* **125**, 296 (1924).
34. Blair, J., and Newbold, G. T., *J. Chem. Soc.* p. 2871 (1955).
35. Boyer, J. H., Krueger, W. E., and Mikol, G. J., *J. Am. Chem. Soc.* **89**, 5504 (1967).
36. Bretherick, L., Gaimster, K., and Wragg, W. R., *J. Chem. Soc.* p. 2919 (1961).
37. British Patent, 859,223 (1961); *Chem. Abstr.* **55**, 15510 (1961).
38. British Patent, 895,430 (1962); *Chem. Abstr.* **57**, 13725 (1962).
39. Brooke, G. M., and Rutherford, R. J. D., *J. Chem. Soc., C* p. 1189 (1967).
40. Brown, F., and Mann, F. G., *J. Chem. Soc.* pp. 847 and 858 (1948).
41. Brown, U. M., Carter, P. H., and Tomlinson, M., *J. Chem. Soc.* p. 1843 (1958).
42. Buchardt, O., *Tetrahedron Letters* p. 6221 (1966).
43. Buchardt, O., Becher, J., and Lohse, D., *Acta Chem. Scand.* **20**, 2467 (1966).
44. Bu'Lock, J. D., and Harley-Mason, J., *J. Chem. Soc.* p. 2248 (1951).
45. Bunnett, J. F., and Hrutfiord, B. F., *J. Am. Chem. Soc.* **83**, 1691 (1961).
46. Burgess, E. M., Carithers, R., and McCullagh, L., *J. Am. Chem. Soc.* **90**, 1923 (1968).
47. Buu-Hoi, N. P., *J. Chem. Soc.* p. 2418 (1958).
48. Buu-Hoi, N. P., and Jacquignon, P., *Compt. Rend.* **251**, 1297 (1960).
49. Buu-Hoi, N. P., Jacquignon, P., and Périn-Roussel, O., *Bull. Soc. Chim. France* p. 2849 (1965).
50. Buu-Hoi, N. P., and Royer, R., *Rec. Trav. Chim.* **66**, 305 (1947).
51. Cadogan, J. I. G., Cameron-Wood, M., Mackie, R. K., and Searle, R. J. G., *J. Chem. Soc.* p. 4831 (1965).
52. Campaigne, E., and Lake, R. D., *J. Org. Chem.* **24**, 478 (1959).
53. Cantrall, E. W., Conrow, R. B., and Bernstein, S., *J. Org. Chem.* **32**, 3445 (1967).
53a. Carlin, R. B., *J. Am. Chem. Soc.* **74**, 1077 (1952).
54. Carlin, R. B., and Carlson, D. P., *J. Am. Chem. Soc.* **81**, 4673 (1959).
55. Carlin, R. B., and Fisher, E. E., *J. Am. Chem. Soc.* **70**, 3421 (1948).
56. Carlin, R. B., Henley, W. O., Jr., and Carlson, D. P., *J. Am. Chem. Soc.* **79**, 5712 (1957).
57. Carlin, R. B., and Larson, G. W., *J. Am. Chem. Soc.* **79**, 934 (1957).
58. Carlin, R. B., Magistro, A. J., and Mains, G. J., *J. Am. Chem. Soc.* **86**, 5300 (1964).
59. Carlin, R. B., Wallace, J. G., and Fisher, E. E., *J. Am. Chem. Soc.* **74**, 990 (1952).
60. Casnati, G., Langella, M. R., Piozzi, F., Ricca, A., and Umani-Ronchi, A., *Gazz. Chim. Ital.* **94**, 1221 (1964).
61. Castro, C. E., Gaughan, E. J., and Owsley, D. C., *J. Org. Chem.* **31**, 4071 (1966).
62. Červinka, O., *Chem. & Ind.* (*London*) p. 1129 (1959).
63. Chastrette, M., *Ann. Chim.* (*Paris*) **7**, 643 (1962); *Chem. Abstr.* **58**, 12493 (1963).
64. Clemo, G. R., and Felton, D. G. I., *J. Chem. Soc.* p. 700 (1951).
65. Clifford, B., Nixon, P., Salt, C., and Tomlinson, M., *J. Chem. Soc.* p. 3516 (1961).
66. Clusius, K., and Weisser, H. R., *Helv. Chim. Acta* **35**, 400 (1952).
67. Corbella, A., Jommi, G., Scolastico, G., and Křepinský, J., *Gazz. Chim. Ital.* **96**, 760 (1966).
68. Cowper, R. M., and Stevens, T. S., *J. Chem. Soc.* p. 1041 (1947).
69. Cromartie, R. I. T., and Harley-Mason, J., *J. Chem. Soc.* p. 2525 (1952).
70. Diels, O., and Reese, J., *Ann. Chem.* **511**, 168 (1934).
71. Domschke, G., and Fürst, H., *Chem. Ber.* **92**, 3244 (1959).
71a. Ebnöthner, A., Niklaus, P., and Suess, R., *Helv. Chim. Acta* **52**, 629 (1969).
72. Ek, A., and Witkop, B., *J. Am. Chem. Soc.* **76**, 5579 (1954).
72a. Elgersma, R. H. C., and Havinga, E., *Tetrahedron Letters* p. 1735 (1969).
73. Elks, J., Elliot, D. F., and Hems, B. A., *J. Chem. Soc.* p. 629 (1944).
74. Erner, W. E., Mills, G. A., and Smith, R. K., U.S. Patent 2,953,575 (1960); *Chem. Abstr.* **55**, 4533 (1961).

75. Feofilaktov, V. V., *Zh. Obshch. Khim.* **17**, 993 (1947); *Chem. Abstr.* **42**, 4537 (1948).
76. Fischer, E., *Ann. Chem.* **236**, 126 (1886).
77. Fitzpatrick, J. T., and Hiser, R. D., *J. Org. Chem.* **22**, 1703 (1957).
78. Forbes, E. J., *J. Chem. Soc.* p. 513 (1956).
79. Frydman, B., Despuy, M. E., and Rapoport, H., *J. Am. Chem. Soc.* **87**, 3530 (1965).
80. Fuson, R. C., Fleming, C. L., and Johnson, R., *J. Am. Chem. Soc.* **60**, 1994 (1938).
81. Gallagher, M. J., and Mann, F. G., *J. Chem. Soc.* p. 5110 (1962).
82. Ginsburg, D., and Pappo, R., *J. Am. Chem. Soc.* **75**, 1094 (1953).
83. Gore, P. H., Hughes, G. K., and Ritchie, E., *Nature* **164**, 835 (1949).
84. Govindachari, T. R., Pillai, P. M., Nagarajan, K., and Viswananathan, N., *Tetrahedron* **21**, 2957 (1965).
85. Grandberg, I. I., Kost, A. N., and Terent'ev, A. P., *J. Gen. Chem. USSR* (*English Transl.*) **27**, 3378 (1957).
86. Green, K. H. B., and Ritchie, E., *J. Proc. Roy. Soc. N. S. Wales* **83**, 120 (1949); *Chem. Abstr.* **45**, 10236 (1951).
87. Grinev, A. N., Ermakova, V. N., Vrotek, E., and Terent'ev, A. P., *J. Gen. Chem. USSR* (*English Transl.*) **29**, 2742 (1959).
88. Grinev, A. N., Florent'ev, V. L., Shevdov, V. I., and Terent'ev, *J. Gen. Chem. USSR* (*English Transl.*) **30**, 2291 (1960).
89. Grinev, A. N., Kul'bovskaya, N. K., and Terent'ev, A. P., *Zh. Obshch. Khim.* **25**, 1355 (1955); *Chem. Abstr.* **50**, 4903 (1956).
89a. Grinev, A. N., Rodzevich, N. E., and Terent'ev, A. P., *J. Gen. Chem. USSR* (*English Transl.*) **27**, 1759 (1957).
90. Grinev, A. N., Shvedov, V. I., and Panisheva, E. K., *J. Org. Chem. USSR* (*English Transl.*) **1**, 2091 (1965).
91. Grinev, A. N., Shvedov, V. I., and Sugrobova, I. P., *J. Gen. Chem. USSR* (*English Transl.*) **31**, 2140 (1961).
92. Grinev, A. N., Shvedov, V. I., and Terent'ev, A. P., *Zh. Obshch. Khim.* **26**, 1449 and 1452 (1956); *Chem. Abstr.* **50**, 14710 and 14711 (1956).
93. Grob, C. A., *Experientia* **13**, 126 (1957).
94. Harley-Mason, J., *J. Chem. Soc.* p. 200 (1953).
95. Marley-Mason, J., and Jackson, A. H., *J. Chem. Soc.* p. 1158 (1954).
96. Hauptmann, S., Blume, H., Hartmann, G., Haendel, D., and Franke, P., *Z. Chem.* **6**, 107 (1966).
97. Heacock, R. A., *Chem. Rev.* **59**, 181 (1959).
98. Heacock, R. A., and Hutzinger, O., *Can. J. Chem.* **43**, 2535 (1965).
99. Heacock, R. A., Hutzinger, O., Scott, B. D., Daly, J. W., and Witkop, B., *J. Am. Chem. Soc.* **85**, 1825 (1963).
100. Heath-Brown, B., and Philpott, P. G., *J. Chem. Soc.* p. 7165 (1965).
101. Heath-Brown, B., and Philpott, P. G., *J. Chem. Soc.* p. 7185 (1965).
102. Hiremath, S. P., and Siddappa, S., *J. Indian Chem. Soc.* **41**, 357 (1964).
103. Hiriyakkanavar, J. G., and Siddappa, S., *Indian J. Chem.* **4**, 188 (1966); *Chem. Abstr.* **65**, 3821 (1966).
104. Hoffmann, E., Ikan, R., and Galun, A. B., *J. Heterocyclic Chem.* **2**, 298 (1965).
105. Holmes, R. R., Untch, K. G., and Benson, H. D., *J. Org. Chem.* **26**, 439 (1961).
106. Huebner, C. F., Troxell, H. A., and Shroeder, D. C., *J. Am. Chem. Soc.* **75**, 5887 (1953).
107. Huffman, J. W., *J. Org. Chem.* **27**, 503 (1962).
108. Hünig, S., Benzing, E., and Lücke, E., *Chem. Ber.* **90**, 2833 (1957).
109. Huisgen, R., and König, H., *Chem. Ber.* **92**, 203 (1959).
110. Huntress, E. H., Bornstein, J., and Hearon, W. M., *J. Am. Chem. Soc.* **78**, 2225 (1956).

111. Huntress, E. H., and Hearon, W. M., *J. Am. Chem. Soc.* **63**, 2762 (1941).
112. Huntress, E. H., Lesslie, T. E., and Hearon, W. M., *J. Am. Chem. Soc.* **78**, 419 (1956).
113. Hyre, J. E., and Bader, A. R., *J. Am. Chem. Soc.* **80**, 437 (1958).
114. Illy, H., and Funderbunk, C., *J. Org. Chem.* **33**, 4283 (1968).
115. Ireland, R. E., *Chem. & Ind. (London)* p. 979 (1958).
116. Ishikawa, M., Yamada, S., and Kaneko, C., *Chem. & Pharm. Bull. (Tokyo)* **13**, 747 (1965).
117. Isomura, K., Kobayashi, S., and Taniguchi, H., *Tetrahedron Letters* p. 3499 (1968).
118. Jackson, A., Gaskell, A. J., Wilson, N. D. V., and Joule, J. A., *Chem. Commun.* p. 364 (1968).
119. Jackson, A., and Joule, J. A., *Chem. Commun.* p. 459 (1967).
120. Jackson, R. W., and Manske, R. H., *J. Am. Chem. Soc.* **52**, 5029 (1930).
121. Janetzky, E. F. J., and Verkade, P. E., *Rec. Trav. Chim.* **64**, 129 (1945).
122. Jardine, R. V., and Brown, R. K., *Can. J. Chem.* **43**, 1293 (1965).
123. Jenisch, G., *Monatsh. Chem.* **27**, 1223 (1906).
124. Jönsson, A., *Svensk Kem. Tidskr.* **67**, 188 (1955); *Chem. Abstr.* **50**, 10709 (1956).
125. Julia, M., and Bagot, J., *Bull. Soc. Chim. France* p. 1924 (1964).
126. Julia, M., Bagot, J., and Siffert, O., *Bull. Soc. Chim. France* p. 1939 (1964).
127. Julia, M., and Gaston-Breton, H., *Bull. Soc. Chim. France* p. 1335 (1966).
128. Julia, M., and Igolen, J., *Bull. Soc. Chim. France* p. 1056 (1962).
129. Julia, M., Igolen, J., and Igolen, H., *Bull. Soc. Chim. France* p. 1060 (1962).
130. Julia, M., and Lenzi, J., *Bull. Soc. Chem. France* p. 2267 (1962).
131. Julia, M., and Lenzi, J., *Bull. Soc. Chim. France* p. 1051 (1962).
132. Julia, M., and Manoury, P., *Bull. Soc. Chim. France* p. 1411 (1965).
132a. Julia, M., Manoury, P., and Voillaume, C., *Bull. Soc. Chim. France* p. 1417 (1965).
133. Julia, M., and Tchernoff, G., *Bull. Soc. Chim. France* p. 741 (1960).
134. Julian, P. L., Meyer, E. W., and Printy, H. C., *in* "Heterocyclic Chemistry" (R. C. Elderfield, ed.), Vol. 3, p. 32. Wiley, New York, 1953.
135. Kakurina, L. N., Kucherova, N. F., and Zagorevskii, V. A., *J. Org. Chem. USSR (English Transl.)* **1**, 1118 (1965); *J. Gen. Chem. USSR (English Transl.)* **35**, 311 (1965).
136. Kanaoka, Y., Ban, Y., Yonemitsu, O., Irie, K., and Miyashita, K., *Chem. & Ind. (London)* p. 473 (1965).
137. Kaneko, C., and Yamada, S., *Chem. & Pharm. Bull. (Tokyo)* **14**, 555 (1966).
137a. Karabatsos, G. J., and Taller, R. A., *J. Am. Chem. Soc.* **85**, 3624 (1963).
138. Kelly, A. H., McLeod, D. H., and Parrick, J., *Can. J. Chem.* **43**, 296 (1965).
139. Kempter, G., Andratschke, P., Heilmann, D., Krausmann, H., and Mietasch, M., *Chem Ber.* **97**, 16 (1964).
140. Kimura, T., and Yamada, S., *Yakugaku Zasshi* **77**, 888 (1957); *Chem. Abstr.* **52**, 1095 (1958).
141. Kissman, H. M., Farnsworth, D. W., and Witkop, B., *J. Am. Chem. Soc.* **74**, 3948 (1952).
142. Kochetkov, N. K., Kucherova, N. F., and Evdakov, V. P., *J. Gen. Chem. USSR (English Transl.)* **27**, 283 (1957).
143. Kochetkov, N. K., Kucherova, N. F., Pronina, L. P., and Petruchenko, M. I., *J. Gen. Chem. USSR (English Transl.)* **29**, 3581 (1959).
144. Komzolova, N. N., Kucherova, N. F., and Zagorevskii, V. A., *J. Gen. Chem. USSR (English Transl.)* **34**, 2396 (1964).
145. Korczynski, A., Brydowna, W., and Kierzek, L., *Gazz. Chim. Ital.* **56**, 903 (1926).
146. Kost, A. N., Sugorova, I. P., and Yakubov, A. P., *J. Org. Chem. USSR (English Transl.)* **1**, 121 (1965).

147. Kost, A. N., Yudin, L. G., and Chiu, Y., *J. Gen. Chem. USSR* (*English Transl.*) **34**, 3487 (1964).
148. Kucherova, N. F., Aksanova, L. A., and Zagorevskii, V. A., *J. Gen. Chem. USSR* (*English Transl.*) **33**, 3331 (1963).
149. Kucherova, N. F., and Kochetkov, N. K., *Zh. Obshch. Khim.* **26**, 3149 (1956); *Chem. Abstr.* **51**, 8721 (1957).
150. Kucherova, N. F., Petruchenko, M. I., and Zagorevskii, V. A., *J. Gen. Chem. USSR* (*English Transl.*) **32**, 3577 (1962).
151. Landquist, J. K., and Marsden, C. J., *Chem. & Ind.* (*London*) p. 1032 (1966).
152. Leete, E., and Marion, L., *Can. J. Chem.* **31**, 1195 (1953).
153. Lesiak, T., *Roczniki Chem.* **31**, 1057 (1957); *Chem. Abstr.* **52**, 8120 (1958).
154. Lesiak, T., *Roczniki Chem.* **32**, 1401 (1958); *Chem. Abstr.* **53**, 10175 (1959).
155. Lewis, A. D., U.S. Patent 3,037,031 (1962); *Chem. Abstr.* **57**, 12439 (1962).
156. Lindwall, H. G., and Mantell, G. J., *J. Org. Chem.* **18**, 345 (1953).
157. Littell, R., and Allen, G. R., Jr., *J. Org. Chem.* **33**, 2064 (1968).
158. Lorenz, R. R., Tullar, B. F., Koelsch, C. F., and Archer, S., *J. Org. Chem.* **30**, 2531 (1965).
159. Lyle, R. E., and Skarlos, L., *Chem. Commun.* p. 644 (1966).
160. Marchand, B., Streffer, C., and Jauer, H., *J. Prakt. Chem.* [4] **13**, 54 (1961).
161. Martynov, V. F., and Kastron, Y. A., *J. Gen. Chem. USSR* (*English Transl.*) **28**, 2119 (1958).
162. Martynov, V. F., and Kastron, Y. A., *Zh. Obshch. Khim.* **24**, 498 (1954); *Chem. Abstr.* **49**, 6150 (1955).
163. Martynov, V. F., and Shchelkunov, A. V., *J. Gen. Chem. USSR* (*English Transl.*) **32**, 2348 (1962).
164. McKay, J. B., Parkhurst, R. M., Silverstein, R. M., and Skinner, W. A., *Can. J. Chem.* **41**, 2585 (1963).
165. McLean, J., McLean, S., and Reed, R. I., *J. Chem. Soc.* p. 2519 (1955).
166. Merchant, J. R., Vaghani, D. V., and Kallianpur, C. S., *J. Indian Chem. Soc.* **40**, 315 (1963).
167. Mkhitaryan, A. V., Kogodovskaya, A. A., Terzyan, A. G., and Tatevosyan, G. T., *Izv. Akad. Nauk Arm. SSR, Khim. Nauk* **15**, 379 (1962); *Chem. Abstr.* **59**, 2753 (1963).
168. Moore, J. A., and Capaldi, E. C., *J. Org. Chem.* **29**, 2860 (1964).
169. Morelli, G., and Stein, M. L., *J. Med. Pharm. Chem.* **2**, 79 (1960).
170. Morin, R. D., Benington, F., and Clark, L. C., Jr., *J. Org. Chem.* **22**, 331 (1957).
171. Murakoshi, I., *Yakugaku Zasshi* **79**, 72 (1959); *Chem. Abstr.* **53**, 10216 (1959).
172. Nakazaki, M., and Maeda, M., *Bull. Chem. Soc. Japan* **35**, 1380 (1962).
173. Neber, P. W., Knöller, G., Herbst, K., and Trissler, A., *Ann. Chem.* **471**, 113 (1929).
174. Nenitzescu, C. D., *Bull. Soc. Chim. Romania* **11**, 37 (1929); *Chem. Abstr.* **24**, 110 (1930).
175. Noland, W. E., and Baude, F. J., *Org. Syn.* **43**, 40 (1963).
176. Noland, W. E., and Sellstedt, J. H., *J. Org. Chem.* **31**, 345 (1966).
177. Ochiai, E., Kataoka, H., Dodo, T., and Takahashi, M., *Chem. & Pharm. Bull.* (*Tokyo*) **10**, 76 (1962).
178. Ochiai, E., and Takahashi, M., *Chem. & Pharm. Bull.* (*Tokyo*) **14**, 1272 (1966).
179. Ochiai, R., Takahashi, M., Tamai, Y., and Kataoka, H., *Chem. & Pharm. Bull.* (*Tokyo*) **11**, 137 (1963).
180. Ockenden, D. W., and Schofield, K., *J. Chem. Soc.* p. 612 (1953).
181. Ockenden, D. W., and Schofield, K., *J. Chem. Soc.* p. 3175 (1957).
182. O'Connor, R., *J. Org. Chem.* **26**, 4375 (1961).
183. O'Connor, R., and Rosenbrook, W., Jr., *J. Org. Chem.* **26**, 5208 (1961).

184. Orr, A. H., and Tomlinson, M., *J. Chem. Soc.* p. 5097 (1957).
185. Owellen, R. J., Fitzgerald, J. A., Fitzgerald, B. M., Welsh, D. A., Walker, D. M., and Southwick, P. L., *Tetrahedron Letters* p. 1741 (1967).
186. Pacheoco, H., *Bull. Soc. Chim. France* p. 633 (1951).
186a. Palmer, M. H., and McIntyre, P. S., *J. Chem. Soc., B* p. 446 (1969).
187. Paragamian, V., Baker, M. B., and Puma, B. M., *J. Org. Chem.* **33**, 1345 (1968).
188. Parmerter, S. M., Cook, A. G., and Dixon, W. B., *J. Am. Chem. Soc.* **80**, 4621 (1958).
189. Pausacker, K. H., *J. Chem. Soc.* p. 621 (1950).
190. Pausacker, K. H., and Schubert, C. I., *J. Chem. Soc.* p. 1814 (1950).
191. Pausacker, K. H., and Schubert, C. I., *Nature* **163**, 289 (1949); *J. Chem. Soc.* p. 1384 (1949).
192. Pelchowicz, Z., and Bergmann, E. D., *J. Chem. Soc.* p. 4699 (1960).
193. Pennington, F. C., Jellinek, M., and Thurn, R. D., *J. Org. Chem.* **24**, 565 (1959).
194. Pennington, F. C., Martin, L. J., Reid, R. E., and Lapp, T. W., *J. Org. Chem.* **24**, 2030 (1959).
195. Pennington, F. C., Trittle, G. L., Boyd, S. D., Bowersox, W., and Aniline, O., *J. Org. Chem.* **30**, 2801 (1965).
196. Phillips, R. R., *Org. Reactions* **10**, 143 (1959).
197. Piers, E., Haarstad, V. B., Cushley, R. J., and Brown, R. K., *Can. J. Chem.* **40**, 511 (1962).
198. Pietra, S., and Tacconi, G., *Farmaco (Pavia), Ed. Sci.* **19**, 741 (1964).
198a. Piozzi, F., and Langella, M. R., *Gazz. Chim. Ital.* **93**, 1392 (1963).
199. Piper, J. R., and Stevens, F. J., *J. Heterocyclic Chem.* **3**, 95 (1966).
200. Plancher, G., *Chem. Ber.* **31**, 1488 (1898).
201. Plancher, G., and Bonavia, A., *Gazz. Chim. Ital.* **32**, 414 (1902).
202. Plieninger, H., *Chem. Ber.* **83**, 268 (1950).
203. Plieninger, H., *Chem. Ber.* **83**, 273 (1950).
204. Plieninger, H., and Lehnert, W., *Chem. Ber.* **100**, 2427 (1967).
205. Plieninger, H., and Nógrádi, I., *Chem. Ber.* **88**, 1961 (1955).
206. Plieninger, H., and Nógrádi, I., *Chem. Ber.* **88**, 1964 (1955).
207. Plieninger, H., and Suhr, K., *Chem. Ber.* **89**, 270 (1956); **90**, 1980 (1957).
208. Prasad, K. B., and Swan, G. A., *J. Chem. Soc.* p. 2024 (1958).
209. Pschorr, R., and Hoppe, G., *Chem. Ber.* **43**, 2543 (1910).
210. Raileanu, D., and Nenitzescu, C. D., *Rev. Roumaine Chim.* **10**, 339 (1965); *Chem. Abstr.* **63**, 9903 (1965).
211. Razumov, A. I., and Gurevich, P. A., *J. Gen. Chem. USSR (English Transl.)* **37**, 1532 (1967).
212. Reissert, A., *Chem. Ber.* **30**, 1030 (1897).
213. Remers, W. A., and Weiss, M. J., *J. Am. Chem. Soc.* **87**, 5263 (1965).
213a. Rice, L. M., Hertz, E., and Freed, M. E., *J. Med. Chem.* **7**, 313 (1964).
214. Robinson, B., *Chem. Rev.* **63**, 373 (1963); **69**, 227 (1969).
215. Robinson, B., *Tetrahedron Letters* p. 139 (1962).
216. Robinson, B., *J. Chem. Soc.* p. 586 (1963).
217. Robinson, F. P., and Brown, R. K., *Can. J. Chem.* **42**, 1940 (1964).
218. Robinson, G. C., U.S. Patent 3,255,205 (1966); *Chem. Abstr.* **65**, 13662 (1966).
219. Robinson, G. M., and Robinson, R., *J. Chem. Soc.* **113**, 639 (1918); **125**, 827 (1924).
220. Rosemund, P., and Haase, W. H., *Chem. Ber.* **99**, 2504 (1966).
221. Rydon, H. N., and Tweddle, J. C., *J. Chem. Soc.* p. 3499 (1955).
222. Schofield, K., and Theobald, R. S., *J. Chem. Soc.* p. 1505 (1950).
223. Schulenberg, J. W., *J. Am. Chem. Soc.* **90**, 7008 (1968).

224. Shagalov, L. B., Sorokina, N. P., and Suvorov, N. N., *J. Gen. Chem. USSR* (*English Transl.*) **34**, 1602 (1964).
225. Shavel, J., Jr., and von Strandtmann, M., U.S. Patent 3,217,029 (1965); *Chem. Abstr.* **64**, 2057 (1966).
226. Shavel, J., Jr., von Strandtmann, M., and Cohen, M. P., *J. Am. Chem. Soc.* **84**, 881 (1962).
227. Shaw, E., and Woolley, D. W., *J. Am. Chem. Soc.* **75**, 1877 (1953).
228. Shine, H. J., "Aromatic Rearrangements," p. 190. Elsevier, Amsterdam, 1967.
229. Shvedov, V. I., Altukhova, L. B., and Grinev, A. N., *J. Org. Chem. USSR* (*English Transl.*) **1**, 882 (1965).
230. Singer, H., and Shive, W., *J. Org. Chem.* **22**, 84 (1957).
231. Smith, W. S., and Moir, R. Y., *Can. J. Chem.* **30**, 411 (1952).
232. Smolinsky, G., *J. Org. Chem.* **26**, 4108 (1961); *J. Am. Chem. Soc.* **83**, 2489 (1961).
233. Smolinsky, G., and Feuer, B. I., *J. Am. Chem. Soc.* **86**, 3085 (1964).
234. Smolinsky, G., and Pryde, C. A., *J. Org. Chem.* **33**, 2411 (1968).
235. Snyder, H. R., and Beilfuss, H. R., *J. Am. Chem. Soc.* **75**, 4921 (1953).
236. Snyder, H. R., Merica, E. P., Force, C. G., and White, E. G., *J. Am. Chem. Soc.* **80**, 4622 (1958).
237. Snyder, H. R., and Smith, C. W., *J. Am. Chem. Soc.* **65**, 2452 (1943).
238. Southwick, P. L., McGrew, G., Engel, R. R., Milliman, G. E., and Owellen, R. J., *J. Org. Chem.* **28**, 3058 (1963).
239. Southwick, P. L., and Owellen, R. J., *J. Org. Chem.* **25**, 1133 (1960).
240. Steck, E. A., Brundage, R. P., and Fletcher, L. T., *J. Org. Chem.* **24**, 1750 (1959).
241. Stephen, H., *J. Chem. Soc.* **127**, 1874 (1925).
242. Stephens, R. D., and Castro, C. E., *J. Org. Chem.* **28**, 3313 (1963).
243. Stetter, H., and Lauterbach, R., *Ann. Chem.* **655**, 20 (1962).
244. Stetter, H., and Siehnhold, E., *Chem. Ber.* **88**, 271 (1955).
245. Stevens, F. J., and Su, H. C., *J. Org. Chem.* **27**, 500 (1962).
246. Stoll, A., Troxler, F., Peyer, J., and Hofmann, A., *Helv. Chim. Acta* **38**, 1452 (1955).
247. Stork, G., and Dolfini, J. E., *J. Am. Chem. Soc.* **85**, 2872 (1963).
248. Streith, J., Darrah, H. K., and Weil, M., *Tetrahedron Letters* p. 5555 (1966).
249. Suh, J. T., and Puma, B. M., *J. Org. Chem.* **30**, 2253 (1965).
250. Sundberg, R. J., *J. Org. Chem.* **30**, 3604 (1965).
251. Sundberg, R. J., *J. Org. Chem.* **33**, 487 (1968).
252. Sundberg, R. J., *J. Am. Chem. Soc.* **88**, 3781 (1966).
253. Sundberg, R. J., Lin, L., and Blackburn, D. E., *J. Heterocyclic Chem.* **6**, 441 (1969).
254. Sundberg, R. J., and Yamazaki, T., *J. Org. Chem.* **32**, 290 (1967).
255. Suvorov, N. N., Fedotova, M. V., Orlova, L. M., and Ogareva, O. B., *J. Gen. Chem. USSR* (*English Transl.*) **32**, 2325 (1962).
256. Suvorov, N. N., Morozovskaya, L. M., and Erskova, L. I., *J. Gen. Chem. USSR* (*English Transl.*) **32**, 2521 (1962).
257. Suvorov, N. N., and Murasheva, V. S., *J. Gen. Chem. USSR* (*English Transl.*) **30**, 3086 (1960).
258. Suvorov, N. N., Preobrazhenskaya, M. N., and Uvarova, N. V., *J. Gen. Chem. USSR* (*English Transl.*) **32**, 1552 (1962).
259. Suvorov, N. N., and Sorokina, N. P., and Sheinker, I. N., *J. Gen. Chem. USSR* (*English Transl.*) **28**, 1058 (1958).
260. Sych, E. D., *Ukr. Khim. Zh.* **19**, 643 (1953); *Chem. Abstr.* **49**, 12429 (1955).
261. Szmuszkovicz, J., Glenn, E. M., Heinzelman, R. V., Hester, J. B., Jr., and Youngdale, G. A., *J. Med. Chem.* **9**, 527 (1966).

262. Teotino, U. M., and Maffii, G., U.S. Patent 3,005,827 (1960); *Chem. Abstr.* **56,** 3460 (1962).
263. Terzyan, A. G., Safrazbekyan, R. R., Sukayan, R. S., Akopyran, Z. G., and Tatevosyan, G. T., *Izv. Akad. Nauk Arm. SSR, Khim. Nauk* **17**, 567 (1964); *Chem. Abstr,* **62**, 11868 (1963).
264. Terzyan, A. G., and Tatevosyan, G. T., *Izv. Akad. Nauk Arm. SSR, Khim. Nauk* **13**, 193 (1960); *Chem. Abstr.* **55**, 7384 (1961).
265. Teuber, M. J., and Glosauer, O., *Chem. Ber.* **98**, 2648 (1965).
266. Towne, E. B., and Hill, H. M., U.S. Patent 2,607,779 (1952); *Chem. Abstr.* **47**, 5452 (1953).
267. Tyson, F. T., *J. Am. Chem. Soc.* **72**, 2801 (1950).
268. Uhle, F. C., *J. Am. Chem. Soc.* **71**, 761 (1949).
269. Velluz, L., Muller, G., Allais, A., and Enezian, J., U.S. Patent, 2,985,659 (1961); *Chem. Abstr.* **55**, 23576 (1961).
270. Voltz, S. E., U.S. Patent 2,886,573 (1959); *Chem. Abstr.* **53**, 18058 (1959).
271. von Strandtmann, M., Cohen, M. P., and Shavel, J., Jr., *J. Med. Chem.* **6**, 719 (1963).
272. von Strandtmann, M., Cohen, M. P., and Shavel, J., Jr., *J. Med. Chem.* **8**, 206 (1965).
273. Walker, G. N., *J. Am. Chem. Soc.* **77**, 3844 (1955); **78**, 3698 (1952).
274. Walton, E., Stammer, C. H., Nutt, R. F., Jenkins, S. R., and Holly, F. W., *J. Med. Chem.* **8**, 204 (1965).
275. Wenkert, E., and Barnett, B. F., *J. Am. Chem. Soc.* **82**, 4671 (1960).
276. Weygand, F., and Richter, E., *Chem. Ber.* **88**, 499 (1955).
277. Wieland, T., and Rühl, K., *Chem. Ber.* **96**, 260 (1963).
278. Wilchek, M., Sample, T. F., Witkop, B., and Milne, G. W. A., *J. Am. Chem. Soc.* **89**, 3349 (1967).
279. Yakhontov, L. N., Uritskaya, M. Y., and Rubstov, M. V., *J. Gen. Chem. USSR* (*English Transl.*) **34**, 1454 (1964).
280. Yamada, S., Chibata, I., and Tsurui, R., *Pharm. Bull.* (*Tokyo*) **1,** 14 (1953); *Chem. Abstr.* **48**, 12078 (1954).
281. Yoneda, F., Miyamae, T., and Nitta, Y., *Chem. & Pharm. Bull.* (*Tokyo*) **15**, 8 (1967).
282. Young, T. E., *J. Org. Chem.* **27**, 507 (1962).

IV SYNTHETIC ELABORATION OF THE INDOLE RING

In this chapter the discussion centers on conversion of preformed indoles to more complex derivatives. Most of the reactions employed in synthetic elaboration have been considered in Chapters I or II in the general discussion of reactions of the indole ring. The organization in this chapter is on the basis of the type of compound which is the objective of the overall synthesis. This organization is useful in correlating the very large amount of synthetic work which has been directed toward a relatively few types of indole derivatives. The approaches to indolealkanoic acids, derivatives of tryptamine, tryptophan derivatives, β-carboline derivatives, and indole alkaloids have been classified in this way.

A. SYNTHETIC APPROACHES TO INDOLEALKANOIC ACIDS

Much work has been done on the synthesis of indole-3-acetic acids both because of the activity of the parent compound as a plant growth stimulant and because substituted indole-3-acetic acids are potential precursors of tryptamines [2-(3-indolyl)ethylamines].

1. Synthesis via Gramine Derivatives

The ease of displacement of secondary or tertiary amines from *N,N*-dialkyl-3-indolylmethylamines and their quaternary salts, respectively, was discussed in Chapter II,B. Displacement with cyanide provides indolyl-3-acetonitriles which can be hydrolyzed to acids (42, 43, 96, 179).

O_2N … $CH_2N(Me)_2$ (A1) —MeI, ^-CN→ O_2N … CH_2CN (A2) (59%)

A1 —(1) $CH_2(CO_2Et)_2$, NaOH; (2) NaOH, H_2O; (3) H^+, heat→ O_2N … $CH_2CH_2CO_2H$ (A4) (45%)

A2 —HCl, EtOH→ O_2N … CH_2CO_2Et (A3)

The use of labeled cyanide permits introduction of ^{14}C into the carboxyl group (43). Alkylation of sodio diethyl malonate by gramine or its derivatives gives indole-3-propionic acids after hydrolysis and decarboxylation (161, 171).

2. Alkylations of Indoles with Ethyl Diazoacetate and Other Diazoalkanes

The decomposition of diazoalkanes in the presence of indole results in the introduction of a 3-alkyl group. Copper catalysts have usually been used but it has been stated that the presence of copper does not affect the course of the reaction (9). The reaction was discovered by Piccinini (132) and confirmed in 1935 by Jackson and Manske (81). It probably involves electrophilic attack at C-3 by a carbene or carbonium ion formed by thermal decomposition of the diazo compound. Table XXIX records representative

TABLE XXIX □ ALKYLATION OF INDOLES WITH ETHYL DIAZOACETATE AND OTHER DIAZOALKANES

Indole substituent	Diazo compounds	Yield (%)	Ref.
None	Ethyl diazoacetate	76	(9, 122)
None	Methyl diazopyruvate	?	(146)
None	Ethyl diazoacetoacetate	62	(146)
None	Ethyl 4-oxo-5-diazovalerate	86	(146)
None	Diazoacetone	?	(128)
None	Diazomethyl ethyl ketone	?	(82)
1-Me	Ethyl diazoacetate	76	(9, 48)
4-Cyanomethyl	Ethyl diazoacetate	?	(138)
6-Bromo-2-methyl	Ethyl diazoacetate	47	(136)
6-Chloro-2-methyl	Ethyl diazoacetate	37	(136)

$$\text{A5} \xrightarrow[{}^{+}CH_2CO_2Et]{:CHCO_2Et \text{ or}} \text{A6} \longrightarrow \text{A7}$$

alkylations of indoles with functionally substituted diazoalkanes.

3. Alkylations with α-Haloacetonitriles

As discussed in Chapter I, G, magnesium derivatives of indoles are alkylated at the 3-position by highly reactive halides. Both chloroacetonitrile and bromoacetonitrile have been used in such alkylation reactions to give indole-3-acetonitriles which can subsequently be hydrolyzed to indole-3-acetic acids. This sequence was originally developed by Majima and Hoshino (111).

$$\text{A8} \xrightarrow{ClCH_2CN} \text{A9}$$

Table XXX records examples of such reactions.

TABLE XXX □ INDOLE-3-ACETONITRILES BY ALKYLATION OF INDOLYLMAGNESIUM HALIDES WITH HALOACETONITRILES

Indole-3-acetonitrile substituent	Acetonitrile	Yield (%)	Ref.
None	2-Bromopropionitrile	29	(48)
4-Chloro	Chloroacetonitrile	19	(58)
5-Benzyloxy	Chloroacetonitrile	?	(163)
6-Chloro	Chloroacetonitrile	26	(58)

4. Direct Formation of Indolealkanoic Acids by Fischer Cyclizations

The Fischer cyclization of arylhydrazones to indoles is discussed in Chapter III,A. When the hydrazone carries a carboxylic acid function on one of the alkyl chains, the cyclization results in the direct formation of indoles having alkanoic acid substitution. This approach has been used extensively. The hydrazones have been prepared from arylhydrazines and keto acids or carboxyaldehydes or by Japp–Klingemann reactions carried out on keto esters. Examples of the first type are recorded in Table XXXI

TABLE XXXI □ INDOLEALKANOIC ACIDS FROM ARYLHYDRAZONES OF KETO ACIDS AND CARBOXYALDEHYDES

Indoleacetic acid substituent	Carbonyl precursor	Cyclization catalyst	Yield (%)	Ref.
None	Succinaldehydic acid	Sulfuric acid	21	(57)
None	Ethyl 4,4-dimethoxy-butyrate	Sulfuric acid	21	(30)
1-Me	Ethyl 4,4-dimethoxy-butyrate	Sulfuric acid	51	(30)
1-Methyl-5-nitro	Succinaldehydic acid	Polyphosphoric acid	?	(6)
2-Me	Levulinic acid	Sulfuric acid	79	(30)
2-Carboxy	2-Oxoglutaric acid	Polyphosphoric acid	64	(149)
2-Methyl-4-chloro-7-methoxy	Ethyl levulinate	Zinc chloride	23	(167, 168)
2-Methyl-5-bromo	Ethyl levulinate	Zinc chloride	58	(167, 168)
2-Methyl-5-methoxy	Ethyl levulinate	Hydrochloric acid, ethanol	86	(154)
2-Methyl-5-carboxy	Ethyl levulinate	Zinc chloride	45	(19)
2,7-Dimethyl	Ethyl levulinate	Zinc chloride	23	(31)
2,7-Dimethyl-4-chloro	Ethyl levulinate	Zinc chloride	46	(31)
5-Benzyloxy	Succinaldehydic acid	Phosphoric acid	?	(118)
5-Fluoro	Ethyl 4,4-dimethoxy-butyrate	Sulfuric acid	53	(30)
5-Nitro	Succinaldehydic acid	Zinc chloride	33	(5)
Indole 3 butyric acid	Methyl 5-formylvalerate	Sulfuric acid	7	(32)

while approaches which involve Japp–Klingemann reactions are summarized in Table XXXII.

TABLE XXXII ☐ INDOLEACETIC ACIDS BY CYCLIZATION OF ARYLHYDRAZONE PREPARED BY JAPP–KLINGEMANN REACTIONS

Indoleacetic acid substituent	β-Keto ester precursor	Cyclization catalyst	Yield (%)	Ref.
2-Carbethoxy	Diethyl 2-acetylglutarate	Hydrochloric acid, ethanol	94	(53)
2-Carbethoxy	2-Carbethoxycyclopentanone	Sulfuric acid	37	(114)
2-Carboxy-5-benzyloxy	2-Carbethoxycyclopentanone	Hydrochloric acid, dioxane	43	(86)
2-Carbethoxy-5-fluoro	Diethyl 2-acetylglutarate	Hydrochloric acid, ethanol	30	(55)
2-Carboxy-5-methoxy	Diethyl 2-acetylglutarate	Hydrochloric acid, ethanol	48	(54)
2-Carbethoxy-7-chloro	Diethyl 2-acetylglutarate	Sulfuric acid	45	(53)
2-Carbethoxy-7-methyl	Diethyl 2-acetylglutarate	Sulfuric acid	95	(53)

Bullock and Hand have compared the effectiveness of various catalyst systems (31). Most cyclizations have been effected with zinc chloride or sulfuric acid as the catalyst.

B. SYNTHESIS OF DERIVATIVES OF TRYPTAMINE

The physiological activity of tryptamine [2-(3-indolyl)ethylamine] and its derivatives, particularly their effect on the nervous system, has led to intense interest in the synthesis of tryptamines. Several approaches have been shown to have general application and these are discussed in this section.

1. Tryptamines from Indole-3-acetonitriles and Indole-3-acetamides

The synthesis of indole-3-acetic acid derivatives was discussed in Section A. The nitriles and amides derived from indole-3-acetic acids are important precursors of tryptamines. Reduction of the nitrile or carboxamide function furnishes tryptamines. The nitrile function has usually been reduced catalytically but lithium aluminum hydride has also been employed. An early synthesis of 5-hydroxytryptamine (serotonin) used lithium aluminum hydride as the reducing agent (166).

Most catalytic methods require high pressures of hydrogen but over rhodium-on-alumina reduction occurs rapidly at 2.5 atm (61). Ammoniacal solutions are usually used to diminish the extent of formation of secondary amines.

CH_2CN —(H)→ $CH_2CH_2NH_2$

B1 B3

(H)

CH_2CONH_2

B2

Lithium aluminum hydride is the reagent of choice for reduction of indole-3-acetamides. Table XXXIII records a few typical examples of reductions of both indoleacetonitriles and indoleacetamides.

2. Tryptamines from Indole-3-glyoxylyl Chlorides

Speeter and Anthony opened an important route to tryptamines from indoles with the demonstration that indole-3-glyoxylyl chlorides (**B6**) could be converted to tryptamines in two steps (165). The glyoxylyl chlorides are usually readily obtained from a substituted indole and oxalyl chloride (93, 165). Treatment of **B6** with ammonia or an amine gives the amides **B8**. By virtue of the fact that carbonyl groups adjacent to the 3-position of

X — B4 + ClCOCOCl (B5) ⟶ X — B6 ($COCOCl$)

B6 + R_2NH (**B7**) ⟶ X — **B8** ($COCONR_2$)

$LiAlH_4$

X — **B9** ($CH_2CH_2NR_2$)

the indole ring are reduced to methylene groups (see Chapter II, C), a single reduction operation gives the desired tryptamine. This synthetic route is very versatile with respect to the substituents which can be introduced on the amino nitrogen (46). Several examples of the successful application of the Speeter–Anthony route are collected in Table XXXIV.

TABLE XXXIII □ REDUCTION OF INDOLE-3-ACETONITRILES AND INDOLE-3-ACETAMIDES TO TRYPTAMINES

Tryptamine substituent	Reducing agent	Yield (%)	Ref.
Indole-3-acetonitriles			
None	Hydrogen, nickel, 90 atm, ammonia, methanol	90	(179)
None	Hydrogen, rhodium on alumina, 2.5 atm, ammonia, ethanol	78	(61)
None	Hydrazine, nickel	95	(176)
1-Benzyl-β-propyl[a]	Lithium aluminum hydride	87	(117)
1-Benzyl-5-methoxy-β-propyl[a]	Lithium aluminum hydride	73	(117)
2-Me	Hydrogen, nickel, 90 atm, ammonia, methanol	?	(90)
5-Benzyloxy	Lithium aluminum hydride	?	(166)
5-Methoxy	Lithium aluminum hydride	70	(174)
5-Benzyloxy-7-methyl	Hydrogen, nickel, 80 atm, ammonia, ethanol	42	(73)
6-Fluoro	Hydrogen, nickel, ammonia methanol	50	(23)
Indole-3-acetamides			
1-Methyl-5-methoxy	Lithium aluminum hydride	47	(154)
2-Methyl-5-benzyloxy	Lithium aluminum hydride	40	(154)
2-Methyl-5-methoxy-*N,N*-dimethyl	Lithium aluminum hydride	71	(154)
5-Benzyloxy	Lithium aluminum hydride	?	(68)

[a] β-refers to side-chain substitution.

When the indole nitrogen carries a substituent, the reduction often terminates at the carbinol stage (4, 36, 194). The reduced tendency for *N*-alkylindoles to undergo hydrogenolysis is discussed in Chapter II, C. Although only recently investigated, diborane appears to be capable of effecting reduction of 1-alkylindole-3-glyoxamides to 1-alkyltryptamines (22).

$$\text{B10 (1-Me-2-Ph-indole-3-}COCON(Me)_2\text{)} \xrightarrow{LiAlH_4} \text{B11 (1-Me-2-Ph-indole-3-}CHOHCH_2N(Me)_2\text{)}$$

The most uniformly successful results have been obtained with the glyoxamide tryptamine synthesis when secondary amines have been used to prepare tertiary amides. The use of benzylamines followed by subsequent benzyl hydrogenolysis has been successfully used to prepare *N*-monoalkytryptamines and unsubstituted tryptamines (4, 152).

A closely related procedure involves synthesis of β-dialkylaminoalkyl 3-indolylketones followed by reduction with lithium aluminum hydride (164).

TABLE XXXIV □ TRYPTAMINES FROM INDOLES VIA INDOLE-3-GLYOXYLYL CHLORIDES

Tryptamine substituent	Percent yield (amide stage)	Percent yield (reduction stage)	Ref.
None	92	?	(156)
N-Hexyl	90	25	(88)
N,N-diMe	89	82	(127)
1-Benzyl-2-methyl-5-methoxy	84	?	(46)
1-Benzyl-2-methyl-5-methoxy-*N,N*-diethyl	93	?	(46)
2-Phenyl-*N,N*-dimethyl	?	?	(4)
2-(*p*-Methoxyphenyl)-*N,N*-dimethyl	?	?	(4)
5-Fluoro	92	62	(35)
5-Benzyloxy-*N,N*-dibenzyl	91	92	(165)
5-Benzyloxy-*N,N*-dimethyl	?	?	(165)
5-Bromo-*N*-methyl	83	20	(35)
5-Methoxy-*N,N*-dimethyl	75	91	(21)
5,6-Dibenzyloxy-*N,N*-dibenzyl	65	75	(152)
6-Trifluoromethyl-*N,N*-dimethyl	?	?	(87)

Derivatives of piperdines and pyrrolidines

Indole substituent	Heterocyclic amine	Percent yield (amide stage)	Percent yield (reduction stage)	Ref.
None	3-Pyrrolindinol	70	78	(194)
None	4-Phenyl-4-piperidinol	55	55	(194)
5-Chloro	4-Phenyl-1,2,5,6-tetrahydropyridine	49	53	(194)

$COCH_2Cl$ $\xrightarrow{R_2NH}$

B12

$COCH_2NR_2$ $\xrightarrow{LiAlH_4}$ $CH_2CH_2NR_2$

B13 B14

MgI + $MeCHCOBr$ (Br) $\longrightarrow$ $COCHMe$ (Br)

B15 B16 B17

R_2NH

Me CH_2CHNR_2 $\xleftarrow{LiAlH_4}$ Me $COCHNR_2$

B19 B18

3. Tryptamines from Indole-3-carboxaldehydes

Indole-3-carboxaldehydes usually can be condensed with nitroalkanes to give 3-(2-nitrovinyl)indoles. The latter compounds can be reduced by various reducing agents to tryptamines. It is also possible to introduce additional substituents at the nitrovinylindole stage. The olefinic linkage substituted by the electron withdrawing nitro group is subject to nucleophilic attack and both carbanions and organometallic reagents have been found to react successfully with 3-(2-nitrovinyl)indoles. The equations in the scheme below illustrate the type of reactions which have been successfully developed to date.

The structure of 3-(2-nitrovinyl)indoles has been discussed by Heinzelman and co-workers who make the statement that spectral data suggest that the compounds are "inner nitronium salts" which they represent by the structure **B32** (74).

This nomenclature has been used by others (73, 206). It is clear the **B32** is a resonance structure of the conventional structure **B31** and, thus, while **B32** may be the more adequate of the two representations, it is meaningless to assign **B32** and exclude **B31**. Structure **B32** clearly represents the drift of electrons from the electron-rich heterocyclic ring to the electron-deficient nitrovinyl moiety.

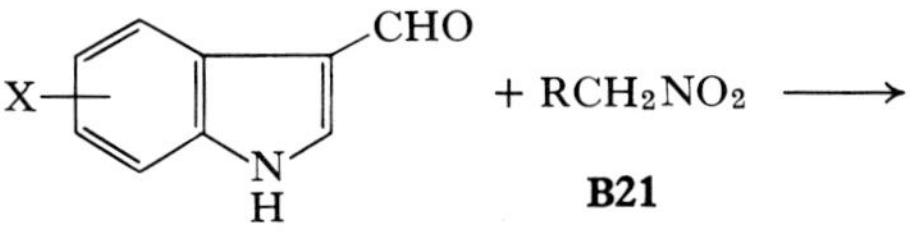

B20 **B21**

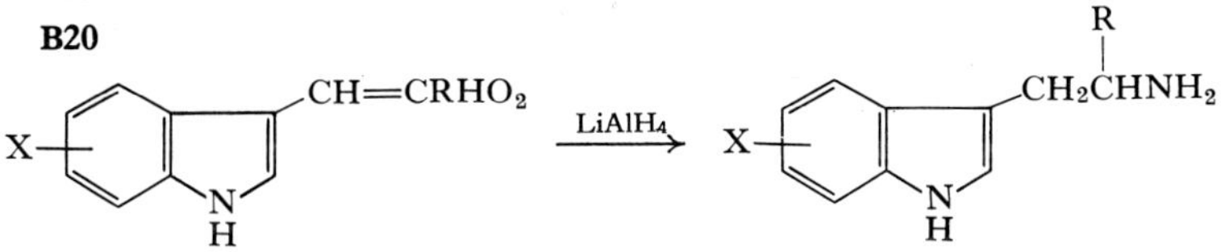

B22 **B23**

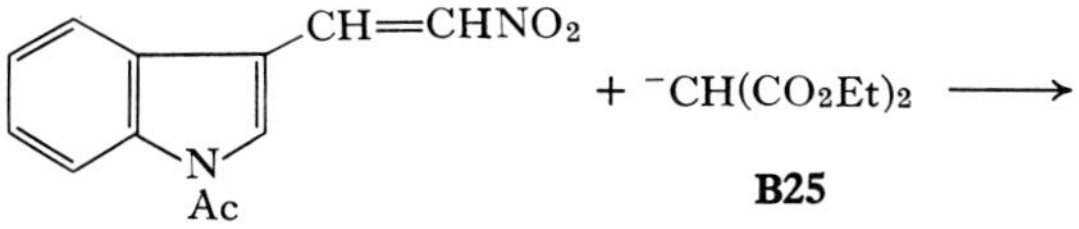

B24 **B25**

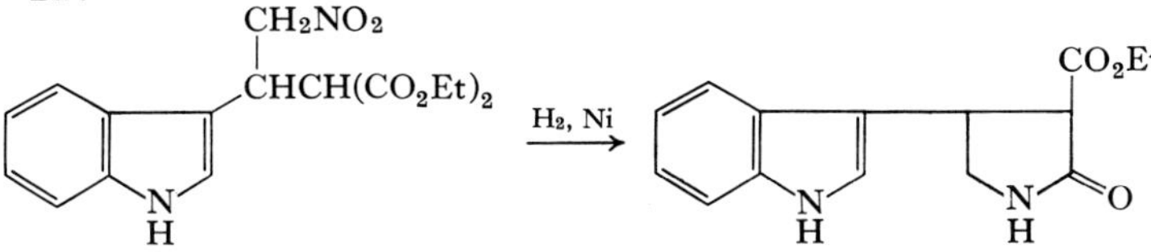

B26 (63%) **B27** (71%) Ref. (112)

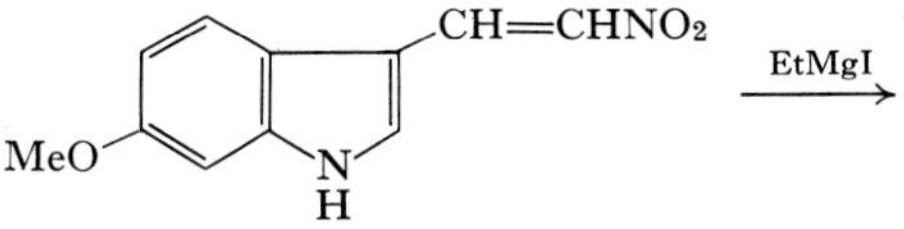

B28

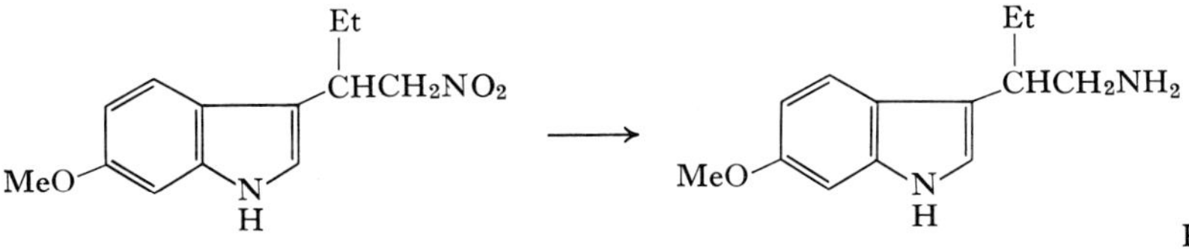

B29 **B30** Ref. (24)

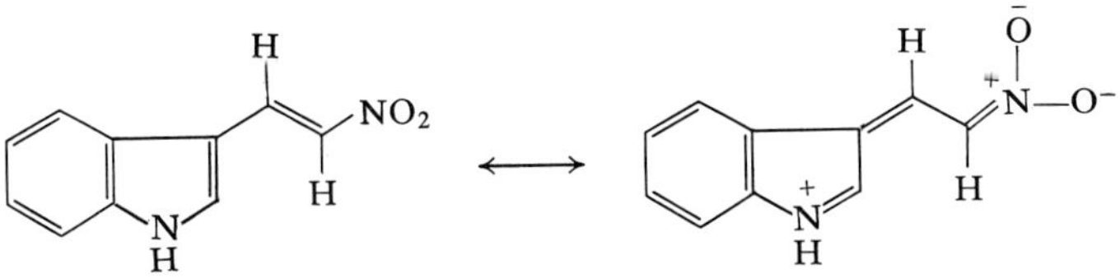

B31 **B32**

The reduction of 3-(2-nitrovinyl)indoles to tryptamines has almost always been effected by lithium aluminum hydride although electrolytic reduction has also been successful in one instance (89). Catalytic methods have been used occasionally (74). Table XXXV catalogs a number of successful applications of the nitrovinyl intermediates in tryptamine synthesis.

TABLE XXXV □ TRYPTAMINES VIA 3-(2-NITROVINYL)INDOLES

Tryptamine substituent[a]	Percent yield condensation step	Percent yield reduction step[b]	Ref.
None	50	87	(130)
α-Methyl	50	71	(74)
α-Ethyl	34	40	(74, 175)
β-Methyl	?	73	(74)
1-Methyl-2-phenyl	45	?	(4)
4-Chloro-α-methyl	85	71	(206)
5-Benzyloxy-α-methyl	59	4	(8)
5-Benzyloxy-1,α-dimethyl	40	84	(74)
5-Bromo-α-methyl	47	63	(44)
5-Chloro	?	?	(211)
5-Methoxy	25, 68	27, 65	(98, 174)
5-Methoxy-α-ethyl	65	47	(73)
6-Benzyloxy	?	?	(211)
6-Chloro-α-methyl	82	29	(206)
7-Chloro-α-ethyl	40	28	(206)
7,α-Dimethyl	66	58	(73)

[a] Greek letter designation refers to side-chain substitution.
[b] Lithium aluminum hydride was the reducing agent in each case.

4. Tryptamines via Gramine Derivatives

The reactivity of gramine derivatives toward nucleophiles was discussed in Chapter II, B. Use of the anions of nitroalkanes as nucleophiles results in the formation of 3-(2-nitroethyl)indoles. Reduction of these intermediates gives tryptamines. This synthetic route was first developed by Snyder and Katz (159) and by Lyttle and Weisblatt (110). Primary nitroalkanes sometimes undergo dialkylation (73, 159). Table XXXVI records typical examples of the method. α-Nitro esters have been frequently alkylated with gramine derivatives (74, 159). This provides a route to substituted tryptophans (see Section C, 1) but the carbonyl group can be removed prior to reduction, providing a route to tryptamines (74).

CH_2NR_2 —(RCH_2NO_2, ^{-}OH)→ CH_2CHRNO_2 → CH_2CHRNH_2

B31 B32 B33

B34 ($CH_2C(NO_2)(Me)CO_2Et$) —NaOH→ B35 ($CH_2C(Me)=NO_2^-$) —HCl→ B36 ($CH_2CH(Me)NO_2$) —$LiAlH_4$→ B37 ($CH_2CH(Me)NH_2$)

PhCH₂O

TABLE XXXVI □ SYNTHESIS OF TRYPTAMINES VIA 3-(2-NITROETHYL)-INDOLES PREPARED FROM GRAMINE DERIVATIVES

Tryptamine substituent[a]	Percent yield alkylation step	Percent yield reduction step	Reducing agent	Ref.
α,α-diMe	80–85	67–75	Hydrogen, nickel	(159)
4,α,α-triMe	76	65	Hydrogen, nickel	(73)
5-Benzyloxy	[b]	[b]	Hydrogen, nickel	(8)
5-Benzyloxy-α,α-dimethyl	70	72[c]	Hydrogen, palladium	(74)
6-Chloro-α,α-dimethyl	76	91	Hydrogen, nickel ammonia	(73)
6-Methoxy-α,α-dimethyl	70	65	Hydrogen, nickel	(73)

[a] Greek letters indicate side-chain substitution.
[b] Overall yield is 40%.
[c] Simultaneous hydrogenolysis of the benzyl group occurs.

A few 3-(2-nitroethyl)amines prepared by the gramine route have also been used to prepare hydroxylamine analogs of tryptamine (37). To prepare the hydroxylamines, zinc and ammonium chloride is employed as the reducing agent.

Phenacylpyridinium ions displace trimethylamine from quaternary salts of gramine (178). Alkaline cleavage removes the benzoyl group and this procedure then provides *N*-[2-(3-indolyl)ethyl]pyridinium ions which should be capable of being reduced to the corresponding piperidines in which the basic nitrogen is part of a heterocyclic ring.

5. Tryptamines from Indoles via Nitroethylation

The alkylation of indole with nitroolefins was discussed in Chapter I, L. This reaction provides 3-(2-nitroethyl)indoles in fair to moderate yields and the nitro compounds can, in turn, be reduced to tryptamines, normally in excellent yields. Table XXXVII records examples of the use of this route to tryptamines.

TABLE XXXVII □ TRYPTAMINES FROM INDOLES VIA NITROETHYLATION

Tryptamine substituent[a]	Percent yield alkylation stage	Percent yield reduction stage	Ref.
β-Ph	66	55	(124)
2-Me	58	?	(126)
2,β-diPh	73	88	(126)
2,α-Dimethyl-β-phenyl	69	60	(126)
5-Benzyloxy	45	84	(125)
5-Benzyloxy-β-phenyl	72	94	(125)
5,6-Dibenzyloxy	20	0	(2)

[a] Greek letters refer to side-chain substitution.

$$\text{B38} + \underset{\textbf{B39}}{RCH{=}CHNO_2} \longrightarrow \text{B40 (3-CHRCH}_2\text{NO}_2\text{)} \longrightarrow \text{B41 (3-CHRCH}_2\text{NH}_2\text{)}$$

X, R, CHCH₂NO₂, N, H

B38 **B39** **B40** **B41**

6. Tryptamines from Indoles and Aziridines

The aziridine ring, particularly when protonated or complexed with a Lewis acid, is a potential electrophile. Successful alkylations of the indole ring have been accomplished with aziridines. The 2-aminoethyl side chain is thereby directly introduced. Indole and 2-methylindole react exothermally with aziridinium tetrafluoroborate (131). The reaction with 2-methylindole gives the corresponding tryptamine in excellent yield but the reaction with indole is less satisfactory. Tryptamine is formed in 40% yield accompanied by 1-(2-aminoethyl)indole. Tryptamine has also been obtained in 46% yield from aziridine and the magnesium derivative of indole (29). Bucourt (28) has reported the preparation of several tryptamines bearing ring substituents by similar reactions of magnesium derivatives with aziridine. Although the reaction has not yet been widely applied, it would seem to have considerable potential when an unbranched side chain is required.

It has been suggested that aziridinium ions may be involved in the alkylation of indolylmagnesium halides by β-haloamines (see Chapter I,G) (62).

7. Tryptamines from 3-(2-Haloethyl)indoles

The reaction of alkyl halides with amines or ammonia is a familiar synthesis of amines and it has been used to some extent in the synthesis of tryptamines. Shaw and Woolley prepared a number of substituted tryptamines from 5-nitro-3-(2-chloroethyl)indole in this way (155). The synthetic transformation **B42** → **B43** provides another example (73). Vitali and

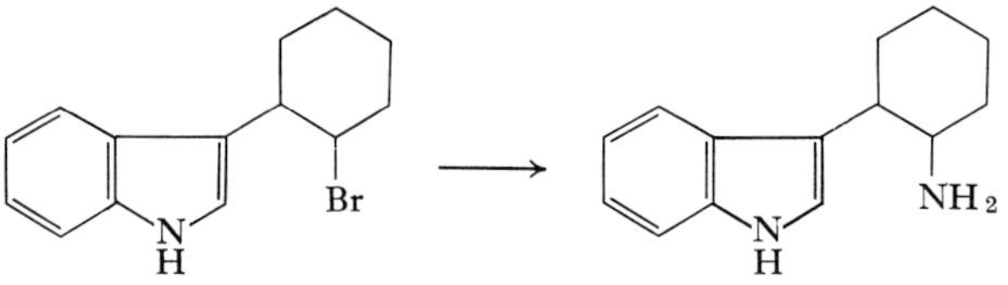

Br → NH2

N H   N H

B42   **B43**

Mossini have reported quite good yields for the alkylation of a series of secondary amines with 3-(2-bromoethyl)indole in sealed-tube reactions at 100–105° (187). Alkylations of pyridines and piperidines with 3-(2-bromoethyl)indoles has been successfully carried out (200, 202, 203) in the course of syntheses directed toward indole alkaloids.

8. Synthesis of Tryptamines by Curtius Rearrangements

The various general degradative syntheses of amines from carboxylic acids, i.e., Hofmann degradation and Curtius rearrangements, seem to have been applied to synthesis of tryptamines only rarely (85, 162).

9. Direct Formation of Tryptamines by Fischer Cyclization

In a number of instances it has been possible to effect Fischer cyclization (Chapter III, A) of arylhydrazones constituted in such a way that a 3-(2-aminoethyl)indole or a simple derivative is formed in the cyclization. The most useful example of this type of approach to tryptamines would seem to be the Abramovitch route to tryptamines starting from 3-carboxy-2-piperidones. This approach was discussed in Chapter III, A, 3 and Table XXII records a number of such syntheses.

An alternative method which has been used occasionally employs γ-aminoketone or aldehyde phenylhydrazones. Cyclization then gives tryptamine derivatives. Transformations **B44** → **B46** and **B47** → **B50** are examples. The factors which govern the direction of cyclization in Fischer reactions were discussed in Chapter III, A, 2. It is clear that tryptamines

NHNH$_2$ Me **B44** + $(EtO)_2CH(CH_2)_3NH_2$ **B45** $\xrightarrow{ZnCl_2}$ $CH_2CH_2NH_2$ N H Me **B46**

MeO N—NH$_2$ CH$_2$ Ph **B47** + Ac(CH$_2$)$_3$N O O **B48** $\xrightarrow{AcOH}$ MeO CH_2CH_2N O O N H Me **B49** $\xrightarrow{NH_2NH_2}$ MeO $CH_2CH_2NH_2$ N H Me **B50**

will be formed from γ-aminoketone phenylhydrazones only when structural features favor cyclization into the branch of the ketone having the amino substituent. The most widely used carbonyl component has been γ-aminobutyraldehyde, employed as its diethyl acetal. Historically, the earliest synthesis of tryptamine itself was carried out in this manner (52). Table XXXVIII records a number of examples of direct formation of tryptamine derivatives by Fischer cyclization.

$$\text{B51 } (C_6H_5NHNH_2) + (EtO)_2CH(CH_2)_3NH_2 \text{ (B52)} \xrightarrow{ZnCl_2} \text{B53 (3-}CH_2CH_2NH_2\text{-indole, N-H)}$$

TABLE XXXVIII □ TRYPTAMINE DERIVATIVES BY FISCHER CYCLIZATION

Tryptamine substituent[a]		Carbonyl precursor	Cyclization catalyst	Yield %	Ref.
None		4-Aminobutyraldehyde diethyl acetal	Zinc chloride	45, 51	(52, 113)
1-Ethyl		4-Aminobutyraldehyde diethyl acetal	Zinc chloride	80	(48)
1-Benzyl		4-Aminobutyraldehyde diethyl acetal	Zinc chloride	40–50	(77)
1-Benzy-2-methyl-5-methoxy		5-Phthalimido-2-pentanone	Acetic acid	80	(157)
7-Methyl		4-Aminobutyraldehyde diethyl acetal	Zinc chloride	?	(47)
Derivatives of Piperidines and Pyrrolidines					
Indole substituent	**Heterocyclic amine**				
1,2,β-triMe	Pyrrolidine	5-Pyrrolidino-4-methyl-2-pentanone	Zinc chloride	?	(63)
1,2-diMe	Piperidine	5-Piperidino-2-pentanone	Zinc chloride	?	(63)

[a] Greek letters refer to side-chain substitution.

10. Syntheses Based on Isatins

The chemical reactivity of isatin (**B54**) is discussed fully in Chapter VII. Substituted isatins are the starting point for several reasonably general synthetic approaches to substituted tryptamines. Isatin can be considered to be a quinone derived from indole and C-3 is particularly reactive toward nucleophilic attack. Advantage of this reactivity has been taken in the syntheses directed toward substituted tryptamines. The general scheme is outlined above. Base-catalyzed addition of a ketone to C-3 of the isatin is followed by oximation. Reduction not only reduces the oximino group to an amino group but also reduces the dioxindole system to an indole ring. Table XXXIX records examples of tryptamines which have been obtained by this method.

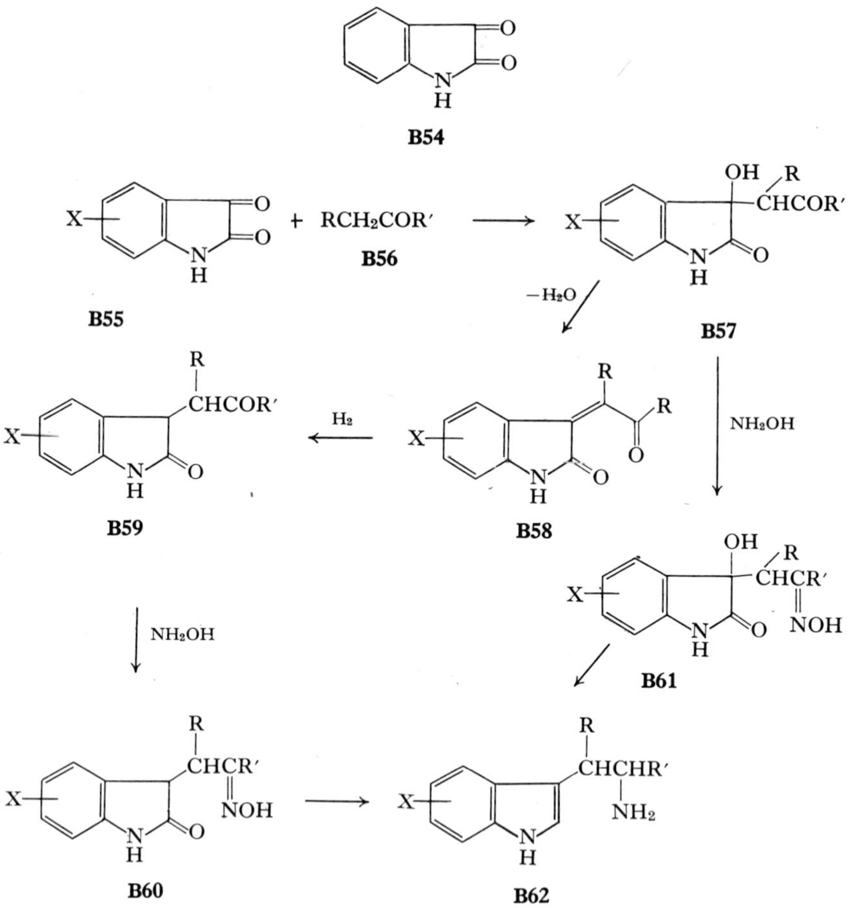

C. SYNTHESIS OF TRYPTOPHAN AND ITS DERIVATIVES

The most general syntheses of tryptophan and its derivatives make use of an alkylation reaction involving a gramine derivative or a Fischer type cyclization of a ketone or aldehyde having the required amino and carboxyl functions in place. These two approaches, along with several other interesting approaches, are considered in the following sections.

TABLE XXXIX □ SYNTHESIS OF TRYPTAMINES FROM ISATINS

Tryptamine substituent[a]	Reducing agent	Percent yield reduction stage	Ref.
α-Me	Sodium borohydride, aluminum chloride	29	(59)
α-Me	Sodium, propanol	58	(133)
α-*t*-Butyl	Sodium, propanol	45	(133)
α-Ph	Sodium, propanol	50	(133)
5-Methoxy-α-methyl	Lithium aluminum hydride	44	(59)
5-Methoxy-α-*t*-butyl	Sodium, propanol	27	(134)
1-Methyl-5-methoxy-α-phenyl	Sodium, propanol	40	(134)
7-Trifluoromethyl-α-methyl	Sodium borohydride, aluminum chloride	?	(59)
3-(2-Aminocyclohexyl)-indole	Sodium, propanol	50	(135)

[a] Greek letters refer to side-chain substitution.

1. Synthesis of Tryptophan Derivatives via Gramine Intermediates

Two types of carbanions have been found suitable for reaction with gramine derivatives to afford precursors of the α-aminopropionic acid side chain of tryptophan. Carbanions derived from diethyl amidomalonate are alkylated by gramine to give diethyl amido(indolylmethyl)malonates (**C3**). Hydrolysis and decarboxylation affords the desired amino acid. This general approach, first developed in the early 1940's (3, 160) has found widespread use. Modifications in which either the free gramine or a quaternary derivative is used have been reported. If substituents are sensitive to the normal

CH_2NR_2 + $^{-}C(CO_2Et)_2$ (NHCOR) → $CH_2C(CO_2Et)_2$ (NHCOR) → $CH_2CH(NH_2)CO_2H$

C1 **C2** **C3** **C4**

$PhCH_2O$ $CH_2N(Me)_2$ + $MeCHCO_2Et$ | NO_2

C5 C6

$PhCH_2O$ $CH_2C(Me)(NO_2)CO_2Et$

C7 (90%)

Ref. (74)

(1) H_2, Pt
(2) NaOH

$PhCH_2O$ $CH_2C(Me)(NH_2)CO_2H$

C8 (75%)

$CH_2N(Me)_2$ + $HC(CO_2Et)(NO_2)CH_2CH{=}C(Me)_2$

C9 C10

$CH_2CH{=}C(Me)_2$ | CH_2CCO_2Et | NO_2

C11 (55%)

H_2, Ni

Ref. (41)

$CH_2CH{=}C(Me)_2$ | CH_2CCO_2Et | NH_2

C12

hydrolytic conditions for removal of the "extra" carboxyl group, the dibenzyl ester of carbobenzyloxyamidomalonic acid can be used (94). Hydrogenolysis then removes the benzyl groups and thermal decarboxylation affords the tryptophan. Table XL summarizes a number of reactions of gramine derivatives with amidomalonate carbanions.

TABLE XL □ PREPARATION OF TRYPTOPHAN DERIVATIVES FROM GRAMINES AND AMIDOMALONATE CARBANIONS[a]

Tryptophan substituent	Gramine derivative	Percent yield alkylation stage	Percent yield hydrolysis and decarboxylation	Ref.
None	Free base	90	81	(78)
None	Methiodide	63–70	81	(160)
None	Ethiodide[b]	95	?	(3)
None	Methosulfate[c]	78	?	(186)
1-Methyl-5-methoxy	Methiodide	?	90	(38)
2-Carbethoxy	Free base[d]	89	76	(94)
2-Phenyl	Free base	70	68	(94)
4-Methoxy	Methosulfate[b]	?	?	(45)
5-Benzyloxy	Free base	78	81	(96)
5-Methoxy	Methosulfate[b]	90	?	(38)
7-Nitro	Free base	?	?	(34)

[a] Diethyl α-acetamidomalonate was employed unless otherwise noted.
[b] Prepared *in situ*.
[c] α-Formamidomalonate was used.
[d] Dibenzyl α-acetamidomalonate was used.

Nitronate anions derived from α-nitrocarboxylate esters are alkylated by gramine derivatives. The resulting α-nitroindole-3-propionic acid derivative can be reduced to a substituted tryptophan (110). The transformations **C5** → **C8** and **C9** → **C12** are illustrative.

$$\textbf{C13}\ (\text{3-}CH_2N(Me)_2\text{-indole}) + O_2NCH(CO_2Et)_2\ (\textbf{C14}) \longrightarrow \textbf{C15}\ (\text{3-}CH_2C(CO_2Et)_2NO_2\text{-indole})$$

A variation on the above approach to tryptophan involves use of the anion of diethyl nitromalonate (**C14**) as the carbon nucleophile. The alkylation product **C15** contains an additional carbethoxy group but it is readily removed by treatment with ethoxide ion. Reduction then gives tryptophan. Tryptophan (193), ethyl 2-carbethoxytryptophan (94), 2-methyltryptophan (7) and 5-methyltryptophan (7) have been prepared via this route.

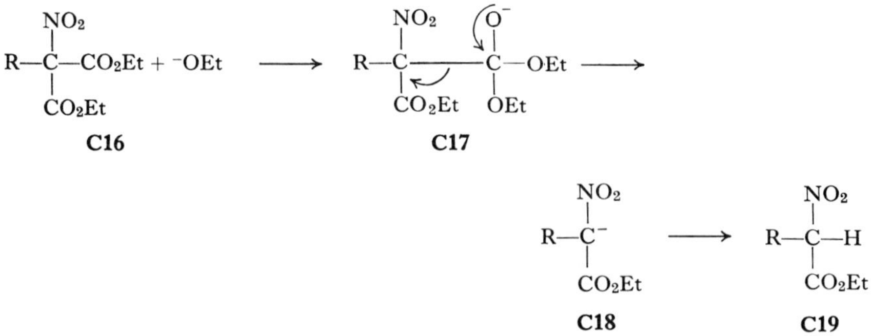

C16 C17 C18 C19

2. Synthesis of Tryptophans via Fischer Cyclizations

As with indole-3-alkanoic acids and tryptamines, it is possible to obtain simple derivatives of tryptophan directly from Fischer cyclizations by choice of appropriate starting materials. For tryptophan derivatives with substituents in the benzene ring, the required precursor is a 4-amino-4-carboxybutyraldehyde derivative. In practice 4-acetamido-4,4-dicarbethoxybutyraldehyde has usually been used and the tryptophan obtained by subsequent

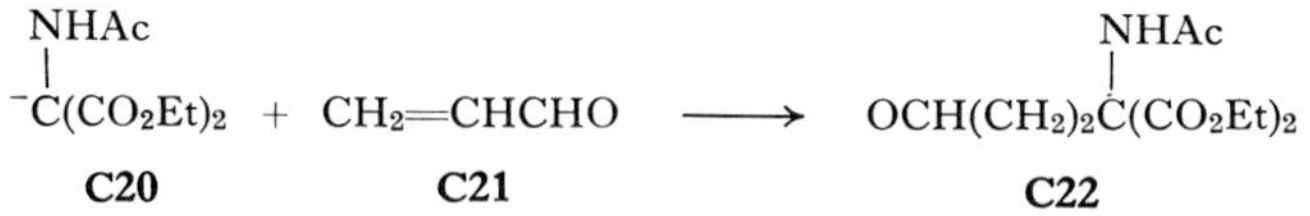

C20 C21 C22

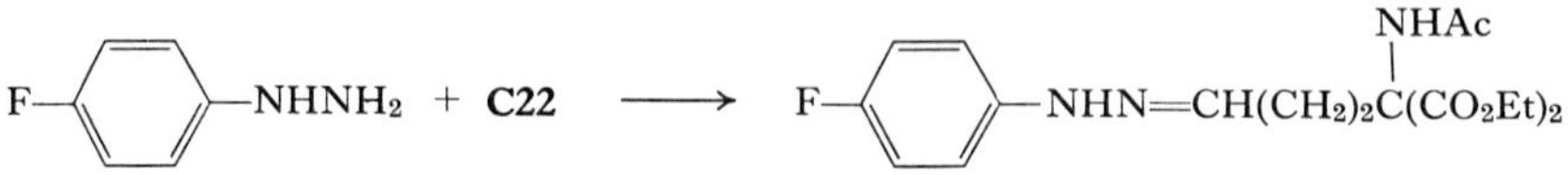

C23 C24 (61%)

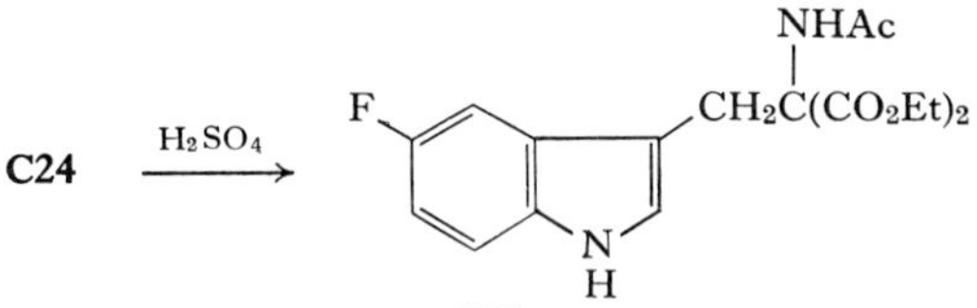

C25

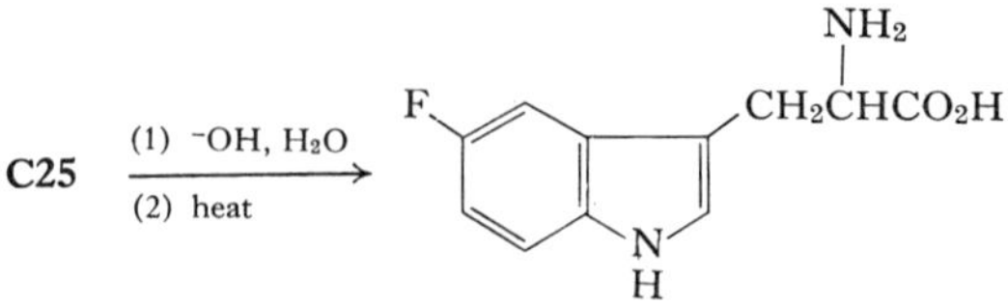

C26

hydrolysis and decarboxylation. The aldehyde is prepared by Michael addition of diethyl α-acetamidomalonate to acrolein (119). The method can be illustrated by a preparation of 5-fluorotryptophan (148). Tryptophan (188), 5-methyltryptophan (158) and a mixture of 4- and 6-methyltryptophan (158) have been prepared by this route.

3. Miscellaneous Approaches to Tryptophan Derivatives

Transformations **C27** → **C28**, **C27** → **C30** and **C31** → **C35** are novel but little-applied routes to tryptophan.

Ac
CH_2CHCO_2Et
N
H
C27

HN_3 →

NHAc
CH_2CHCO_2Et
N
H
C28 (83%)

↓

CH_2CCO_2H
NOH
N
H
C29

H_2, Ni / NH_3 →

NH_2
CH_2CHCO_2H
N
H
(72%)
C30 Ref. (76)

Good yields are reported for each of the steps in the latter sequence (18).

X
N
Mg Br
C31

+

ClCH
O
N
O
Ph
C32

→

H
X
O
N
O
N
H
Ph
C33 Ref. (18)

↓ 1% NaOH

NHCOPh
CH_2CHCO_2H
X
N
H
C35

H_2 ←

H
CO_2H
X
NHCOPh
N
H
C34

D. SYNTHESIS OF DERIVATIVES OF β-CARBOLINE

The β-carboline nucleus (**D1**) has been the object of considerable synthetic effort in indole chemistry. The ring forms a portion of the skeleton of

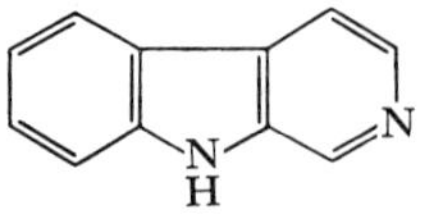

D1

many indole alkaloids and this fact, along with interest in the compounds from a pharmacological point of view, has spurred the synthetic work. Many of the important derivatives of the β-carboline system contain the pyridine ring in a reduced state and the emphasis here will be on tetrahydro-β-carbolines.

1. Tetrahydro-β-Carbolines from Tryptamines

Tryptamines and aldehydes condense to give tetrahydro-β-carbolines. The mechanism which has been generally accepted for this cyclization suggests Schiff base formation followed by an electrophilic attack at C-2 of the indole ring (1).

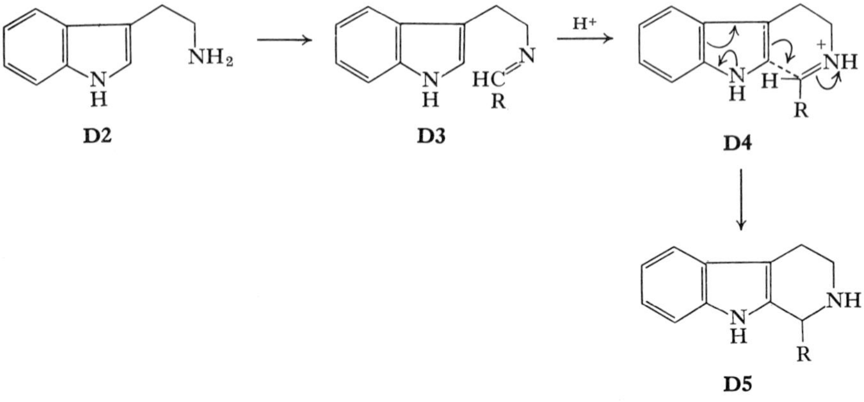

D2 → **D3** $\xrightarrow{H^+}$ **D4** → **D5**

Jackson and Smith have suggested that, in line with other electrophilic substitutions of 3-alkylindoles, the reaction should be formulated as involving initial substitution at the 3-position of the indole ring followed by rearrangement of the resulting indolenine (80). Efforts to obtain experimental support for this proposal by isolating indolenines have failed but indirect evidence for the intermediacy of indolenines was obtained by an alternative

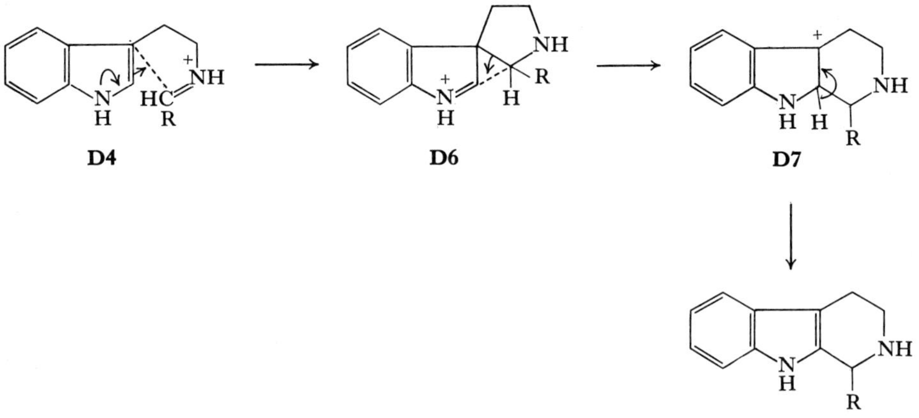

D4 D6 D7 D8

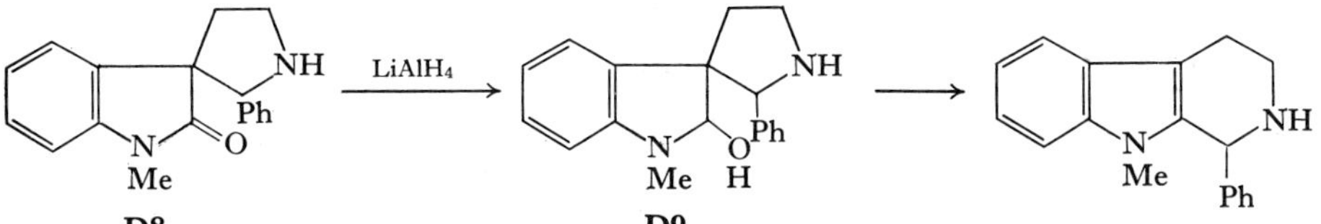

D8 D9 D10

$LiAlH_4$

scheme. It was demonstrated that the carbinolamine **D9**, a potential precursor of the desired indolenine system, was rearranged under extremely mild acidic conditions to the tetrahydro-β-carboline **D10**. This result is consistent with the mechanism proposed by Jackson and Smith but, of course, does not prove that initial attack is at C-3.

Recent intramolecular cyclizations of 1-[2-(3-indolyl)ethyl]tetrahydropyridines provide examples of cyclizations which clearly occur by attack at the 3-position of the indole ring. In these cases, subsequent rearrangement is blocked by the presence of a substituent at C-2. The cyclization of **D14** in which the initial product **D15** is stabilized by tautomerization to **D16**, is very efficient (197). On the other hand, it appears that **D11** may exist in equilibrium with **D12** since reduction of the crude product mixture gives the reduction products of both **D11** and **D12** (197).

The cyclization of **D17** to **D19**, reported by van Tamelen, Dolby and Lawton (181), is apparently another example of cyclization at C-3. This cyclization may be driven to completion by subsequent intramolecular aldol condensation of the indolenine **D18**. One of the intermediate steps in Woodward's synthesis of strychnine (210) also provides an example of cyclization occurring at the 3-position when the 2-position is blocked. None of these results permits an unequivocal choice of mechanism in the normal type of cyclization to tetrahydro-β-carbolines, but the Jackson mechanism (80)

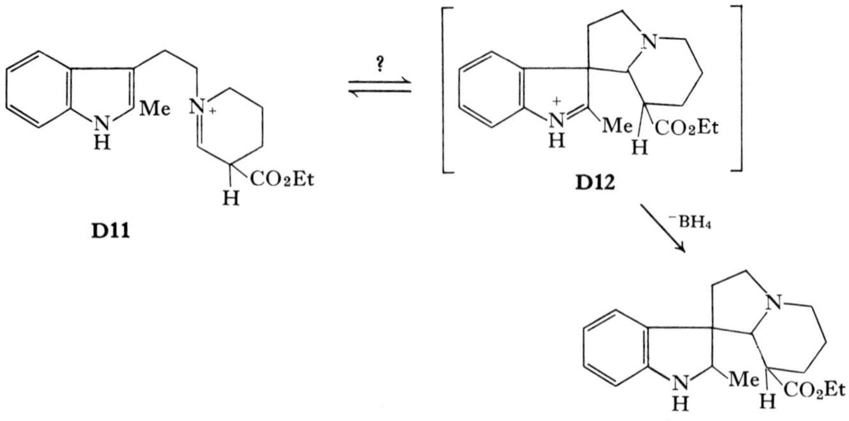

N H Me N+ H CO_2Et D11

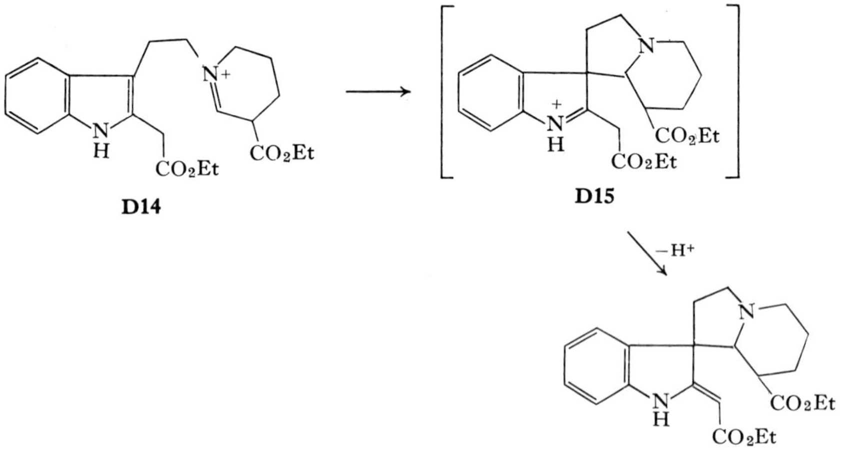

?

N+ H Me CO_2Et H N D12

$^-BH_4$

N H Me H CO_2Et N D13

N H CO_2Et N+ CO_2Et D14

N+ H CO_2Et CO_2Et N D15

$-H^+$

N H CO_2Et CO_2Et N D16

is in accord with results in the study of other electrophilic substitutions in 3-alkylindoles (see Chapter I, N).

The formation of tetrahydro-β-carbolines from tryptamines is a special version of the Pictet–Spengler synthesis of tetrahydroisoquinolines and Whaley and Govindachari have tabulated a number of such cyclizations carried out prior to 1950 in their review of the Pictet–Spengler reaction (205). The transformations **D22** → **D24, D25** → **D27** and **D22** → **D29** are recent examples of the application of the reaction. The latter reaction demonstrates the use of functionally substituted aldehydes for further synthetic elaboration. The compound **D29** served as an intermediate in the total synthesis of the alkaloid vincamine **D30** (99).

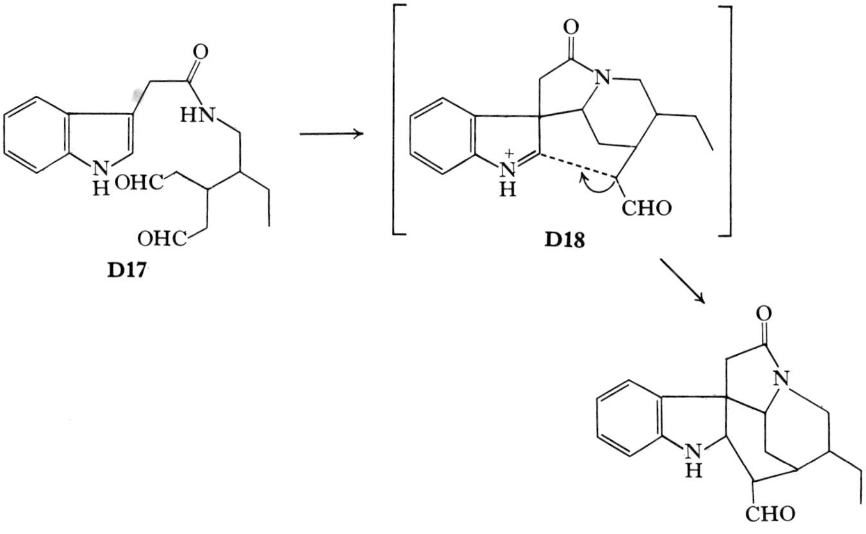

D17 → [D18] → D19

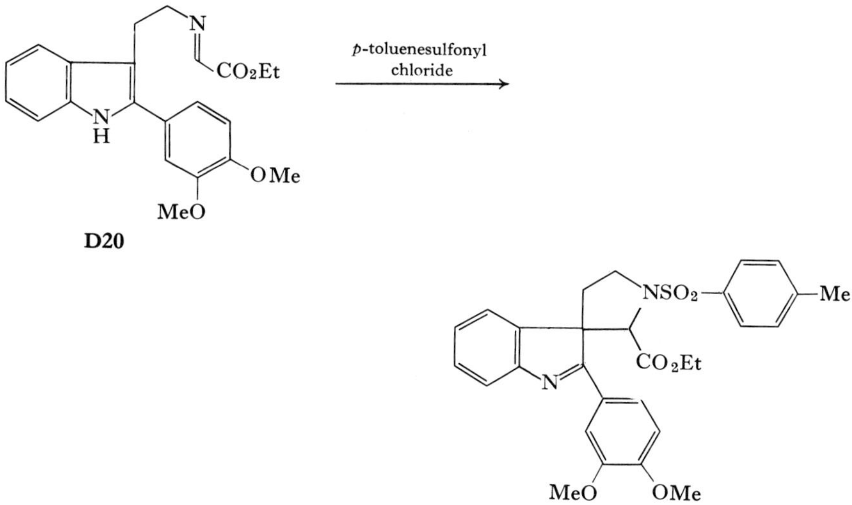

D20 —*p*-toluenesulfonyl chloride→ D21

Pyruvic acid derivatives are suitable carbonyl components in the cyclization of tryptamines to tetrahydro-β-carbolines. Carbethoxypyruvic acid and tryptamine react to give **D33** in 59% yield (65, 95).

The use of pyruvic acids in such condensations was apparently first reported by Hahn and Werner (66). The method has been applied particularly effectively for the conversion of tryptamines to yohimbanes via the

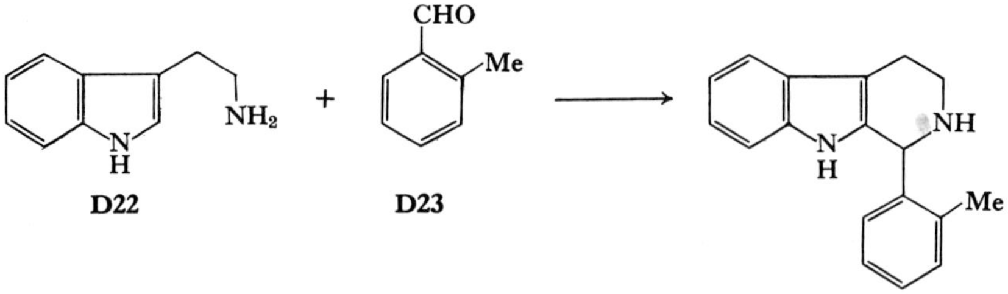

NH_2 + CHO Me → NH N H Me

D22 D23 D24 Ref. (190)

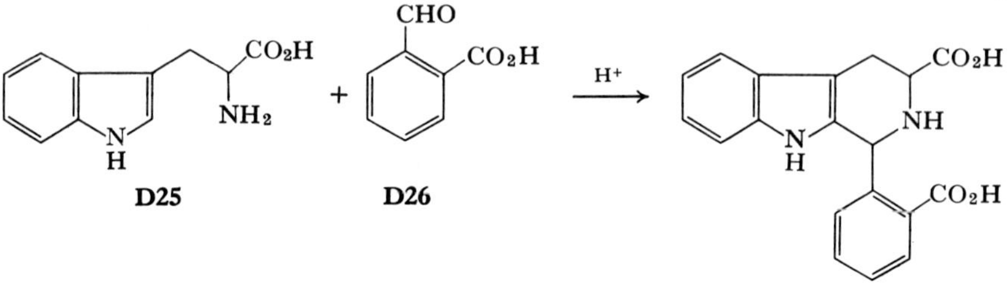

CHO CO_2H NH_2 + CO_2H H^+ → CO_2H NH CO_2H

D25 D26 D27

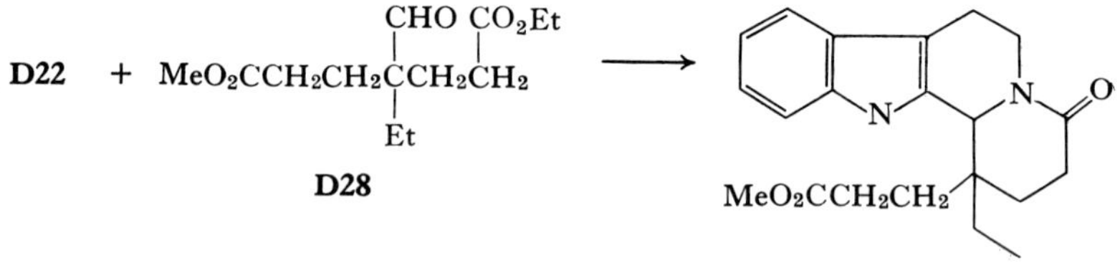

D22 + $MeO_2CCH_2CH_2C(CHO)(Et)CH_2CH_2CO_2Et$ → N N O $MeO_2CCH_2CH_2$

D28 D29

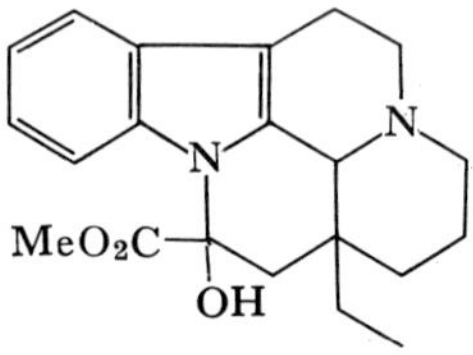

N N MeO_2C OH

D30

NH_2 + $EtO_2CCH_2COCO_2H$ → NH N H CH_2CO_2Et

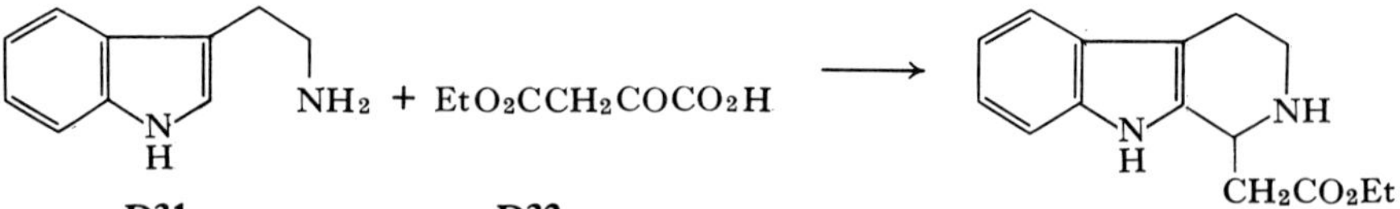

D31 D32 D33

route **D34** → **D36** (137). The final cyclization to the pentacyclic system can be effected by formaldehyde when the potential E-ring carries electron releasing substituents. The facile decarboxylation which accompanies cyclization is reminiscent of the decarboxylation of iboga alkaloids (Chapter II, D) and a similar acid-catalyzed mechanism can be suggested. Hahn and Werner (66) reported the isolation of the intermediate acid **D38** in one of their early examples of the reaction.

D31 + X–C_6H_4–$COCO_2H$ ⟶ **D35** —CH_2O→ **D36**

D34

D35

D36

Further elaboration of the β-carbolines is again possible with the use of appropriately substituted pyruvic acid derivatives.

A report that tryptamine condenses with *p*-*N*,*N*-bis(2-chloroethyl)amino benzaldehyde to give the Schiff base and not the tetrahydro-β-carboline is unsubstantiated by the data presented (51). Very careful control of reaction conditions permitted the isolation of a Schiff base from 6-methoxytryptamine and cyclohexanecarboxaldehyde derivative during the course of the synthesis of reserpine (208).

The report (212) that **D48** accompanies the reasonable products **D46** and **D47** in the condensation of *N*-methyltryptamine with formaldehyde is difficult to accept in view of the severe steric strain which would result from incorporation of a planar trans system into a seven-membered ring.

Table XLI contains a few examples of condensations of tryptamines with pyruvic acids. The reader is referred to the review of Whaley and Govindachari (205) and a recent thorough review (1) of both synthesis and reactions of carbolines for other examples.

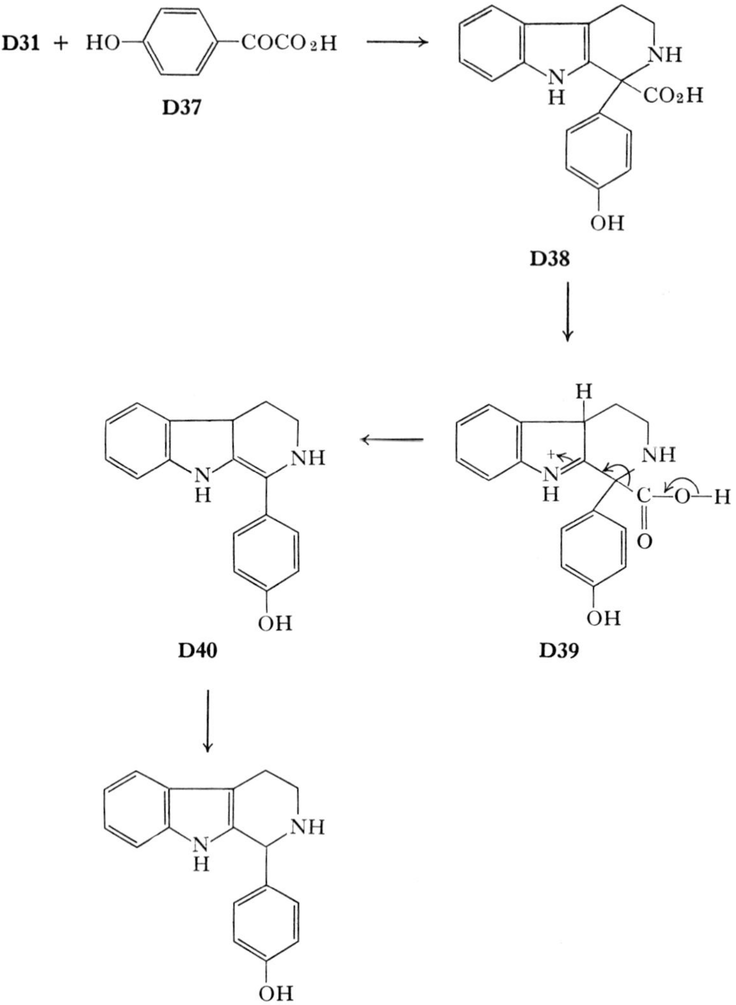

D31 + HO–C_6H_4–$COCO_2H$ ⟶ D38

D37

NH
N
H
CO_2H
OH

D38

↓

H
NH
+N
H
C–O–H
O
OH

D39 ⟶ D40

NH
N
H
OH

D40

↓

NH
N
H
OH

D41

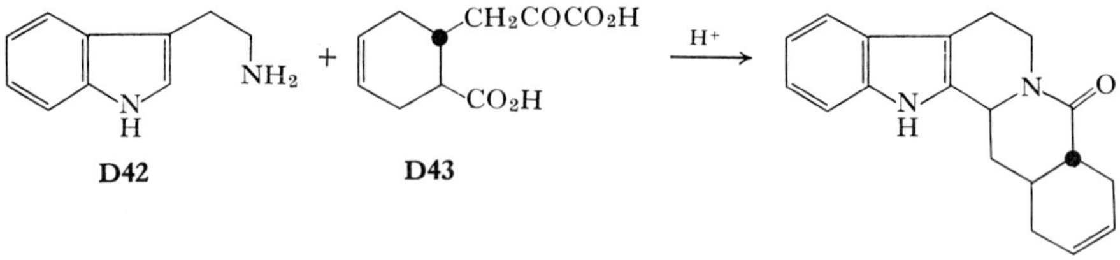

NH
N
H
NH_2

D42

+

CH_2COCO_2H
CO_2H

D43

$\xrightarrow{H^+}$

N
H
N
O

D44 Ref. (39)

TABLE XLI □ TETRAHYDRO-β-CARBOLINES BY CONDENSATION OF TRYPTAMINES AND PYRUVIC ACIDS

Tryptamine substituent	Pyruvic acid substituent	Yield (%)	Ref.
None	*m*-Methoxyphenyl	?	(173)
None	3,4-Dimethoxy-5-chlorophenyl	37	(33)
None	3,5-Dimethoxyphenyl	44	(137)
None	3,5-Dimethyl-4-hydroxyphenyl	?	(109)
5-Chloro	3,4-Dimethoxyphenyl	22	(33)
5-Methoxy	4-Hydroxy-3-methylphenyl	?	(108)

Condensations between ketones and tryptamine are less common than those involving aldehydes. The cyclization of **D49** to **D50**, however, can be effected by use of dilute methanolic hydrochloric acid. Similar reactions have been carried out by Wawzonek and co-workers.

Ref. (207)

The fact that these condensations result in the tetracyclic ring systems may be important to their success. Hester (75) has studied the reactions of tryptamine and its 5- and 6-methoxy derivatives with acetone. 6-Methoxytryptamine condenses with acetone at slightly acidic pH but the other two compounds are inert to cyclization under such mild conditions. This result

D51 → D52 Ref. (189)

D53 + $RCOCH_2CH_2CO_2H$ (D54) → D55 Ref. (191)

R = Me or Ph

D56 —acetone, pH 4.7→ D57 (95%)

has been explained in terms of an electron releasing effect of the 6-methoxy group. The exceptional magnitude of this effect is further emphasized by the fact that the Schiff bases of tryptamine and 5-methoxytryptamine were prepared in benzene under acid catalysis without any cyclization occurring. This unusually large substituent effect would seem to merit further investigation.

2. β-Carboline Derivatives from Bischler–Napieralski Cyclizations

Amides of tryptamines are cyclized to dihydro-β-carbolines under the conditions of the Bischler–Napieralski synthesis. A review published in 1951 (204) tabulates a number of successful cyclizations. A more recent review (1) has discussed additional aspects of the reaction. The cyclization has normally been effected by phosphorus oxychloride or phosphorus pentoxide. The mechanism outlined below is seen to be fundamentally similar to the intermolecular acylation of indoles by amides in the presence of phosphorus oxychloride (Chapter I, K). The possibility that initial attack is at C-3 should again be kept in mind.

The dihydro-β-carbolines obtained in the cyclization can be reduced to the tetrahydro derivatives. The Bischler–Napieralski cyclization was used

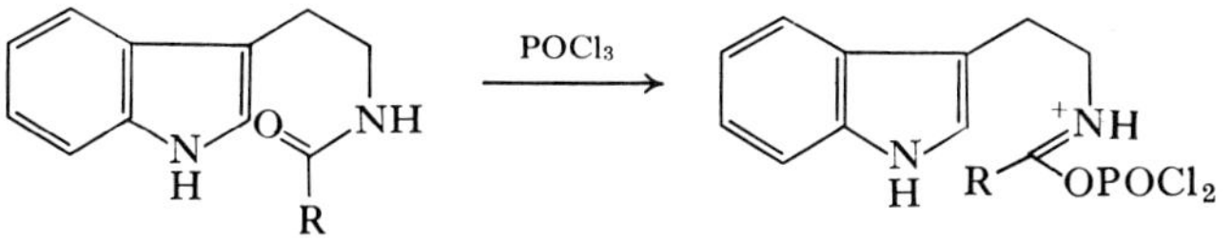

$POCl_3$

D58 **D59**

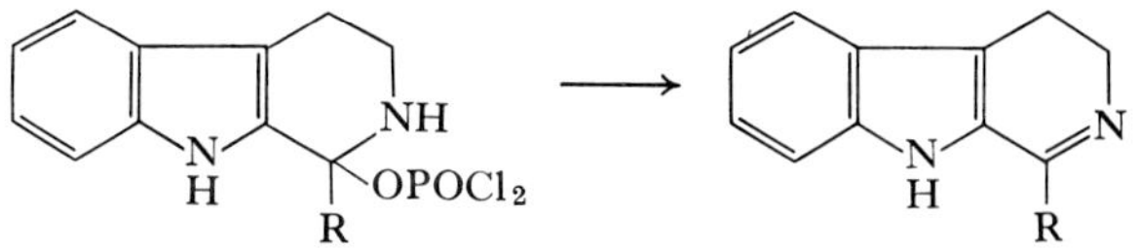

D60 **D61**

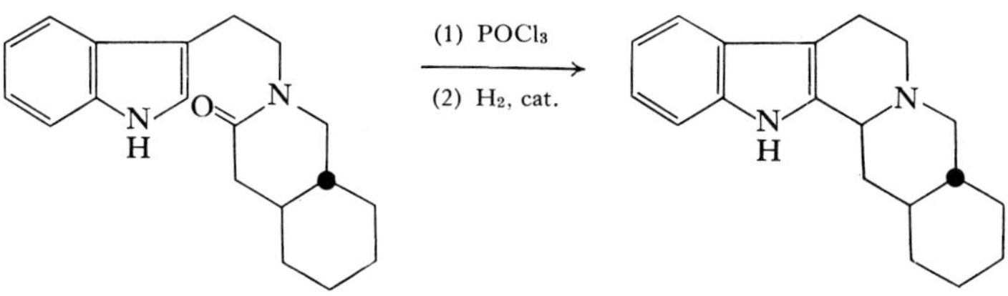

(1) $POCl_3$
(2) H_2, cat.

D62 **D63**

to advantage in synthesis of stereoisomers of yohimbane (**D63**) (170, 184). In certain cases it has been possible to isolate and characterize the intermediate dihydro-β-carboline. Wenkert and Wickberg (203) reported the the isolation of the perchlorate **D65** after the cyclization of **D64**.

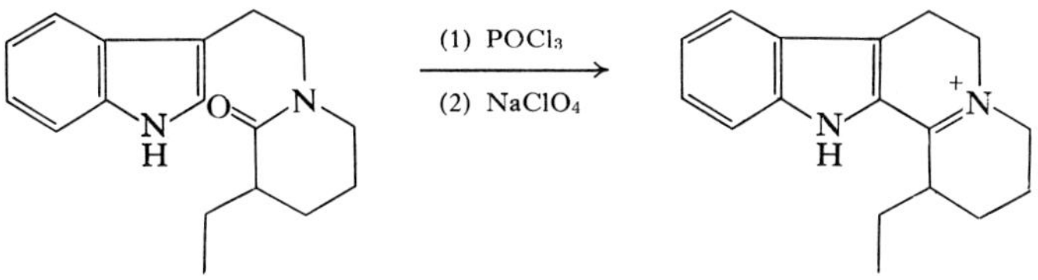

(1) $POCl_3$
(2) $NaClO_4$

D64 **D65**

The chloroform-soluble phosphate polymer "polyphosphate ester" has recently been recommended as an alternative cyclization catalyst (91). Yields with polyphosphate ester were reported to have been generally superior to those of comparable cyclizations with phosphorus pentoxide. Table XLII tabulates a number of examples of Bischler–Napieralski cyclizations of acyltryptamines.

A recently developed method of cyclization of tryptamine derivatives, which resembles the Bischler–Napieralski method to some extent, has been

TABLE XLII □ BISCHLER–NAPIERALSKI CYCLIZATION OF ACYL TRYPTAMINES

Tryptamine substituent[a]	Acyl group	Cyclization reagent	Yield (%)	Ref.
None	Formyl	Polyphosphate ester	73	(91)
None	Benzoyl	Polyphosphate ester	68	(91)
None	Cyclohexaneacetyl	Phosphorus pentoxide	low	(92)
None	*o*-Carboxyphenylacetyl	Phosphorus oxychloride	36	(151)
None	*o*-Methylphenylacetyl	Phosphorus pentoxide	63	(83)
α-Carboxy	Formyl	Polyphosphate ester	82	(91)
5-Methoxy	Phenylacetyl	Phosphorus oxychloride	?	(141)
7-Methoxy	Phenylacetyl	Phosphorus oxychloride	?	(141)

[a] Greek letters refer to side-chain substitution.

HgCl$_2$, MeCN

D66 → D67 (90%)

reported to give quite good yields of 3,4-dihydro-β-carbolines (129). The procedure involves cyclodesulfurization of thioamides with mercuric chloride. The example shown above is illustrative. Similar cyclizations can be catalyzed by acids. 3,4-Dihydro-β-carboline is formed in excellent yield from 3-(2-thioformamidoethyl)indole in acidic solution (56).

3. Formation and Cyclization of 1-[2-(3-Indolyl)ethyl] tetrahydropyridines

The treatment of a 1-[2-(3-indolyl)ethyl]tetrahydropyridine **D68** with acid generates, by C-protonation, an iminium bond which is favorably disposed for cyclization by electrophilic attack on the indole ring.

This general sequence has been very valuable in the construction of alkaloid skeletons. The main synthetic problem involves the generation of

H$^+$

D68 → D69 → D70

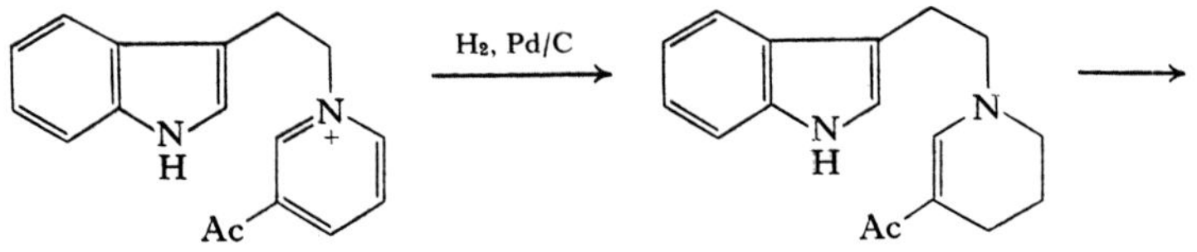

H_2, Pd/C

D71 D72

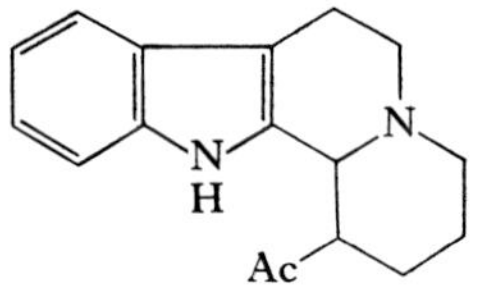

D73

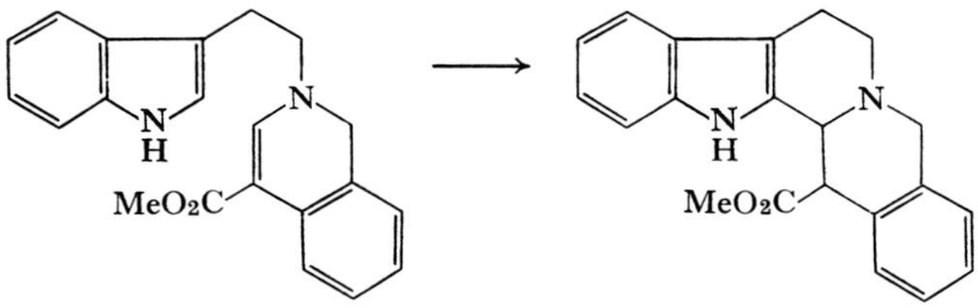

D74 D75

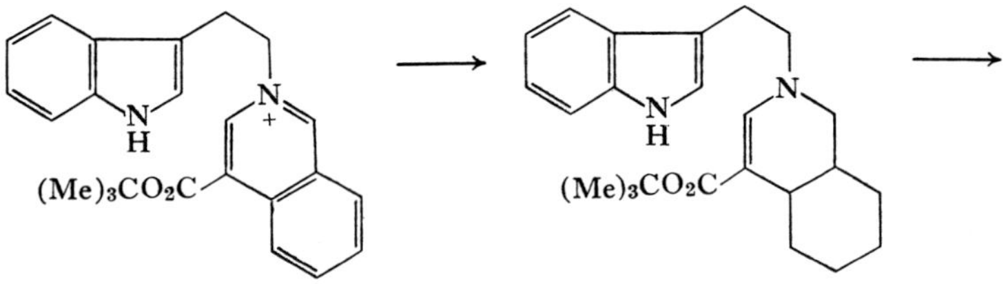

D76 D77

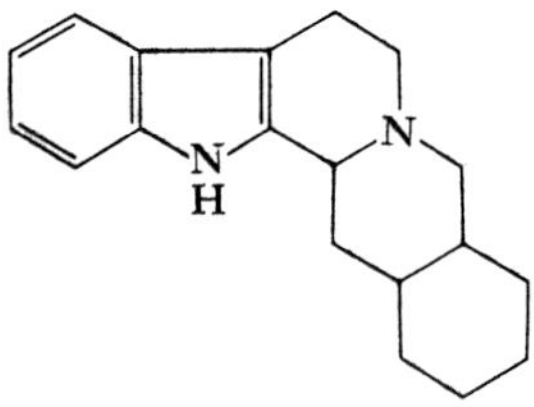

D78

the pyridine ring in the proper state of oxidation. This has been approached both via oxidation of piperidines and via selective reduction of pyridines. Very significant advances have been made in the reductive approach recently by Wenkert and co-workers. When a 3-acyl or 3-carbalkoxy group is present on a pyridine ring, reduction can be stopped at the tetrahydro stage (60, 142, 203). For example, reduction of **D71** over palladium–charcoal gives **D72**. Treatment of **D72** with acid effects cyclization to **D73** via a Pictet–Spengler cyclization. Wenkert and co-workers (198, 199) have since demonstrated the generality of both the reduction and cyclization steps. The syntheses of *d,l*-dihydrogambirtannine (**D75**) and *d,l*-epialloyohimbane (**D78**) are illustrative (197, 198).

The significance of the loss of the *t*-butoxycarbonyl group concomitant with cyclization is discussed by Wenkert and co-workers. A general summary of the work is available (196). (See also Chapter I, K.)

Prior to the development of the catalytic method, chemical reduction by lithium aluminum hydride had been only occasionally successful in effecting cyclization of indolylethylpyridinum salts. The reduction of **D79** to **D80** was documented in 1955 (140). An earlier report of a similar reductive cyclization (84) was found to be invalid when it was shown that the starting material had been assigned an incorrect structure (20).

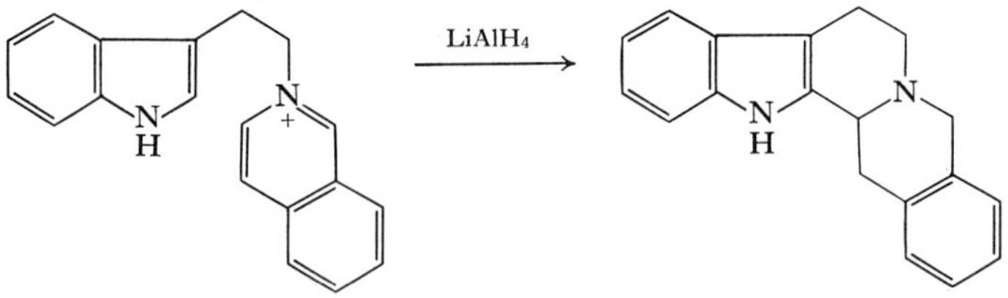

$LiAlH_4$

D79 **D80**

Early attempts to effect reductive cyclization of the simple pyridine **D81** failed, an uncyclized tetrahydro derivative **D82** being formed (50). Elderfield and Fischer found, however, that the isoquinolinium salt **D83** could be successfully cyclized by lithium aluminum hydride but not by sodium

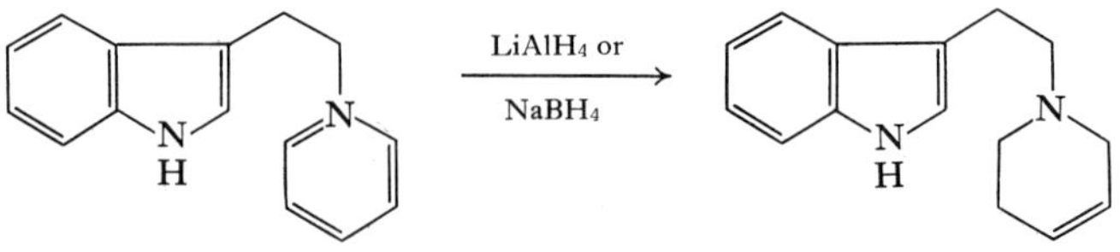

$LiAlH_4$ or $NaBH_4$

D81 **D82**

borohydride (49). Successful reductive cyclizations of indolylethylpyridines have been effected by Wenkert, Massy-Westrop, and Lewis (201). *N*-Substituted indoles were found to be cyclized effectively and lithium tri-*t*-butoxy aluminum hydride was used with moderate success in systems not substituted on the indole nitrogen. Potts and Liljegren observed that reduc-

D83 D84

tion with $LiAlH_4$ in tetrahydrofuran was more effective for formation of cyclized products than was reduction in ether (139).

D85 —$LiAlH_4$, ether-reflux→ D86 (33%)

D85 —$LiAlH_4$, tetrahydrofuran→ D87 (56%)

As Wenkert has pointed out (201), cyclization results when reduction ceases at the dihydro stage. Subsequent treatment with acid effects cyclization. If reduction proceeds to the tetrahydro stage, subsequent cyclization is not feasible, since the 1,2,3,6-tetrahydropyridines formed do not generate an iminium system when protonated.

D88 —H^+→ D89 → D90

An alternative route to the dihydropyridine system has been examined by Lyle and co-workers. Reduction of **D91** and **D92** with sodium dithionite gave the dihydropyridines **D93** and **D96** which were converted by mild acid to **D95** and **D96**, respectively (172).

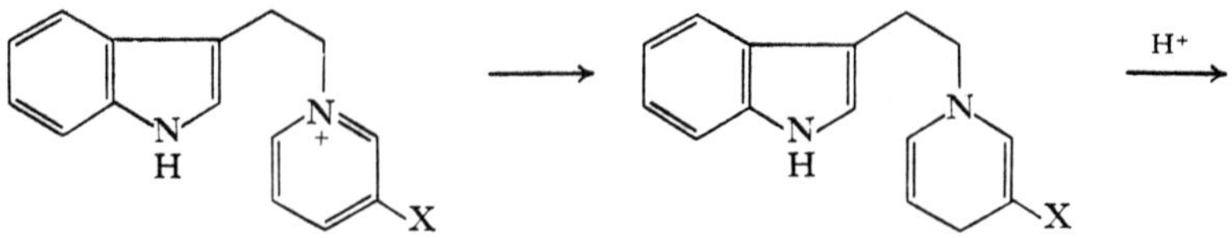

H+

D91 X = PhCO
D92 X = CN

D93
D94

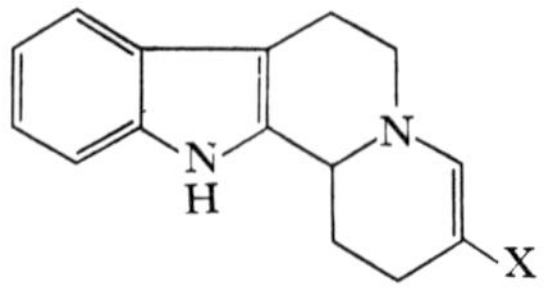

D95
D96

A brief report of a catalytic reduction of an indolylethylpyridinium salt to a tetrahydropyridine which cyclized on standing has appeared (177).

Oxidation of piperidines to tetrahydropyridines by mercuric acetate has been the principal oxidative approach to cyclization of 1-(2-indolylethyl)-piperidines. Wenkert and Wickberg (202) examined the oxidation of **D97** and **D100**. The isolated products of principal interest are shown below:

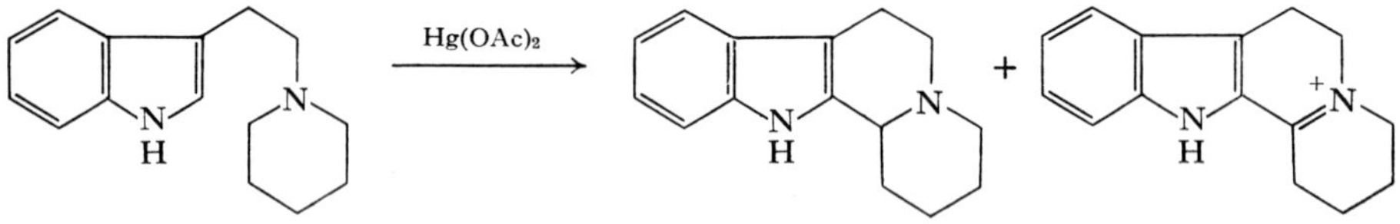

$Hg(OAc)_2$

D97 **D98** **D99**

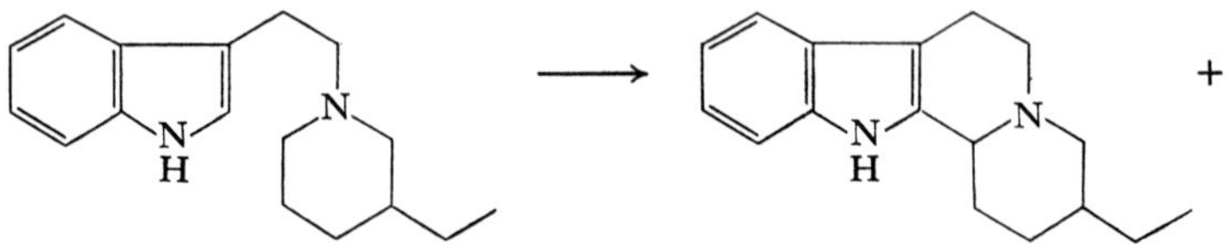

D100 **D101**

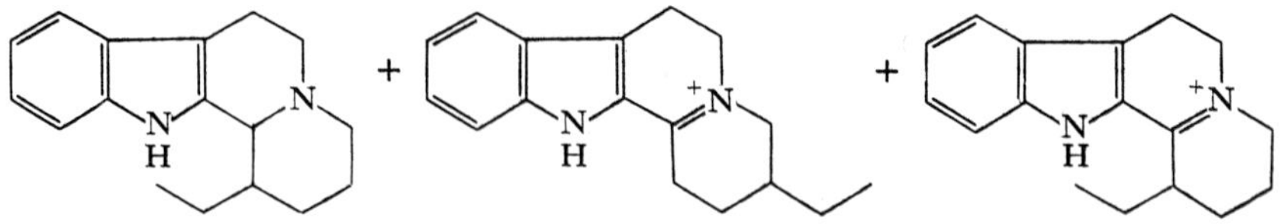

D102 **D103** **D104**

The overoxidation products which are produced can be reduced with sodium borohydride to give the desired cyclization products. Morrison and co-workers have carried out the related oxidative cyclization **D105** → **D106**

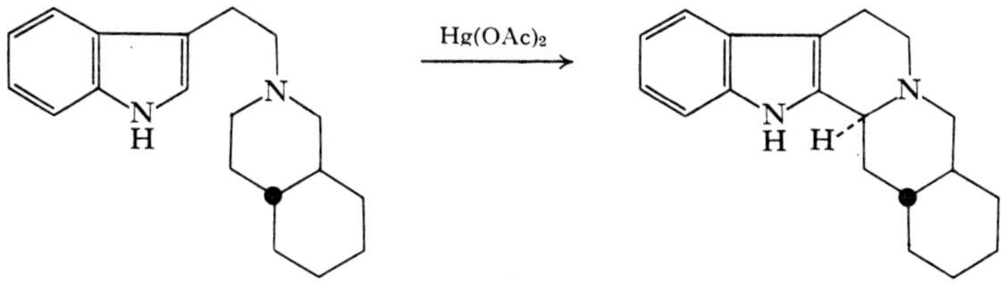

Hg(OAc)$_2$

D105 **D106** (31%)

in 31% yield (120). The stereochemical outcome of these reactions is critical to their synthetic usefulness in the alkaloid area, and has been discussed by Wenkert (202) and Morrison (120).

E. SYNTHESIS OF INDOLE ALKALOIDS

While a complete discussion of synthetic approaches to the indole alkaloids is clearly beyond the scope of this book, it seems pertinent to discuss some of the recent advances in this area, particularly those reactions which directly involve the indole ring. It is in the area of alkaloid synthesis that the synthetic techniques for elaboration of the indole ring have met the most exacting requirements to date and much interesting chemistry is found in the area. We direct our attention here to the reactions used to generate the fundamental alkaloid skeletons, leaving the often elegant synthetic chemistry involved in securing appropriate starting materials largely undiscussed.

1. Iboga Alkaloids

The iboga skeleton is represented by ibogamine (**E1**) and is seen to consist of an indole ring and an isoquinuclidine ring fused by ring C, a seven-membered ring. One successful approach to the system involved synthesis of the ketone **E2**. Fischer cyclization then gave ibogamine (150).

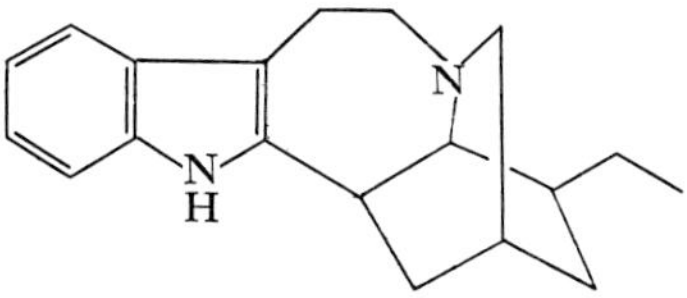

E1

The available details do not indicate that cyclization to a 3*H*-indole was a competing process. The vinylhydrazine tautomer required for cyclization to the more substituted α-carbon is sterically strained as it contains a bridgehead multiple bond and this factor is presumably involved in directing the cyclization in the desired direction.

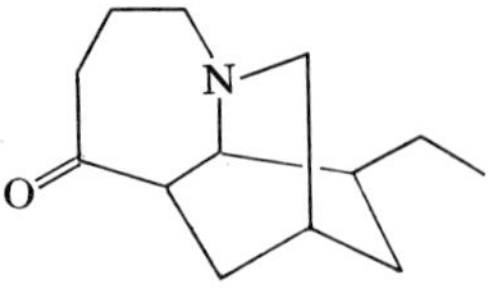

E2 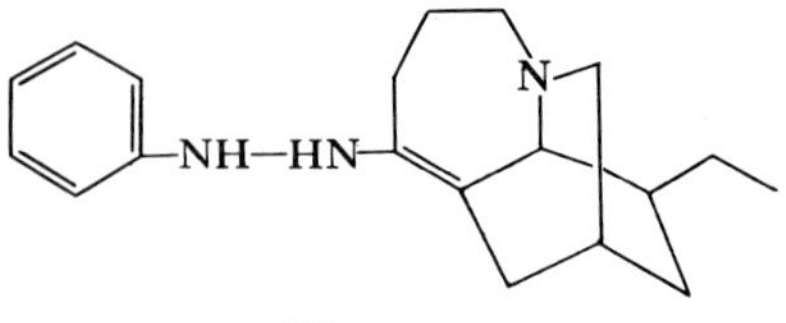 **E3**

An alternative approach has been to generate intermediates of structure **E4** in which an electrophilic center is generated in the isoquinuclidine ring. The 7-membered ring can then be formed by an intramolecular alkylation of C-2 of the indole ring. Thus, Huffman, Rao, and Kamiya (79) employed the cyclization **E6** → **E7** as a key step in their synthesis of desethylibogamine.

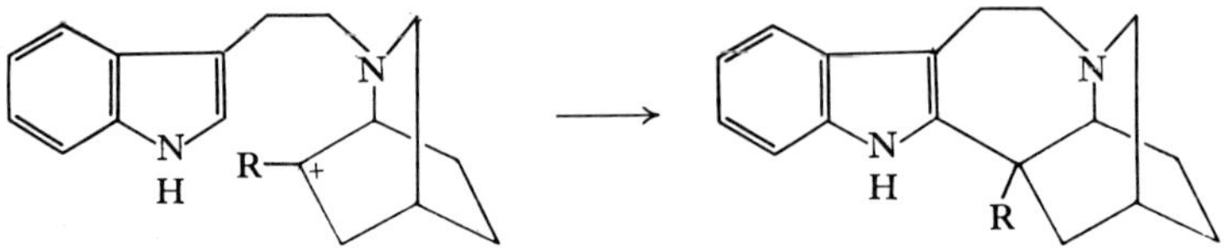

E4 → **E5**

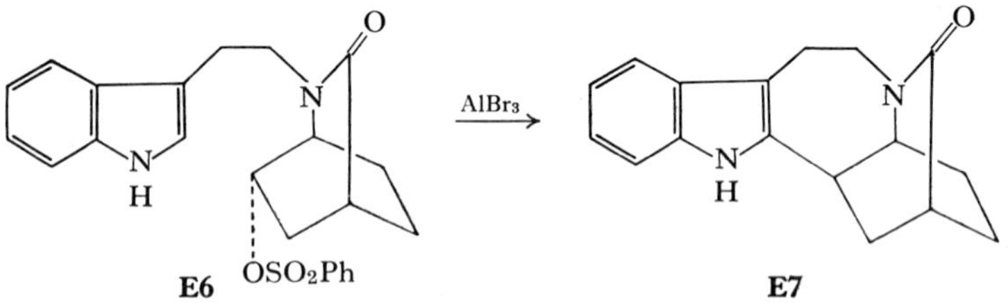

E6 $\xrightarrow{AlBr_3}$ **E7**

Although the cyclization is formally accommodated by the general scheme **E4** → **E5**, it has been pointed out that $AlBr_3$ may have catalyzed the cyclization by a process more complex than simple ionization of the tosyl group. A two-step conversion of **E6** → **E7** was effected under mechanistically less ambiguous circumstances but in low yield.

E6 → **E8** → **E7**

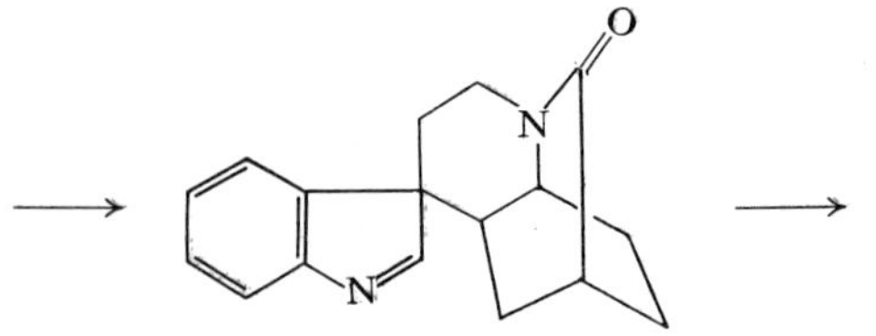

The first total synthesis of a completely elaborated iboga alkaloid was carried out in the laboratory of Büchi (25). Here the synthesis was accompanied by a fascinating series of rearrangements. The isoquinuclidone **E9** was treated with *p*-toluenesulfonic acid in acetic acid and unexpectedly gave the rearranged cyclization product **E10**.

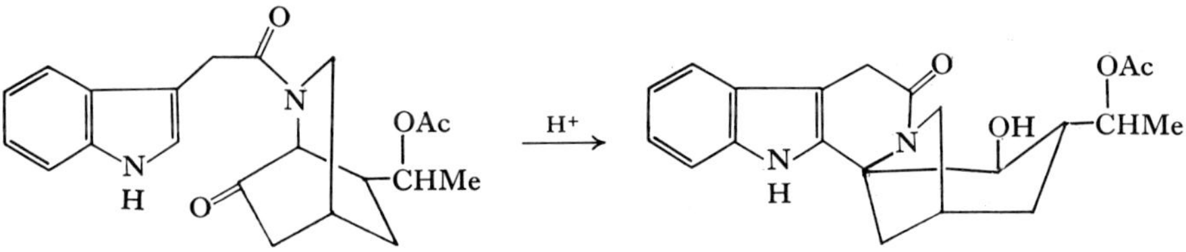

$\xrightarrow{H^+}$

E9 **E10**

The amine **E11** also cyclized with rearrangement.

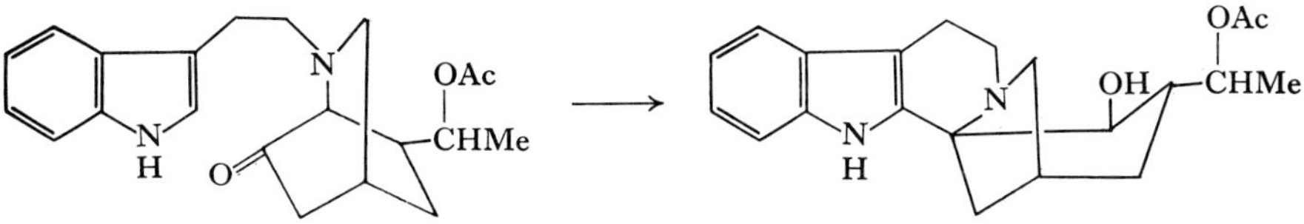

E11 **E12**

The iboga skeleton was reestablished when the ketone **E13**, derived from **E12**, was reduced with zinc and acetic acid.

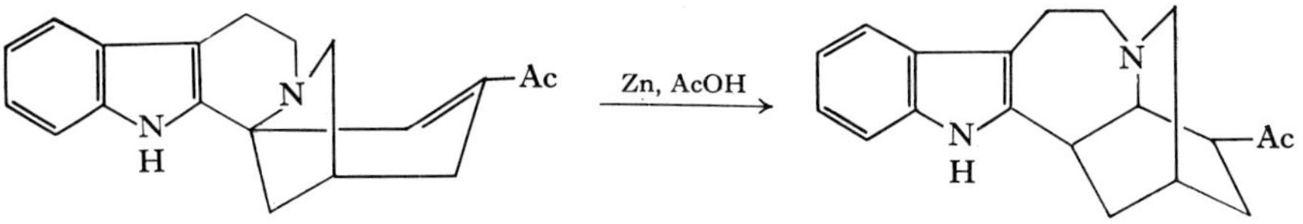

Zn, AcOH

E13 **E14**

The rearrangement which occurs during the cyclization of **E11** to **E12** may involve ionization of a leaving group from an indolylcarbinyl carbon.

The alcohol **E20** cyclized to **E22** without rearrangement. An intermediate analogous to **E18** presumably cyclized intramolecularly to give **E22**. The hydroxyl group would be expected for steric reasons to attack the aziridinium ion **E21** to form a five-membered ether ring.

The ketone **E23** also apparently cyclizes without rearrangement (121). Büchi and co-workers effected the cyclization of the ketal **E25** with *p*-toluenesulfonic acid in refluxing benzene. An unrearranged iboga skeleton was also formed in this case.

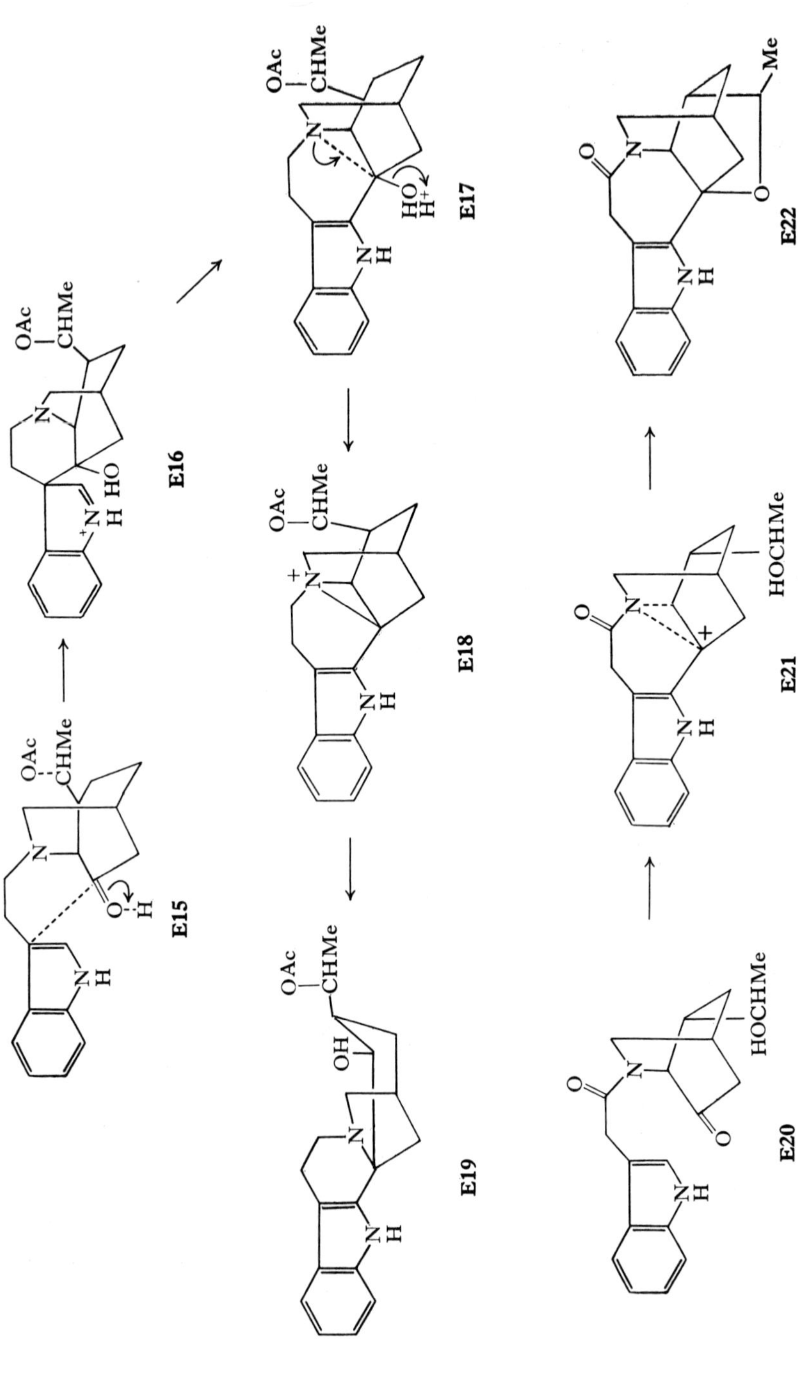

E15

E16

E17

E18

E19

E20

E21

E22

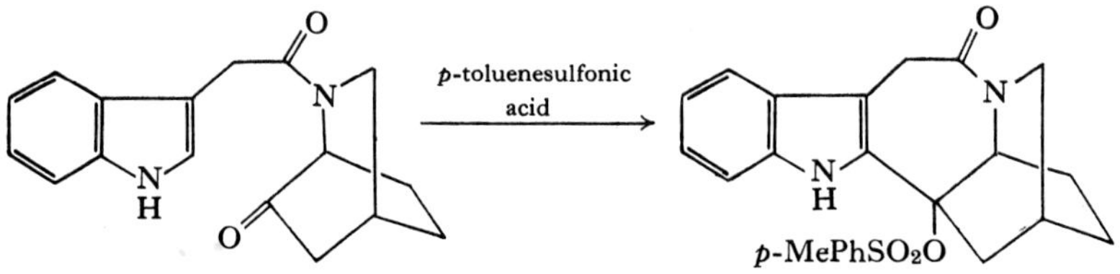
O
N
p-toluenesulfonic
acid
N
H
O
p-MePhSO₂O

E23 E24

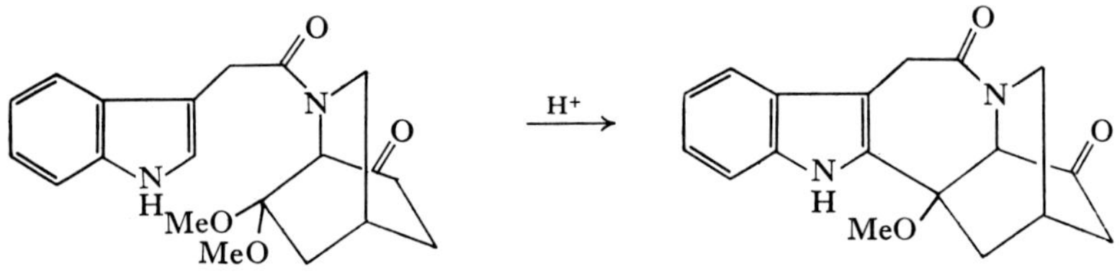
O
N
O
H+
N
H
MeO
MeO

E25 E26

Each of the preceding cyclizations involves an electrophilic substitution on a 3-alkylindole. It is likely that, as in other electrophilic cyclizations, the initial electrophilic attack occurs at C-3 and that rearrangement to the iboga skeleton follows. In this regard it can be noted that the indolenine **E8** prepared by Huffman, Rao, and Kamiya (79) rearranged under acidic conditions to **E7** with formation of the iboga skeleton. Formation of a six-membered ring in the initial stage of the cyclization could help to explain the relative efficiency with which the seven-membered C-rings are eventually formed.

A rather different approach to the iboga skeleton which parallels the Wenkert hypothesis (195) for biosynthesis was effected by Kutney and co-workers (101). The intermediate **E27** is converted by mercuric acetate into an epimeric mixture of compounds **E28** along with other products which are discussed in Section E, 2. The igoba skeleton arises by addition of the methoxycarbonyl-substituted carbon to the imine linkage generated by oxidation of the piperidine ring.

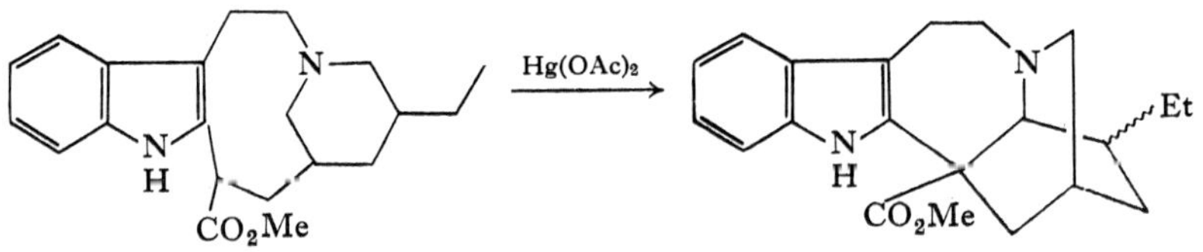
N
N
H
CO₂Me
Hg(OAc)₂
Et

E27 E28

The synthesis of the important intermediate **E27** is outlined in Section E, 2.

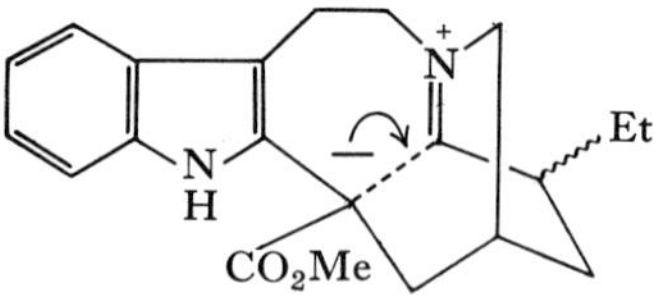

E29

2. Monomeric Vinca and Aspidosperma Alkaloids

The tetracyclic compound **E33** has served as a key intermediate in the synthesis of a number of the monomeric vinca alkaloids (102). It is synthesized from the aldehyde **E31** and tryptamine in a reaction involving cyclization to a β-carboline followed by closure of a lactam ring, reduction, and intramolecular alkylation.

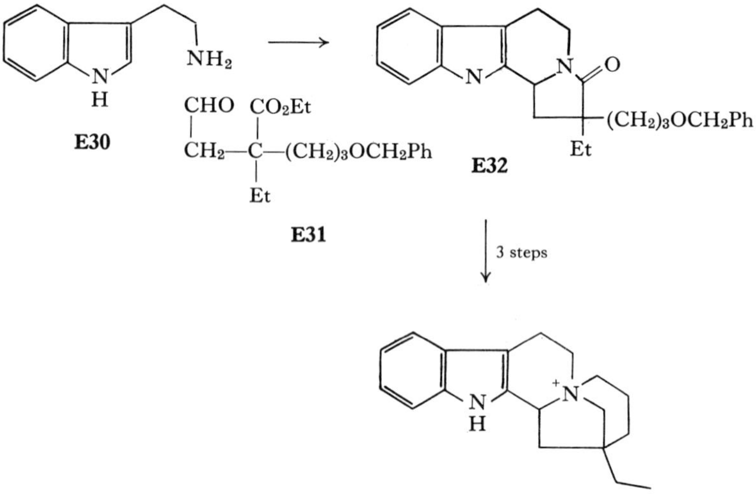

E30 E31 E32 E33

3 steps

Conversions of **E33** into various alkaloids are illustrated in the scheme which follows. Many of these reactions serve to emphasize the reactivity of the bond between C-3 and the nonindolic nitrogen toward nucleophilic attack.

The displacement reaction with cyanide ion was used earlier by Harley-Mason (69) in the synthesis of **E39**. It seems to be more efficient than an alternative route for the same transformation applied by Kutney and co-workers in the conversion **E38** → **E39** (104).

LiAlH$_4$ or Li/NH$_3$ ← E33 → $^-$CN

E37
dl-quebrachamine

E34

(1) H_2O, ^-OH
(2) CH_2N_2

Hg(OAc)$_2$ or O_2, Pt

CO_2Me

E36
vincadifformine

CO_2Me

E35
dl-vincadine and
dl-epivincadine

$^-$CN

E38

CN

E39

E38 —Na/NH$_3$→

E40

$(Me)_3COCl$

Cl

E39 ←$^-$CN—

E41

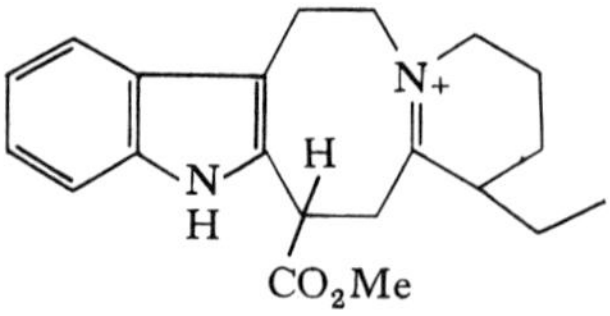

E42

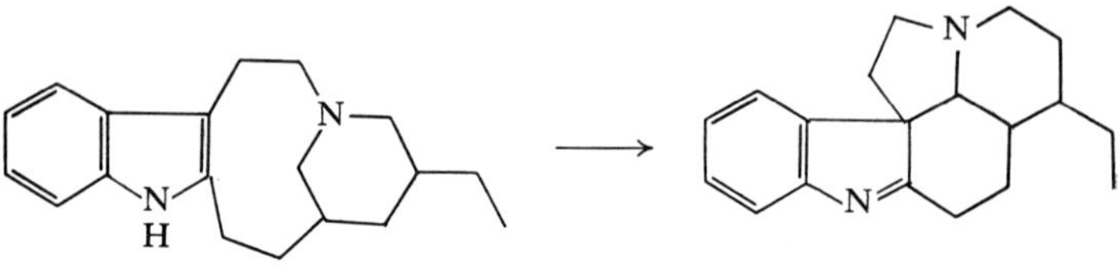

E43 → E44

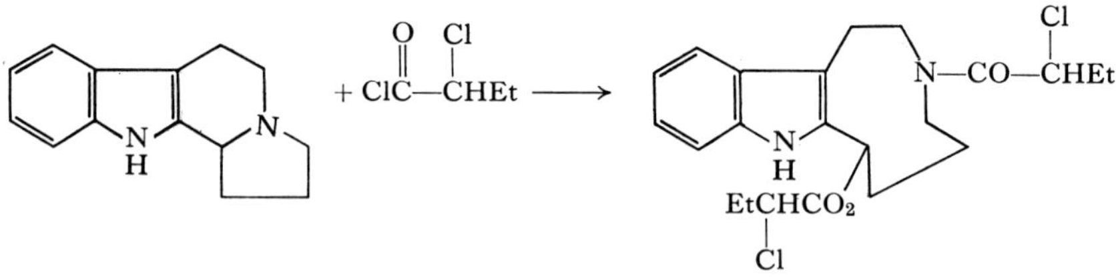

E45 + E46 ($ClC(=O)—CHClEt$) → E47

E47 → (1) ^{-}OH (2) MnO_2 → E48

E48 → $(Me)_3CCO^-$ → E49

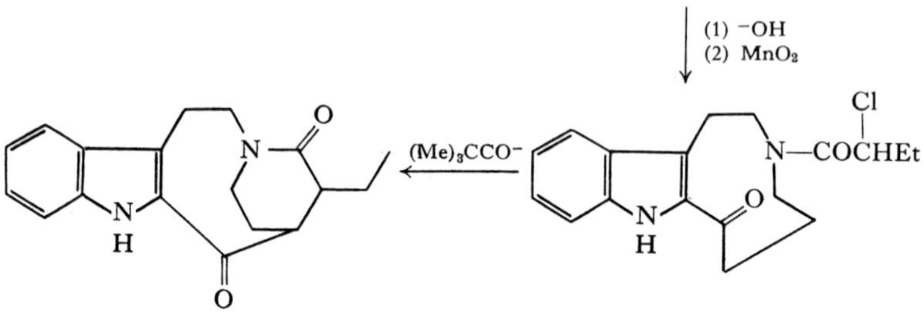

E49 → (1) NH_2NH_2, ^{-}OH (2) $LiAlH_4$ → E50

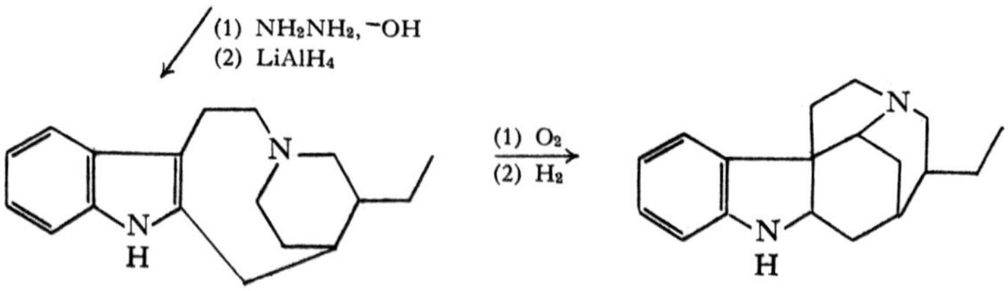

E50 → (1) O_2 (2) H_2 → E51

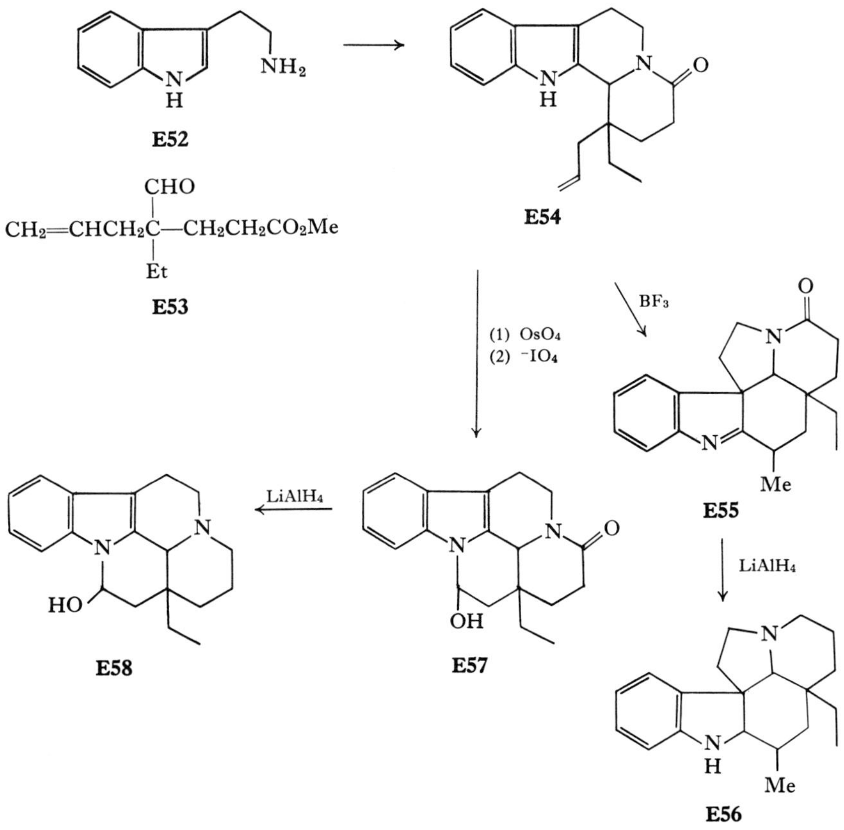

E52

$CH_2{=}CHCH_2C(CHO)(Et){-}CH_2CH_2CO_2Me$

E53

E54

(1) OsO_4
(2) $^{-}IO_4$

BF_3

E55

LiAlH₄

E56

LiAlH₄

E57

E58

The transformation of the vincadine (**E35**) to vincadifformine (**E36**) involves cyclization of the dehydro compound **E42**. The reaction is related to the oxidative approach to tetrahydro-β-carbolines, but cyclization occurs at C-3 rather than C-2 of the indole ring. Kutney and his group have used similar transannular cyclization to effect the oxidative cyclization of dihydrocleavamine (**E43**) to **E44** (105) and an analogous cyclization of carbomethoxydihydrocleavamine (100). This type of transannular cyclization is proving to be an exceedingly useful approach to the syntheses of pentacyclic indole alkaloids and model systems. This approach was also used, for example, in the synthesis of tubifolidine (**E51**) (40). In this synthesis the iminium species needed for transannular electrophilic substitution was generated by catalytic oxidation of the piperidine **E50** (153).

Harley-Mason and co-workers have synthesized the β-carboline **E54** from the aldehyde ester **E53**. Conversion of **E54** to eburnamine (**E58**) and 3-methylaspidospermidine (**E56**) is outlined in the scheme above (11).

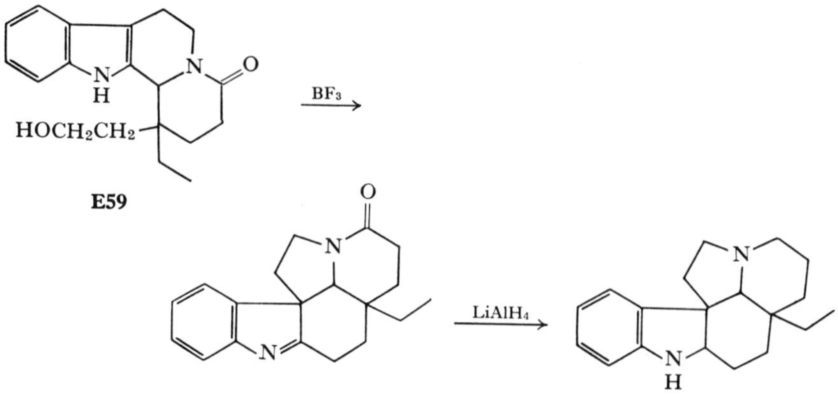

BF_3

E59

$LiAlH_4$

E60 **E61**

Aspidospermidine (**E61**) was later obtained by a route that parallels the synthesis of **E56** (71). In both of the latter syntheses a skeletal rearrangement has been used to advantage. The rearrangement is presumably triggered by formation of a carbonium ion in each case. Two subsequent paths for the rearrangement can be considered but specific proposals on the details of the process have not yet been put forward. These rearrangements involve the

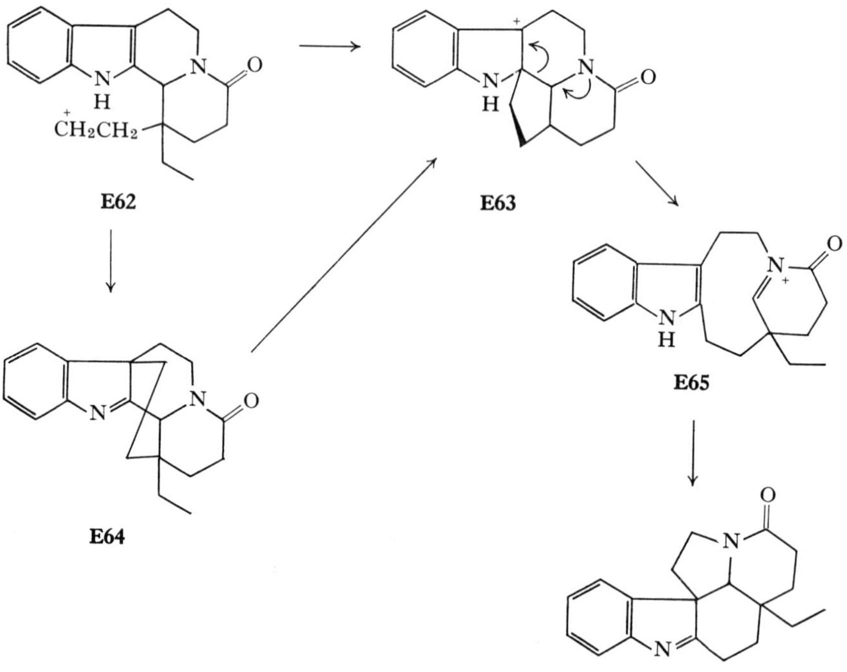

E62 **E63**

E65

E64

E66

question of C-2 versus C-3 as the initial site of electrophilic attack in indoles. If initial attack is at C-3, **E64** results and subsequent rearrangement to **E63** would initiate one possible path to the aspidospermine skeleton. If the initial alkylation is at C-2, **E63** is formed directly.

3. Dimeric Vinca Alkaloids

The "dimeric" vinca alkaloids such as vinblastine (**E67**) have attracted attention because of their anticancer activity (123). No total synthesis of such substances has yet been accomplished but partial syntheses have been effected. The reaction **E68** + **E69** → **E70** can serve as an example for discussion (70).

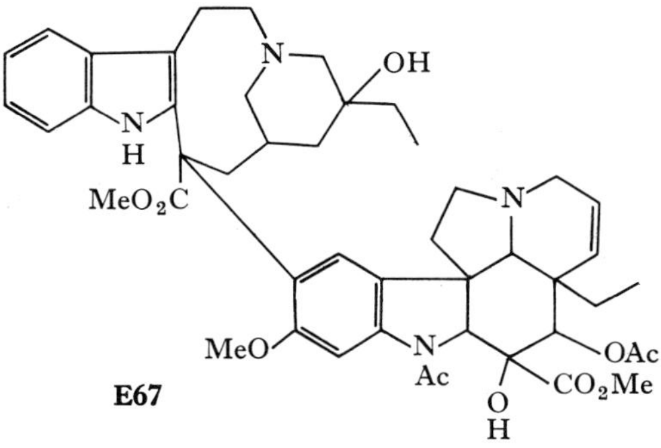

N OH N H MeO_2C N MeO N Ac OAc CO_2Me O H

E67

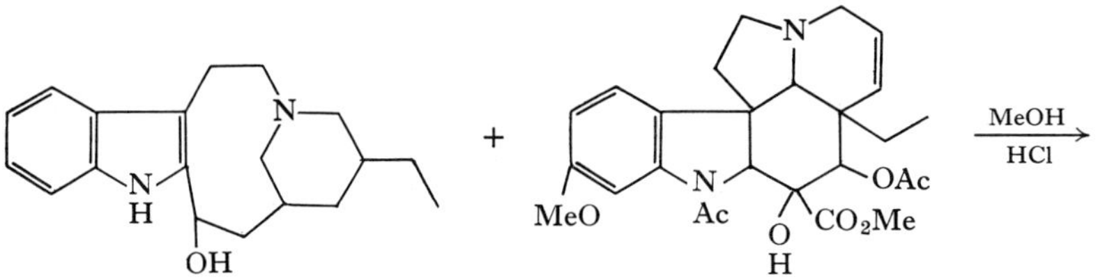

N N H OH

E68

+

N MeO N Ac OAc CO_2Me O H

E69

$\xrightarrow[\text{HCl}]{\text{MeOH}}$

N N H N MeO N Ac OAc O CO_2Me H

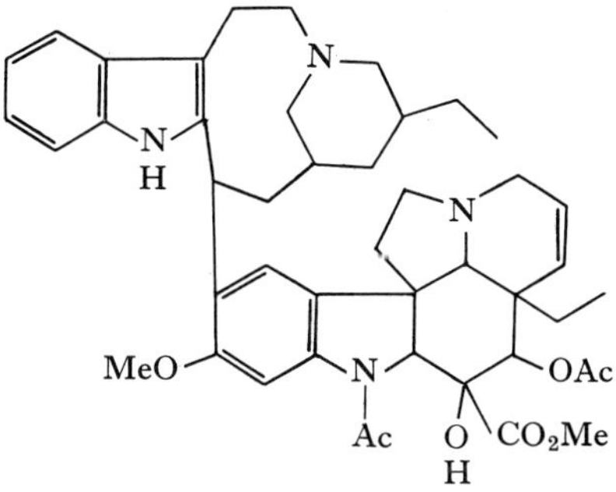

E70

The "dimeric" vinca alkaloids are characterized by bonds between C-16 and a methoxyindoline ring such as that in vindoline (**E69**). Such bonds have been formed (27, 147) by treatment of hydroxy compounds such as **E68** with acid, which generates positive ions such as **E71**. A bond is formed by subsequent electrophilic attack by the ion **E71** on a reactive methoxy-substituted aromatic ring. Related reactions are presumably the source of such alkaloids in the plant.

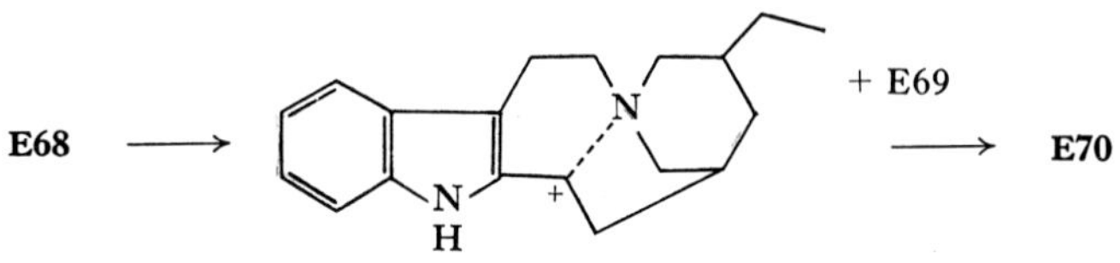

E68 ⟶ **E71** $\xrightarrow{+\ E69}$ **E70**

E71

Reduction of vinblastine in acidic media cleaves the molecule into velbanamine (**E72**) and deacetylvindoline. Velbanamine has been synthesized by Büchi and co-workers (26).

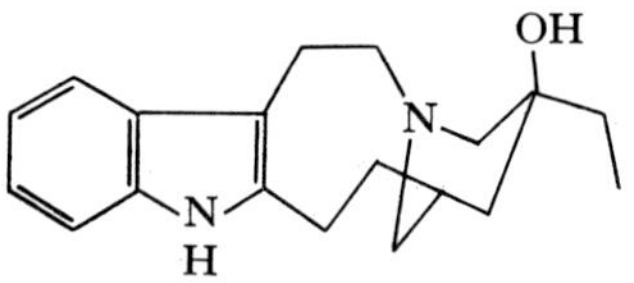

E72

4. Alkaloids of the Indoloquinolizidine and Sarpagine Families

The indoloquinolizidine skeleton is common to a large group of indole alkaloids such as corynantheidine (**E73**). The tetrahydro-β-carboline

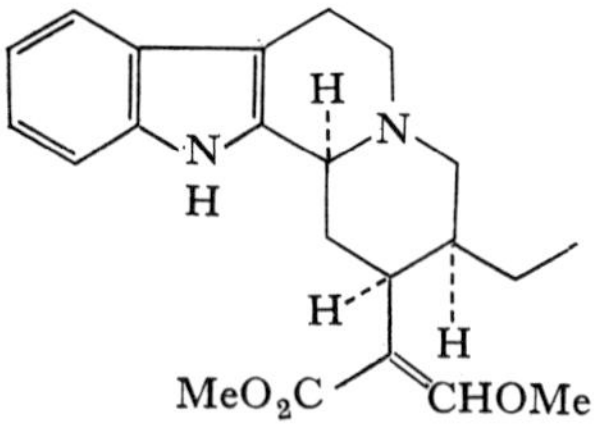

E73

skeleton is present and most of the successful syntheses of members of this group have used reactions which form the tetrahydro-β-carboline nucleus as a key step. Thus, Wenkert's synthesis of corynantheidine employs the reduction of a pyridine to a tetrahydropyridine followed by acid-catalyzed

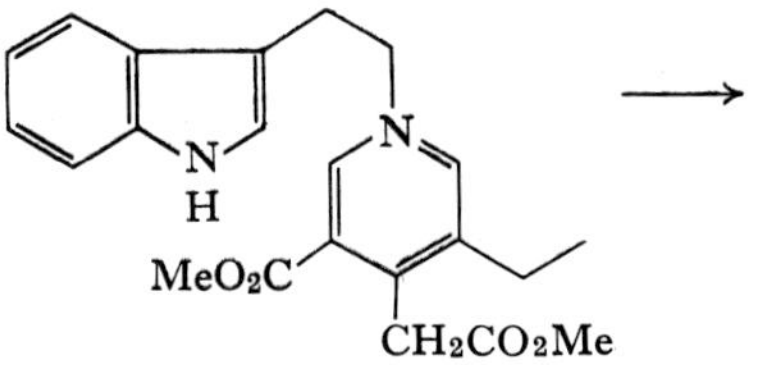

E74

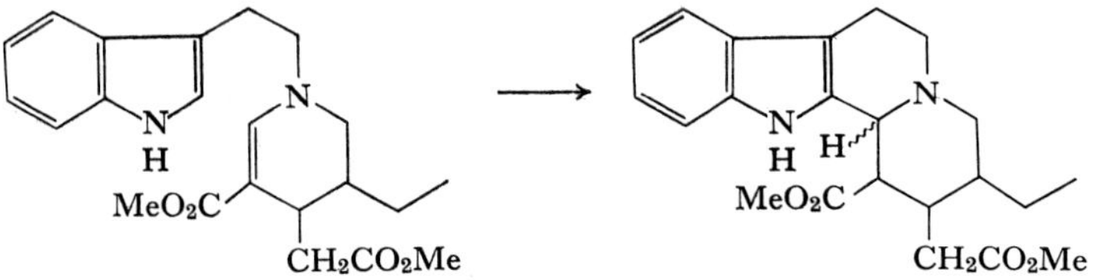

E75 → **E76**

several steps

E73

cyclization (200). An alternative synthesis of corynantheidine starts with the β-carboline system intact (192). The synthesis of the stereoisomeric dihydrocorynantheine used a Bischler-Napieralski cyclization to form the

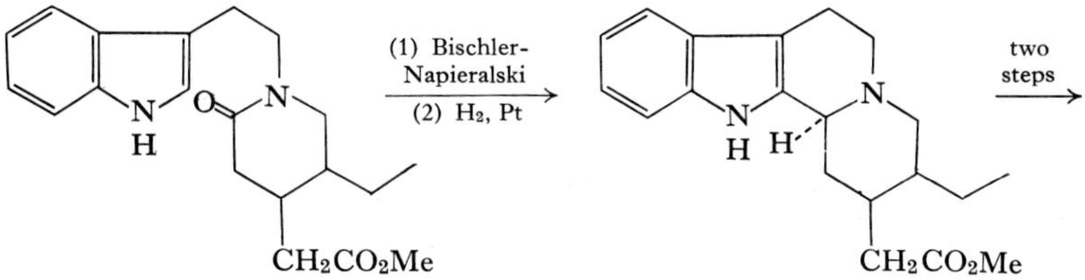

(1) Bischler-Napieralski
(2) H_2, Pt

two steps → **E73**

basic skeletal system (182). A Bischler-Napieralski cyclization was also employed in the synthesis of ajmalicine (**E77**) (183). In the Woodward synthesis of reserpine (**E78**), after the critical substitution and stereochemistry of potential ring E was established, the synthesis was brought near consumation by a Bischler-Napierelski cyclization (208).

The alkaloid eburnamonine (**E90**) has been obtained by a Bischler–Napieralski cyclization and two modifications of the Pictet–Spengler type tetrahydro-β-carboline synthesis (10, 203).

In the synthesis of yohimbine (**E100**) the β-carboline ring was established by briefly heating the dialdehyde **E94** with dilute phosphoric acid (185). The dialdehyde is generated from **E93** by cleavage with periodate.

The early stages of the synthesis (115) of ajmaline are similar in some respects. The cyclopentene **E101** was cleaved by osmium tetroxide and sodium metaperiodate to the carbinolamine **E102**. Cyclization to the 2-carbon occurred, generating **E103** which was interrelated with ajmaline.

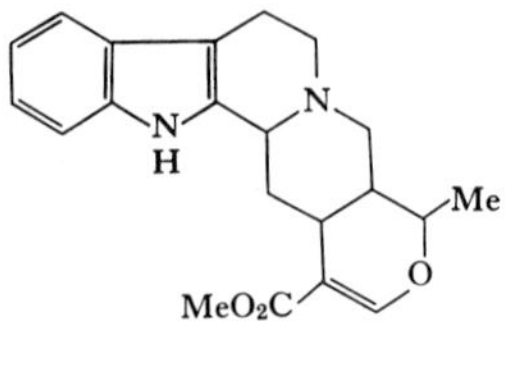

E77

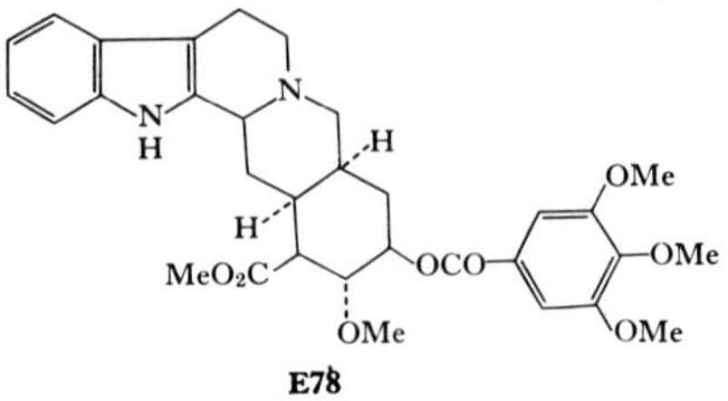

E78

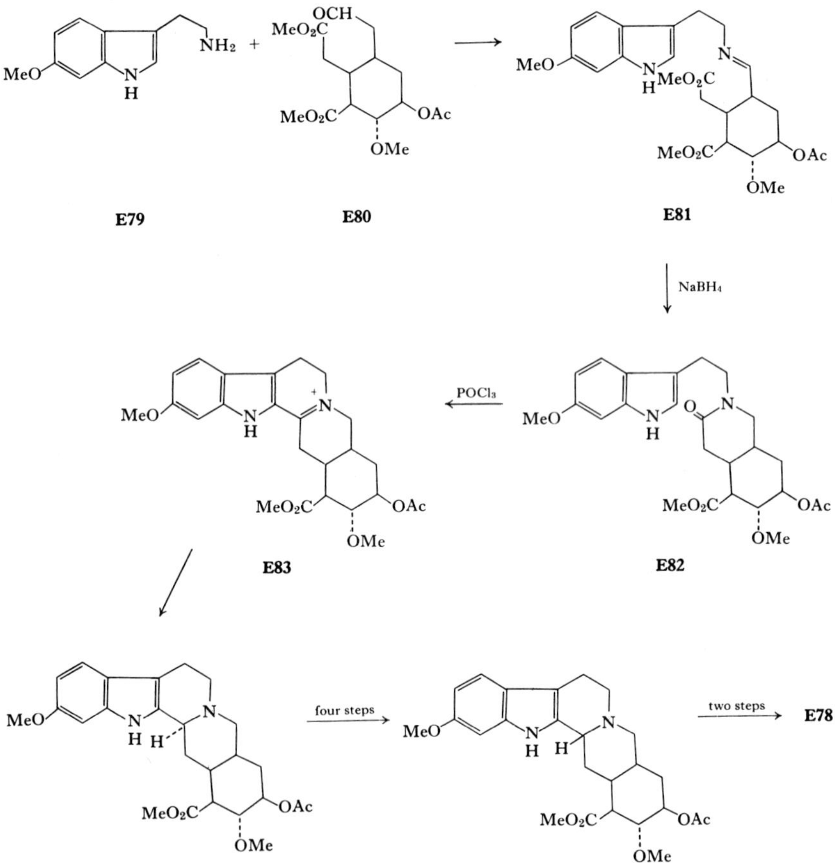

E79 + **E80** → **E81**

$NaBH_4$

E82

$POCl_3$

E83

E84

four steps

E85

two steps → **E78**

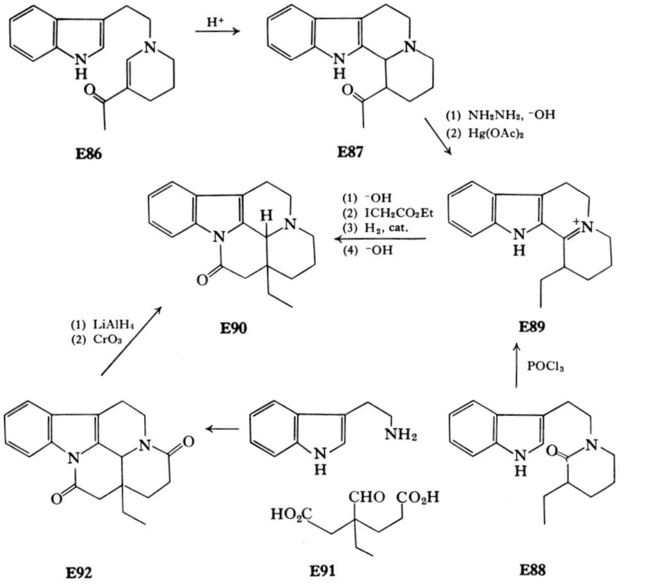

H^+

E86 E87

(1) NH_2NH_2, ^-OH
(2) $Hg(OAc)_2$

(1) ^-OH
(2) ICH_2CO_2Et
(3) H_2, cat.
(4) ^-OH

E90 E89

(1) $LiAlH_4$
(2) CrO_3

$POCl_3$

NH_2

CHO CO_2H
HO_2C

E92 E91 E88

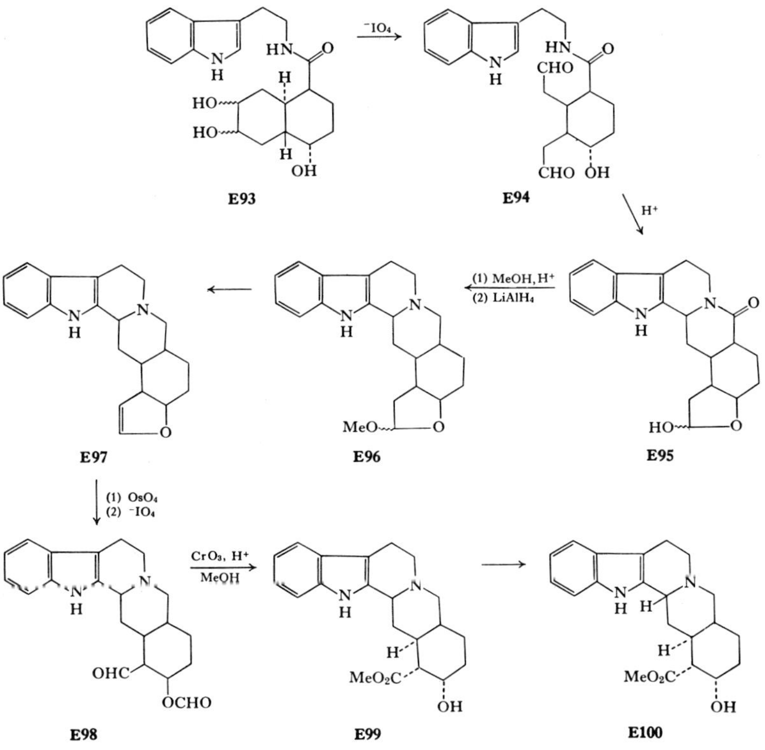

$^-IO_4$

HO HO OH CHO CHO OH

E93 E94

H^+

(1) MeOH, H^+
(2) $LiAlH_4$

MeO HO

E97 E96 E95

(1) OsO_4
(2) $^-IO_4$

CrO_3, H^+
MeOH

OHC OCHO MeO_2C OH

E98 E99 E100

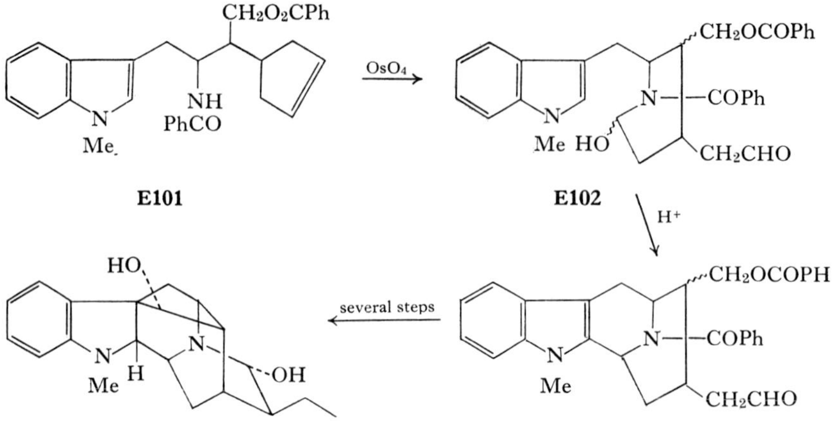

CH2O2CPh; NH PhCO; N Me; E101 —OsO4→ E102 (CH2OCOPh, N—COPh, CH2CHO, Me HO) —H+→ E103 (CH2OCOPH, N—COPh, CH2CHO, N Me) —several steps→ E104 (HO, N, Me H, OH)

5. Alkaloids of the Strychnine Structural Family

Synthetic approaches to strychnine (**E105**) and related alkaloids in which the β-carbon atom of the dihydroindole nucleus is disubstituted can involve alkylation of a disubstituted indole. The Woodward approach to strychnine

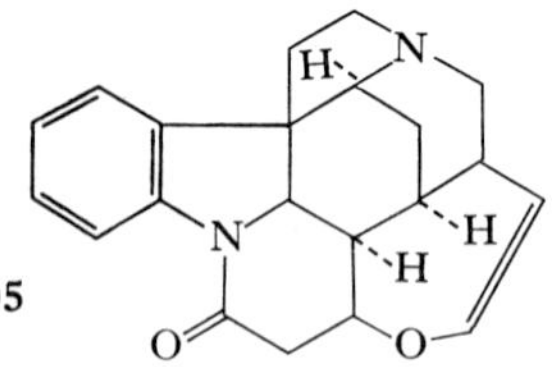

E105

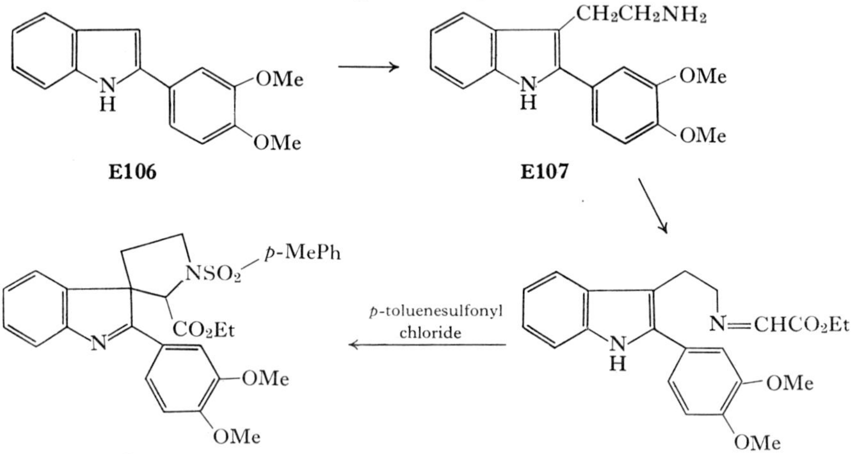

E106 (N H, OMe, OMe) → E107 (CH2CH2NH2, N H, OMe, OMe) → E108 (N=CHCO2Et, N H, OMe, OMe) —p-toluenesulfonyl chloride→ E109 (NSO2 p-MePh, CO2Et, N, OMe, OMe)

used such a pattern (209, 210). The indole **E106** was converted by a gramine-type sequence (Section B) to the tryptamine **E107** which gave the Schiff base **E108** with ethyl glyoxalate. Treatment of **E108** with *p*-toluenesulfonyl chloride gave **E109**, presumably by intramolecular alkylation by the sulfonylimine. The brilliant synthetic chemistry which followed served to

$^{-}IO_4$

E110

E111

E112

E113

E114

E115

(1) $NaBH_4$
(2) $LiAlH_4$

E116

construct the remainder of the nonindolic portion of the molecule chiefly from the dimethoxyphenyl ring of **E109**.

Van Tamelen, Dolby, and Lawton (181) obtained a large part of the strychnine-type skeleton by generating the dialdehyde **E111** by periodate oxidation. The carbinolamine **E112** is undoubtedly an intermediate in the cyclization. Addition of one of the aldehyde chains to the indolenine generated by Mannich-type alkylation of the 3-position completes the reaction.

6. Ergot Alkaloids

The synthesis of the skeleton of the ergot alkaloids presents the problem of construction a six-membered ring fused to the c and d sides of the indole ring. The total synthesis of lysergic acid (**E118**) was finally achieved by building the skeleton upon an indoline nucleus. Oxidation of the indoline to the indole stage was accomplished as the last step of the synthesis (97).

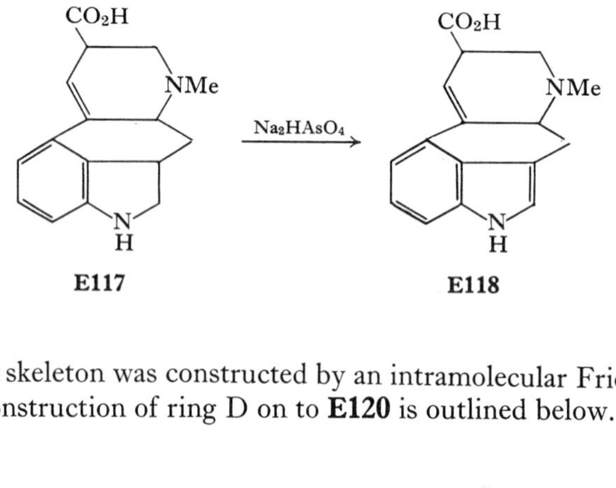

CO_2H NMe N H **E117** $\xrightarrow{Na_2HAsO_4}$ CO_2H NMe N H **E118**

The tricyclic skeleton was constructed by an intramolecular Friedel–Crafts acylation. Construction of ring D on to **E120** is outlined below. Numerous

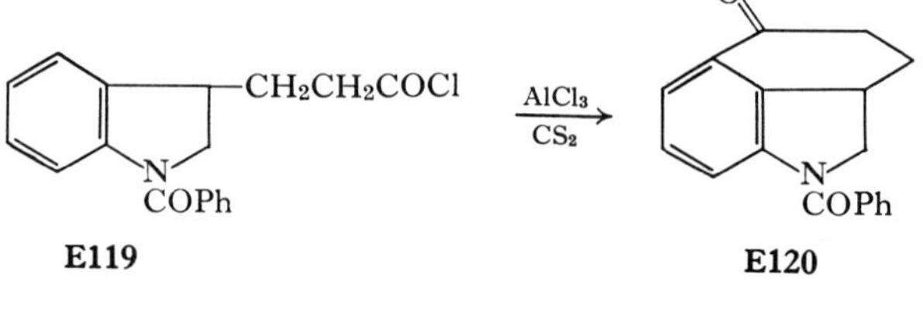

CH_2CH_2COCl N COPh **E119** $\xrightarrow[CS_2]{AlCl_3}$ O N COPh **E120**

references to other synthetic approaches to the lysergic acid skeleton have been compiled by Stoll and Hofmann (169).

E120 → (1) $ClCH_2CO_2Et$, $(Me)_3CO^-$ (2) NaOH (3) $NH_2NHCONH_2$ (4) $MeCOCO_2H$ → E121

E121 → (1) H_2O_2, ^-OH (2) $NaBH_4$ → E122

E122 → (1) $MeNCH_2$—C—Me (H; dioxolane) (2) $^-IO_4$ → E123

E123 → (1) H^+ (2) MeO^- → E124

E124 → (1) $NaBH_4$ (2) $SOCl_2$, SO_3 (3) CN^- → E125

E125 → (1) MeOH, H^+ (2) H_2O, H^+ → E117

7. The Biosynthesis of Indole Alkaloids

The indole portion of indole alkaloids is derived from tryptophan (12). The origin of the nonindolic portion of the various indole alkaloid skeletons has been of great interest and recent work has shown that this portion of the alkaloid skeleton is derived from the terpene precursor mevalonic acid (12). The incorporation of labeled mevalonic acid (13–15, 64, 67, 107, 116), geraniol (15, 107), and loganin (106) into various members of the important structural families of indole alkaloids has been observed. Double labeling experiments with the substance vincoside (**E144**) which incorporates both the indolic and terpenoid structures have shown that it is incorporated into representative alkaloids (15a). Some of the key labeling results are shown in the scheme below. The percentages indicate the percent of the total specific activity found at the designated carbon atom.

Me

$HOCH_2CH_2C—^{*}CH_2CO_2H$ ⟶ Ref. (14)

OH

(100%)

E126

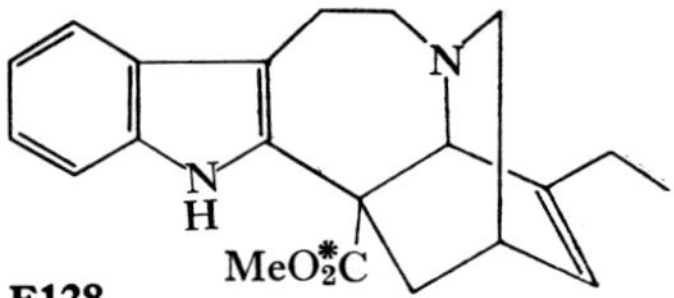

$MeO_2^{*}C$

(24%)

E127

E126 ⟶ Ref. (14)

$MeO_2^{*}C$

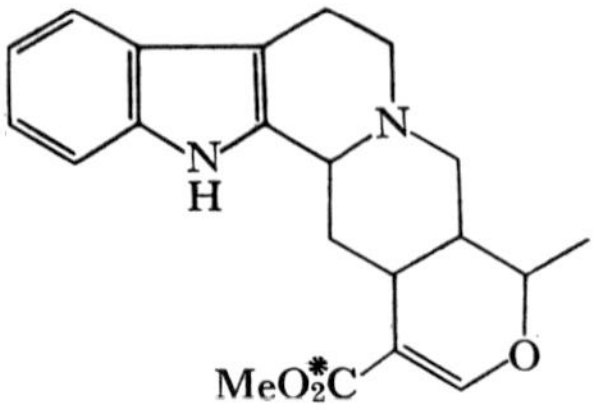

E128 (23%)

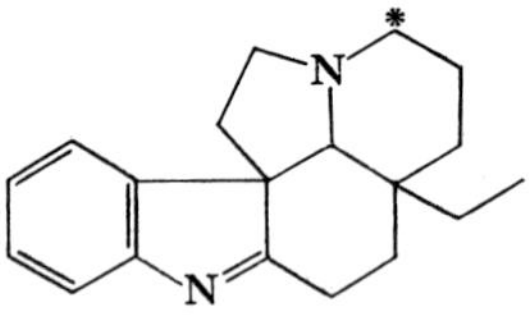

(65%)

E126 ⟶ Ref. (14)

E129

E126 ⟶ Ref. (64, 116)

MeO OAc

Me OH

$^{*}CO_2Me$

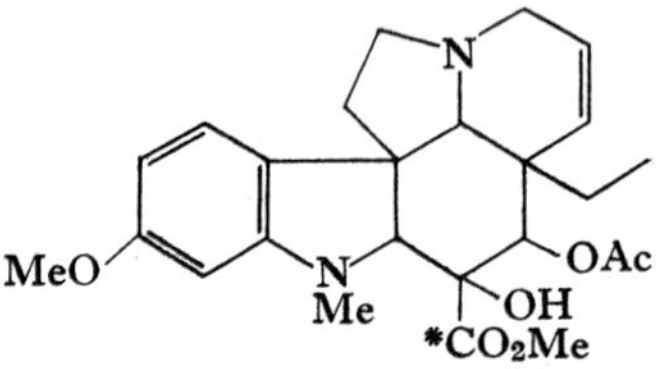

(23%)

E130

E126 ⟶ Ref. (64)

MeO

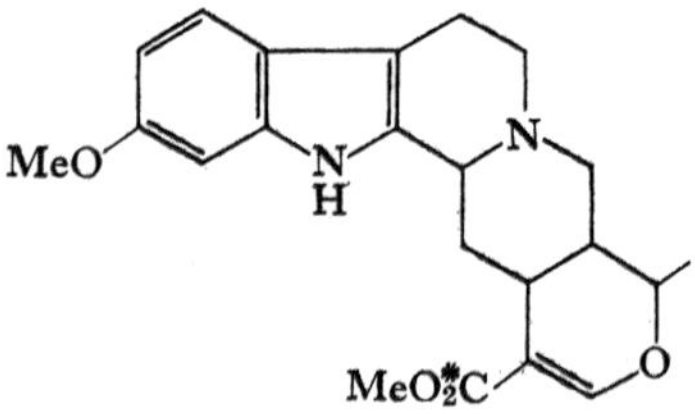

$MeO_2^{*}C$

E131

$HOCH_2CH_2$—C*(Me)(OH)—CH_2CO_2H ⟶ Ref. (14)

(100%) **E132** (47%) **E133**

E132 ⟶ Ref. (15, 107)

MeO_2C

(42%) **E134**

$(Me)_2C$=$CHCH_2CH_2C$(Me)=*$CHCH_2OH$ ⟶ Ref. (15)

(100%) **E135**

MeO_2C

(100%) **E136**

E135 ⟶ Ref. (15, 107)

N Me OAc OH CO_2Me

(100%) **E137**

Me OH D D O O ⟶ Ref. (67)

(100%) **E138**

D D MeO N Me OAc OH CO_2Me

E139

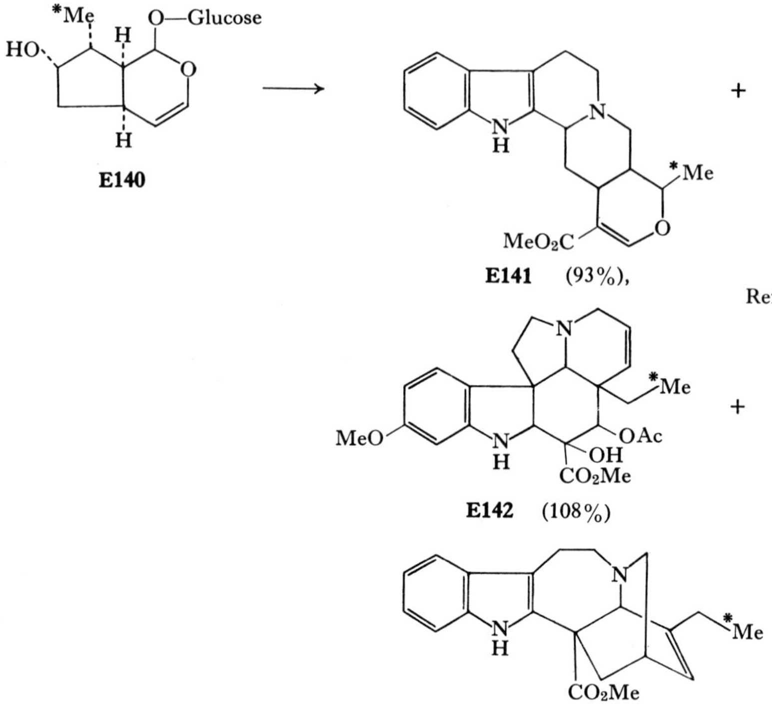
*Me
O—Glucose
HO
H
O
H
N
H
N
*Me
MeO2C
O
+
N
*Me
MeO
N
H
OAc
OH
CO2Me
+
N
N
H
*Me
CO2Me

E140

E141 (93%),

Ref. (106)

E142 (108%)

E143 (107%)

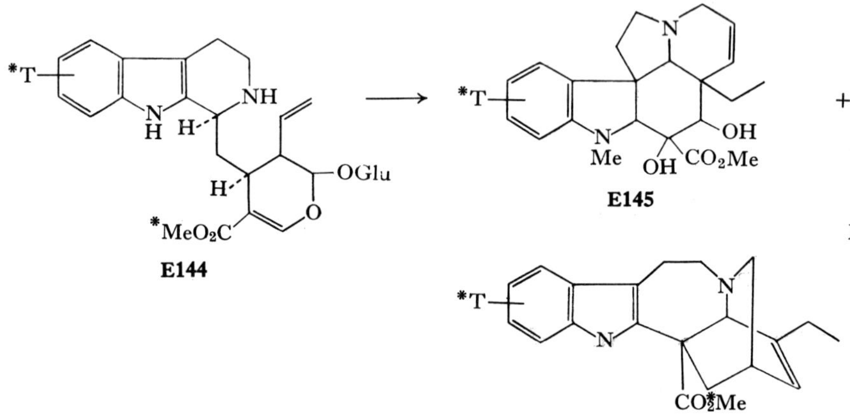
*T
N
H
NH
H
H
OGlu
*MeO2C
O
*T
N
N
Me
OH
OH
CO2Me
+
*T
N
N
CO2*Me

E144

E145

Ref. (15a)

E146

The results provide strong evidence that the nonindolic portion of indole alkaloids is of terpenoid origin, confirming hypotheses advanced by Thomas (180) and Wenkert (195). Further work (17, 106) has shown that loganin (**E140**), can be biosynthesized from geraniol (**E135**) and that it is incorporated into typical alkaloids. The results can be combined with pathways demonstrated (103, 143–145) or hypothesized (145) to develop a working hypothesis (15a, 16, 145) for elaboration of the various types of alkaloidal skeletons *in vivo*. This biogenetic scheme is outlined below:

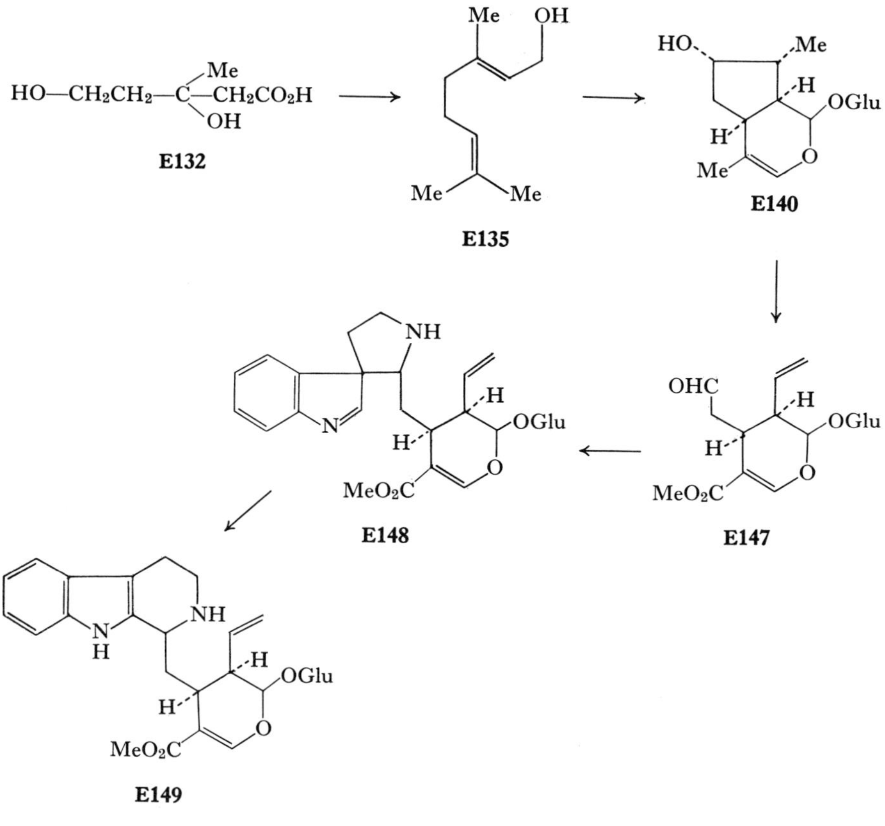

$HO—CH_2CH_2—C(Me)(OH)—CH_2CO_2H$

E132 → **E135** → **E140** → **E147** → **E148** → **E149**

The steps **E147** → **E148** → **E149** (tetrahydro-β-carboline formation), and **E155** → **E159** → **E160** → **E156** (substituent cleavage and recombination) can be recognized as having counterparts in *in vitro* indole chemistry.

One of the proposed routes (15a) for formation of **E150** involves the sequence **E149** → **E161** → **E162** → **E150**. The closest analogy to the latter process is probably the conversion of **E163** to **E166** by acid (72).

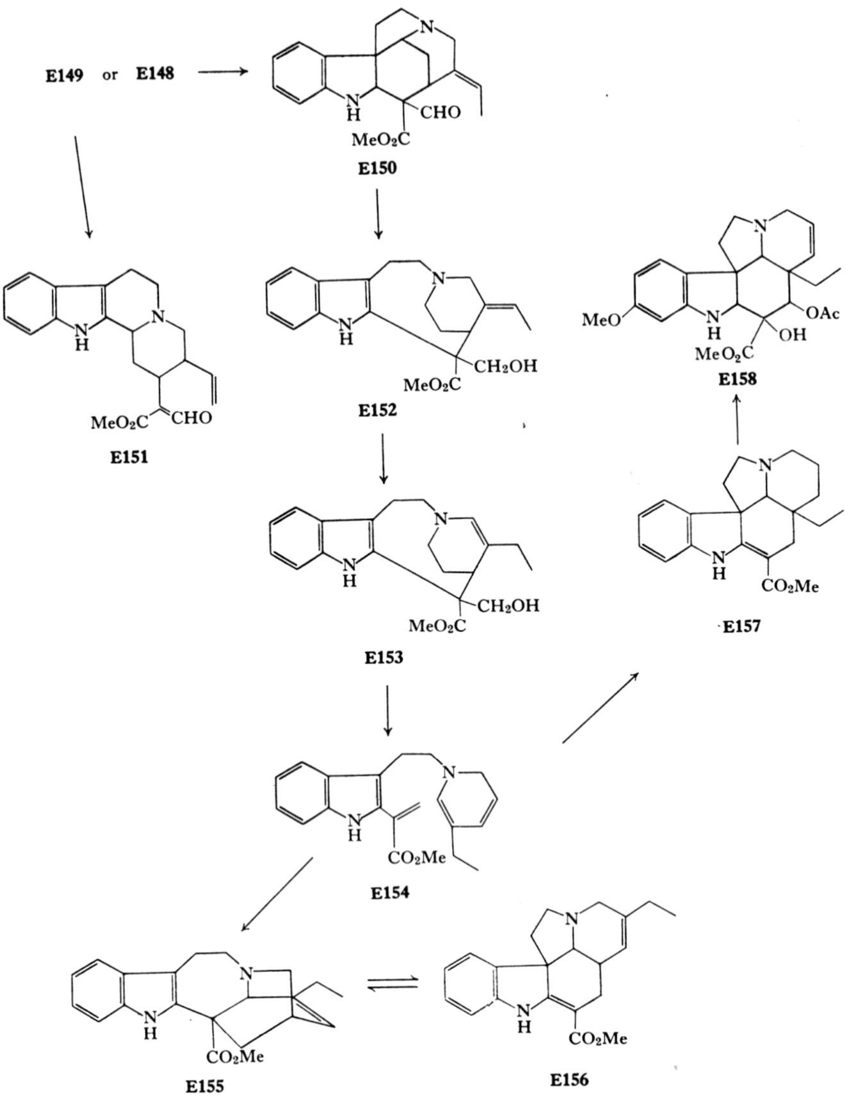

E149 or E148 → E150

CHO
MeO₂C
E150

E151
MeO$_2$C CHO

E152
MeO$_2$C CH$_2$OH

E153
MeO$_2$C CH$_2$OH

E154
CO$_2$Me

E155
CO$_2$Me

E156
CO$_2$Me

E157
CO$_2$Me

E158
MeO OAc OH MeO$_2$C

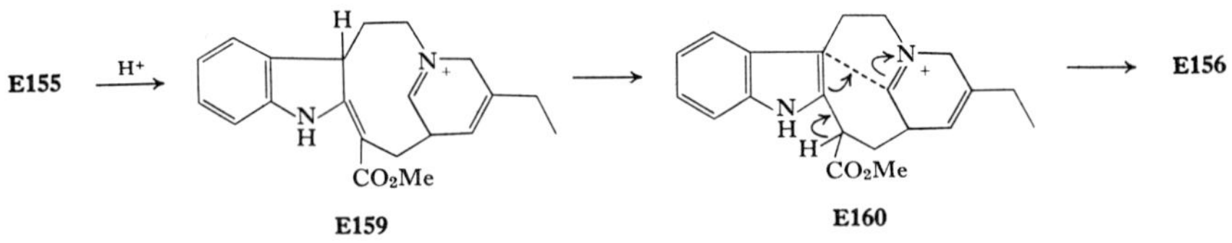

E155 $\xrightarrow{H^+}$ E159 → E160 → E156

E159
CO$_2$Me

E160
CO$_2$Me

E149 → E161 → E162 → E150

The mechanism of indole alkaloid biosynthesis remains a very active area of investigation. Present efforts are directed toward more precisely determining the points at which the branching toward the various skeletal patterns takes place.

E163 →(H^+) E164 → E165 → E166

REFERENCES

1. Abramovitch, R. A., and Spenser, I. D., *Advan. Heterocyclic Chem.* **3**, 79 (1964).
2. Acheson, R. M., Hands, A. R., and King, L. J., *J. Chem. Soc.* p. 1134 (1962).
3. Albertson, N. F., Archer, S., and Suter, C. M., *J. Am. Chem. Soc.* **66**, 500 (1944); **67**, 36 (1945).
4. Ames, A. F., Ames, D. E., Coyne, C. R., Grey, T. F., Lockhart, I. M., and Ralph, R. S., *J. Chem. Soc.* p. 3388 (1959).
5. Amorosa, M., and Lipparini, L., *Ann. Chim.* (*Rome*) **46**, 451 (1956); *Chem. Abstr.* **51**, 3558 (1957).

6. Amorosa, M., and Lipparini, L., *Ann. Chim.* (*Rome*) **49**, 322 (1959); *Chem. Abstr.* **53**, 20036 (1959).
7. Arai, I., *J. Pharm. Soc. Japan* **71**, 677 (1951); *Chem. Abstr.* **45**, 10301 (1951).
8. Ash, A. S. F., and Wragg, W. R., *J. Chem. Soc.* p. 3887 (1958).
9. Badger, G. M., Christie, B. J., Rodda, H. J., and Pryke, J. M., *J. Chem. Soc.* p. 1179 (1958).
10. Bartlett, M. F., and Taylor, W. I., *J. Am. Chem. Soc.* **82**, 5941 (1960).
11. Barton, J. E. D., and Harley-Mason, J., *Chem. Commun.* p. 298 (1965).
12. Battersby, A. R., *Pure Appl. Chem.* **14**, 117 (1967).
13. Battersby, A. R., Brown, R. T., Kapil, R. S., Knight, J. A., Martin, J. A., and Plunkett, A. O., *Chem. Commun.* p. 810 (1966).
14. Battersby, A. R., Brown, R. T., Kapil, R. S., Plunkett, A. O., and Taylor, J. B., *Chem. Commun.* p. 46 (1966).
15. Battersby, A. R., Brown, R. T., Knight, J. A., Martin, J. A., and Plunkett, A. O., *Chem. Commun.* p. 346 (1966).
15a. Battersby, A. R., Burnett, A. R., and Parsons, R. G., *J. Chem. Soc., C* p. 1193 (1969).
16. Battersby, A. R., Byrne, J. C., Kapil, R. S., Martin, J. A., and Payne, T. G., *Chem. Commun.* p. 951 (1968).
17. Battersby, A. R., Kapil, R. S., Martin, J. A., and Mo, L. *Chem. Commun.* p. 133 (1968).
18. Behringer, H., and Taul, H., *Chem. Ber.* **90**, 1398 (1957).
19. Beitz, H., Stroh, H., and Fiebig, H., *J. Prakt. Chem.* [4] **36**, 304 (1967).
20. Belleau, B., *Chem. & Ind.* (*London*) p. 229 (1955).
21. Benington, F., Morin, R. D., and Clark, L. C., Jr., *J. Org. Chem.* **23**, 1977 (1958).
22. Biswas, K. M., and Jackson, A. H., *Tetrahedron* **24**, 1145 (1968).
23. British Patent, 846,675 (1960); *Chem. Abstr.* **55**, 11436 (1961).
24. British Patent, 899,549 (1962); 910,992 (1962); *Chem. Abstr.* **57**, 15076 (1962); **58**, 9029 (1963).
25. Büchi, G., Coffen, D. L., Kocsis, K., Sonnet, P. E., and Ziegler, F. E., *J. Am. Chem. Soc.* **87**, 2073 (1965); **88**, 3099 (1966).
26. Büchi, G., Kulsa, P., and Rosati, R. L., *J. Am. Chem. Soc.* **90**, 2448 (1968).
27. Büchi, G., Manning, R. E., and Monti, S. A., *J. Am. Chem. Soc.* **86**, 4631 (1964).
28. Bucourt, R., Valls, J., and Joly, R., U.S. Patent 2,920,080 (1960); *Chem. Abstr.* **54**, 13018 (1960).
29. Bucourt, R., and Vignau, M., *Bull. Soc. Chim. France* p. 1190 (1961).
30. Bullock, M. W., and Fox, S. W., *J. Am. Chem. Soc.* **73**, 5155 (1951).
31. Bullock, M. W., and Hand, J. J., *J. Am. Chem. Soc.* **78**, 5852 (1956).
32. Bullock, M. W., and Hand, J. J., *J. Am. Chem. Soc.* **78**, 5854 (1956).
33. Buzas, A., Hoffmann, C., and Régnier, G., *Bull. Soc. Chim. France* p. 645 (1960).
34. Casini, G., and Goodman, L., *Can. J. Chem.* **42**, 1235 (1964).
35. Chapman, N. B., Clarke, K., and Hughes, H., *J. Chem. Soc.* p. 1424 (1965).
36. Chapman, N. B., Clarke, K., and Iddon, B., *J. Med. Chem.* **9**, 819 (1966).
37. Cohen, A., and Heath-Brown, B., *J. Chem. Soc.* p. 7179 (1965).
38. Cook, J. W., Loudon, J. D., and McCloskey, P., *J. Chem. Soc.* p. 1203 (1951).
39. Corsano, S., Romeo, A., and Panizzi, L., *Ric. Sci.* **28**, 2274 (1958); *Chem. Abstr.* **53**, 20107 (1959).
40. Dadson, B. A., Harley-Mason, J., and Foster, G. H., *Chem. Commun.* p. 1233 (1968).
41. David, S., and Fischer, J.-C., *Bull. Soc. Chim. France* p. 2306 (1965).
42. DeGraw, J., and Goodman, L., *J. Org. Chem.* **27**, 1728 (1962).
43. Delvigs, P., Isaac, W. M., and Taborsky, R. G., *J. Biol. Chem.* **240**, 348 (1965).

44. deStevens, G., Lukaszewski, H., Sklar, M., Halmandaris, A., and Blatter, H. M., *J. Org. Chem.* **27**, 2457 (1962).
45. Doig, G. G., Loudon, J. D., and McCloskey, P., *J. Chem. Soc.* p. 3912 (1952).
46. Domschke, G., and Fürst, H., *Chem. Ber.* **94**, 2353 (1961).
47. Eiter, K., and Nezval, A., *Monatsh. Chem.* **81**, 404 (1950).
48. Eiter, K., and Svierak, O., *Monatsh. Chem.* **83**, 1453 (1952).
49. Elderfield, R. C., and Fischer, B. A., *J. Org. Chem.* **23**, 949 (1958).
50. Elderfield, R. C., Fischer, B., and Lagowski, J. M., *J. Org. Chem.* **22**, 1376 (1957).
51. Elderfield, R. C., and Wood, J. R., *J. Org. Chem.* **27**, 2463 (1962).
52. Ewins, A. J., *J. Chem. Soc.* **99**, 270 (1911).
53. Feofilaktov, V. V., and Semenova, N. K., *Zh. Obshch. Khim.* **23**, 644 (1953); *Chem. Abstr.* **48**, 7600 (1954).
54. Findley, S. P., and Dougherty, G., *J. Org. Chem.* **13**, 560 (1948).
55. Finger, G. C., Gortatowski, M. J., Shirley, R. H., and White, R. H., *J. Am. Chem. Soc.* **81**, 94 (1959).
56. Fleming, I., and Harley-Mason, J., *J. Chem. Soc., C* p. 425 (1966).
57. Fox, S. W., and Bullock, M. W., *J. Am. Chem. Soc.* **73**, 2754 (1951).
58. Fox, S. W., and Bullock, M. W., *J. Am. Chem. Soc.* **73**, 2756 (1951).
59. Franklin, C. S., and White, A. C., *J. Chem. Soc.* p. 1335 (1963).
60. Freifelder, M., *J. Org. Chem.* **29**, 2895 (1964).
61. Freifelder, M., *J. Am. Chem. Soc.* **82**, 2386 (1960).
62. Ganellin, C. R., Hollyman, D. R., and Ridley, H. F., *J. Chem. Soc., C* p. 2220 (1967).
63. German Patent, 878,802 (1953); *Chem. Abstr.* **52**, 10202 (1958).
64. Goeggel, H., and Arigoni, D., *Chem. Commun* p. 538 (1965).
65. Groves, L. H., and Swan, G. A., *J. Chem. Soc.* p. 650 (1952).
66. Hahn, G., and Werner, H., *Ann. Chem.* **520**, 123 (1935).
67. Hall, E. S., McCapra, F., Money, T., Fukumoto, K., Hanson, J. R., Mootoo, B. S., Phillips, G. T., and Scott, A. I., *Chem. Commun.* p. 348 (1966).
68. Hamlin, K. E., and Fischer, F. E., *J. Am. Chem. Soc.* **73**, 5007 (1951).
69. Harley-Mason, J., and Atta-ur-Rahman, *Chem. Commun.* p. 208 (1967).
70. Harley-Mason, J., and Atta-ur-Rahman, *Chem. Commun.* p. 1048 (1967).
71. Harley-Mason, J., and Kaplan, M., *Chem. Commun.* p. 915 (1967).
72. Harley-Mason, J., and Waterfield, W. R., *Tetrahedron* **19**, 65 (1963).
73. Heath-Brown, B., and Philpott, P. G., *J. Chem. Soc.* p. 7165 (1965).
74. Heinzelman, R. V., Anthony, W. C., Lyttle, D. A., and Szmuszkovicz, J., *J. Org. Chem.* **25**, 1548 (1960).
75. Hester, J. B., Jr., *J. Org. Chem.* **29**, 2864 (1964).
76. Holland, D. O., and Nayler, J. H. C., *J. Chem. Soc.* p. 280 (1953).
77. Hörlein, U., *Chem. Ber.* **87**, 463 (1954).
78. Howe, E. E., Zambito, A. J., Snyder, H. R., and Tishler, M., *J. Am. Chem. Soc.* **67**, 38 (1945).
79. Huffman, J. W., Rao, C. B. S., and Kamiya, T., *J. Am. Chem. Soc.* **87**, 2288 (1965); *J. Org. Chem.* **32**, 697 (1967).
80. Jackson, A. H., and Smith, A. E., *Tetrahedron* **24**, 403 (1968).
81. Jackson, R. W., and Manske, R. H., *Can. J. Res.* **B13**, 170 (1935).
82. Julia, M., Igolen, H., and Lenzi, J., *Bull. Soc. Chim. France* p. 2291 (1966).
83. Julian, P. L., Karpel, W. J., Magnani, A., and Meyer, E. W., *J. Am. Chem. Soc.* **70** 180 (1948).
84. Julian, P. L., and Magnani, A., *J. Am. Chem. Soc.* **71**, 3207 (1949).
85. Justoni, R., and Pessina, R., British Patent 770,370 (1957); *Chem. Abstr.* **51**, 14822 (1957).

86. Justoni, R., and Pessoma, R., *Farmaco (Pavia), Ed. Sci.* **10**, 356 (1955); *Chem. Abstr.* **49**, 13968 (1955).
87. Kalir, A., Pelah, Z., and Balderman, D., *Israel J. Chem.* **5**, 101 (1967); *Chem. Abstr.* **68**, 68815 (1968).
88. Kalir, A., Szara, S., *J. Med. Chem.* **9**, 341 (1966).
89. Kamentani, T., and Fukumoto, K., *Yakugaku Kenkyu* **33**, 83 (1961); *Chem. Abstr.* **55**, 19897 (1961).
90. Kanaoka, Y., Ban, Y., Oishi, T., Yonemitsu, O., Terashima, M., Kimura, T., and Nakagawa, M., *Chem. & Pharm. Bull. (Tokyo)* **8**, 294 (1960).
91. Kanaoka, Y., Sato, E., and Ban, Y., *Chem. & Pharm. Bull. (Tokyo)* **15**, 101 (1967).
92. Kao, Y., and Robinson, R., *J. Chem. Soc.* p. 2865 (1955).
93. Kharash, M. S., Kane, S. S., and Brown, H. C., *J. Am. Chem. Soc.* **62**, 2242 (1940).
94. Kissman, H. M., and Witkop, B., *J. Am. Chem. Soc.* **75**, 1967 (1953).
95. Kline, G. B., *J. Am. Chem. Soc.* **81**, 2251 (1959).
96. Koo, J., Avakian, S., and Martin, G. J., *J. Org. Chem.* **24**, 179 (1959).
97. Kornfeld, E. C., Fornefeld, E. J., Kline, G. B., Mann, M. J., Morrison, D. E., Jones, R. G., and Woodward, R. B., *J. Am. Chem. Soc.* **78**, 3087 (1956).
98. Kralt, T., Asma, W. J., Haeck, H. H., and Mold, H. D., *Rec. Trav. Chim.* **80**, 313 (1961).
99. Kuehne, M. E., *J. Am. Chem. Soc.* **86**, 2946 (1964).
100. Kutney, J. P., Brown, R. T., and Piers, E., *J. Am. Chem. Soc.* **86**, 2286 (1964).
101. Kutney, J. P., Brown, R. T., and Piers, E., *J. Am. Chem. Soc.* **86**, 2287 (1964); Kutney, J. P., Brown, R. T., Piers, E., and Hadfield, J. R., *J. Am. Chem. Soc.*, **92**, 1708 (1970).
102. Kutney, J. P., Chan, K. K., Failli, A., Fromson, J. M., Gletsos, C., and Nelson, V. R., *J. Am. Chem. Soc.* **90**, 3891 (1968).
103. Kutney, J. P., Cretney, W. J., Hadfield, J. F., Hall, E. S., Nelson, V. R., and Wigfield, D. C., *J. Am. Chem. Soc.* **90**, 3566 (1968).
104. Kutney, J. P., Cretney, W. J., Le Quensne, P., McKague, B., and Piers, E., *J. Am. Chem. Soc.* **88**, 4756 (1966).
105. Kutney, J. P., and Piers, E., *J. Am. Chem. Soc.* **86**, 953 (1964); Kutney, J. P., Piers, E., and Brown, R. T., *J. Am. Chem. Soc.*, **92**, 1700 (1970).
106. Loew, P., and Arigoni, D., *Chem. Commun.* p. 137 (1968).
107. Loew, P., Goeggel, H., and Arigoni, D., *Chem. Commun.* p. 347 (1966).
108. Logeman, W., Almirante, L., Caprio, L., and Meli, A., *Chem. Ber.* **88**, 1952 (1955).
109. Logeman, W., Caprio, L., Almirante, L., and Meli, A., *Chem. Ber.* **89**, 1043 (1956).
110. Lyttle, D. A., and Weisblat, D. I., *J. Am. Chem. Soc.* **69**, 2118 (1947).
111. Majima, R., and Hoshino, T., *Chem. Ber.* **58**, 2042 (1923).
112. Mamaev, V. P., and Rodina, O. A., *Izv. Sibirsk. Otd. Akad. Nauk SSSR* No. 8, p. 72 (1962); *Chem. Abstr.* **58**, 12499 (1963).
113. Manske, R. H. F., *Can. J. Res.* **B5**, 592 (1931).
114. Manske, R. H. F., and Robinson, R., *J. Chem. Soc.* p. 240 (1927).
115. Masamune, S., Ang, S. K., Egli, C., Nakatsuka, N., Saikar, S. K., and Yasunari, Y., *J. Am. Chem. Soc.* **89**, 2506 (1967).
116. McCapra, F., Money, T., Scott, A. I., and Wright, I. G., *Chem. Commun.* p. 537 (1965).
117. Meier, J., *Bull. Soc. Chim. France* p. 290 (1962).
118. Mentzer, C., Beaudet, C., and Bory, M., *Bull. Soc. Chim. France* p. 421 (1953).
119. Moe, O. A., and Warner, D. T., *J. Am. Chem. Soc.* **70**, 2763 (1948).
120. Morrison, G. C., Cetenko, W., and Shavel, J., Jr., *J. Org. Chem.* **32**, 4089 (1967).
121. Nagata, W., Hirai, S., Kawata, K., and Okumura, T., *J. Am. Chem. Soc.* **89**, 5046 (1967).

122. Nametkin, S. S., Mel'nikov, N. N., and Bokarev, K. S., *Zh. Prikl. Khim.* **29**, 459 (1956); *Chem. Abstr.* **50**, 13867 (1956).
123. Neuss, N., Gorman, M., Hargrove, W., Cone, N. J., Biemann, K., Büchi, G., and Manning, R. E., *J. Am. Chem. Soc.* **86**, 1440 (1964).
124. Noland, W. E., Christensen, G. M., Sauer, G. L., and Dutton, G. G. S., *J. Am. Chem. Soc.* **77**, 456 (1955).
125. Noland, W. E., and Hovden, R. A., *J. Org. Chem.* **24**, 894 (1959).
126. Noland, W. E., and Lange, R. F., *J. Am. Chem. Soc.* **81**, 1203 (1959).
127. Noland, W. E., and Sundberg, R. J., *J. Org. Chem.* **28**, 884 (1963).
128. Novák, J., Ratuský, J., Šneberg, V., and Šorm, F., *Chem. Listy* **51**, 479 (1957); *Chem. Abstr.* **51**, 10508 (1957).
129. Omar, A. M. E., and Yamada, S., *Chem. & Pharm. Bull. (Tokyo)* **14**, 856 (1966).
130. Onda, M., Kawanishi, M., and Sasamoto, M., *J. Pharm. Soc. Japan* **76**, 409 (1956); *Chem. Abstr.* **50**, 13930 (1956).
131. Pfeil, E., and Harder, U., *Angew. Chem. Intern. Ed. Engl.* **6**, 178 (1967).
132. Piccinni, A., *Gazz. Chim. Ital.* **29**, 363 (1899).
133. Pietra, S., and Tacconi, G., *Farmaco (Pavia), Ed. Sci.* **13**, 893 (1958); *Chem. Abstr.* **53**, 21874 (1959).
134. Pietra, S., and Tacconi, G., *Farmaco (Pavia), Ed. Sci.* **14**, 854 (1959); *Chem. Abstr.* **54**, 9880 (1960).
135. Pietra, S., and Tacconi, G., *Farmaco (Pavia), Ed. Sci.* **16**, 492 (1961); *Chem. Abstr.* **58**, 5613 (1963).
136. Piper, J. R., and Stevens, F. J., *J. Heterocyclic Chem.* **3**, 95 (1966).
137. Plieninger, H., and Kiefer, B., *Chem. Ber.* **90**, 617 (1957).
138. Plieninger, H., and Suhr, K., *Chem. Ber.* **90**, 1984 (1957).
139. Potts, K. T., and Liljegren, D. R., *J. Org. Chem.* **28**, 3066 (1963).
140. Potts, K. T., and Robinson, R., *J. Chem. Soc.* p. 2675 (1955).
141. Protiva, M., Jílek, J. O., Hachová, E., Novák, L., Vejdělek, Z. J., and Adlerova, E., *Chem. Listy* **51**, 1915 (1957); *Chem. Abstr.* **52**, 4666 (1958).
142. Quan, P. M., and Quin, L. D., *J. Org. Chem.* **31**, 2487 (1966).
143. Qureshi, A. A., and Soctt, A. I., *Chem. Commun.* p. 945 (1968).
144. Qureshi, A. A., and Scott, A. I., *Chem. Commun.* p. 947 (1968).
145. Qureshi, A. A., and Scott, A. I., *Chem. Commun.* p. 948 (1968).
146. Ratusky, J., and Sorm, F., *Chem. Listy* **51**, 1091 (1957); *Chem. Abstr.* **51**, 13843 (1957).
147. Renner, U., and Fritz, H., *Tetrahedron Letters* p. 283 (1964).
148. Rinderknecht, H., and Niemann, C., *J. Am. Chem. Soc.* **72**, 2296 (1950).
149. Robinson, J. R., and Good, N. E., *Can. J. Chem.* **35**, 1578 (1957).
150. Sallay, S. I., *J. Am. Chem. Soc.* **89**, 6762 (1967).
151. Schlittler, E., and Allemann, T., *Helv. Chim. Acta* **31**, 128 (1948).
152. Schlossberger, H. G., and Kuch, H., *Chem. Ber.* **93**, 1318 (1960).
153. Schumann, D., and Schmid, H., *Helv. Chim. Acta* **46**, 1996 (1963).
154. Shaw, E., *J. Am. Chem. Soc.* **77**, 4319 (1955).
155. Shaw, E., and Woolley, D. W., *J. Am. Chem. Soc.* **75**, 1877 (1953).
156. Shaw, K. N. F., McMillan, A., Gudmundson, A. G., and Armstrong, M. D., *J. Org Chem.* **23**, 1171 (1958).
157. Sletzinger, M., Gaines, W. A., and Ruyle, W. V., *Chem. & Ind. (London)* p. 1215 (1957).
158. Snyder, H. R., Beilfuss, H. R., and Williams, J. K., *J. Am. Chem. Soc.* **75**, 1873 (1953).
159. Snyder, H. R., and Katz, L., *J. Am. Chem. Soc.* **69**, 3140 (1947).
160. Snyder, H. R., and Smith, C. W., *J. Am. Chem. Soc.* **66**, 350 (1944).
161. Snyder, H. R., Smith, C. W., and Stewart, J. M., *J. Am. Chem. Soc.* **66**, 200 (1944).

162. Spanish Patent, 227,606 (1957); *Chem. Abstr.* **52**, 2923 (1958).
163. Speeter, M. E., U.S. Patent 2,728,778 (1955); *Chem. Abstr.* **50**, 10786 (1956).
164. Speeter, M. E., U.S. Patent 2,814,625 (1957); *Chem. Abstr.* **52**, 11948 (1958).
165. Speeter, M. E., and Anthony, W. C., *J. Am. Chem. Soc.* **76**, 6208 (1954).
166. Speeter, M. E., Heinzelman, R. V., and Weisblat, D. I., *J. Am. Chem. Soc.* **73**, 5514 (1951).
167. Stevens, F. J., Ashby, E. C., and Downey, W. E., *J. Am. Chem. Soc.* **79**, 1680 (1957).
168. Stevens, F. J., and Higginbotham, D. H., *J. Am. Chem. Soc.* **76**, 2206 (1954).
169. Stoll, A., and Hofmann, A., *Alkaloids* **8**, 742–746 (1965).
170. Stork, G., and Hill, R. K., *J. Am. Chem. Soc.* **76**, 949 (1954); **79**, 494 (1957).
171. Stork, G., and Singh, G., *J. Am. Chem. Soc.* **73**, 4742 (1951).
172. Supple, J. H., Nelson, D. A., and Lyle, R. E., *Tetrahedron Letters* p. 1645 (1963).
173. Swan, G. A., *J. Chem. Soc.* p. 1534 (1950).
174. Szmuskovicz, J., Anthony, W. C., and Heinzelman, R. V., *J. Org. Chem.* **25**, 857 (1960).
175. Szmuszkovicz, J., and Thomas, R. C., *J. Org. Chem.* **26**, 960 (1961).
176. Terent'ev, A. P., Preobrazhenskaya, M. N., and Ban-Lun Ge, *Khim. Nauka i Promy* **4**, 281 (1959); *Chem. Abstr.* **53**, 21879 (1959).
177. Thesing, J., and Festag, W., *Experientia* **15**, 127 (1959).
178. Thesing, J., Ramloch, H., and Willersinn, C.-H., *Chem. Ber.* **89**, 2896 (1956).
179. Thesing, J., and Schülde, F., *Chem. Ber.* **85**, 324 (1952).
180. Thomas, R., *Tetrahedron Letters* p. 544 (1961).
181. van Tamelen, E. E., Dolby, L. J., and Lawton, R. G., *Tetrahedron Letters* No. 19, p. 30 (1960).
182. van Tamelen, E. E., and Hester, J. B., Jr., *J. Am. Chem. Soc.* **81**, 3805 (1959); **91**, 7342 (1969).
183. van Tamelen, E. E., and Placeway, C., *J. Am. Chem. Soc.* **83**, 2594 (1961); E. E. van Tamelen, C. Placeway, G. P. Schiemenz and I. G. Wright, *J. Am. Chem. Soc.* **91**, 7359 (1969).
184. van Tamelen, E. E., and Shamma, M., *J. Am. Chem. Soc.* **76**, 950 (1954).
185. van Tamelen, E. E., Shamma, M., Burgstahler, A. W., Wolinsky, J., Tamm, R., and Aldrich, P. E., *J. Am. Chem. Soc.* **80**, 5006 (1958); **91**, 7315 (1969).
186. Vejdelek, Z. J., *Chem. Listy* **44**, 73 (1950); *Chem. Abstr.* **45**, 8004 (1951).
187. Vitali, T., and Mossini, F., *Boll. Sci. Fac. Chim. Ind. Bologna* **17**, 84 (1959); *Chem. Abstr.* **54**, 19644 (1960).
188. Warner, D. T., and Moe, O. A., *J. Am. Chem. Soc.* **70**, 2765 (1948).
189. Wawzonek, S., and Maynard, M. M., *J. Org. Chem.* **32**, 3618 (1967).
190. Wawzonek, S., and Nelson, G. E., *J. Org. Chem.* **27**, 1377 (1962).
191. Wawzonek, S., and Nordstrom, J. D., *J. Med. Chem.* **8**, 265 (1965).
192. Weisbach, J. A., Kirkpatrick, J. L., Williams, K. R., Anderson, E. L., Yim, N. C., and Douglas, B., *Tetrahedron Letters* p. 3457 (1965).
193. Weisblat, D. I., and Lyttle, D. A., *J. Am. Chem. Soc.* **71**, 3079 (1949).
194. Welstead, W. J., Jr., DaVanzo, J. P., Helsley, G. C., Lunsford, C. D., and Taylor, C. R., Jr., *J. Med. Chem.* **10**, 1015 (1967).
195. Wenkert, E., *J. Am. Chem. Soc.* **84**, 98 (1962).
196. Wenkert, E., *Accounts Chem. Res.* **1**, 78 (1968).
197. Wenkert, E., Dave, K. G., Gnewuch, C. T., and Sprague, P. W., *J. Am. Chem. Soc.* **90**, 5251 (1968).
198. Wenkert, E., Dave, K. G., and Haglid, F., *J. Am. Chem. Soc.* **87**, 5461 (1965).
199. Wenkert, E., Dave, K. G., Haglid, F., Lewis, R. G., Oishi, T., Stevens, R. V., and Terashima, M., *J. Org. Chem.* **33**, 747 (1968).

200. Wenkert, E., Dave, K. G., Lewis, R. G., and Sprague, P. W., *J. Am. Chem. Soc.* **89**, 6741 (1967).
201. Wenkert, E., Massy-Westrop, R. A., and Lewis, R. G., *J. Am. Chem. Soc.* **84**, 3732 (1962).
202. Wenkert, E., and Wickberg, B., *J. Am. Chem. Soc.* **84**, 4914 (1962).
203. Wenkert, E., and Wickberg, B., *J. Am. Chem. Soc.* **87**, 1580 (1965).
204. Whaley, W. M., and Govindachari, T. R., *Org. Reactions* **6**, 74 (1951).
205. Whaley, W. M., and Govindachari, T. R., *Org. Reactions* **6**, 151 (1951).
206. Whittle, B. A., and Young, E. H. P., *J. Med. Chem.* **6**, 378 (1963).
207. Winterfeldt, E., *Chem. Ber.* **97**, 2463 (1964).
208. Woodward, R. B., Bader, F. E., Bickel, H., Frey, A. J., and Kierstead, R. W., *Tetrahedron* **2**, 1 (1958).
209. Woodward, R. B., Cava, M. P., Ollis, W. D., Hunger, A., Daeniker, H. U., and Schenker, K., *J. Am. Chem. Soc.* **76**, 4749 (1954).
210. Woodward, R. B., Cava, M. P., Ollis, W. D., Hunger, A., Daeniker, H. U., and Schenker, K., *Tetrahedron* **19**, 247 (1963).
211. Young, E. H. P., *J. Chem. Soc.* p. 3493 (1958).
212. Yurashevskii, N. K., *Zh. Obshch. Khim.* **24**, 729 (1954); *Chem. Abstr.* **49**, 6271 (1955).

V
OXIDATION, DEGRADATION, AND METABOLISM OF THE INDOLE RING

A. THE REACTIONS OF INDOLES WITH MOLECULAR OXYGEN AND PEROXIDES

The nature of the reaction of indoles with oxygen was put on a firm basis in the early 1950's by the determination of the structure of the hydroperoxides formed from oxygen and tetrahydrocarbazoles (7) and by a series of papers (101–110) by Witkop and co-workers which considered the paths and mechanisms of oxidation for a number of indole derivatives. Beer, McGrath, and Robertson (7) isolated crystalline hydroperoxides from tetrahydrocarbazole and certain derivatives. It was shown that the hydroperoxide had the structure **A2**. Witkop's work demonstrated that indolenyl hydroperoxides were usually the primary products from reaction of indoles with oxygen, and that subsequent reactions of these hydroperoxides could account for the structures of other oxidation products.

$$\textbf{A1} \xrightarrow{O_2} \textbf{A2}$$

A1 **A2**

Since the characterization of the hydroperoxide of tetrahydrocarbazole by Beer, McGrath, and Robertson (7) the formation of such derivatives from 2,3-disubstituted indoles has come to be recognized as a quite general reaction. Table XLIII records a number of the 3-hydroperoxyindolenines isolated by oxidation of indoles. Attempts to prepare hydroperoxides from

TABLE XLIII □ 3-HYDROPEROXYINDOLENINES FROM INDOLES

Indole substituent	Mode of oxidation	Yield (%)	Ref.
2,3-diMe	Air, peroxide catalyst	25	(6)
2,3-Diethyl	Air	71	(59)
1,2,3,4-Tetrahydrocarbazole	Air	Variable	(7)
3-Methyl-2-phenyl	Air	10	(108)
2-(*p*-Methoxyphenyl)-3-methyl	Oxygen, platinum	30–60	(109)
2,3-Dimethyl-5-methoxy	Air, peroxide catalyst	50	(6)
5-Carbomethoxy-1,2,3,4-tetrahydrocarbazole	Air	90	(5)
6-Bromo-1,2,3,4-tetrahydrocarbazole	Air	55	(5)

2-methyl-3-phenyl and 2,3-diphenylindole failed, and these indoles are said to be relatively resistant to oxygen (6) although subsequent work by Chen and Leete indicates that 3-phenylindoles are not completely resistant to oxidation (11).

The indolenyl peroxides are relatively reactive substances and the products isolated from oxidations are often the result of subsequent reactions. The peroxide linkage is subject to reductive cleavage. Reduction with dithionite (7) or by hydrogenation (106) over platinum oxide gives **A3**.

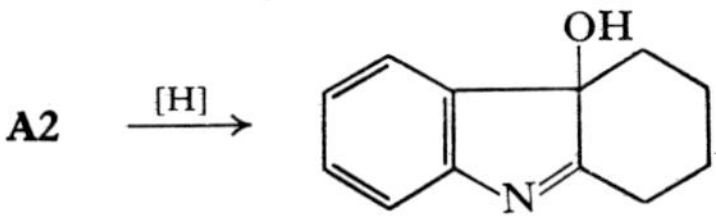

A3

Treatment of **A3** with alkali or acid gives the indolinone **A4** (73, 106).

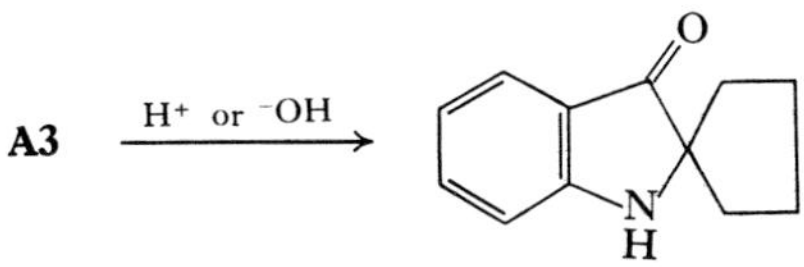

A4

The hydroperoxide **A2** is converted to the lactam **A5** under neutral or slightly acidic conditions (101, 106).

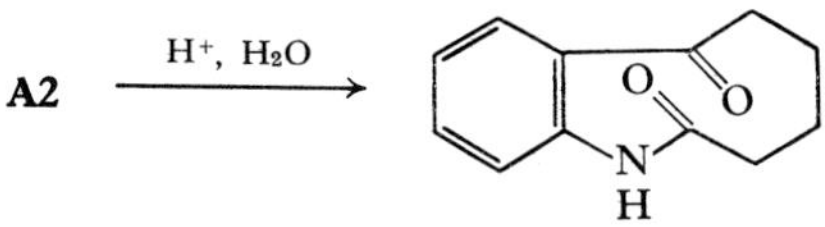

A5

Witkop and Patrick (107) proposed the mechanism below to account for the formation of **A5**. They also found that **A3** was converted to **A5** by perbenzoic acid and formulated the reaction as proceeding through the intermediate **A8**.

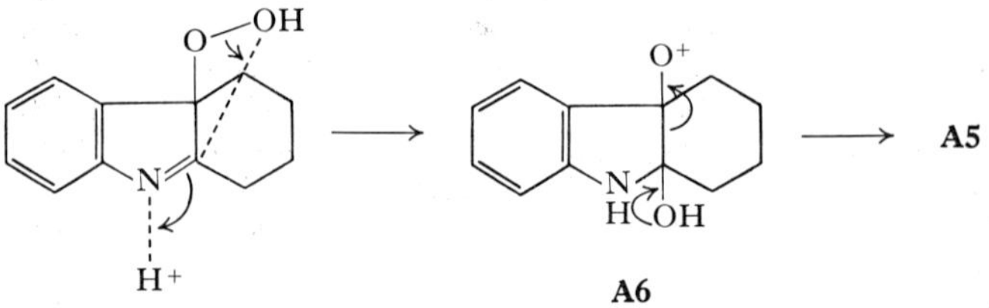

A2 → **A6** → **A5**

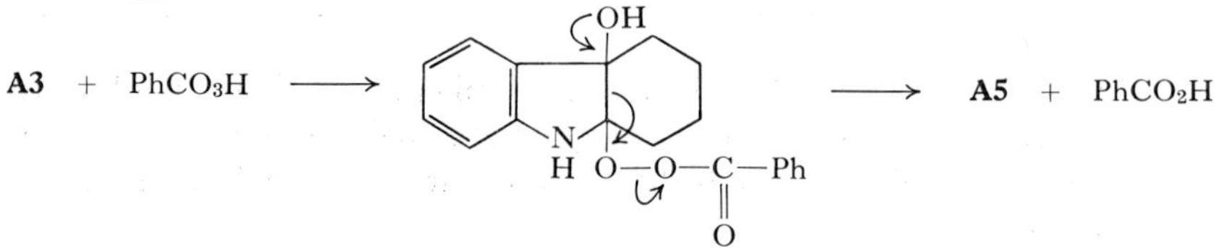

A3 + $PhCO_3H$ → **A8** → **A5** + $PhCO_2H$

A7

The effect of ring size on the course of these reactions was investigated (110). The indole **A9** gives the lactam **A10** very readily on catalytic oxidation. The seven-membered ring analog **A11** apparently reacts similarly but the lactam **A12** was not isolated; its intramolecular condensation product **A13** was found instead.

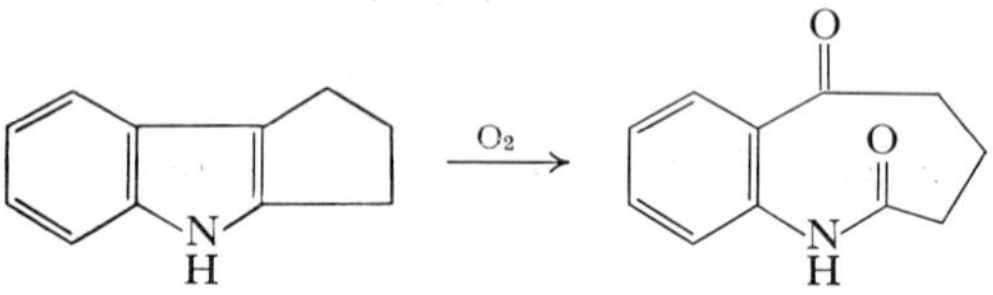

A9 $\xrightarrow{O_2}$ **A10**

Related intramolecular condensations have been found to follow the oxidation of *N*-substituted tetrahydro-β-carbolines such as **A14** (104).

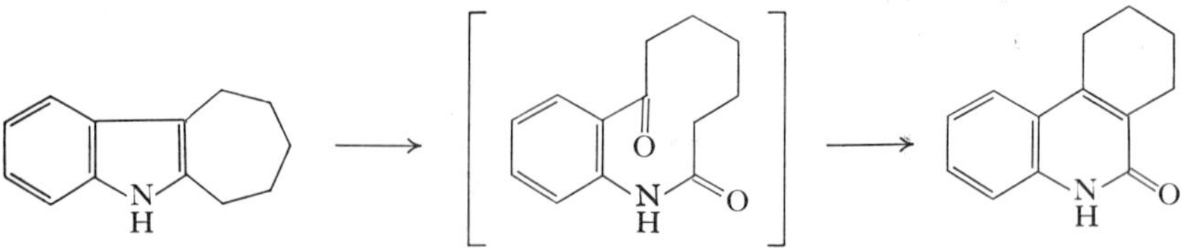

A11 → [**A12**] → **A13**

Similar condensations also occurred when yohimbine and yohimbane were ozonized and then treated with base.

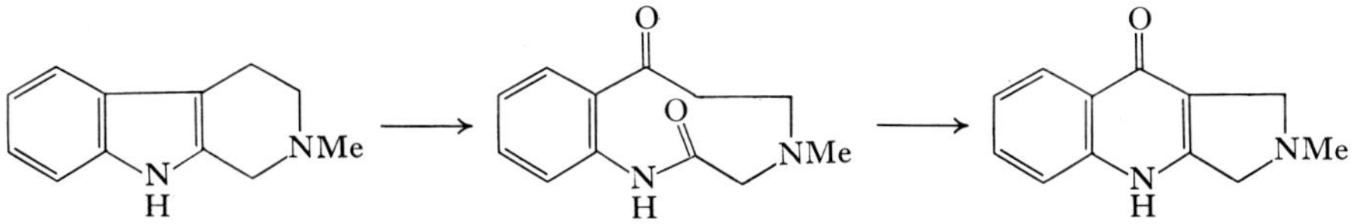

A14 A15 A16

With the introduction of an eight-membered ring, the reaction takes a different course and leads to the 2-acylindole **A18**. The structure of **A18**

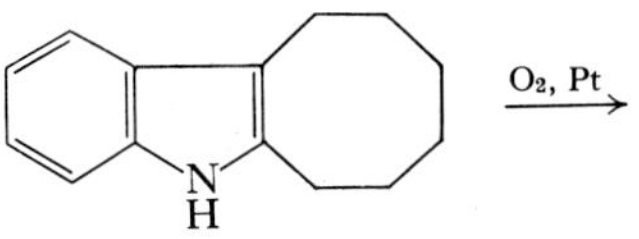

O_2, Pt

A17

EtOH, H^+

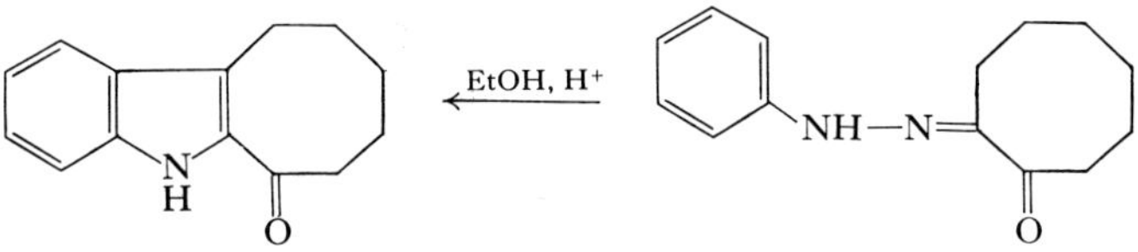

A18 A19

was proven by synthesis (110). This relatively unusual pattern of oxidation, that is, apparent attack at the 2-alkyl substituent, has been observed subsequently in the other systems. Leete (59) observed that autoxidation of 2,3-diethylindole gave 2-acetyl-3-ethylindole and Taylor (92) found that 2,3-dimethylindole gave small amounts of 2-formyl-3-methylindole along with the major product **A24**. Leete (59) provided evidence that the initial

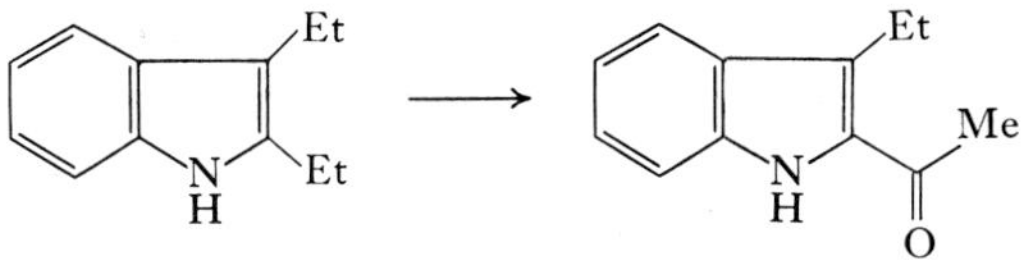

A20 A21

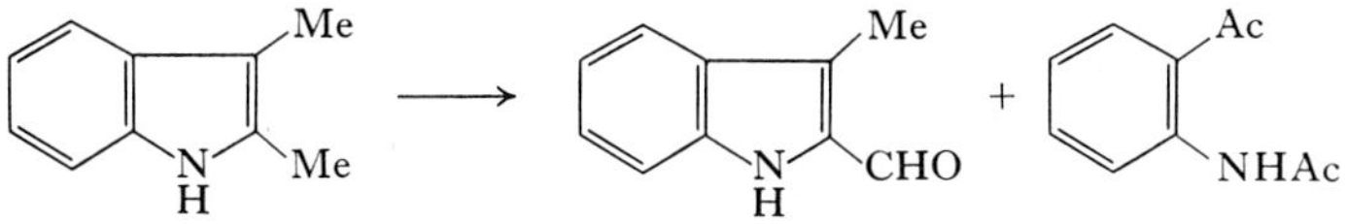

A22 A23 A24

stage of this type of oxidation was attack of oxygen at the 3-position to generate the hydroperoxide **A25**, analogous to the usual course of indole oxidation. The hydroperoxide was, in fact, isolated and shown to give **A21** on heating.

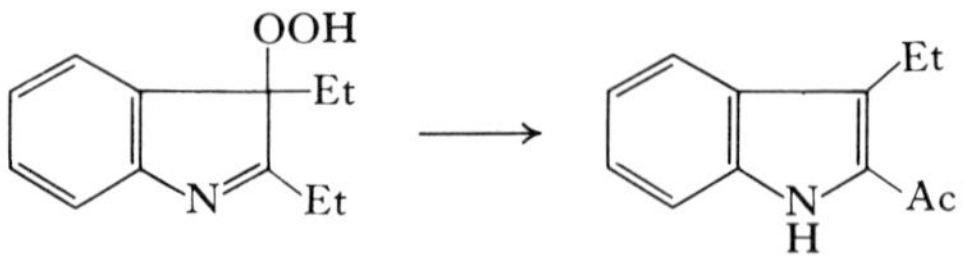

A25 → **A21**

Taylor (92) has offered a mechanism for these oxidation reactions. He suggests an equilibrium between **A25** and **A26**, the enamine form **A26** being favored by alkyl substitution. Allylic rearrangement of **A26** to **A27** which decomposes to **A28** completes Taylor's proposal.

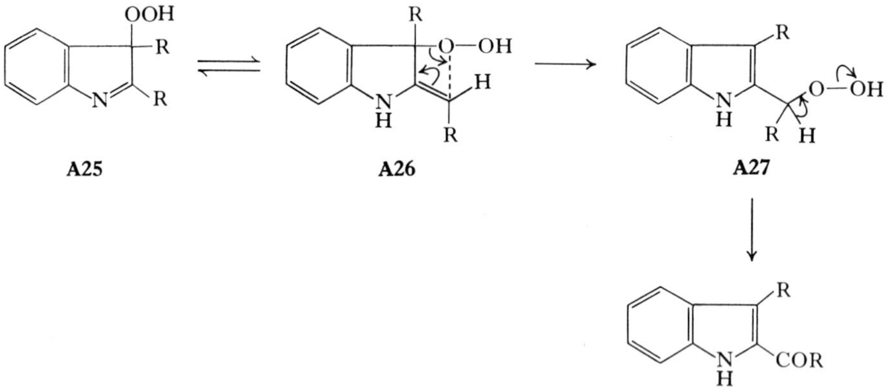

A25 ⇌ **A26** → **A27** → **A28**

Wasserman and Floyd (97) have obtained **A35** and **A36** by catalytic oxidation of 2-isopropyl-3-methylindole and proposed the mechanism shown below.

Wasserman has proposed the scheme **A38** → **A40** to account for the formation of 2-acetyl-3-ethylindole from 2,3-diethylindole. Chen and Leete (11) studied the oxidation of 2-benzyl-3-phenylindole. Autoxidation gave 3-phenyldioxindole **A43** but catalytic oxidation gave the 2-acylindole **A42**. These results can be accommodated by the Wasserman scheme, although the reason for the change in the nature of products from the two modes of oxidation is not evident. Thus, while initial attack by oxygen at C-3 is implicated in all current interpretations of indole oxidations, there are divergent pathways for decomposition of indolenyl peroxides and understanding of the factors which govern the formation of subsequent products is incomplete. Thermal decomposition of the hydroperoxides usually gives *o*,*N*-diacylanilines. Even the decomposition of the hydroperoxide of 2,3-

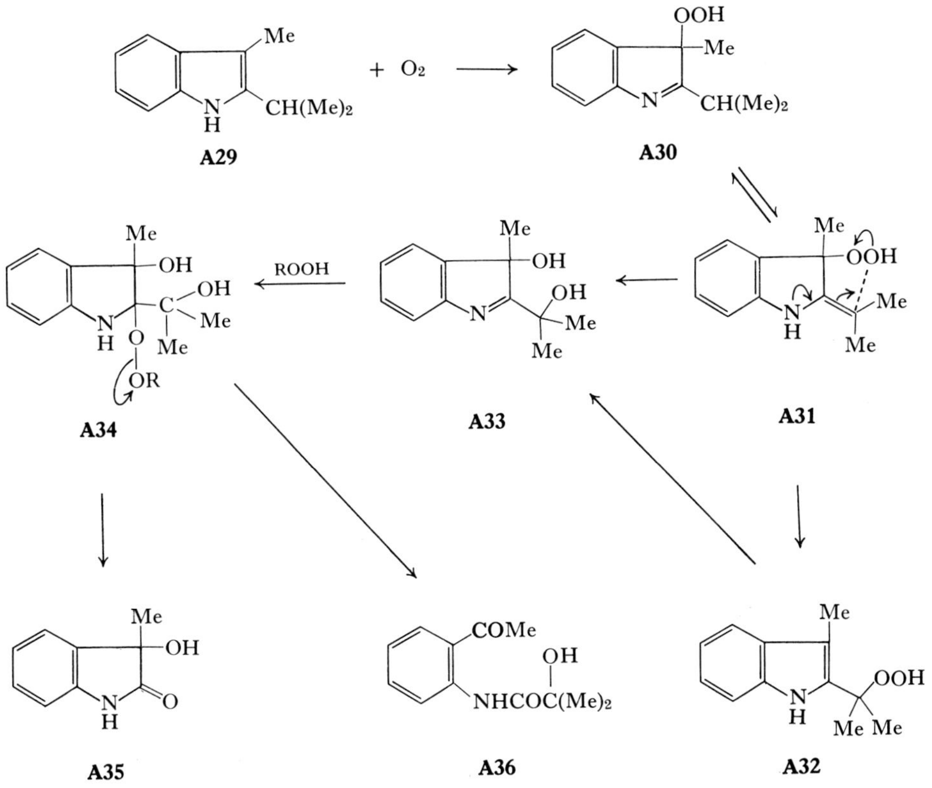
Me
N
H
CH(Me)2
A29
+ O2
OOH
Me
N
CH(Me)2
A30
Me
OOH
N
H
Me
Me
A31
Me
OH
N
OH
Me
Me
A33
ROOH
Me
OH
OH
C
Me
N
H
O
Me
OR
A34
Me
OH
N
H
O
A35
COMe
OH
NHCOC(Me)2
A36
Me
N
H
OOH
Me Me
A32

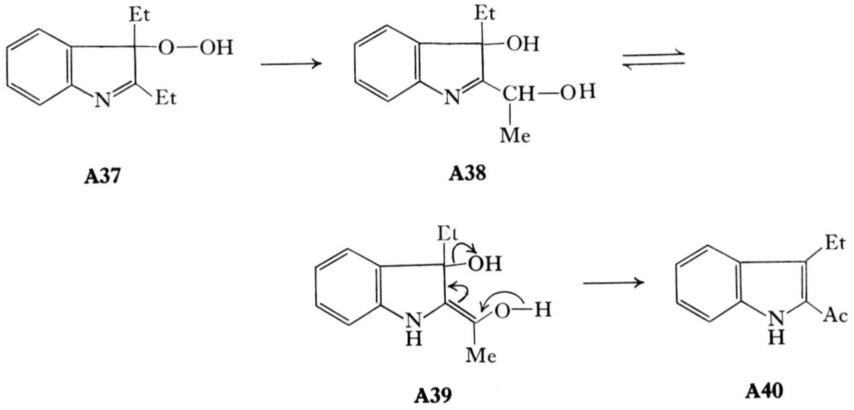
Et
O—OH
N
Et
A37
Et
OH
N
CH—OH
Me
A38
OH
N
H
O—H
Me
A39
Et
N
H
Ac
A40

A41 —(O_2, Pt)→ A42

A41 —(air)→ A43

A44 —(OH, heat)→ A45 Ref. (6)

A46 —(H_2O, heat)→ A47 (50%) Ref. (6, 92)

A48 —(H^+, H_2O)→ A49 Ref. (108)

diethylindole follows this path in water (59). Addition of water to the indolenine may promote this mode of decomposition. When allylic rearrangement of the peroxide competes successfully with ring opening, net oxidation of the 2-substituent is observed.

A50 —(H_2O)→ A51

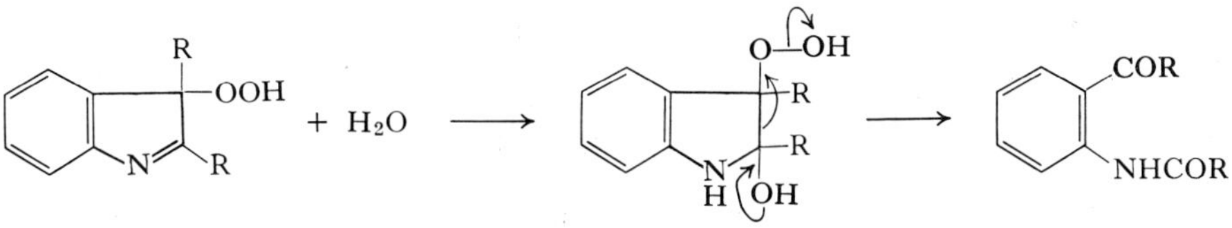

A52 + H_2O ⟶ A53 ⟶ A54

The oxidation of the 2-methylindole by a variety of oxidizing agents including oxygen and peracetic acid gives **A56** which is clearly formed from two molecules of 2-methylindole. The structure of the oxidation product was established by Witkop and Patrick (105).

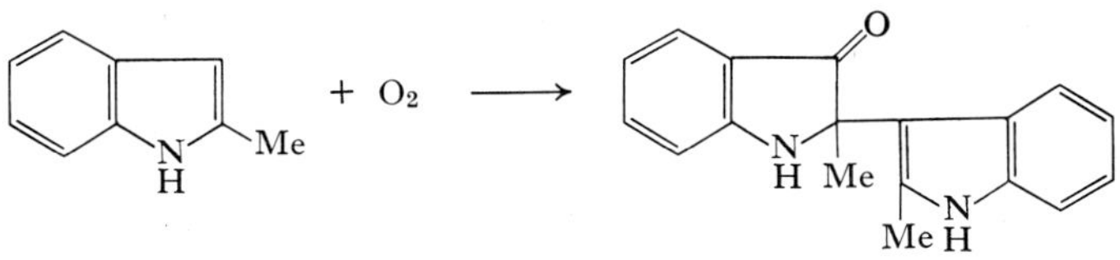

A55 + O_2 ⟶ A56

The initially formed hydroperoxide **A57** may decompose to **A58**. The strongly electrophilic C=N bond in **A58** should be capable of effecting 3-alkylation of an unreacted molecule of 2-methylindole. The formation of **A56** is then easily reconciled with the general patterns of reactivity found for 2,3-disubstituted indoles.

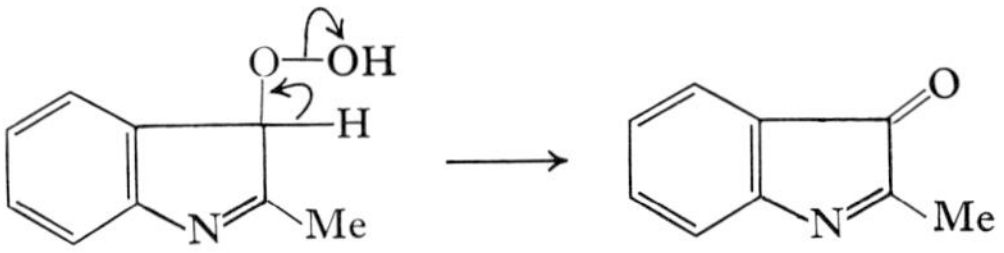

A57 ⟶ A58

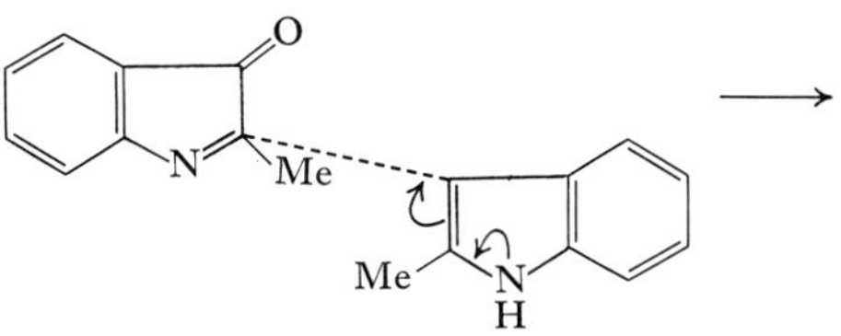

A58 + A55 ⟶ A56

The base-catalyzed decomposition of 3-hydroperoxindolenines in aprotic solvents is chemiluminescent. The light-emitting species is believed to be the anion of the *o*-acylaminoketone formed by cleavage of the cyclic hydroperoxide **A61** (63, 90).

A59 —$(Me)_3CCO^-$→ A60 → A61 → A62 → A63 $+ h\nu$ —H^+→ A64

Sugiyama and co-workers have investigated a large number of indoles and have detected strong chemiluminescence in 2,3-dialkylindoles and 3-aminovinylindoles (89, 90). Electron releasing substituents enhance the intensity of chemiluminescence of 2,3-dimethylindoles (89).

Electron releasing substituents at C-2 and C-3 would be expected to activate the indole nucleus toward oxidation and available data tend to confirm this expectation. 3-Aminoindoles are very susceptible to autoxidation. 3-Amino-1-methyl-2-phenylindole is reported to give **A66** or its ring-chain tautomer **A67**, while 3-amino-1,2-diphenylindole has been shown to give **A69** (8a). The structures of the oxidation products of **A65** and **A68** suggest that initial attack of oxygen is probably at C-2 of the indole ring although it is possible to write mechanisms leading to **A67** or **A69** involving initial attack at C-3. The tricyclic 3-aminoindole **A70** is reported to be extremely sensitive to air oxidation (40). It is converted to **A71** in

A65 —O_2→ A66 ⇌ A67

A68 —O_2→ A69

70% yield by exposure to air in ethyl acetate solution for 18 hours. The product again bears the oxygen substituent at C-2 of the indole ring. The *N*-ethyl derivative of **A70** is reported to be less readily oxidized.

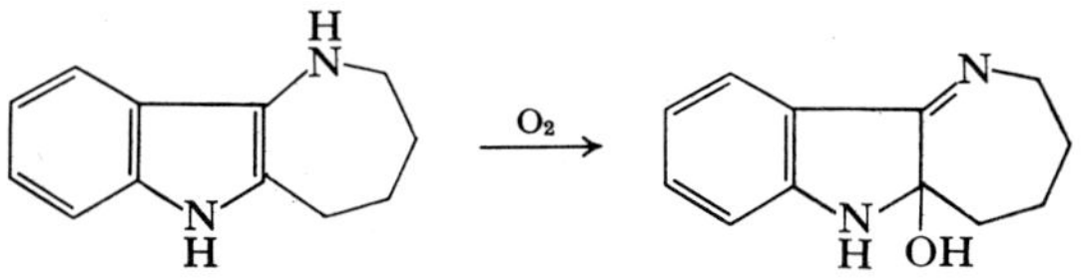

A70 $\xrightarrow{O_2}$ **A71**

Attempts to isolate the 2-aminoindolenine **A72** from its hydrobromide salts have failed because of the extraordinary susceptibility of **A72** to auto-oxidation. The hydroxy compound **A74** is isolated instead (43). Although there is no apparent reason for **A72** to be highly susceptible to oxidation, its tautomer **A73**, a 2-aminoindole, would be expected to be rapidly attacked

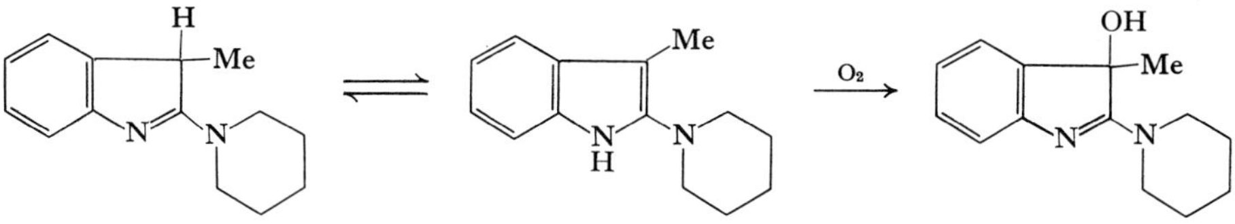

A72 ⇌ **A73** $\xrightarrow{O_2}$ **A74**

by oxygen at C-3. 2-Ethoxy-3-methylindole and 2-ethylthio-3-methylindole are also auto-oxidized to the corresponding 3-hydroxy derivatives (44). The substituted 2-ethoxyindole **A75** is rapidly transformed into the hydroperoxide **A76** on exposure to air (78). The irradiation of tryptophan and related indoles in the presence of oxygen results in disappearance of the indole but the numerous products have not yet been adequately characterized (36).

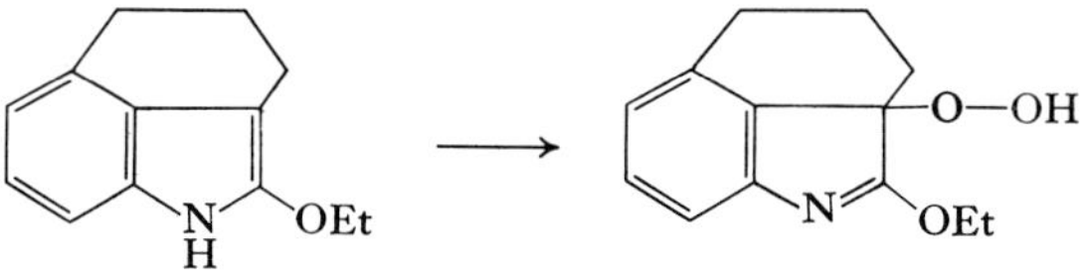

A75 ⟶ **A76**

Indole alkaloids are also susceptible to oxidation. The hydroxyindolenine **A77** has been isolated from *Evatamia dichotoma* (83). The structure of **A77** was proven by synthesis from voacristine (**A78**) by photochemically induced oxidation It is not known if **A77** is actually present in the plant since it may be an artifact derived from **A78** by autoxidation.

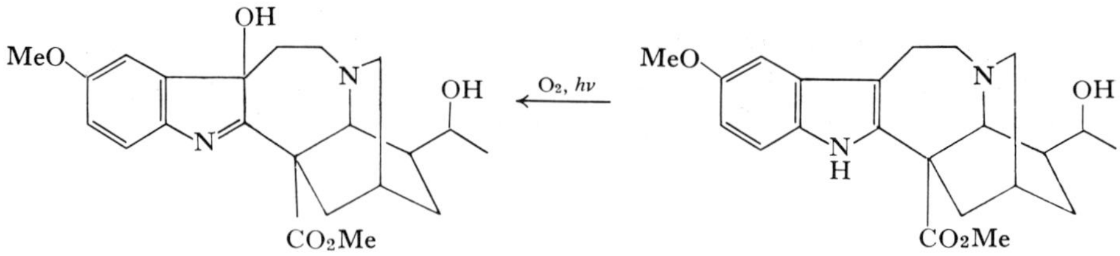

OH MeO N OH O_2, $h\nu$ MeO N OH

CO_2Me CO_2Me

A77 **A78**

Ibogaine (**A79**) is converted to iboluteine (**A80**) by aeration of an irradiated benzene solution followed by treatment with base (3). Interestingly, the hydroperoxy derivative of ibogaine (**A81**) is obtained by oxidation with chromic oxide in pyridine. The catalytic oxidation of voacangine (**A82**) takes a different course, the isoquinuclidine ring being the site of oxidation (35).

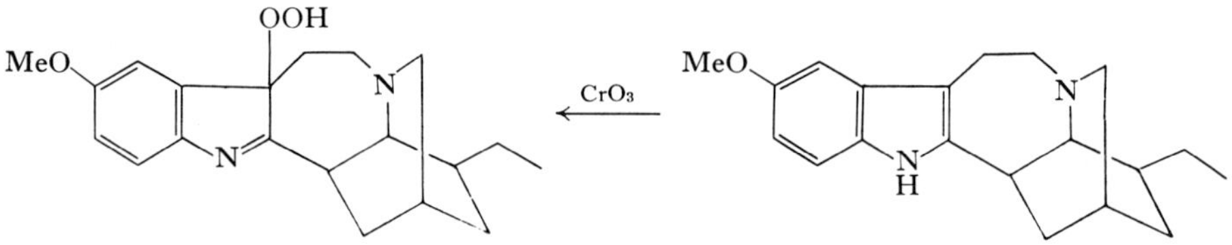

OOH MeO N CrO_3 MeO N

A81 **A79**

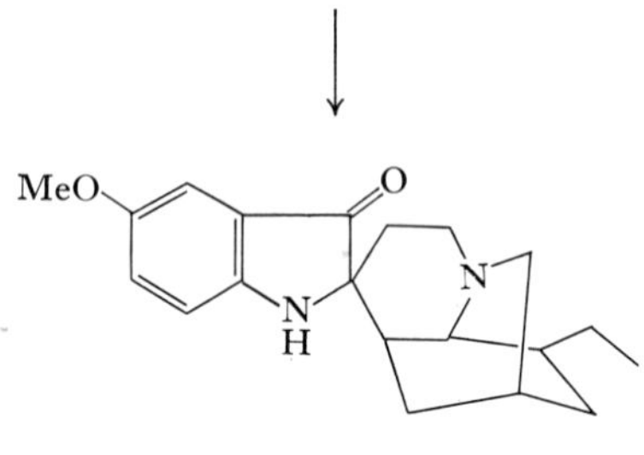

MeO O N

A80

MeO N O_2, Pt MeO O N

CO_2Me CO_2Me

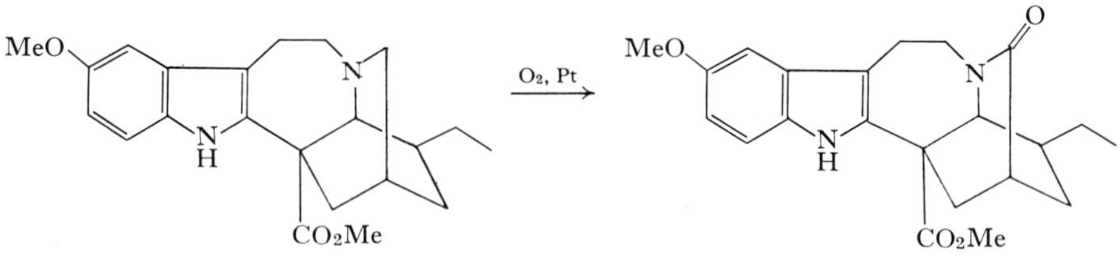

A82 **A83**

An alternative mode of oxidation of indoles has been investigated. This involves oxygenation of the Grignard derivative. The products of this procedure are 3-hydroxyindolenines (35) or subsequent transformation products.

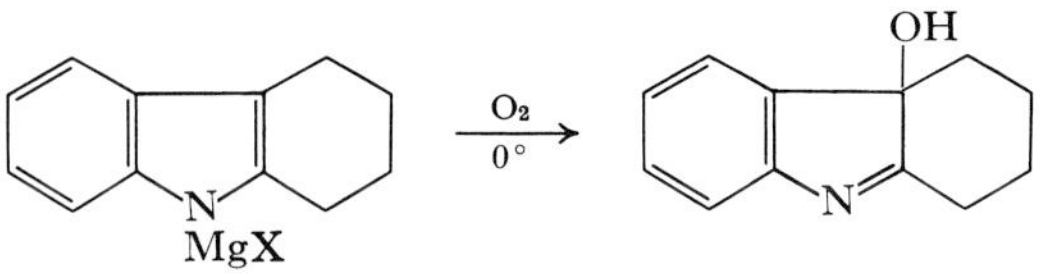

A84 $\xrightarrow[0°]{O_2}$ A85 (65%) Ref. (35)

A86 $\xrightarrow[0°]{O_2}$ A87 (30%) Ref. (35)

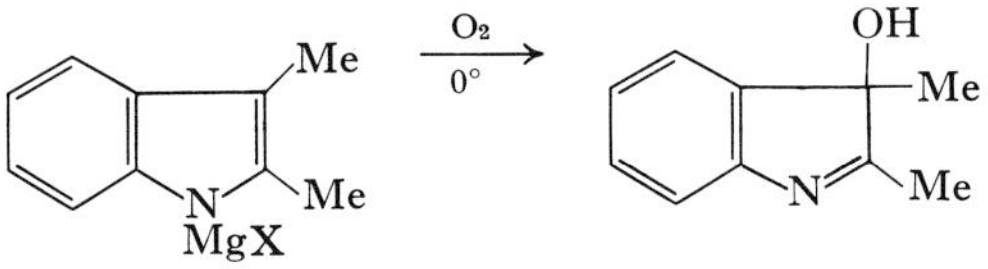

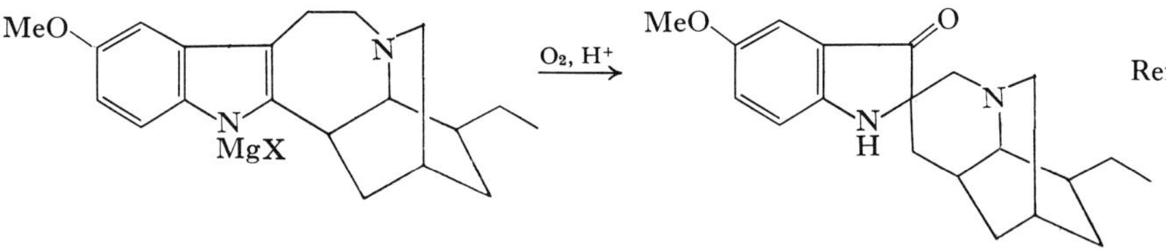

A88 $\xrightarrow{O_2, H^+}$ A89 Ref. (35)

The oxidation of Grignard derivatives of disubstituted indoles with hydrogen peroxide has been investigated occasionally but usually without definitive results (35, 106). Oxidation of the Grignard derivative of indole with *p*-nitroperbenzoic acid gives 3-bromoindole (68a).

Reaction of 3-methylindole with perbenzoic acid gives *o*-formamidoacetophenone (103). The alkaloid derivatives, tetrahydroisoyobrin and

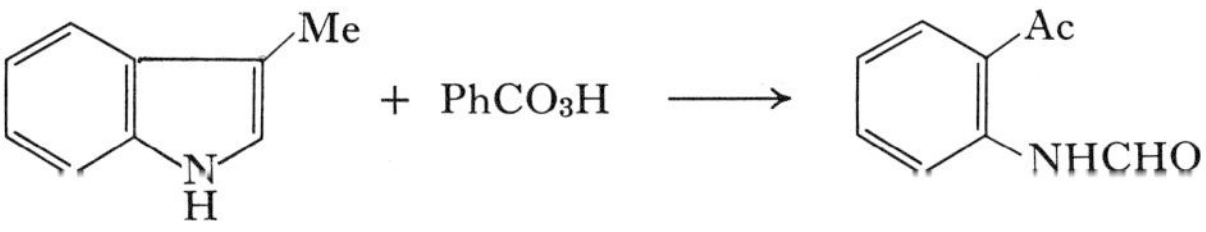

A90 + $PhCO_3H$ (A91) ⟶ A92

yohimbone are cleaved in a similar manner with perbenzoic acid (103). In contrast, the alkaloid cinchonamine gives the corresponding 3-hydroxyindolenine (quinamine) on peracetic acid oxidation (101a). A solution of

N H Me $\xrightarrow[\text{AcOH}]{H_2O_2}$ O N H Me Me N H

A93 A94

hydrogen peroxide oxidizes 2-methylindole to **A94** (100, 105). Under similar conditions, indole gives 2,2-di-3-indolyl-3-indolinone (105). The reaction of 2-alkylindoles with hydrogen peroxide in acetic acid has been investigated subsequently by Piozzi and Langella (75). These workers confirmed the formation of products analogous to **A94** from 2-alkylindoles in concentrated solutions, but in more dilute solutions the symmetric coupling products **A98** were observed. The formation of **A98** might occur if the indolinone **A97** is also formed under these conditions. It also seems possible that **A98** might be formed from **A96** by subsequent oxidation.

A93 ⟶ [O N Me] + O N H H Me

A95 A97

↓ + **A93**

O N H Me Me N H $\xrightarrow{?}$ O O N H Me Me H N

A96 A98

3-Methylindole and tryptophan give the corresponding oxindole under these conditions but not in high yield (100). *N*-Pivaloylanthranilic acid has been obtained by the oxidation of 2-*t*-butylindole, along with other incompletely characterized neutral products (17).

R N H $\xrightarrow[\text{AcOH}]{H_2O_2}$ H R N H O

A99 A100

Hydrogen peroxide selectively cleaves the 2-3 bond of the indole ring in the presence of ammonium molybdate (12, 65). 2-Substituted indoles give 2-acylanthranilic acids and 2,3-disubstituted indoles give *o*-aminophenyl ketones. This method of oxidation has found some use in degradative structural determinations (87, 88).

Potassium persulfate has been found to oxidize 3-methylindole to the oxindole (16) (38% yield) and other products (39) but this reagent has thus far found little synthetic use.

B. OZONOLYSIS OF INDOLES

Early studies on ozonolysis of simple substituted indoles showed that the indole ring could be selectively cleaved between C-2 and C-3 (66, 99). *o*-Formamidophenyl ketones were isolated from 3-substituted indoles and *o*-acylaminophenyl ketones were formed from 2,3-disubstituted indoles. 2-Substituted indoles usually give *N*-acyl derivatives of anthranilic acid although Witkop (99) isolated some *o*-acetamidobenzaldehyde from 2-methylindole. Besides providing an important degradative tool, the ozonolysis of indoles has attracted attention because it is often possible to isolate quite stable ozonides, a matter which will be discussed below. Ockenden and Schofield ozonized several indoles and 1-acylindoles and isolated acylaminophenylketones. Yields of 40% were typical (72). In some cases these cleavages provided good synthetic routes to the ketones.

A number of workers have isolated quite stable ozonides from ozonolysis of indoles (8, 13, 52, 66, 67, 72, 108, 109). Witkop and co-workers (108, 109) investigated rather extensively the chemistry of the ozonide obtained from 2-phenyl-3-methylindole. The ozonide **B1** was found to react with acetic anhydride to give **B3**. Heat or acid caused its decomposition to the *o*-acylamino ketone **B2**. From ultraviolet spectral data and reduction potentials, Witkop and co-workers (109) concluded that the ozonide **B1** is in equilibrium (ring-chain tautomerism) with the oxazine **B4**.

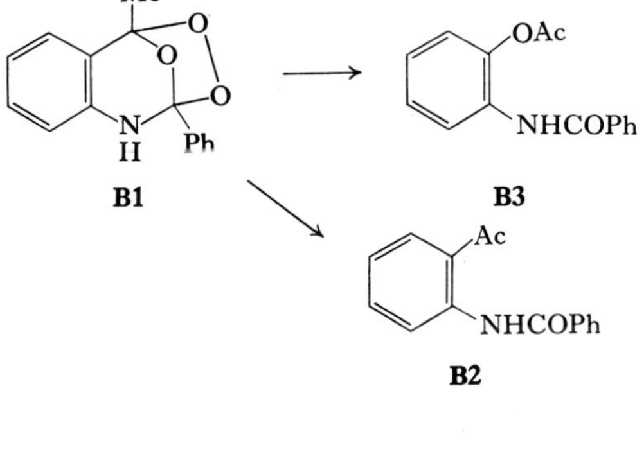

Me
O
O
O
N
H
Ph
B1

OAc
NHCOPh
B3

Ac
NHCOPh
B2

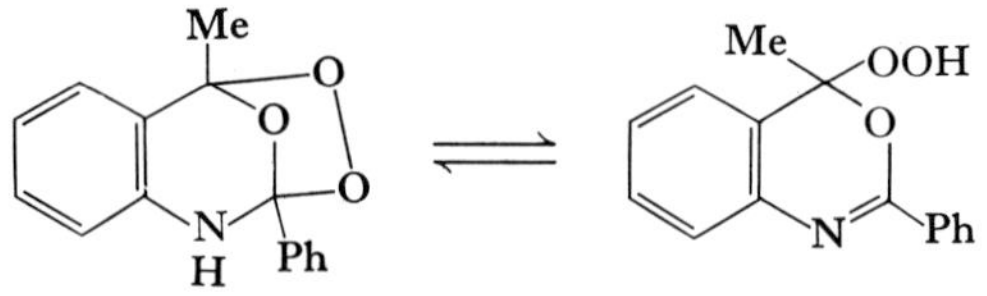

B1 ⇌ B4

Protonation of the nitrogen atom favors the closed form. Little more has been reported concerning the chemistry of such ozonides since the work of Witkop (108, 109).

C. OXIDATION OF INDOLES WITH OTHER OXIDIZING AGENTS

Most reagents which can be generally classed as oxidizing agents have some effect on the indole ring. The various reagents show divergent behavior, however, both as to site of attack and extent of oxidation. Chromium trioxide in acetic acid can effect the cleavage of indoles between C-2 and C-3. The reaction has most often been applied to indoles in which the benzenoid ring is substituted with electron withdrawing groups. In cases where poor yields are obtained, prior acetylation of the indole nitrogen may be beneficial (70). Oxidative dimers are among the products which have been obtained in addition to products of ring cleavage. Thus, Noland and coworkers isolated and characterized **C3** as an important product of oxidation of **C1** with chromic acid. 6-Nitroisatin is a by-product of the oxidation of

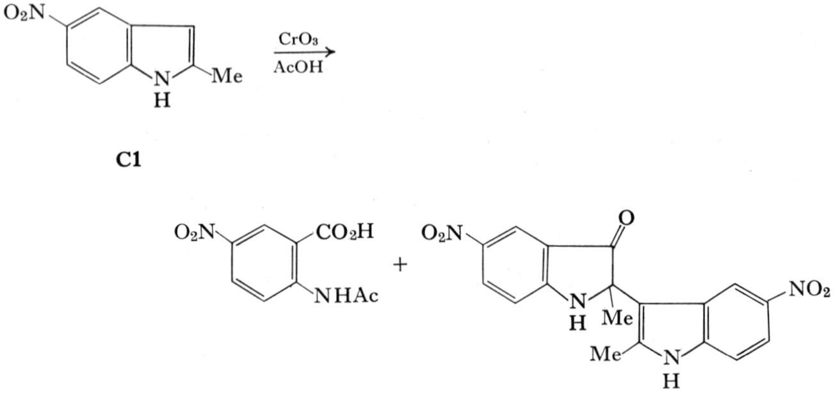

C1 $\xrightarrow[\text{AcOH}]{\text{CrO}_3}$ C2 + C3

3-acetyl-2-methyl-6-nitroindole (71). Oxidation of 2-aminomethylindoles leads to benzodiazepines, presumably via cyclization of the aminoketone **C8** formed by the oxidative ring cleavage (112).

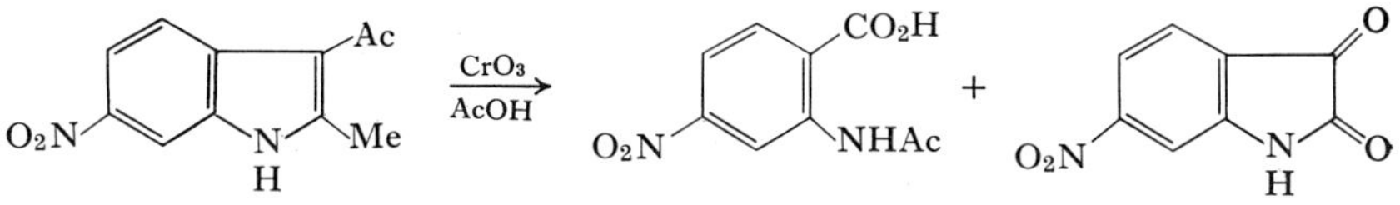

C4 → (CrO_3, AcOH) → C5 + C6

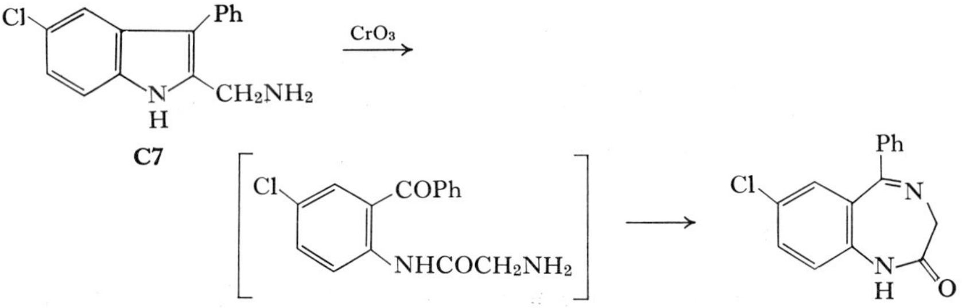

C7 → (CrO_3) → [C8] → C9

Oxidative cleavage of the indole ring with potassium permanganate has also been reported (10). 2-Methylindole is reported to give a 40% yield of *N*-acetylanthranilic acid. When the benzenoid ring is substituted by methyl groups, oxidation of the methyl groups to carboxyl groups occurs (10). Most of the yields reported for such reactions are below 20%. Oxida-

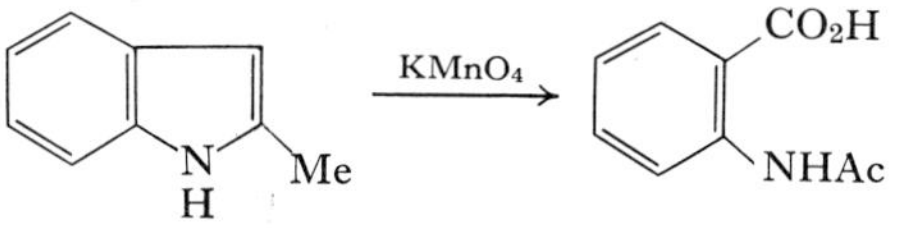

C10 → ($KMnO_4$) → C11 (40%) Ref. 10

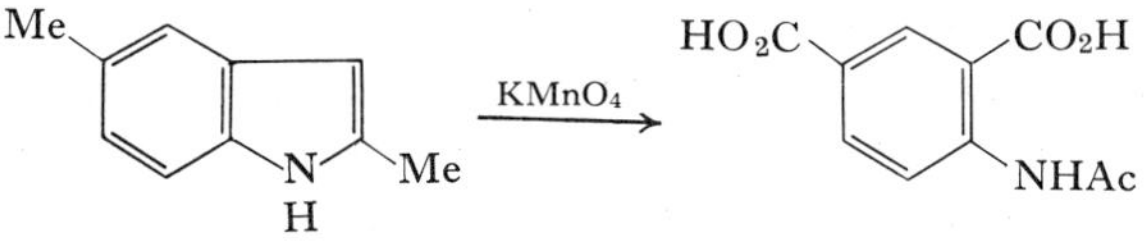

C12 → ($KMnO_4$) → C13 (15%) Ref. 10

tion of the 3-nitroindole **C14** with alkaline permanganate leaves the indole ring intact, giving **C15** (71). The presence of the nitro groups is no doubt instrumental in diminishing the susceptibility of the pyrrole ring to oxidative attack. Selective oxidation of alkyl substituents is not generally successful.

The reactivities of a number of the more selective oxidizing agents toward the indole ring have been examined. Osmium tetroxide appears usually to react by hydroxylation of the C-2 and C-3 carbons of the indole ring, giving 2,3-dihydroxyindolines. Ockenden and Schofield (72) found that such glycols could be isolated only from *N*-substituted indoles.

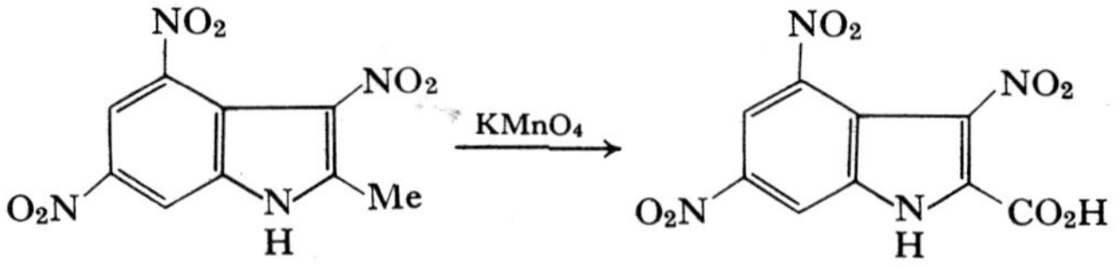

$KMnO_4$

C14 → C15

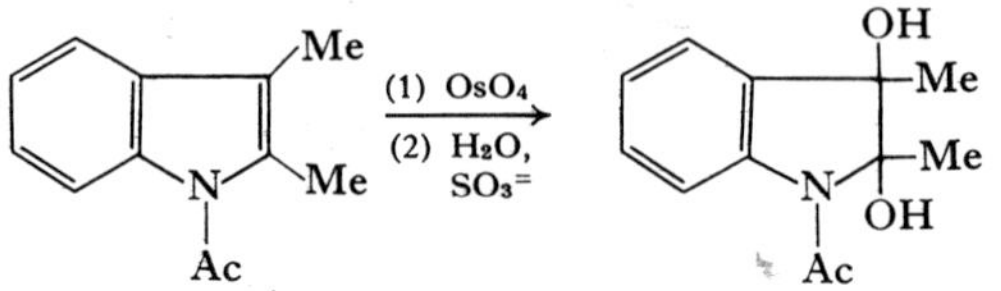

(1) OsO_4 (2) H_2O, $SO_3^=$

C16 → C17 Ref. (72)

Several 2-*t*-butylindoles have been found to give 3-*t*-butyldioxindoles on oxidation with osmium tetroxide followed by treatment with base (74). These products may arise by subsequent rearrangement of glycols formed in the normal manner, but the usual course of base-catalyzed rearrangements of such glycols results in the formation of 2,2-disubstituted-3-indolinones (82, 96).

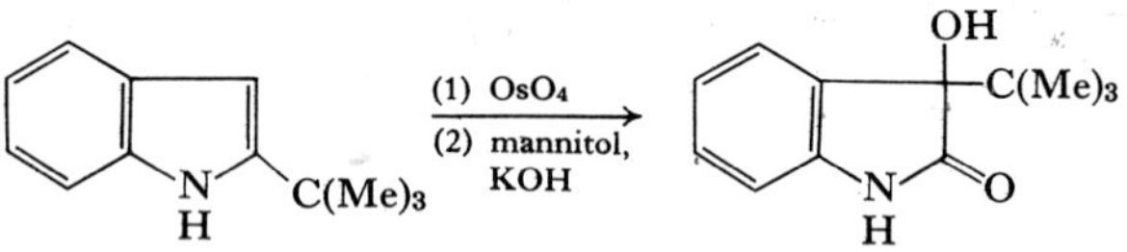

(1) OsO_4 (2) mannitol, KOH

C18 → C19

2,3-Dihydroxylindolines have also been isolated as by-products of nitration of 2,3-disubstituted indoles (76, 77). A recent reinvestigation (2) of these reactions has confirmed that substantial amounts of glycols do accompany the normal substitution products. Thus 1-acetyl-2,3-dimethylindole affords a 23% yield of the glycol **C23**, and **C24** gives **C25** which is easily converted to the glycol **C26**.

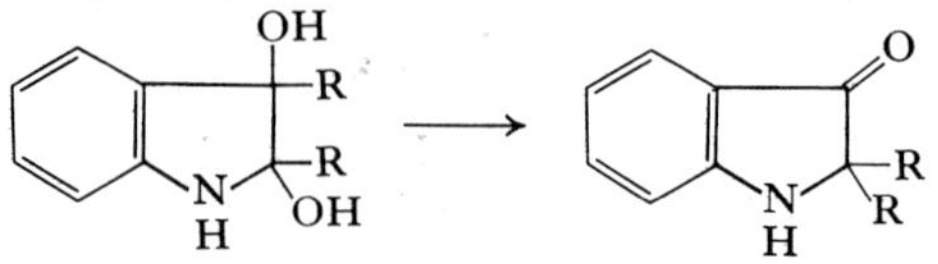

C20 → C21

The stereochemistry of the glycols **C23** and **C26** has been assigned as trans on the basis of slow reaction with periodate and the fact that the glycol **C26** differs from an isomeric glycol prepared by osmium tetroxide oxidation

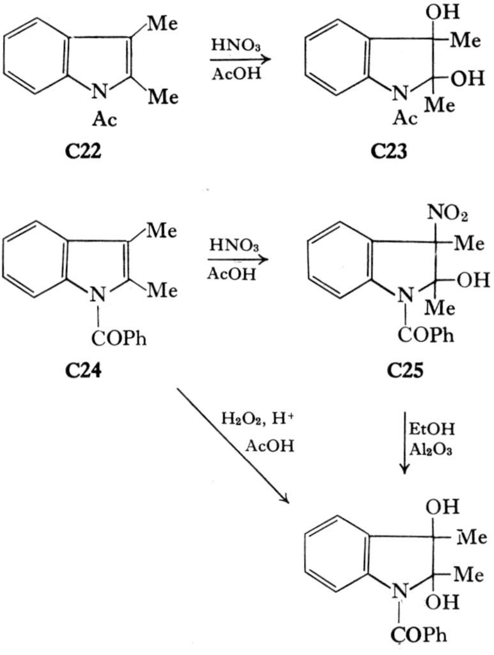

C22 $\xrightarrow[AcOH]{HNO_3}$ **C23**

C24 $\xrightarrow[AcOH]{HNO_3}$ **C25**

C24 $\xrightarrow[AcOH]{H_2O_2,\ H^+}$ **C26**

C25 $\xrightarrow[Al_2O_3]{EtOH}$ **C26**

from **C24** (72). The glycol from the osmium tetroxide reaction is assumed to have the cis configuration. The formation of **C25** is presumably the result of the addition of water to the intermediate 3-nitroindolenium ion **C27**.

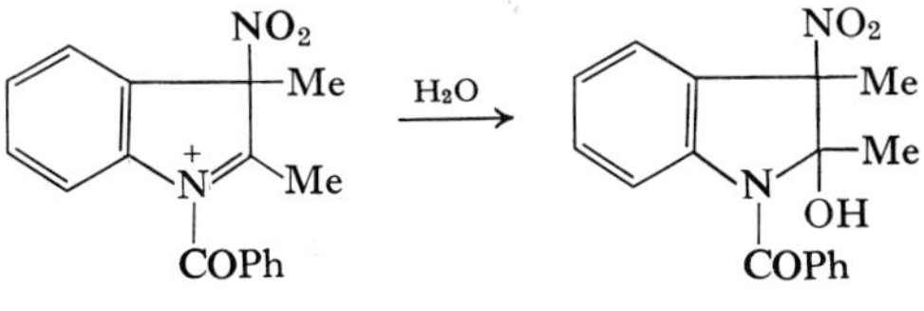

C27 $\xrightarrow{H_2O}$ **C28**

The easy displacement of the nitro group may be attributed to the fact that it is tertiary and in an *o*-aminobenzylic position.

The oxidizing action of manganese dioxide toward indole has been investigated. In benzene, *N*-formylanthranilic acid is formed in good yield (46). The oxidation of tetrahydrocarbazole with the same reagent was found to be considerably more complex. Carbazole and 2-oxotetrahydrocarbazole were produced from an oxidation run under nitrogen. When oxidation was carried out in the presence of air, some oxidation occurred via the indolenyl hydroperoxide.

Dolby and Booth have investigated the effect of periodic acid and periodate on indoles. Mechanistic interpretation is quite incomplete at present, but sodium periodate was found to cleave the pyrrole ring of several 3-substituted and 2,3-disubstituted indoles. For example, an 85% yield of *o*-acetamidoacetophenone was obtained from 2,3-dimethylindole (19, 20). Periodic acid oxidation takes an alternate course. The alkyl substituent attached to C-2 is the site of oxidation. Thus, 2,3-dimethylindole gives a low yield of 2-formyl-3-methylindole and tetrahydrocarbazole gives the 2-acylindole **C33**.

Me, Me, N, H — $NaIO_4$ / MeOH → Ac, NHAc

C29 **C30**

C29 — H_5IO_6 → Me, CHO, N, H

C31 (10%)

N, H — H_5IO_6 → N, H, O

C32 **C33** (62%)

When 2-benzyl-3-phenylindole is heated with lead tetraacetate, 1-acetyl-2-benzoyl-3-phenylindole is obtained (11). A mechanism has been put forward for this reaction which involves initial attack at C-3 of the indole ring.

The correctness of this scheme remains to be demonstrated but it is significant to note that it suggests that apparent oxidation at the C-2 substituent may be the result of rearrangements of 3*H*-indole systems. The same idea has been put forward in interpretation of autoxidation of indoles (Section A).

Table XLIV presents a summary of various oxidations of indoles with inorganic oxidizing agents.

The effect of lead tetraacetate on certain of the tetrahydro-β-carboline indole alkaloids has been investigated (25). When one equivalent of lead tetraacetate is used, 3-acetoxyindolenines can be isolated. These compounds are proposed to arise via an unstable intermediate in which the lead is bound to C-3. Acid can catalyze loss of acetic acid leading to dehydro derivatives

TABLE XLIV □ OXIDATION OF INDOLES

Indole substituent	Oxidizing agent	Oxidation product	Yield (%)	Ref.
1-Acetyl-2-methyl-5-nitro	Chromic acid	*N*-Acetyl-5-nitroanthranilic acid	46	(70)
1,6-Diacetyl-2,3-dimethyl	Chromic acid	2,5-Diacetylacetanilide	32	(31)
1-Acetyl-6-bromo-2,3-diphenyl	Chromic acid	2-(*N*-Acetylbenzamido)-4-bromobenzophenone	55	(55)
2-Methyl-5-nitro	Chromic acid	*N*-Acetyl-5-nitroanthranilic acid	1	(70)
2,3-Dimethyl-5-nitro	Chromic acid	2-Acetamido-5-nitroacetophenone	25	(84)
2,3-Diphenyl-6-nitro	Chromic acid	2-Benzamido-4-nitrobenzophenone	70	(85)
3-Acetyl-2-methyl-6-nitro	Chromic acid	*N*-Acetyl-4-nitroanthranilic acid	47	(71)
3,5-Dinitro-2-methyl	Chromic acid	*N*-Acetyl-5-nitroanthranilic acid	15	(71)
Ethyl 2,6-dicarbethoxyindole-3-propionate	Chromic acid	Ethyl 3-(4-carbethoxy-2-ethoxalylamidobenzoyl)-propionate	83	(56)
2-Me	Potassium permanganate	*N*-Acetylanthranilic acid	40	(10)
2,5-diMe	Potassium permanganate	4-Acetamidoisophthalic acid	15	(10)
1-Acetyl-2,3-dimethyl	Osmium tetroxide	1-Acetyl-2,3-dihydroxy-2,3-dimethylindoline	69	(72)
1-Acetyl-2,3-diphenyl	Osmium tetroxide	1-Acetyl-2,3-dihydroxy-2,3-diphenylindoline	24, 29	(72, 82)
N,*N'*-Diacetyl-1,2,3,4-tetrahydro-β-carboline	Osmium tetroxide	Corresponding dihydroxy indoline	?	(96)
None	Manganese dioxide	*N*-Formyl anthranilic acid	75	(46)
3-Me	Sodium periodate	2-Formamidoacetophenone	82	(19)
1,2,3,4-Tetrahydrocarbazole	Sodium periodate	1-Aza-8,9-benzcyclonona-2,7-dione	99	(19)
1,2,3,4-Tetrahydrocarbazole	Periodic acid	1-Oxo-1,2,3,4-tetrahydrocarbazole	62	(19)
2-Benzyl-3-phenyl	Lead tetraacetate	1-Acetyl-2-benzoyl-3-phenylindole	?	(11)

Pb(OAc)4 Ph N H CH2Ph
C34

OAc Ph N+ H CH2Ph
C35

OAc Ph N H CHPh −OAc
C36

Ph N H CHPh OAc
C37

OAc Ph N CHPh OAc
C38

OAc Ph N H Ph OAc
C39

Ph N H C(OAc)2 Ph
C40

Ph N Ac COPh
C41

N H N
C42

Pb(OAc)4

OAc N N
C43

OAc N N
C44

O N H N
C46

H+

H+

N H +N
C45

OR

N N H O
C47

(**C45**) or, if the stereochemistry of the D-E ring junction is cis, to oxindoles (**C47**). The mechanistic basis for the stereochemical requirement has been discussed (25). Under basic conditions the acetoxyindolenines undergo an alternative rearrangement giving 2,2-disubstituted 3-indolinones (**C46**).

A similar series of transformations is initiated by reaction of indoles with *t*-butyl hypochlorite. Treatment of tetrahydro-β-carboline alkaloids with this reagent gives 3-chloroindolenines which can subsequently be converted to oxindoles (26, 32, 86, 113).

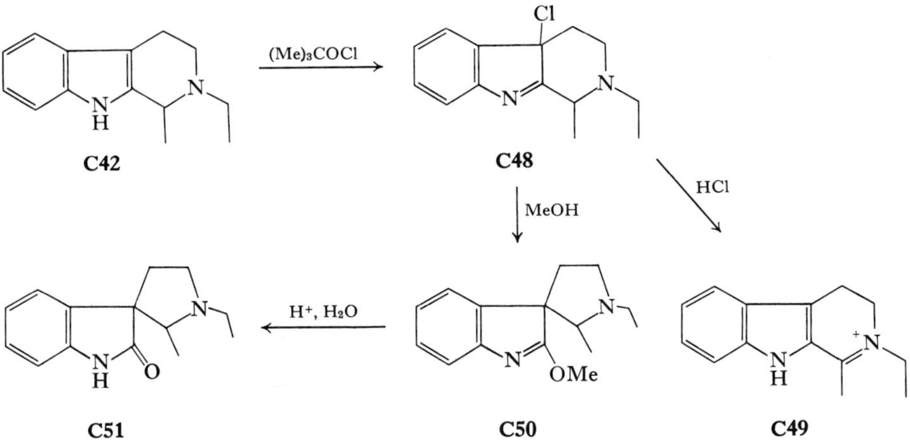

$(Me)_3COCl$; **C42**; **C48**; MeOH; HCl; H^+, H_2O; **C51**; **C50**; **C49**

Another electrophilic reagent which can effect net oxidation of the indole ring is *N*-bromosuccinimide. Treatment of indole-3-propionic acid in acetic acid with *N*-bromosuccinimide followed by hydrogenolysis over palladium gives a 50% yield of oxindole-3-propionic acid (**C54**) (58). Indole-3-acetic acid is oxidized in a similar fashion by this procedure. The oxidation of indole-3-propionic acid proceeds by way of the bromolactone

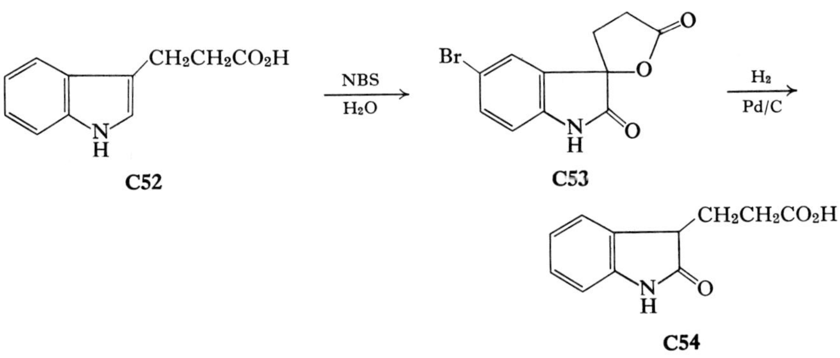

$CH_2CH_2CO_2H$; NBS, H_2O; H_2, Pd/C; **C52**; **C53**; **C54**

C53. Simultaneous hydrogenolysis of the C–Br and the benzylic C–O bond yields **C54**. The mechanism shown below has been advanced to account for the formation of **C53** (57). Alternatively, **C57** might be formed from a 3-bromoindolenine arising by attack by bromine at the 3-position.

$CH_2CH_2CO_2H$ + Br_2 →

C52 **C55**

$-HBr$

C57 **C56**

H_2O (O)

NBS

C58 **C53**

3-Methylindole reacts with *N*-bromophthalimide in benzene to give 2,6-dibromo-3-methylindole which can be hydrolyzed to 6-bromo-3-methyloxindole (57). In aqueous acetic acid 3-methylindole and *N*-bromosuccinimide form 5-bromo-3-methyloxindole (57). The 5-bromo derivative presumably arises from 3-methyloxindole formed by hydrolysis of the primary product, 2-bromo-3-methylindole. Facile hydrolysis to oxindoles is a general property of 2-haloindoles (Chapter I, D). Hinman and Bauman (41) have investigated the reaction of 3-substituted indoles with *N*-bromosuccinimide in *t*-butyl alcohol. When 1:1 mole ratios of the indole and the brominating agent are used, followed by treatment with water, yields of up to 50% of the corresponding oxindole can be obtained. Use of a second mole of *N*-bromosuccinimide results in the formation of 3-bromoxindoles. Subsequent bromination in the 5-position is apparently greatly retarded in *t*-butyl alcohol relative to water.

CH_2CO_2H N H **C59** —1 mole NBS, $(Me)_3COH$→ CH_2CO_2H N H O **C60** (42%)

C59 —2 moles NBS, $(Me)_3COH$→ Br CH_2CO_2H N H O **C61** (75%)

Bromination of 3-substituted indoles in pyridine results in the formation of *N*-(2-indolyl)pyridinium bromides which can subsequently be hydrolyzed to oxindoles (54a).

The reaction of peptides with *N*-bromosuccinimide has been used to selectively cleave the peptide chain at tryptophan positions (102). The scheme **C62** → **C65** rationalizes the selective cleavage (72a).

$CH_2CHCONHR$ O C NH R N H **C62** → NR O NHCOR N H Br **C63** —H_2O (O)→ NR O NHCOR N H O **C64** —H_2O→ O O NHCOR N H O **C65**

Foglia and Swern (27) have investigated the reactions of *N,N*-dichloro-urethane with indoles. Like *N*-bromosuccinimide, *N,N*-dichlorourethane is a source of positive halogen. Both indole-2-carboxylic acid and indole-3-carboxylic acid give 3,3,5-trichlorooxindole, as does indole itself, on reaction with dichlorourethane in aqueous solution. The carboxyl substituents are

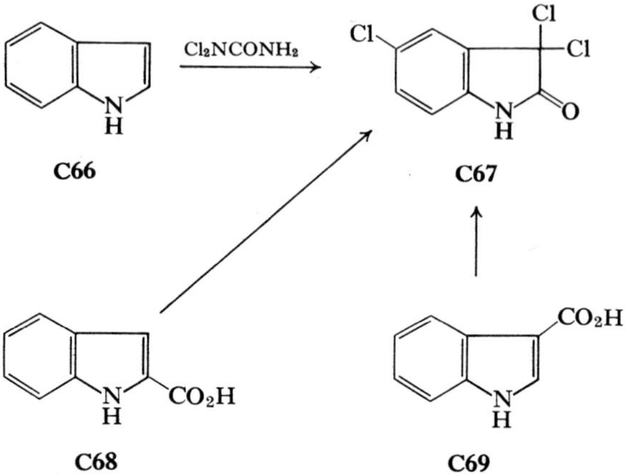

Cl_2NCONH_2

C66 **C67** **C68** **C69**

not lost when the methyl esters are subjected to oxidation. Rational mechanisms for these reactions involve attack by Cl^+ on the indole ring.

In connection with his studies of the reaction of potassium nitrosodisulfonate which are discussed in more detail in Chapter VII, Teuber (93) has had occasion to study the reactions of simple indoles with this reagent. 2-Methylindole in acidic solutions gave the indolinone **C82** which has been previously isolated from 2-methylindole on reaction with other oxidizing reagents. More extensive oxidation to **C83** occurs under neutral conditions. 2-Phenylindole is oxidized to **C85** which is a well-known product of nitrosation of 2-phenylindole.

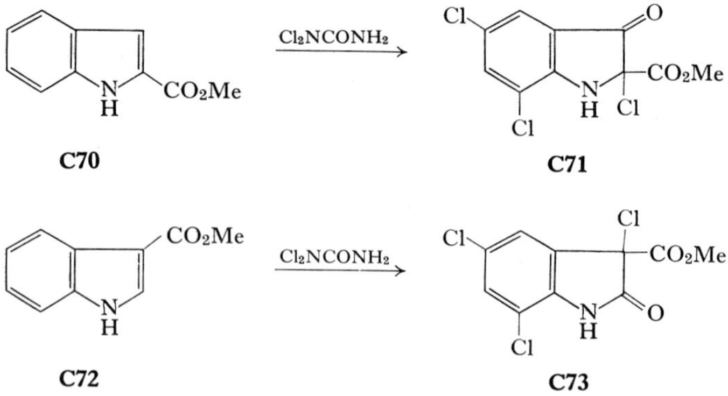

Cl_2NCONH_2

C70 **C71**

Cl_2NCONH_2

C72 **C73**

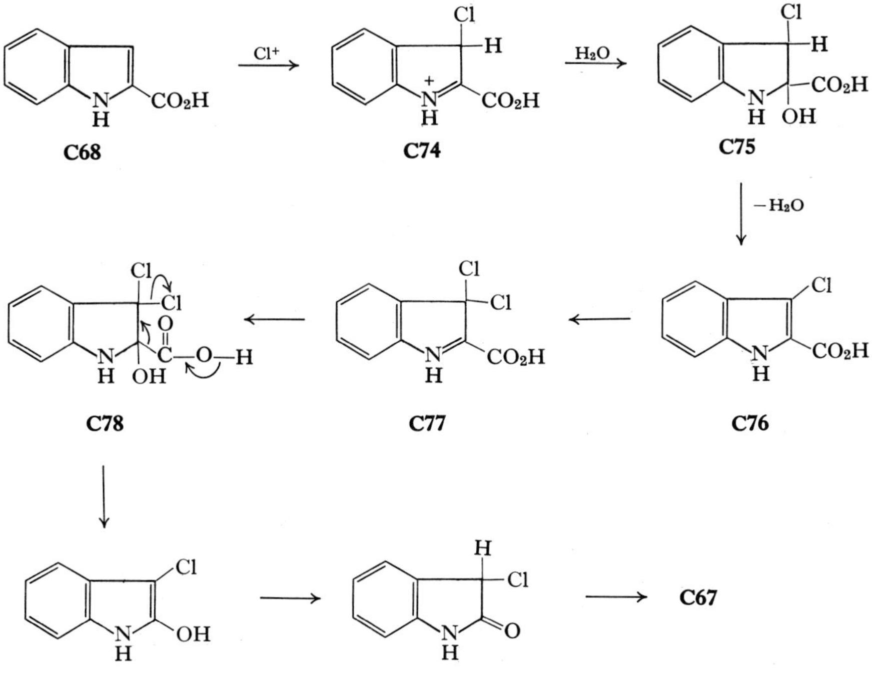

Cl^+ H_2O $-H_2O$

Cl H CO_2H N H OH

C68 C74 C75

C78 C77 C76

C79 C80 C67

$ON(SO_3K)_2$

Me H O N H Me

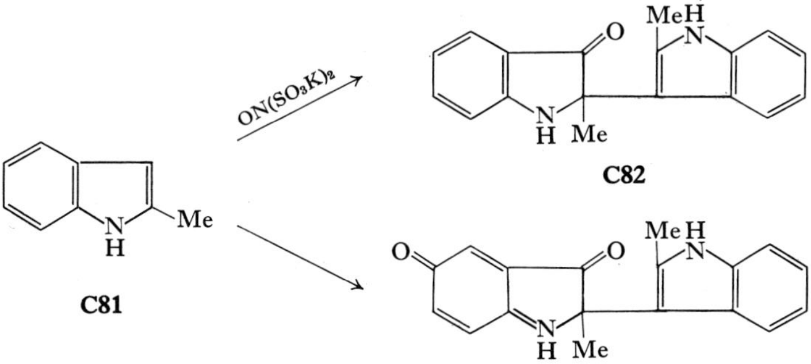

C81 C82 C83

NOH Ph

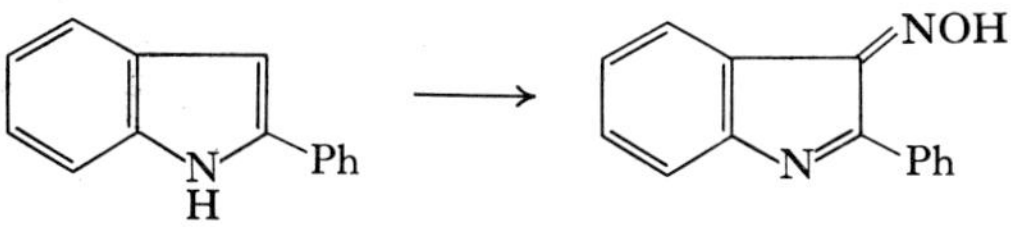

C84 C85

D. OXIDATION OF INDOLE DERIVATIVES IN BIOLOGICAL SYSTEMS AND CHEMICAL MODELS OF BIOLOGICAL SYSTEMS

The physiological importance of the indole ring has generated considerable interest in the biological transformations of indole derivatives. This has been particularly true of the tryptamine and tryptophan systems which have attracted attention because of the possibility that defects in metabolism of tryptamines, particularly serotonin (5-hydroxytryptamine) may be involved in certain mental disorders (111).

A recent study of the metabolism of indole-2-^{14}C has led to evidence for the formation of 3-hydroxindole, oxindole, isatin, *N*-formylanthranilic acid, anthranilic acid, and derivatives of 5-hydroxyoxindole as metabolites (53). Previously much evidence had accumulated that 3-hydroxylation was a primary metabolic pathway for indole (79, 98). Hydroxylation in the 6-position can also be a metabolic reaction in humans as established by the identification of 6-hydroxyskatole as a metabolite (45). Oxindoles are found, on the other hand, to be hydroxylated at the 5-position of the aromatic ring in rats, guinea pigs, and rabbits (4) and when incubated with rat liver preparations (53).

Tryptamine is apparently not hydroxylated to serotonin (5-hydroxylation). Serotonin is believed to be formed from tryptophan by 5-hydroxylation followed by decarboxylation (22, 94). A major pathway for metabolism of tryptamine involves oxidative deamination of the amino-substituted side chain, a metabolic pathway which is very general for amines (30). Like indole, tryptamine and related indoles, e.g., dimethyltryptamine and indole-3-acetic acid, are hydroxylated at the 6-position by liver microsomes (48) and *in vivo* in rats (90a). 3-(2-Acetoxyethyl)-5-methoxyindole is reported to be metabolized in the rat by 7-hydroxylation on the basis of the nonidentity of the metabolite with 6-hydroxy and 4-hydroxy-5-methoxyindole-3-acetic acid (18).

It is interesting to note that, except for the last example and the important case of tryptophan, biological hydroxylation of 3-substituted indoles usually takes place at the site predicted (C-6) for electrophilic aromatic substitution (see Chapter I, O). The notion that 5- and 7- are the reactive positions for electrophilic substitution on the indole ring is clearly not in accord with the current picture of orientation of electrophilic substitution in indoles and has led to incorrect structural assignments for products of metabolic hydroxylations (48).

The hydroxylation of tryptophan at the 5-position has been observed in a number of biological systems (33, 60, 68, 69) and with enzyme preparations (29, 81). It has been shown (80) that hydroxylation is accompanied by hydrogen migration from C-5 to C-4 of the indole ring. The preferential introduction of the hydroxyl group at C-5 tends to argue against direct

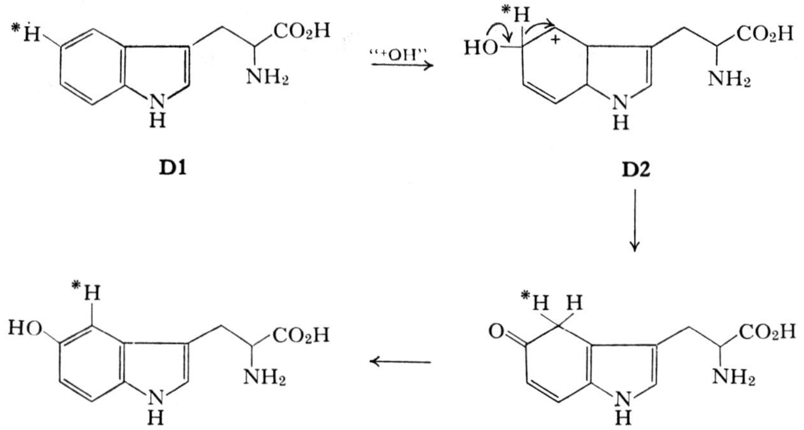

"+OH"

D1

D2

D4

D3

electrophilic substitution as the mechanism of hydroxylation since extensive substitution at the 6-position would be predicted on the basis of current knowledge of electrophilic substitution in indoles. The possibility that arene oxides could be involved as intermediates in biological hydroxylation of aromatic substates is under active investigation (49–51). The intermediacy of an arene oxide such as **D6** could provide a rationale for selective 5-hydroxylation of tryptophan. The selective hydroxylation could also arise from steric factors at the enzyme active site.

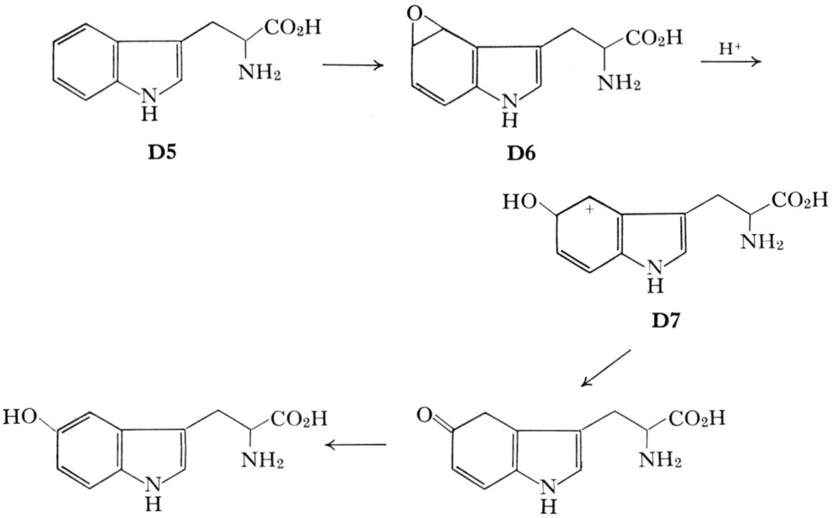

H^+

D5

D6

D7

D9

D8

Relatively little is known about the enzyme or enzymes responsible for hydroxylation (30, 36a).

A chemical system which, to some extent, models biological hydroxylation has been developed and applied to indole derivatives (9, 15). The system consists of ascorbic acid, ferrous sulfate, and Versene in phosphate buffer. Oxygen is also required. This system is apparently nonselective with regard to position of substitution in that 3-methylindole has been shown to give all four benzenoid hydroxylation products as well as 3-methyloxindole, *o*-aminoacetophenone and *N*-formylacetophenone (38). Indole apparently shows similar behavior (21). Hydroxylation at C-5 and C-6 has been reported for indole-3-carboxylic acid (1).

In addition to hydroxylation, oxidative cleavage of the indole ring is an important reaction of indoles in biological systems. Tryptophan undergoes oxidative cleavage between C-2 and C-3, giving formylkynurenine. Structural studies on the catalytic enzyme (tryptophan pyrrolase) have shown that it contains a ferroporphyrin (91) and probably copper (61).

CO_2H, NH_2, N, H $\xrightarrow[\text{tryptophan pyrrolase}]{O_2}$ $COCH_2CHNH_2$ (CO_2H), NCHO

D9 **D10**

The enzyme has been resolved into an apoenzyme and a ferroporphyrin (23, 34) cofactor. The apoenzyme is inactive but it is reactivated by ferriprotoporphyrin. It is believed that cyclic oxidation–reduction of at least part of the iron in the enzyme occurs during the process of catalytic oxidation (62). Spectral evidence has been reported (47) which indicates that tryptophan and oxygen form a complex with the enzyme which may be an intermediate in the oxidation process.

The mechanism of oxidation has been expressed as shown below (24). The precise nature of the "Tryp^{ox}" and "$O_2^{-(\text{Red})}$" are regarded as open to question. Analogy with autoxidation of indoles would suggest that

$$\text{Tryp} + Fe^{+3}_{(H)} \longrightarrow \text{Tryp}^{(ox)} + Fe^{+2}_{(H)}$$

$$Fe^{2}_{(H)} + O_2 \longrightarrow Fe^{+3}_{(H)} + O_2^{-}\ (\text{red})$$

$$\text{Tryp}^{(ox)} + O_2^{-}\ (\text{red}) \longrightarrow \text{formyl kynurenine}$$

$(H) = \text{hematin}$

3*H*-indolylhydroperoxides might be intermediates in the formation of formyl kynurenine. Reviews discussing oxidation of tryptophan in the presence of tryptophan pyrrolase are available (14, 37, 54, 64, 95).

The oxidation of the plant growth hormone, indole-3-acetic acid, has also received substantial attention. This oxidation is catalyzed by horseradish peroxidase. The major ultimate product of the reaction has been identified as 3-methyleneoxindole (42). The net transformation is unusual when compared with normal modes of oxidative cleavage. Hinman and Lang have put forward a mechanism which envisages intramolecular participation of the acetic acid side chain in the decomposition of a 3-hydroperoxyindolenine intermediate. Fox, Purves, and Nakada (28) have shown that spectra changes

CH_2CO_2H

D11

[O]

CH_2CO_2H

D12

CH_2CO_2H

D13

D14

CH_2

D15

CH_2

D16

CH_2OH

D17

CH_2

D18

accompany interaction of indole-3-acetic acid and horseradish peroxidase in the presence of oxygen. Their result would require modification of the early stages of the Hinman–Lang mechanism but require no change in the later, product-determining steps.

REFERENCES

1. Acheson, R. M., and King, L. J., *Biochim. Biophys. Acta* **71**, 643 (1963).
2. Atkinson, C. M., Kershaw, J. W., and Taylor, A., *J. Chem. Soc.* p. 4426 (1962).
3. Bartlett, M. F., Dickel, D. F., and Taylor, W. I., *J. Am. Chem. Soc.* **80**, 126 (1958).
4. Beckett, A. H., and Morton, D. M., *Biochem. Pharmacol.* **15**, 937 (1966).
5. Beer, R. J. S., Broadhurst, T., and Robertson, A., *J. Chem. Soc.* p. 4946 (1952).
6. Beer, R. J. S., Donavanik, T., and Robertson, A., *J. Chem. Soc.* p. 4139 (1954).
7. Beer, R. J. S., McGrath, L., and Robertson, A., *J. Chem. Soc.* pp. 2118 and 3283 (1950).
8. Berti, G., Da Settimo, A., and Segnini, D., *Tetrahedron Letters* No. 26 p. 13 (1960).

8a. Bird, C. W., *J. Chem. Soc.* p. 3490 (1965).

9. Brodie, B. B., Axelrod, J., Shore, P. A., and Udenfriend, S., *J. Biol. Chem.* **208**, 741 (1954).
10. Cardani, C., and Piozzi, F., *Gazz. Chim. Ital.* **86**, 849 (1956); *Chem. Abstr.* **52**, 5373 (1958).
11. Chen, F. Y., and Leete, E., *Tetrahedron Letters* p. 2013 (1963).
12. Clerc-Bory, G., Clerc-Bory, M., Pacheco, H., and Mentzger, C., *Bull. Soc. Chim. France* p. 1229 (1955).
13. Clerc-Bory, M., *Bull. Soc. Chim. France* p. 88, (1955).
14. Crandall, D. I., *in* "Oxidases and Related Redox Systems" (T. King, H. S. Mason, and M. Morrison, eds.), p. 269. Wiley, New York, 1965.
15. Dalgliesh, C. E., *Arch. Biochem. Biophys.* **58**, 214 (1955).
16. Dalgliesh, C. E., and Kelly, W., *J. Chem. Soc.* p. 3726 (1958).
17. David, S., and Monnier, J., *Bull. Soc. Chim. France* p. 1333 (1959).
18. Delvigs, P., and Taborsky, R. G., *Biochem. Pharmacol.* **16**, 579 (1967).
19. Dolby, L. J., and Booth, D. L., *J. Am. Chem. Soc.* **88**, 1049 (1966).
20. Dolby, L. J., and Gribble, G. W., *J. Org. Chem.* **32**, 1391 (1967).
21. Eich, E., and Rochelmeyer, H., *Pharm. Acta Helv.*, **41**, 109 (1966); *Chem. Abstr.* **64**, 14159 (1966).
22. Erspamer, V., *Progr. Drug Res.* **3**, 151 (1961).
23. Feigelson, P., and Greengard, O., *Biochim. Biophys. Acta* **50**, 200 (1961).
24. Feigelson, P., Ishimura, Y., and Hayaishi, O., *Biochim. Biophys. Acta* **96**, 283 (1965).
25. Finch, N., Gemenden, G. W., Hsu, I. H., and Taylor, W. I., *J. Am. Chem. Soc.* **85**, 1520 (1963).
26. Finch, N., and Taylor, W. I., *J. Am. Chem. Soc.* **84**, 3871 (1962).
27. Foglia, T. A., and Swern, D., *J. Org. Chem.* **33**, 4440 (1968).
28. Fox, L. R., Purves, W. K., and Nakada, H. I., *Biochemistry* **4**, 2754 (1965).
29. Freedland, R. A., *Biochim. Biophys. Acta* **73**, 71 (1963).
30. Garanttini, S., and Valzelli, L., "Serotonin," pp. 27–73. Elsevier, Amsterdam, 1965.
31. Gaudion, W. J., Hood, W. H., and Plant, S. G. P., *J. Chem. Soc.* p. 1631 (1947).
32. Godtfredsen, W. O., and Vangedal, S., *Acta Chem. Scand.* **10**, 1414 (1956).
33. Grahame-Smith, D. G., *Biochem. Biophys. Res. Commun.* **16**, 586 (1964).
34. Greengard, O., and Feigelson, P., *J. Biol. Chem.* **237**, 1903 (1962).

35. Guise, G. B., Ritchie, E., and Taylor, W. C., *Australian J. Chem.* **18**, 1279 (1965).
36. Gwinani, S., Arifuddin, M., and Augusti, K. T., *Photochem. Photobiol.* **5**, 495 (1966).
36a. Hagen, P. B., and Cohen, L. H., *Handbuch Exptl. Pharmakol.* **19**, 182 (1966).
37. Hayaishi, O., "Oxygen in the Animal Organism," p. 151. Macmillan, New York, 1964.
38. Heacock, R. A., and Mahon, M. E., *Can. J. Biochem. Physiol.* **41**, 2381 (1963).
39. Heacock, R. A., and Mahon, M. E., *Can. J. Biochem.* **43**, 1985 (1965); *Chem. Abstr.* **64**, 931 (1966).
40. Hester, J. B., Jr., *J. Org. Chem.* **32**, 3804 (1967).
41. Hinman, R. L., and Bauman, C. P., *J. Org. Chem.* **29**, 1206 (1964).
42. Hinman, R. L., and Lang, J., *Biochemistry* **4**, 144 (1965).
43. Hino, T., Nakagawa, M., Wakatsuki, T., Ogawa, K., and Yamada, S., *Tetrahedron* **23**, 1441 (1967).
44. Hino, T., Nakagawa, M., and Akaboshi, S., *Chem. Commun.* p. 656 (1967).
45. Horning, E. C., Sweeley, C. C., Dalgliesh, C. E., and Kelly, W., *Biochim. Biophys. Acta* **32**, 566 (1959).
46. Hughes, B., and Suschitzky, H., *J. Chem. Soc.* p. 875 (1965).
47. Ishimura, Y., Nozaki, M., Hayaishi, O., Tamura, M., and Yamazaki, I., *J. Biol. Chem.* **242**, 2574 (1967).
48. Jepson, J. B., Zaltzman, P., and Udenfriend, S., *Biochim. Biophys. Acta* **62**, 91 (1962).
49. Jerina, D. M., Daly, J. W., and Witkop, B., *J. Am. Chem. Soc.* **90**, 6523 (1968).
50. Jerina, D. M., Daly, J. W., Witkop, B., Zaltzman-Nirenberg, P., and Udenfriend, S., *J. Am. Chem. Soc.* **90**, 6525 (1968).
51. Jerina, D., Daly, J. W., Witkop, B., Zaltzman-Nirenberg, P., and Udenfriend, S., *Arch Biochem. Biophys.* **128**, 126 (1969).
52. Karrer, P., and Enslin, P., *Helv. Chim. Acta* **32**, 1390 (1949).
53. King, L. J., Parke, D. V., and Williams, R. T., *Biochem. J.* **98**, 266 (1966).
54. Knox, W. E., and Tokyama, K., *in* "Oxidases and Related Redox Systems," (T. King, H. S. Mason, and M. Morrison, eds.), p. 514. Wiley, New York, 1965.
54a. Kobayashi, T., and Inokuchi, N., *Tetrahedron* **20**, 2055 (1964).
55. Koelsch, C. F., *J. Am. Chem. Soc.* **66**, 1983 (1944).
56. Koelsch, C. F., *J. Org. Chem.* **8**, 295 (1943).
57. Lawson, W. B., Patchornik, A., and Witkop, B., *J. Am. Chem. Soc.* **82**, 5918 (1960).
58. Lawson, W. B., and Witkop, B., *J. Org. Chem.* **26**, 263 (1961).
59. Leete, E., *J. Am. Chem. Soc.* **83**, 3645 (1961).
60. Lovenberg, W., Levine, R. J., and Sjoerdsma, S., *Biochem. Pharmacol.* **14**, 887 (1965).
61. Maeno, H., and Feigelson, P., *Biochem. Biophys. Res. Commun.* **21**, 297 (1965).
62. Maeno, H., and Feigelson, P., *J. Biol. Chem.* **242**, 596 (1967).
63. McCapra, F., and Chang, Y. C., *Chem. Commun.* p. 522 (1966).
64. Mehler, A., *in* "Oxygenases" (O. Hayaishi, ed.), p. 87. Academic Press, New York, 1962.
65. Mentzer, C., and Berguer, Y., *Bull. Soc. Chim. France* p. 218 (1952).
66. Mentzer, C., Molho, D., and Berguer, Y., *Bull. Soc. Chim. France* pp. 555 and 782 (1950).
67. Mills, B., and Schofield, K., *J. Chem. Soc.* p. 5558 (1961).
68. Mitoma, C., Weissbach, H., and Udenfriend, S., *Arch. Biochem. Biophys.* **63**, 122 (1956).
68a. Mousseron-Canet, M., and Boca, J.-P., *Bull. Soc. Chim. France* p. 1294 (1967).
69. Nakamura, S., Ichiyama, A., and Hayaishi, O., *Federation Proc.* **24**, 604 (1965).
70. Noland, W. E., Smith, L. R., and Johnson, D. C., *J. Org. Chem.* **28**, 2262 (1963).
71. Noland, W. E., Smith, L. R., and Rush, K. R., *J. Org. Chem.* **30**, 3457 (1965).

72. Ockenden, D. W., and Schofield, K., *J. Chem. Soc.* pp. 612 and 3440 (1953).
72a. Patchornik, A., Lawson, W .B., Gross, E., and Witkop, B., *J. Am. Chem. Soc.* **82,** 5923 (1960).
73. Patrick, J. B., and Witkop, B., *J. Am. Chem. Soc.* **72,** 633 (1950).
74. Piozzi, F., and Cecere, M., *Atti Accad. Nazl. Lincei, Rend., Classe Sci. Fis., Mat. Nat.* [8] **28,** 639 (1960); *Chem. Abstr.* **55,** 9372 (1961).
75. Piozzi, F., and Langella, M. R., *Gazz. Chim. Ital.* **93,** 1373 (1963); *Chem. Abstr.* **60,** 9232 (1964).
76. Plant, S. G. P., and Tomlinson, M. L., *J. Chem. Soc.* p. 955 (1933).
77. Plant, S. G. P., and Whitaker, W. B., *J. Chem. Soc.* p. 283 (1940).
78. Plieninger, H., Bauer, H. Bühler, W., Kurze, J., and Lerch, U., *Ann. Chem.* **680,** 69 (1964).
79. Posner, H. S., Mitoma, C., and Udenfriend, S., *Arch. Biochem. Biophys.* **94,** 269 (1961).
80. Renson, J., Daly, J., Weissbach, H., Witkop, B., and Udenfriend, S., *Biochem. Biophys. Res. Commun.* **25,** 504 (1966).
81. Renson, J., Weissbach, H., and Udenfriend, S., *J. Biol. Chem.* **237,** 2261 (1962).
82. Sarel, S., and Klug, J. T., *Israel J. Chem.* **2,** 143 (1964); *Chem. Abstr.* **62,** 6452 (1965).
83. Schnoes, H. K., Thomas, D. W., Aksormvitaya, R., Schleigh, W. R., and Kupchan, S. M., *J. Org. Chem.* **33,** 1225 (1968).
84. Schofield, K., and Theobald, R. S., *J. Chem. Soc.* p. 796 (1949).
85. Schofield, K., and Theobald, R. S., *J. Chem. Soc.* p. 1505 (1950).
86. Shavel, J., Jr., and Zinnes, H., *J. Am. Chem. Soc.* **84,** 1320 (1962).
87. Suehiro, T., *Chem. Ber.* **100,** 905 and 915 (1963).
88. Suehiro, T., and Nakagawa, A., *Bull. Chem. Soc. Japan* **40,** 2919 (1967).
89. Sugiyama, N., Akutagawa, M., Gasha, T., Saiga, Y., and Yamamoto, H., *Bull. Chem. Soc. Japan.* **40,** 347 (1967).
90. Sugiyama, N., Yamamoto, H., and Omote, Y., *Bull. Chem. Soc. Japan* **41,** 1917 (1968).
90a. Szára, S., Hearst, E., and Putney, F., *Federation Proc.* **19,** 23 (1960).
91. Tanaka, T., and Knox, W. E., *J. Biol. Chem.* **234,** 1162 (1959).
92. Taylor, W. I., *Proc. Chem. Soc.* p. 247 (1962).
93. Teuber, H.-J., and Staiger, G., *Chem. Ber.* **88,** 1066 (1955).
94. Udenfriend, S., Creveling, C. R., Posner, H., Redfield, B. G., Daly, J., and Witkop, B., *Arch. Biochem. Biophys.* **83,** 501 (1959).
95. Ullrich, V., and Staudinger, H. J., *in* "Biological and Chemical Aspects of Oxygenases" (K. Bloch and O. Hayaishi, eds.), p. 235. Maruzen Co., Ltd., Tokyo, 1966.
96. van Tamelen, E. E., Siebrasse, K. V., and Hester, J. B., *Chem. & Ind.* (*London*) p. 1145 (1956).
97. Wasserman, H. H., and Floyd, M. B., *Tetrahedron Letters* p. 2009 (1963).
98. Williams, R. T., "Detoxication Mechanisms," 2nd ed., p. 668. Chapman & Hall London, 1959.
99. Witkop, B., *Ann. Chem.* **556,** 103 (1944).
100. Witkop, B., *Ann. Chem.* **558,** 98 (1947).
101. Witkop, B., *J. Am. Chem. Soc.* **72,** 1428 (1950).
101a. Witkop, B., *J. Am. Chem. Soc.* **72,** 2311 (1950).
102. Witkop, B., *Advan. Protein Chem.* **16,** 221 (1961).
103. Witkop, B., and Fiedler, H., *Ann. Chem.* **558,** 91 (1947).
104. Witkop, B., and Goodwin, S., *J. Am. Chem. Soc.* **75,** 3371 (1953).
105. Witkop, B., and Patrick, J. B., *J. Am. Chem. Soc.* **73,** 713 (1951).
106. Witkop, B., and Patrick, J. B., *J. Am. Chem. Soc.* **73,** 2188 (1951).
107. Witkop, B., and Patrick, J. B., *J. Am. Chem. Soc.* **73,** 2196 (1951).

108. Witkop, B., and Patrick, J. B., *J. Am. Chem. Soc.* **74**, 3855 and 3861 (1952).
109. Witkop, B., Patrick, J. B., and Kissman, H. M., *Chem. Ber.* **85**, 949 (1952).
110. Witkop, B., Patrick, J. B., and Rosenblum, M., *J. Am. Chem. Soc.* **73**, 2641 (1951).
111. Woolley, D. W., "The Biochemical Basis of Psychoses." Wiley, New York, 1962.
112. Yamamoto, H., Inaba, S., Hirohashi, T., and Ishizumi, K., *Chem. Ber.* **101**, 4245 (1968).
113. Zinnes, H., and Shavel, J., Jr., *J. Org. Chem.* **31**, 1765 (1966).

VI

REARRANGEMENT, RING EXPANSION, AND RING OPENING REACTIONS OF INDOLES

A. SIMPLE SUBSTITUENT MIGRATIONS

The equilibria below express in a general way the type of simple substituent migrations which can be observed in the indole system. As is discussed in

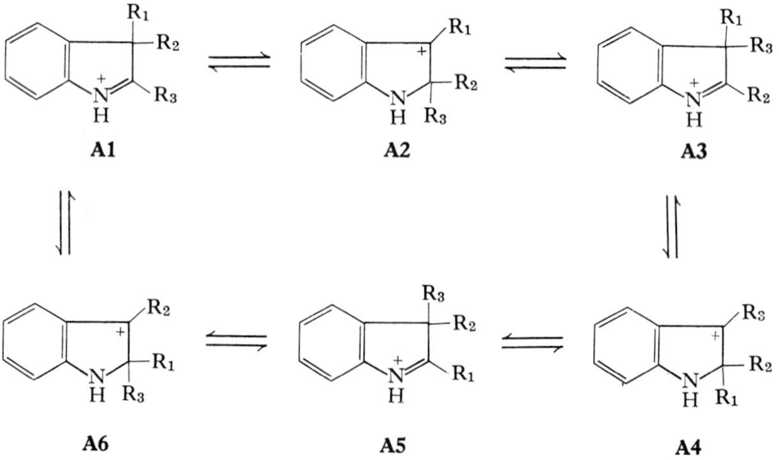

A1 ⇌ A2 ⇌ A3

A6 ⇌ A5 ⇌ A4

Chapter I, B, the ion **A3** is expected to be more stable than **A2**. Nevertheless, skeletal migrations occurring through ions of general structure **A2** have been observed. In this section we discuss migrations of 2- and 3-substituents which can be considered to comprise parts of the general scheme above. It has been found that at 150°C polyphosphoric acid catalyzes the interconversion of **A7** and **A8** (10). Interconversion was also effected by hot 47% hydrobromic acid and by heating various Lewis acid salts of the 3*H*-indoles

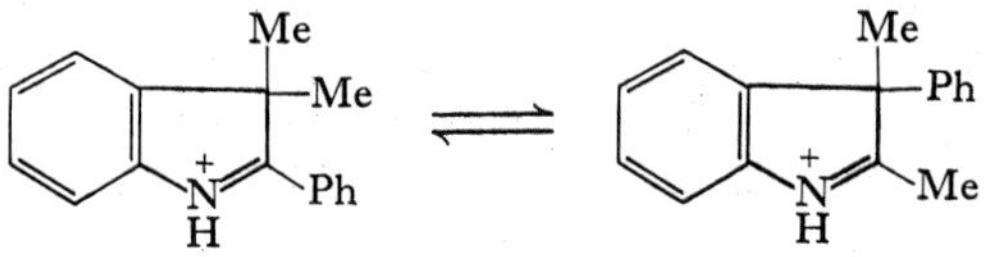

A7 (30%) ⇌ A8 (70%)

above 150°C. The position of the equilibrium is interesting. The conjugation present in **A7** might have suggested that it would be the more stable species. Steric interference with the adjacent methyl substituents may largely disrupt the conjugation. In acidic solution the relative basicities of the two compounds will also effect the equilibrium. Evans and Lyle (10a) further studied the rearrangement using both deuterium and ^{14}C labeling techniques. Equilibration of **A9** and **A10** was effected by heating the BF_3 salt in xylene for 24 hours. The equilibrium **A11** ⇌ **A12** was shown not to be established by heating in polyphosphoric acid at 150°C, indicating that the indole skeleton had retained its integrity under these conditions. These results

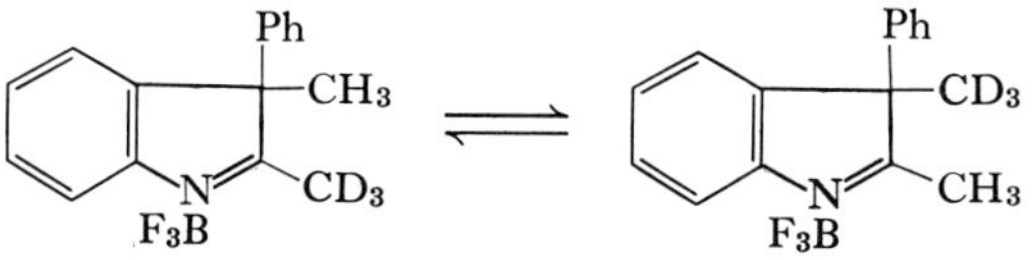

A9 ⇌ A10

are consistent with formulating the equilibration of **A7** and **A8** as occurring

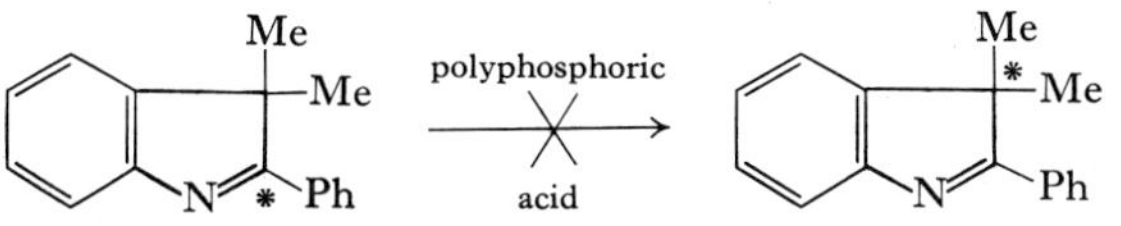

A11 —(polyphosphoric acid, ✕)→ A12

via a double Wagner–Meerwein rearrangement. Nakazaki and co-workers have observed other rearrangements which can be accommodated by the general scheme. Thus **A13** and **A15** gave the same indolenine **A14**, estab-

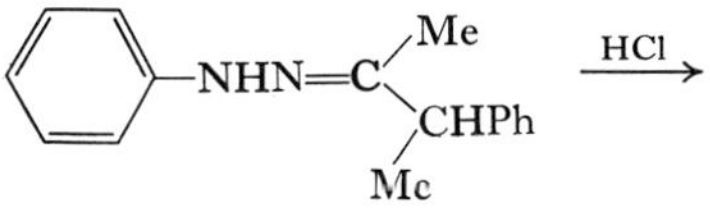

A13 —(HCl)→ A14 ←(polyphosphoric acid)— A15

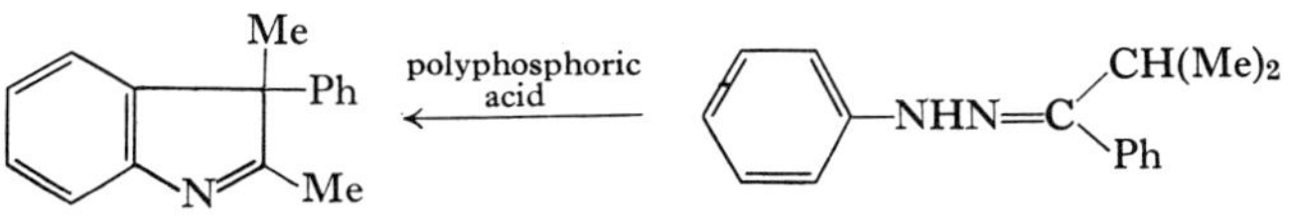

lishing that migration occurred in one instance (25), although an erroneus conclusion was apparently drawn as to which reaction proceeded with migration (10). The transformation of **A16** to **A17** with polyphosphoric acid was also observed (25).

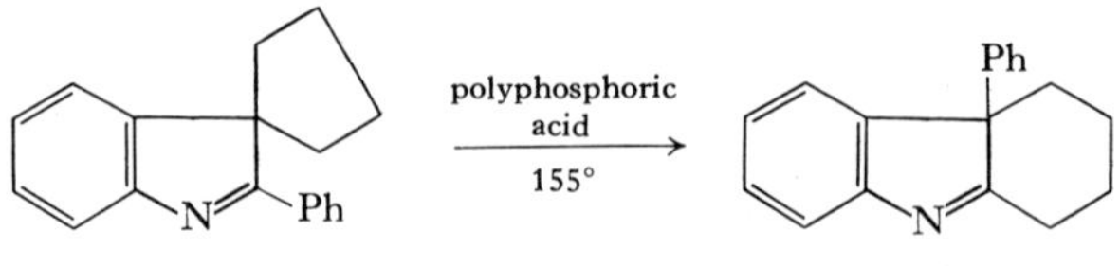

polyphosphoric acid, 155°

A16 **A17**

2-Methyl-3-phenylindole is isomerized to 3-methyl-2-phenylindole by heating with aluminum chloride (23). Similar treatment of **A18** gives **A20**. The earliest example of such a rearrangement was the conversion of 3-phenylindole to 2-phenylindole which can be effected with zinc chloride

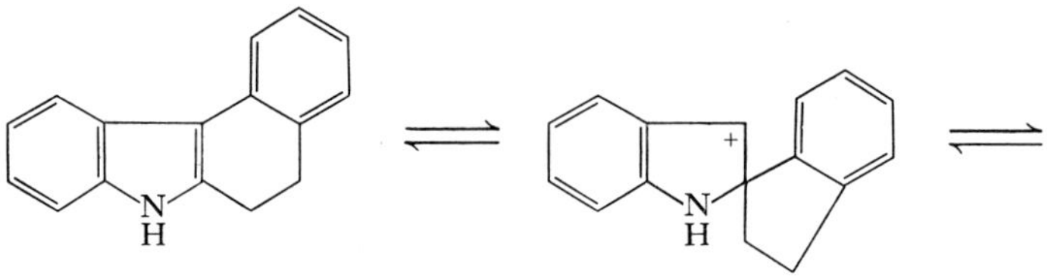

A18 **A19** 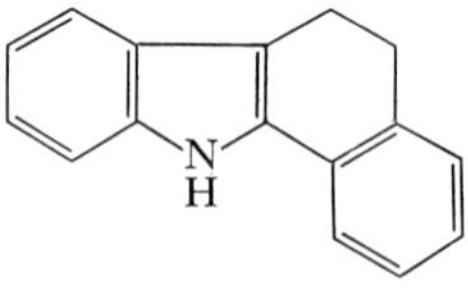 **A20**

(14) or aluminum chloride (8). These migrations have been viewed as two-fold Wagner–Meerwein shifts initiated by complexation of the indole with the Lewis acid.

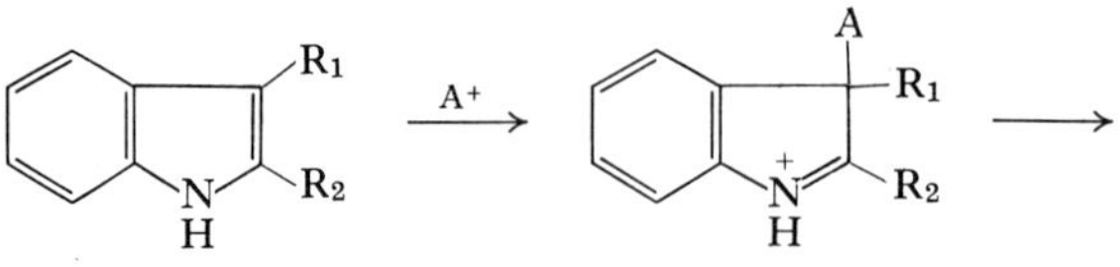

A^+

A21 **A22** 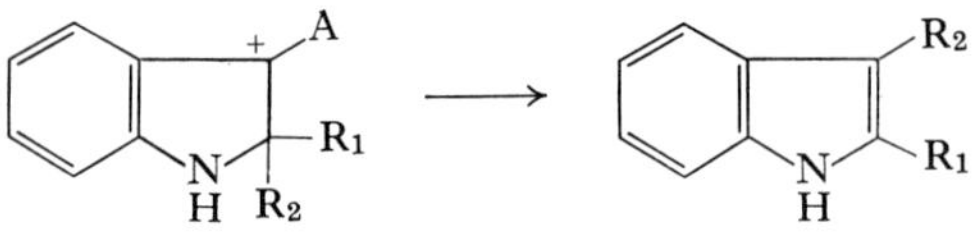 **A23** **A24**

Skeletal rearrangements are also observed when quaternary derivatives of 3*H*-indoles are heated. The reaction of 2-phenylindole with methyl iodide gives 1,2,3-trimethyl-3-phenylindolenium iodide. Unequivocal proof of structure was effected, demonstrating (7) that phenyl migration occurred during the course of the alkylation. The rearrangements **A27** → **A28** has

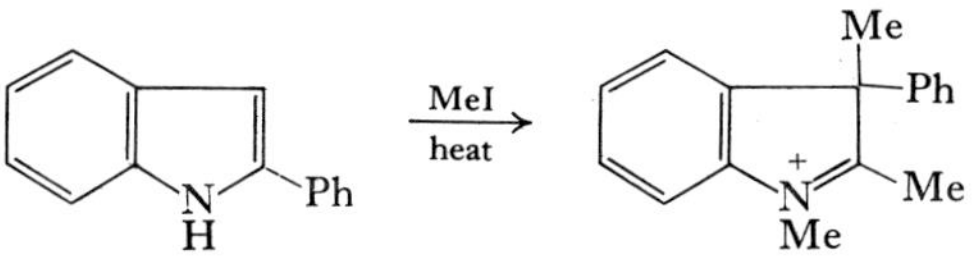

MeI, heat

A25 **A26**

been observed by Nakazaki (24). A second reported example, **A29** → **A30**, is equivocal (10).

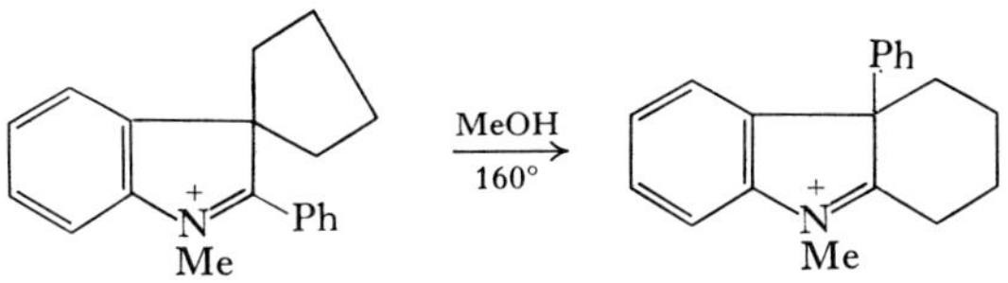

MeOH, 160°

A27 **A28**

MeOH, 120–150°

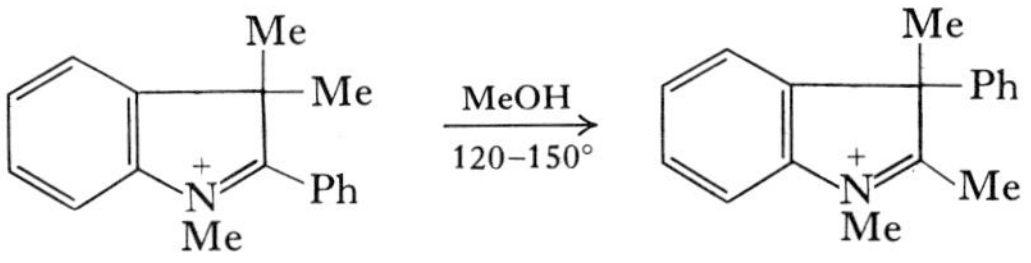

A29 **A30**

A final class of related skeletal rearrangements which should be considered here is the conversion of 3,3-disubstituted indolenines to 2,3-disubstituted indoles. Among the transformations which can be cited are those shown below.

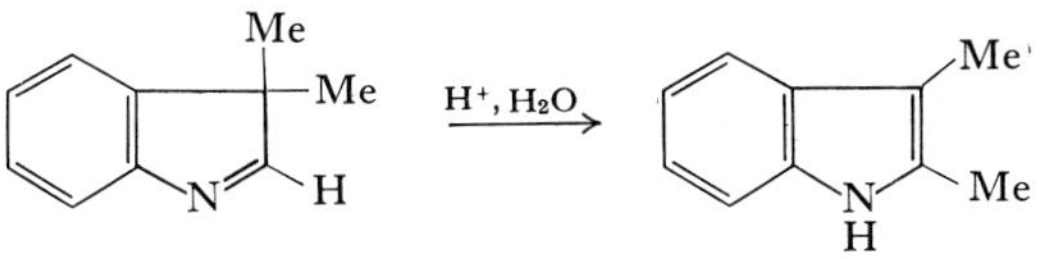

H^+, H_2O — Ref. (30)

A31 **A32**

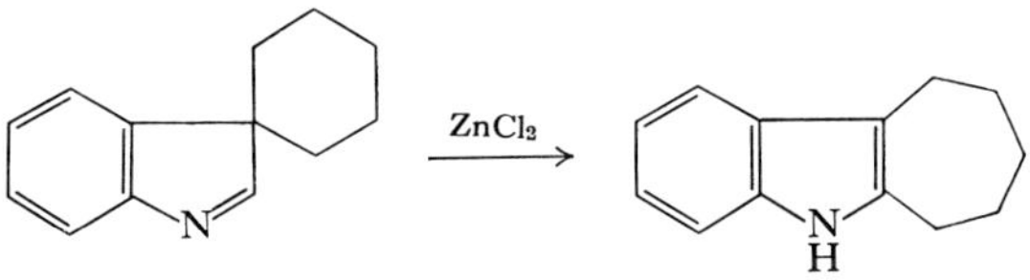

$ZnCl_2$ — Ref. (20)

A33 **A34**

A particularly elegant and significant case is found in the experiments of Jackson and Smith (18). The labeled butanol **A35** was cyclized to **A37** by the action of boron trifluoride etherate. The label was found distributed as indicated. The redistribution of label can be explained by formation and subsequent rearrangement of the 3*H*-indole **A36**. Jackson and Smith (16)

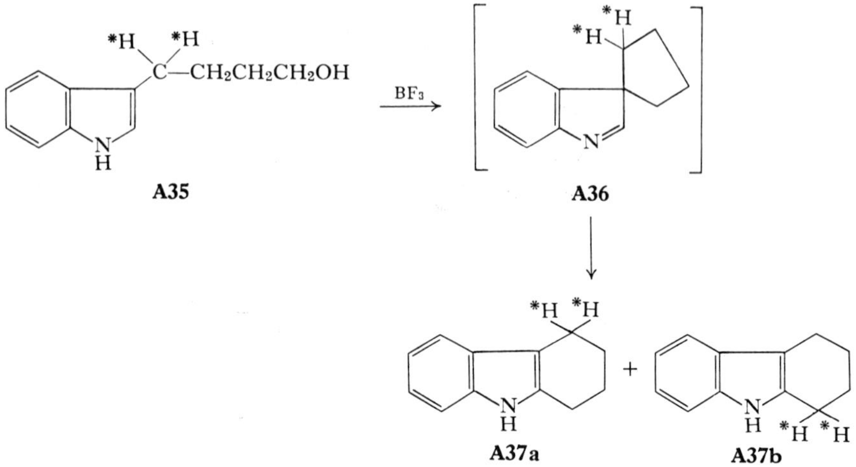

*H *H C—$CH_2CH_2CH_2OH$ N H **A35** $\xrightarrow{BF_3}$ [*H *H N **A36**] → *H *H N H **A37a** + N H *H *H **A37b**

have observed very facile acid-catalyzed migration of allyl groups in 3,3-disubstituted 3*H*-indoles. Migrations of crotyl and 3,3-dimethylallyl

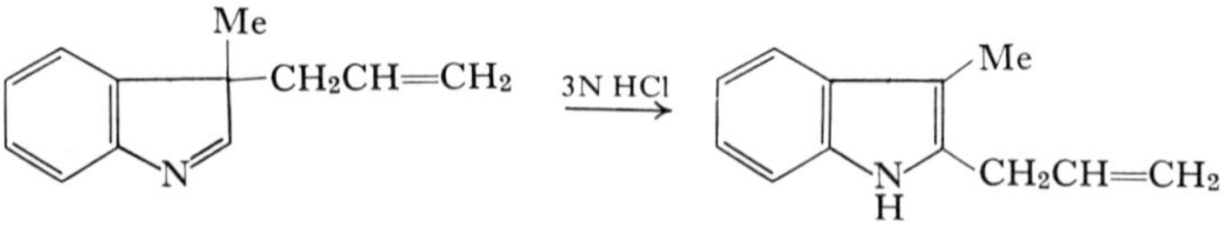

Me $CH_2CH{=}CH_2$ N **A38** $\xrightarrow{3N\ HCl}$ Me N H $CH_2CH{=}CH_2$ **A39**

groups are apparently even more facile. No rearrangement within the allyl groups was noted during the migration. Subsequent extension (19) of the investigation of migrations in 3,3-disubstituted-3*H*-indoles has indicated the following order of migratory aptitude in ethanolic hydrochloric acid: methyl < ethyl < propyl < i-propyl < allyl < benzyl < α-aminobenzyl. Alkyl groups with α-amino substituents migrate with great facility (17). It has been demonstrated by crossover experiments that the rearrangement is intramolecular.

The behavior of 3,3-disubstituted-3*H*-indoles reported to date can be summarized by stating that migration of one of the 3-substituents to C-2 is quite facile and the migration is catalyzed by acid. A driving force for such migration clearly exists as the aromaticity of the indole ring is thereby established. The group which migrates will be that with the greatest migrat-

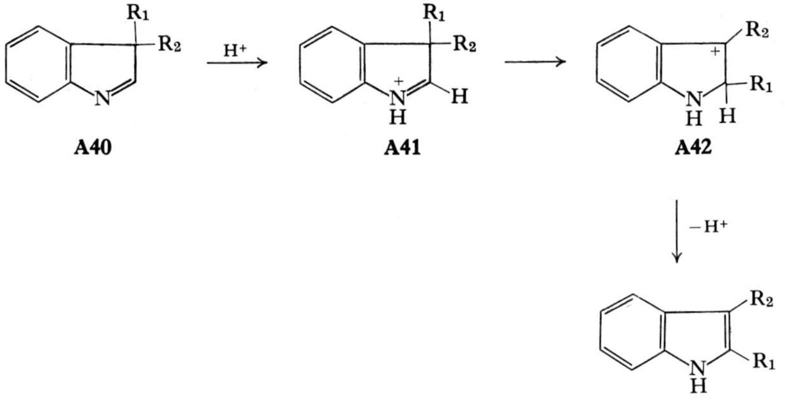

A40 —H^+→ A41 → A42 —$-H^+$→ A43

ory aptitude toward electron-deficient centers; the relative rate of rearrangement for a given indole also is determined by the migrating ability of the 3-substituents. The structure of the resulting indole is then controlled by kinetic factors. There is no evidence that equilibrium is established between the two possible indoles (18) under the mild conditions necessary to effect rearrangements of 3*H*-indoles lacking 2-substituents.

The details of the thermal and acid-catalyzed rearrangements of 2,3,3-trisubstituted-3*H*-indoles are much less clear. While the available evidence indicates that equilibrium is established between **A44** and **A45** in polyphosphoric acid, no satisfactory explanation of the position of the equilibrium has been advanced (10). It is clear that equilibrations of indoles catalyzed

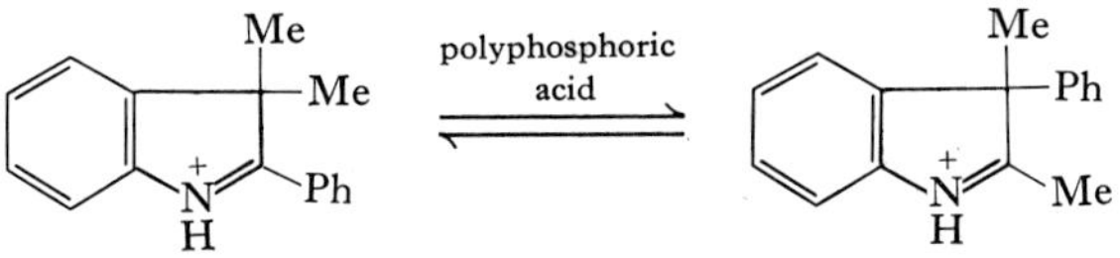

A44 (30%) ⇌ (polyphosphoric acid) A45 (70%)

by Lewis acids actually involve establishment of equilibrium between the Lewis acid complexes of the indoles when large amounts of the catalyst are present. For example, the equilibrium presumed to be established in isomerization of 2-methyl-3-phenylindole to 3-methyl-2-phenylindole (23) is properly considered as the equilibrium **A46** ⇌ **A47** and the structure of the complex must be considered if one is to understand the position of the equilibrium under such circumstances.

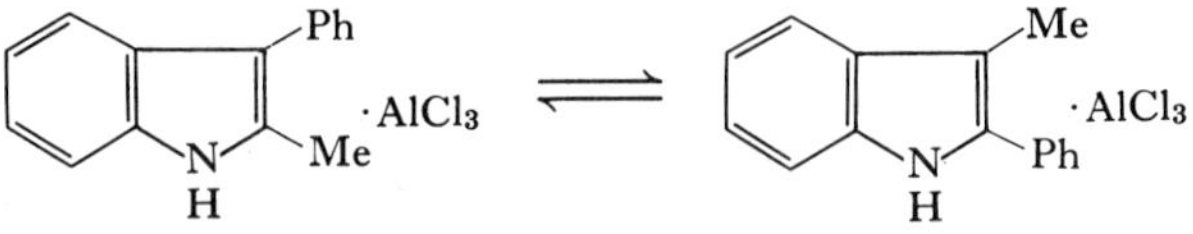

A46 ⇌ A47

An interesting special case of migration of 3,3-disubstituted 3*H*-indoles has been reported by Noland and Baude (26). Compounds of structure **A48** are rearranged by heat or by acid to the isomer **A49**. A 3*H*-indolium ion is formed by elimination of the phenolic group from C-2 of the indole skeleton and migration of the phenyl group then gives rise to the observed product.

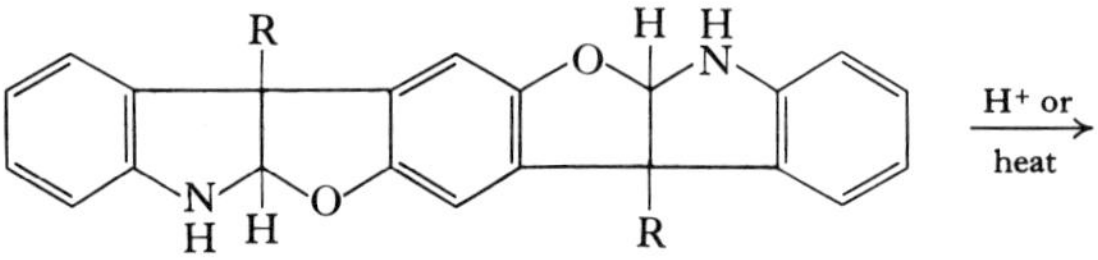

A48 —(H^+ or heat)→ A49

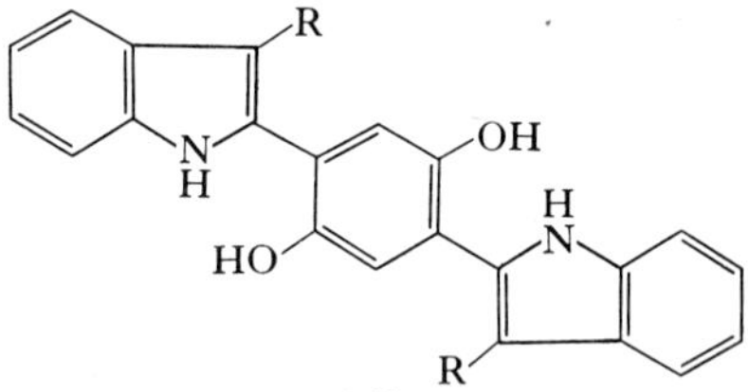

B. REARRANGEMENTS OF INDOLINONES AND INDOLINOLS

The presence of hetero atoms, usually oxygen, bonded to the indole nucleus in intermediates such as **B1**, **B4**, **B8** or **B10** can have a profound effect on the ease of migration of substituent groups on the indole ring. The structural changes illustrated below are the most common types.

Reaction of *N*-methylisatin with phenylmagnesium bromide gives a mixture of 2,2-diphenyl-1-methyl-3-indolinone and 3,3-diphenyl-1-methyloxindole (39). Reduction of the former compound with lithium aluminum hydride gives the carbinol **B13** which, when treated with hydrochloric acid, gives 1-methyl-2,3-diphenylindole. The formation of the indole clearly takes place via a pinacol-type rearrangement of **B13** followed by deprotonation of **B14** to give the indole **B15**. Other reactions which follow this pattern include the conversion of **B16** to **B17**, and **B18** to **B19** (37, 40). The reduction product of **B16** rearranges rapidly under very slightly acidic conditions. In the case of **B18**, the rearrangement occurs during the course of the reduction,

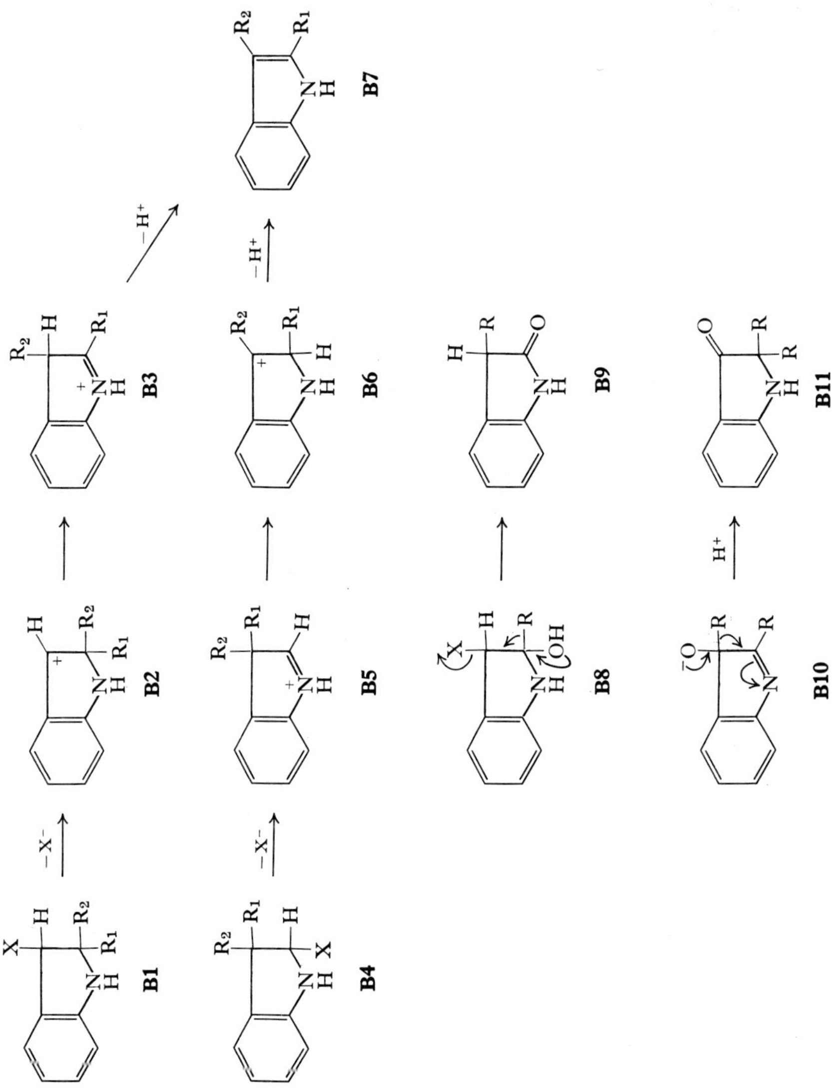

B1 → B2 → B3

B4 → B5 → B6 → B7

B8 → B9

B10 → B11

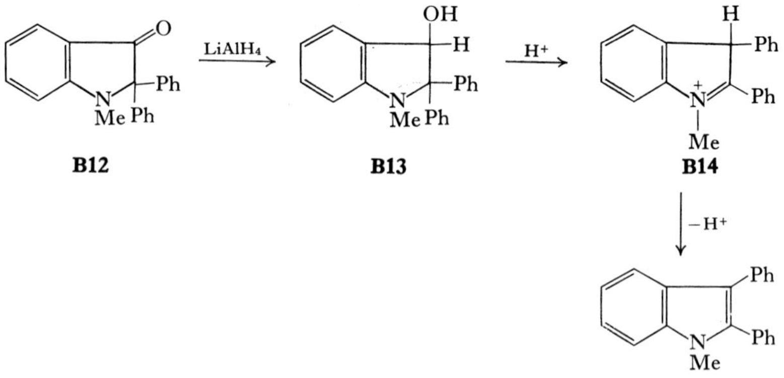

O Ph N Me Ph — $LiAlH_4$ → OH H Ph N Me Ph — H^+ → H Ph Ph N Me

B12 B13 B14

$-H^+$

Ph Ph N Me

B15

O N H — (1) $LiAlH_4$ (2) H^+ → N H

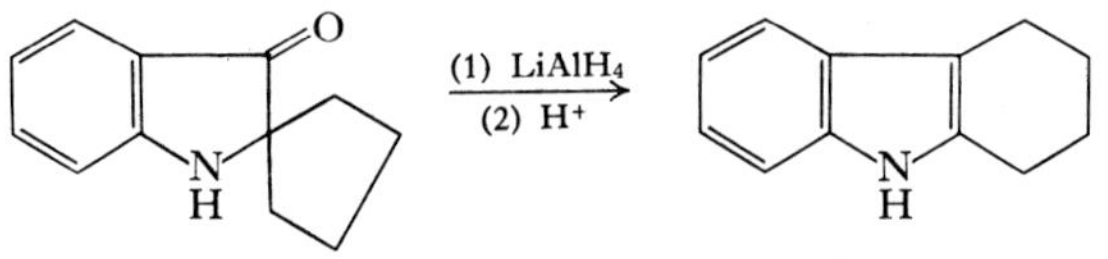

B16 B17

O N H Me Me N H → N H Me Me N H

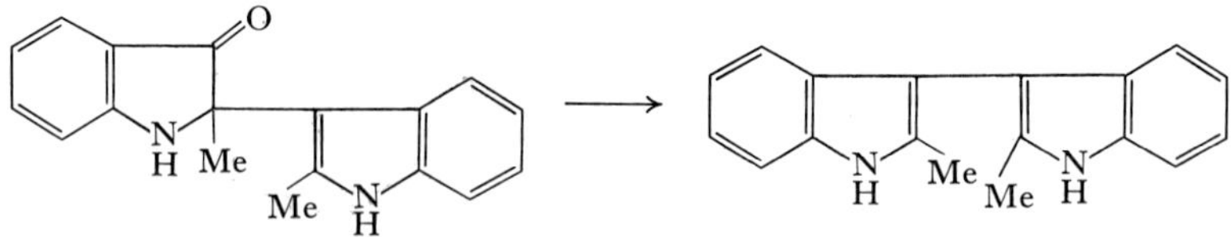

B18 B19

with an aluminum compound serving as a Lewis acid to effect the migration (40).

The formation of the rearranged product **B22** from the reaction of indolinone **B20** with methyllithium (42) is a related reaction.

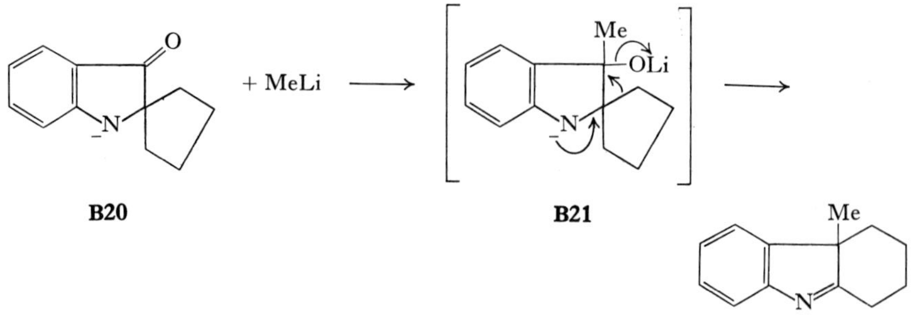

O N + MeLi → [Me OLi N] →

B20 B21

Me N

B22 (10%)

Rearrangement can also be initiated by departure of a leaving group from C-2 of a 3,3-disubstituted indoline ring. The indolinol **B23** when heated with concentrated hydrochloric acid or polyphosphoric acid gives the tetrahydrocarbazole (**B26**) (43). The indolium ion **B24** is presumably an intermediate. The conversion of the oxindole **B27** into **B28** by lithium aluminum hydride (5a) fits into the same pattern of reactivity.

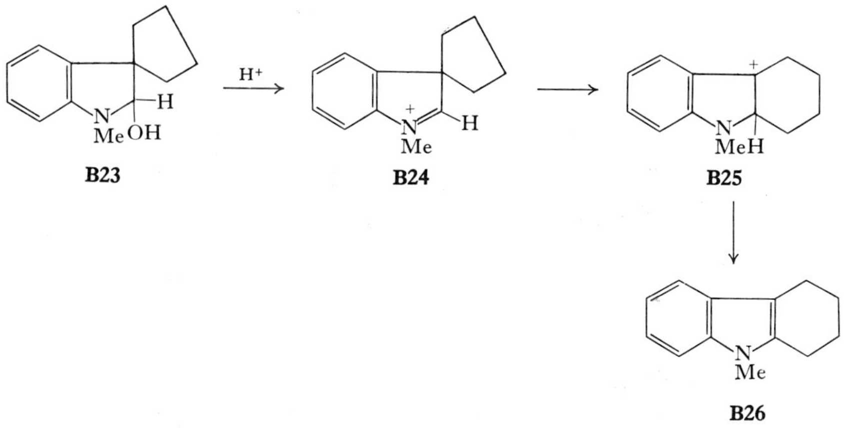

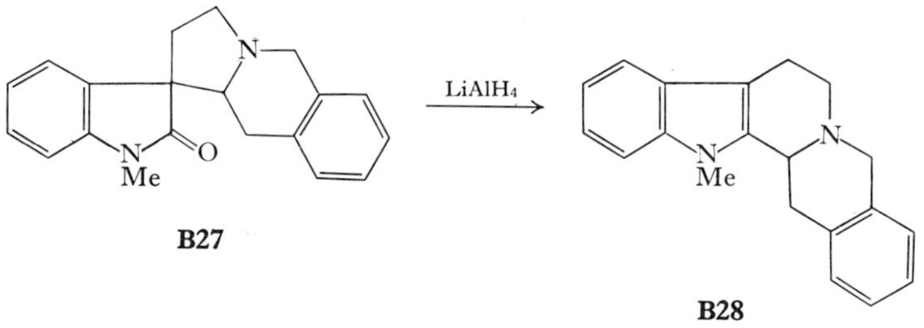

The conversion of tetrahydro-β-carboline type alkaloids into oxindole systems via 3-chloroindolenines (13, 44) involves substituent migration. In the procedure used by Finch and Taylor (13), the chloroindolenines are converted by methanol to iminoethers of general structure **B31**, which are in turn hydrolyzed to oxindoles. The rearrangements may be of the pinacol type with migration being initiated by cleavage of the carbon–chlorine bond and assisted by the indolinyl nitrogen atom. Zinnes and Shavel (44) have effected the same net transformation in one step by refluxing the chloroindolenines in slightly acidic aqueous methanol. These workers suggest that migration proceeds by a fragmentation–recombination process.

Cl N N B29 → Cl H N N MeO B30 →

N OMe B31 → N H O N B32

Cl N N B33 $\xrightarrow{H_2O}$ Cl N H OH N B34

↓ $-Cl^-$

N H O N + B36 ← N H OH N + B35

↓

N H O N B37

The rearrangement of 2,3-dihydroxyindoline derivatives can be effected with either acid or base. The glycol **B38** on treatment with acid gives the oxindole **B39** (27). Likewise **B40** gives **B41** on treatment with acetic anhydride (35).

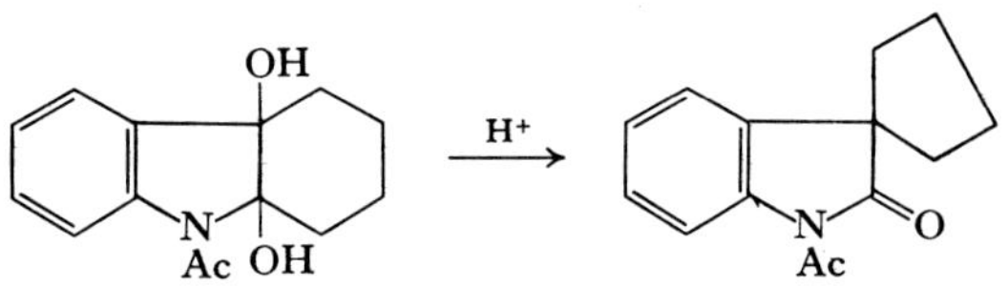

OH
H+
N
Ac OH
N
Ac
O

B38 **B39**

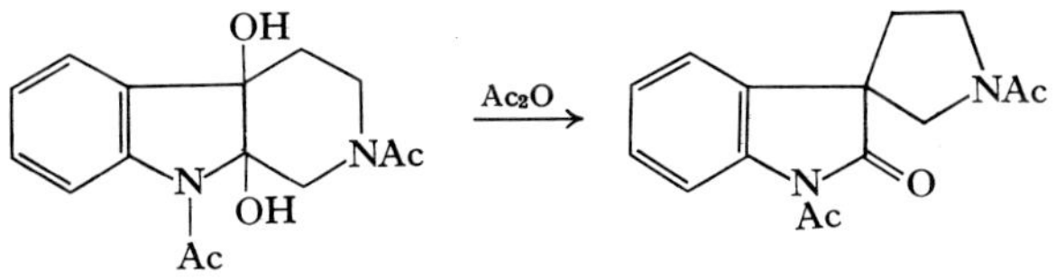

OH
Ac_2O
NAc
NAc
N
OH
Ac
N
Ac
O

B40 **B41**

In these reactions a pinacol rearrangement pattern is again evident. With base, similar glycols undergo a different rearrangement. Thus, with base **B42** gives **B43**. Compounds **B44** and **B46** behave similarly. The base-catalyzed rearrangement of **B48** to **B49** is also analogous (41).

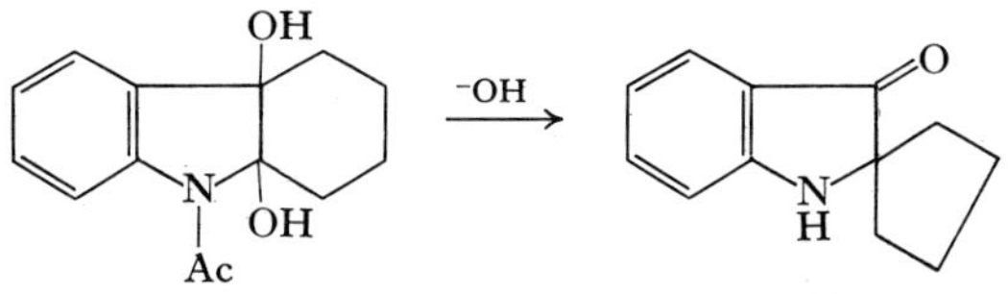

OH
^{-}OH
O
N
OH
Ac
N
H
Ref. (37)

B42 **B43**

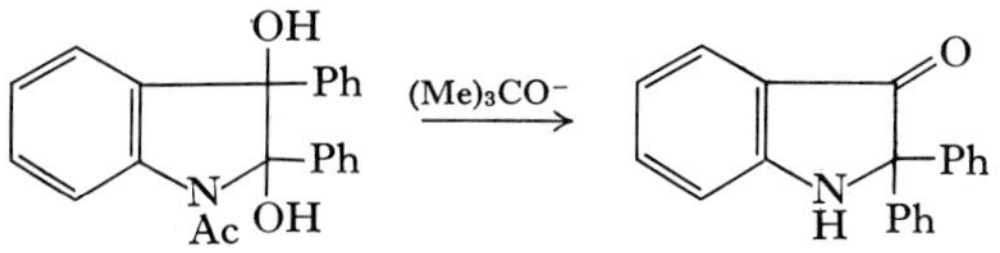

OH
Ph
$(Me)_3CO^-$
O
Ph
Ph
N
Ac OH
N
H Ph
Ref. (31)

B44 **B45**

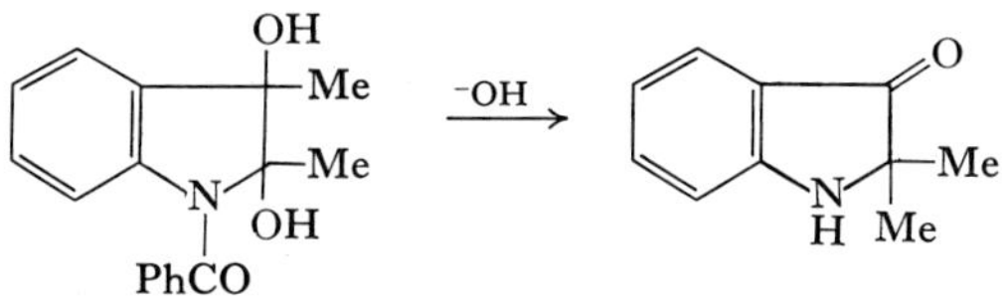

OH
Me
^{-}OH
O
Me
Me
N
OH
PhCO
N
H Me
Ref. (21)

B46 **B47**

OH
^{-}OH
O
N
N
H

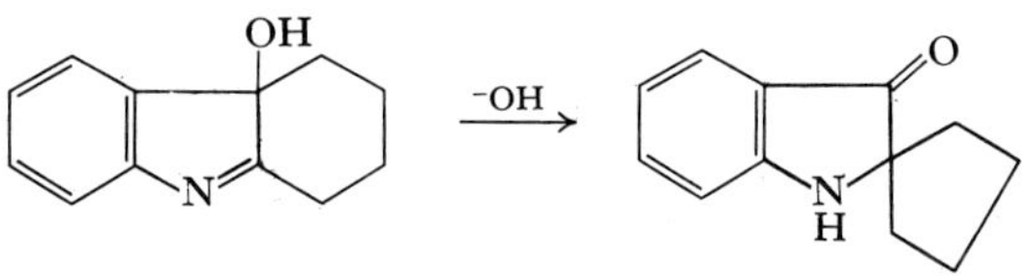

B48 **B49**

Kershaw and Taylor have reported the isolation of the stereoisomeric glycols **B51** and **B53**. These workers reported that the two stereoisomers behaved quite differently toward base. The isomer from the osmium tetrox-

OH Me Me N PhCOOH **B51** $\xrightarrow{^-OH}$ O Me N H Me **B52**

Me Me N PhCO **B50** — OsO_4 → B51; (1) HNO_3 (2) Al_2O_3 → B53

Refs. (5, 21)

OH Me Me N PhCO OH **B53** →

O Ph N H Me + OH Me Ph N

B54 **B55**

ide oxidation, assigned cis stereochemistry, gave 2,2-dimethyl-3-indolinone but the trans isomer gave **B54** and **B55** in which a phenyl substituent has been introduced. The authors put foward no mechanism for the rearrangement and no very conventional pathway is evident to the writer. The phenyl ring in the rearranged product must be derived from the benzoyl group and cleavage of the C-2–C-3 bond of the indole ring obviously must have taken place.

The dichotomy in behavior of the indoline-2,3-diol system toward acid and base must reflect quite different mechanisms of rearrangement. The base-catalyzed rearrangement seems to be initiated by hydroxide attack at the C-3 hydroxyl group whereas in the acid-catalyzed reaction the migration of the alkyl substituent to C-3 suggests that the rearrangement is initiated by ionization of the C-3 hydroxyl group.

The acetoxyindolenines formed from tetrahydro-β-carboline type alkaloids by lead tetraacetate oxidation show similar reactivity patterns (11, 12). Basic catalysts such as methoxide ion induce rearrangement by attack at the acetoxy group. However, oxindole formation results when the acetoxyindole-

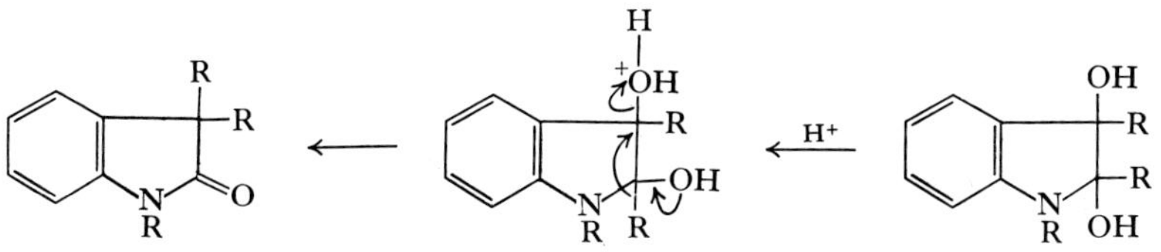

B58 B57 B56

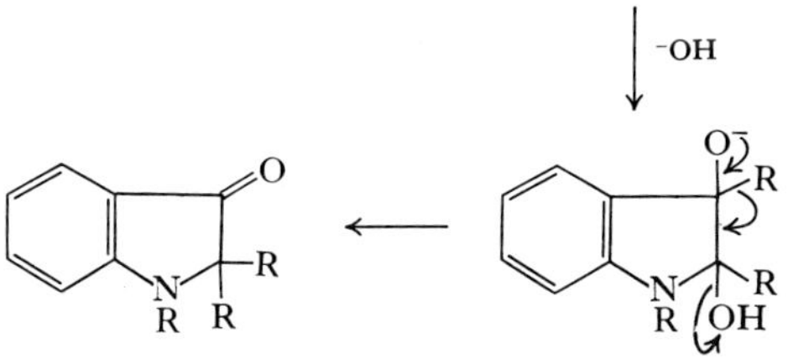

B60 B59

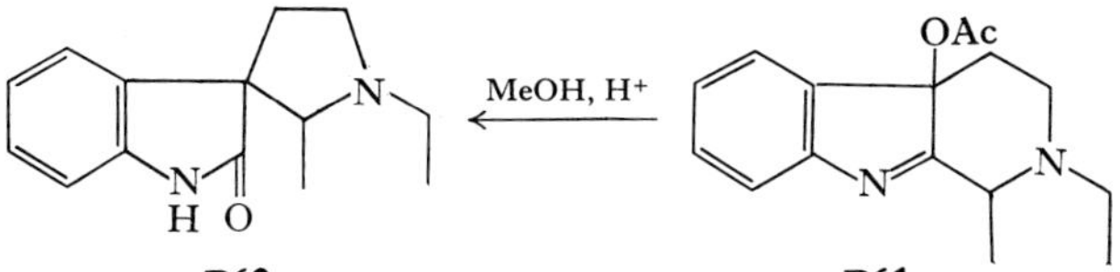

B62 B61

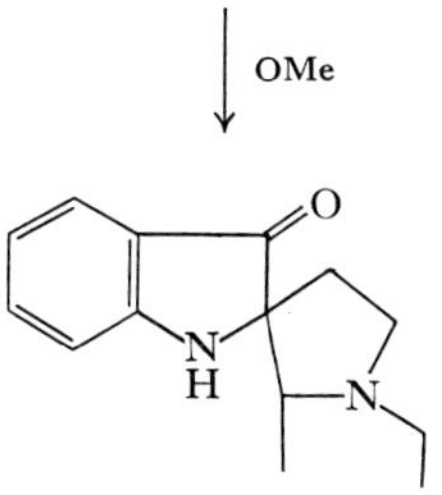

B63

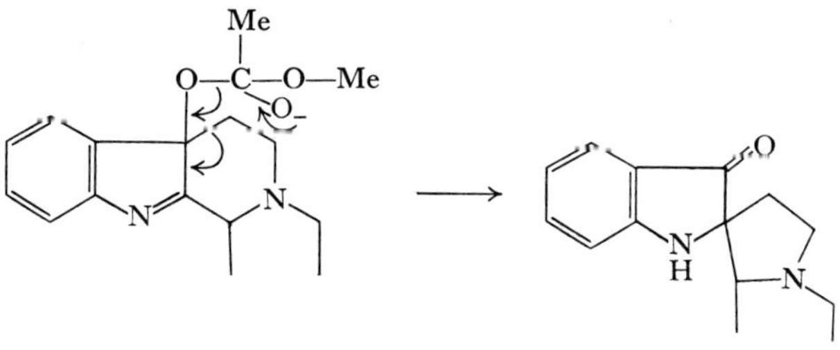

B64 B65

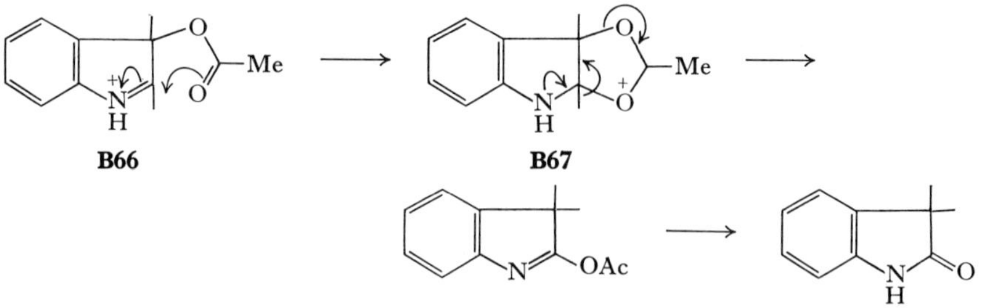

B66 B67

B68 B69

nines are treated with methanol and traces of acetic acid. The mechanism put forward for the acid-catalyzed rearrangement is depicted above. As in other acid-catalyzed rearrangements to oxindoles, departure of a leaving group from C-3 is considered to initate rearrangement.

The conversion of **B70** to **B72** by base appears to involve migration of a carbethoxy substituent (1). Electrophilic attack by bromine has been observed to initiate a similar rearrangement (2).

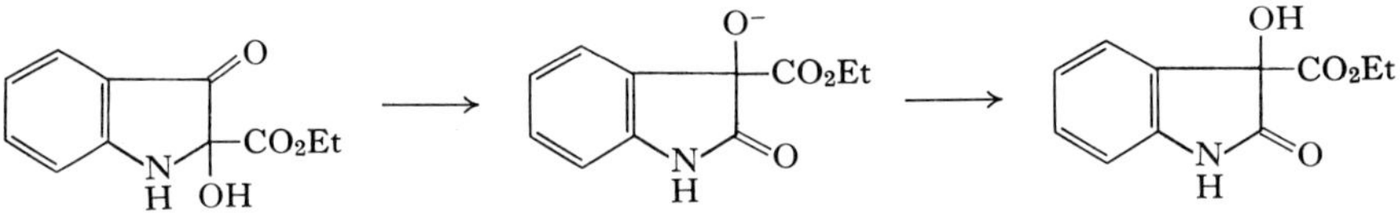

B70 B71 B72

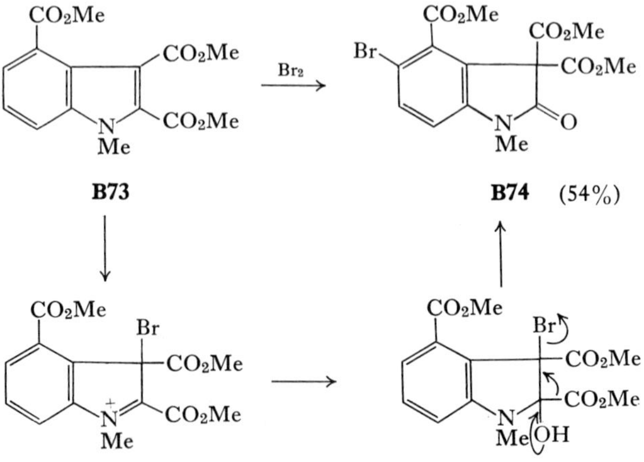

B73 B74 (54%) Ref. (2)

B75 B76

2,2-Disubstituted 3-indolinones can be rearranged to 3,3-disubstituted oxindoles by the action of Lewis acids. Witkop and Ek (39) have formulated the reaction as shown below.

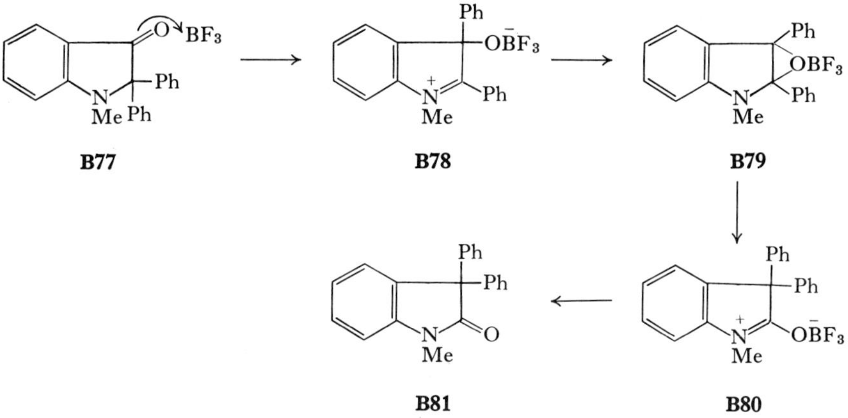

The reaction apparently involves a series of three substituent migrations. Witkop has summarized his investigations of rearrangements of indolinones and indolinols in a brief review (38).

C. RING EXPANSION AND RING OPENING REACTIONS OF INDOLES

Most ring opening reactions of indoles are initiated by protonation or other electrophilic attack at C-3 generating a 3*H*-indolium ion. Nucleophilic addition (intramolecular or intermolecular) can then occur at C-2 and this may lead eventually to a ring opening. Simple indoles are not subject to hydrolytic ring opening but the strained system in **C1** has been opened under hydrolytic conditions (6). The importance of the strain caused by the

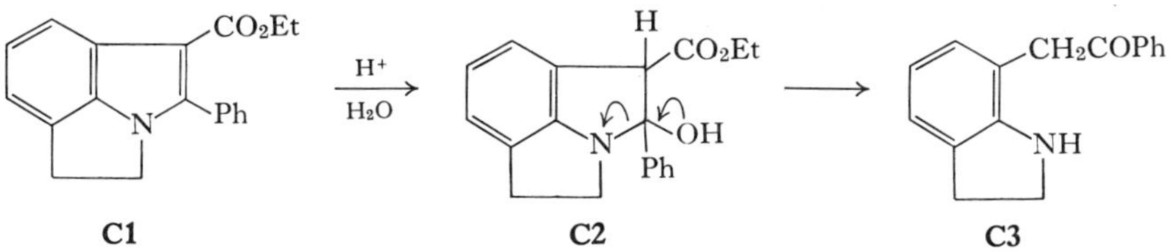

N-C-7 bridge present in **C1** in facilitating the ring opening is demonstrated by the stability of **C4** to similar reaction conditions. Although it would be expected that **C4** could be hydrated to **C5**, it reverts to **C4** by loss of water rather than undergoing ring opening.

Treatment of 2-methyltryptophan with two equivalents of alkaline hypochlorite results in the formation of 4-acetylquinoline (34). The ring expansion, which is formulated below, is probably mechanistically similar to biochemical processes which convert indoles to quinolines.

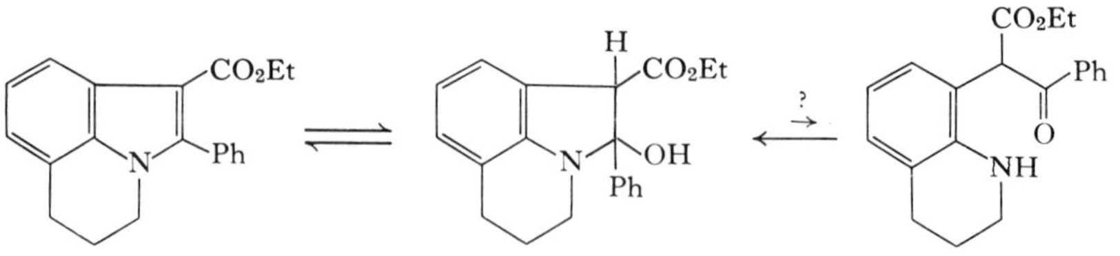

CO_2Et Ph N — **C4** ⇌ H CO_2Et N OH Ph — **C5** $\xleftarrow{?}$ CO_2Et Ph O NH — **C6**

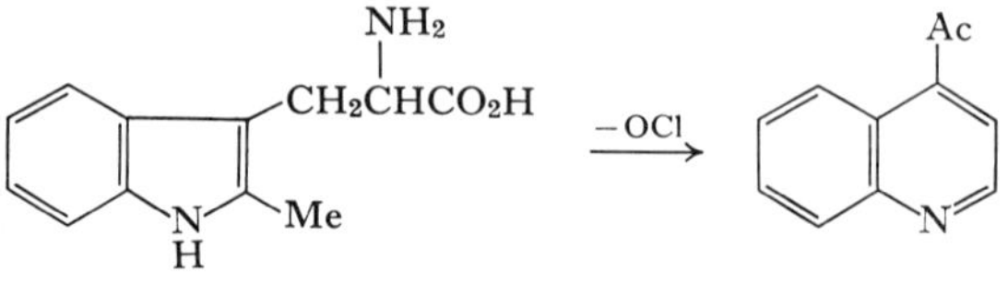

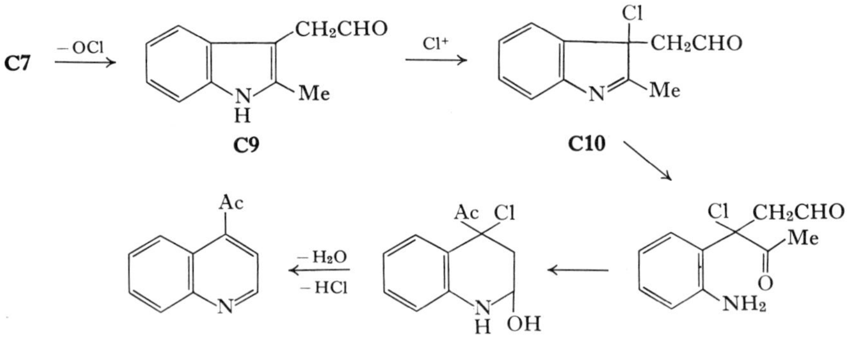

NH_2 CH_2CHCO_2H N H Me — **C7** $\xrightarrow{-OCl}$ Ac N — **C8**

C7 $\xrightarrow{-OCl}$ CH_2CHO N H Me — **C9** $\xrightarrow{Cl^+}$ Cl CH_2CHO N Me — **C10** → Cl CH_2CHO Me O NH_2 — **C11** → Ac Cl N H OH — **C12** $\xrightarrow{-H_2O,\ -HCl}$ Ac N — **C8**

A somewhat similar pattern can be perceived in the conversion of **C13** to **C14** and **C15** with hot polyphosphoric acid (33).

Mechanisms for these transformations have been suggested.

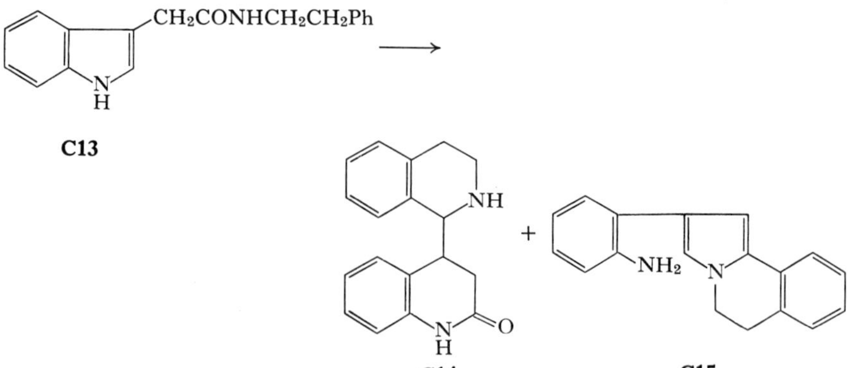

$CH_2CONHCH_2CH_2Ph$ N H — **C13** →

NH N H O — **C14** + NH_2 N — **C15**

C13 $\xrightarrow{H^+}$

H
$\overset{+}{N}$
H
HN
O
CH_2CH_2Ph
C16

→

H
$\overset{+}{N}$
N
O
H H
CH_2CH_2Ph
C17

H
$\overset{+}{N}$
N
H H
C20

↓

C15

NH_2
$\overset{+}{N}$
O
C18

←

NH_2
N
O
C19

C14 ←

$CH_2CONHCH_2Ph$

C21

$\xrightarrow[AcN(Me)_2]{POCl_3}$

Me +
C=N(Me)$_2$
$CH_2CONHCH_2Ph$

C22

Ac
$CH_2CONHCH_2Ph$

C23

Ac
$CONHCH_2Ph$
H

C24

MeC—O$^-$
$CONHCH_2Ph$

C25

OH
Me
NCH_2Ph
O

C26

The key step leading to the opening of the indole ring is the intramolecular nucleophilic addition of the amido group to the 3*H*-indolium ion **C16**.

Dolby and Furukawa (9) have proposed a carbanion addition to an indolenine as a key step in a mechanism proposed to account for the formation of **C26** from **C21** (10% yield). The formation of the cyclopropane **C24** is a most novel feature of the mechanism.

The 3-acylindole **C26** gives a mixture of **C29** and **C30** on treatment with acid (15). Again, nucleophilic addition to an indolenine can satisfactorily account for the ring opening.

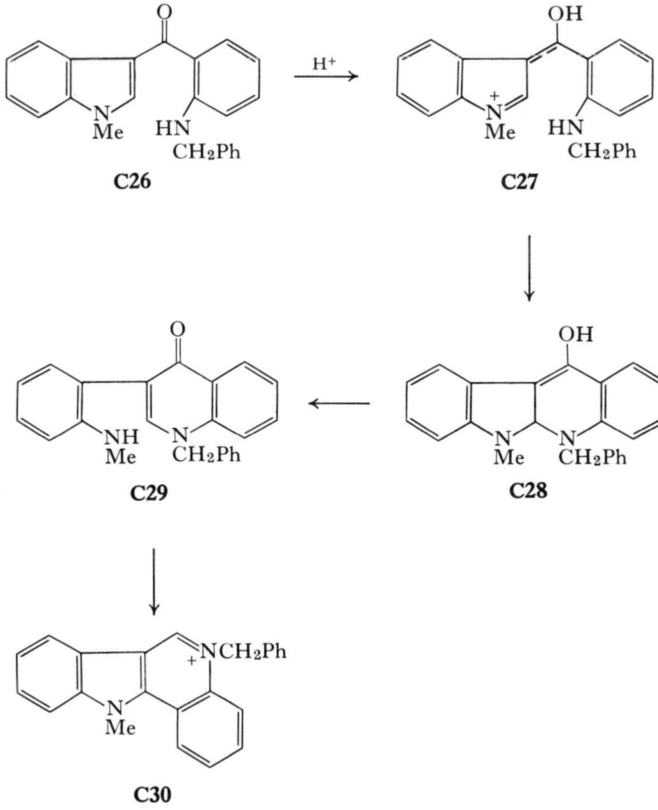

O
N Me HN CH2Ph
C26

H^+ →

OH
$\overset{+}{N}$ Me HN CH2Ph
C27

↓

OH
N Me N CH2Ph
C28

←

O
NH Me N CH2Ph
C29

↓

$\overset{+}{N}CH_2Ph$
N Me
C30

The conversion of 3-acylindoles into 4-(*o*-aminophenyl)pyrazoles by hydrazine is a reaction of the same type (3).

Wenkert and co-workers observed a somewhat related process (36). It was found that oximes of 3-acyloxindoles gave 2-substituted indoles on reaction with sodium acetate. Yields were in the range of 15–35%. The mechanism shown below outlines the steps involved in indole formation.

COR + NH_2NH_2 → C—R, N—NH_2

C31 C32

H^+

R H R

NH_2 N N H N N H N H H H

C34 C33

NOH

R AcO⁻ N H O N H R

C35 C36

R =NOH CO₂H R

C35 → NH_2 CO_2H → O NH_2 → C36

C37 C38

O R R ⁻OH H_2O R R

N R O N R R

C39 C40

^-O_2C O R R NH R CO_2^- OH R N R R → C40

C41 C42

The sequence of steps is undefined but must involve ring opening, decarboxylation, and cyclization.

A rearrangement described by Anthony (4) also converts oxindole derivatives into indoles but with a net skeletal change. Treatment of oxindoles **C39** with hot aqueous base results in rearrangement and formation of 2,3-disubstituted indoles. Hydrolytic opening of the oxindole ring is proposed, followed by opening of the epoxide ring to give **C42**.

There is no experimental evidence concerning the timing of the rearrangement and decarboxylation steps. Perhaps the most reasonable view would be to regard the rearrangement as being of the 2,2-disubstituted indolinol type. Facile decarboxylation of an intermediate such as **C43** has ample precedent.

C42 ⟶ C43 ⟶ C40

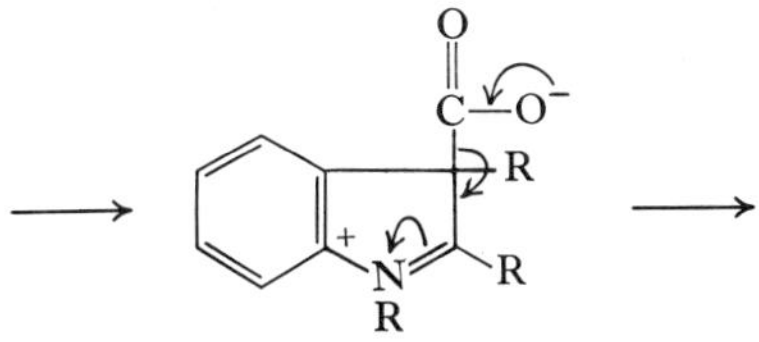

C43

In 1965, Teuber reported that the indole **C44** was converted in good yield to the rearranged indoline **C45** on treatment with methanolic acid (32). The mechanism outlined below was proposed to account for the reaction. Nucleophilic addition of the ketonic chain to an indolenium ion generated

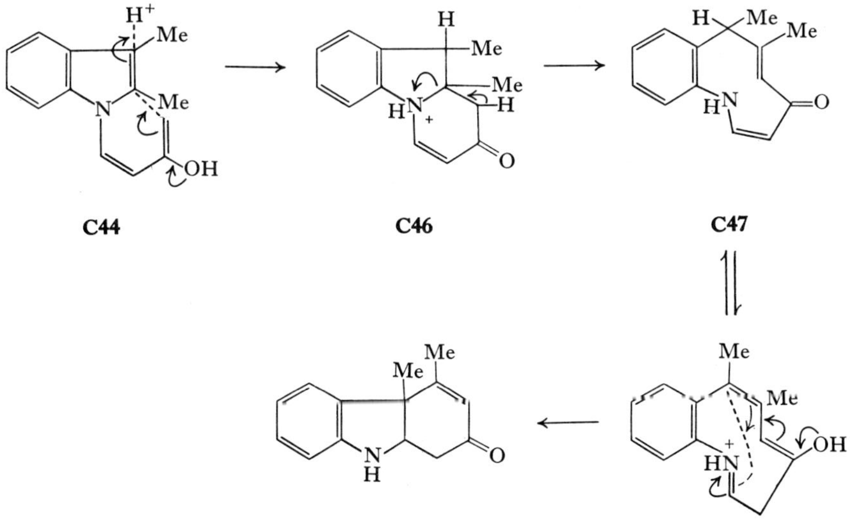

C44 ⟶ **C46** ⟶ **C47** ⇌ **C48** ⟶ **C45**

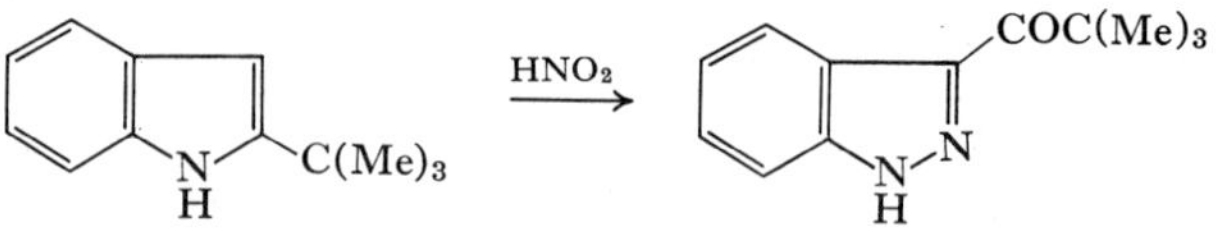

C49 $\xrightarrow{HNO_2}$ C50

by protonation at C-3 is followed by ring opening via β-elimination. The final step which recloses the indole ring is an addition of the γ-atom of an α,β-unsaturated carbonyl system to an iminum ion generated by protonation of the enamine. The initiating step in the rearrangement is again a nucleophilic addition to a 3*H*-indole intermediate.

Indoles substituted at C-2 by tertiary butyl groups are converted by nitrous acid into 3-acylindazoles (28, 29). With primary or secondary substituents at C-2, nitrosation takes place at C-3 and α- to the indole ring

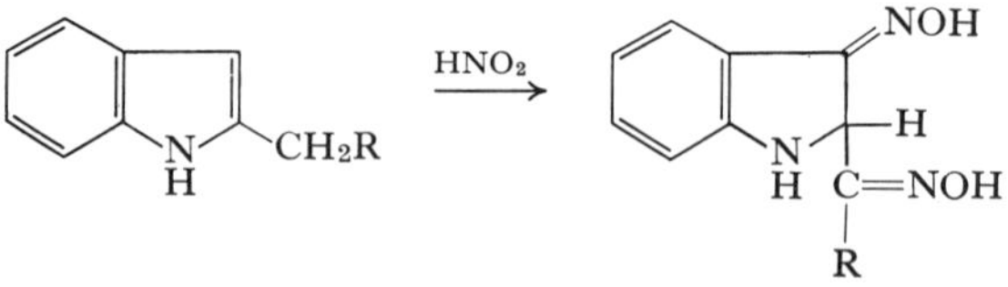

C51 $\xrightarrow{HNO_2}$ C52

(22, 28, 29). The mechanism shown below has been suggested to explain these reactions.

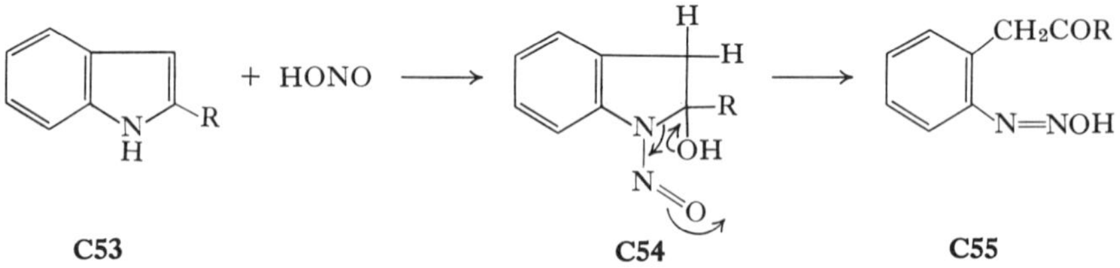

C53 + HONO ⟶ C54 ⟶ C55 ⟶ C56 ⟶ C57

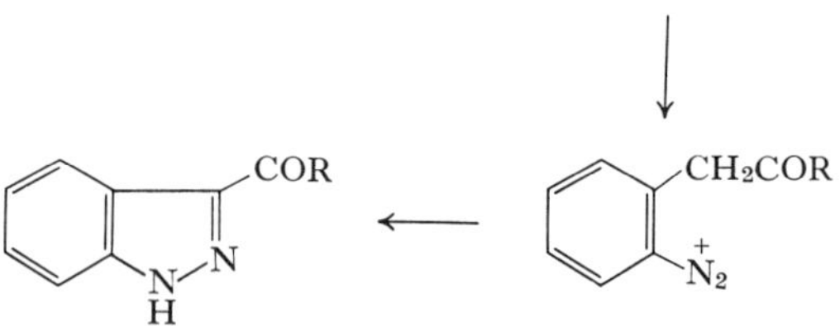

The mechanism bypasses the normally reactive 3-position as the primary site for nitrosation. Possibly a steric effect of the tertiary group in the 2-position may account for this unusual feature of the mechanism. It is, however, possible that initial nitrosation occurs at C-3. A pathway such as **C58** → **C62** could then account for indazole formation.

C53 + HNO_2

C58

C59

C61

C60

C57

C62

REFERENCES

1. Acheson, R. M., and Booth, S. R. G., *J. Chem. Soc., C* p. 30 (1968).
2. Acheson, R. M., Snaith, R. W., and Vernon, J. M., *J. Chem. Soc.* p. 3229 (1964).
3. Alberti, C., *Gazz. Chim. Ital.* **87**, 720 and 736 (1957); **85**, 245 (1955); *Chem. Abstr.* **52**, 15532 (1958); **50**, 9389 (1956).
4. Anthony, W. C., *J. Org. Chem.* **31**, 77 (1966).
5. Atkinson, C. M., Kershaw, J. W., and Taylor, A., *J. Chem. Soc.* p. 4426 (1962).

5a. Belleau, B., *Chem. & Ind. (London)* p. 229 (1955).

6. Blake, J., Tretter, J. R., Juhasz, G. J., Bonthrone, W., and Rapoport, H., *J. Am. Chem. Soc.* **88**, 4061 (1966).
7. Boyd-Barrett, H. S., *J. Chem. Soc.* p. 321 (1932).
8. Buu-Hoi, N. P., and Jacquignon, P., *Bull. Soc. Chim. France* p. 1104 (1967).
9. Dolby, L. J., and Furukawa, S., *J. Org. Chem.* **28**, 2512 (1963).
10. Evans, F. J., Lyle, G. G., Watkins, J., and Lyle, R. E., *J. Org. Chem.* **27**, 1553 (1962).

10a. Evans, F. J., and Lyle, R. E., *Chem. & Ind. (London)* p. 986 (1963).

11. Finch, N., Gemenden, C. W., Hsu, I. H., Kerr, A., Sim, G. A., and Taylor, W. I., *J. Am. Chem. Soc.* **87**, 2229 (1965).
12. Finch, N., Gemenden, C. W., Hsu, I. H., and Taylor, W. I., *J. Am. Chem. Soc.* **85**, 1520 (1963).
13. Finch, N., and Taylor, W. I., *J. Am. Chem. Soc.* **84**, 3871 (1962).
14. Fischer, E., and Schmidt, T., *Chem. Ber.* **21**, 1811 (1888).
15. Garcia, E. E., Riley, J. G., and Fryer, R. I., *J. Org. Chem.* **33**, 2868 (1968).
16. Jackson, A. H., and Smith, A. E., *Tetrahedron* **21**, 989 (1965).
17. Jackson, A. H., and Smith, A. E., *Tetrahedron* **24**, 403 (1968).

18. Jackson, A. H., and Smith, P., *Chem. Commun.* p. 264 (1967); Jackson, A. H., Naidoo, B., and Smith, P., *Tetrahedron* **24**, 6119 (1968).
19. Jackson, A. H., and Smith, P., *Tetrahedron* **24**, 2227 (1968).
20. Jones, G., and Stevens, T. S., *J. Chem. Soc.* p. 2344 (1953).
21. Kershaw, J. W., and Taylor, A., *J. Chem. Soc.* p. 4320 (1964).
22. Mann, F. G., and Haworth, R. C., *J. Chem. Soc.* p. 670 (1944).
23. Nakazaki, M., *Bull. Chem. Soc. Japan.* **33**, 461 (1960).
24. Nakazaki, M., *Bull. Chem. Soc. Japan* **33**, 472 (1960).
25. Nakazaki, M., Yamamota, K., and Yamagami, K., *Bull. Chem. Soc. Japan* **33**, 466 (1960).
26. Noland, W. E., and Baude, F. J., *J. Org. Chem.* **31**, 3321 (1966).
27. Patrick, J. B., and Witkop, B., *J. Am. Chem. Soc.* **72**, 633 (1950).
28. Piozzi, F., and Dubini, M., *Gazz. Chim. Ital.* **89**, 638 (1959); *Chem. Abstr.* **54**, 12116 (1960).
29. Piozzi, F., and Umani-Ronchi, A., *Gazz. Chim. Ital.* **94**, 1248 (1964); *Chem. Abstr.* **62**, 9122 (1965).
30. Robinson, R., and Suginome, H., *J. Chem. Soc.* p. 298 (1932).
31. Sarel, S., and Klug, J. T., *Israel J. Chem.* **2**, 143 (1964); *Chem. Abstr.* **62**, 6452 (1965).
32. Teuber, H.-J., Reinehr, U., and Cornelius, D., *Tetrahedron Letters* p. 1703 (1965).
33. Thesing, J., and Funk, F. H., *Chem. Ber.* **89**, 2498 (1956); **91**, 1546 (1967).
34. van Tamelen, E. E., and Haarstad, V. B., *Tetrahedron Letters* p. 390 (1961).
35. van Tamelen, E. E., Siebrasse, K. V., and Hester, J. B., *Chem. & Ind.* (*London*) p. 1145 (1956).
36. Wenkert, E., Bernstein, B. S., and Udelhofen, J. H., *J. Am. Chem. Soc.* **80**, 4899 (1958).
37. Witkop, B., *J. Am. Chem. Soc.* **72**, 614 (1950).
38. Witkop, B., *Bull. Soc. Chim. France* p. 423 (1954).
39. Witkop, B., and Ek, A., *J. Am. Chem. Soc.* **73**, 5664 (1951).
40. Witkop, B., and Patrick, J. B., *J. Am. Chem. Soc.* **73**, 713 (1951).
41. Witkop, B., and Patrick, J. B., *J. Am. Chem. Soc.* **73**, 2188 (1951).
42. Witkop, B., and Patrick, J. B., *J. Am. Chem. Soc.* **73**, 1558 (1951).
43. Witkop, B., and Patrick, J. B., *J. Am. Chem. Soc.* **75**, 2572 (1953).
44. Zinnes, H., and Shavel, J., Jr., *J. Org. Chem.* **31**, 1765 (1966).

VII

HYDROXYINDOLES AND DERIVATIVES INCLUDING OXINDOLE, 3-INDOLINONE, AND ISATIN

In this chapter the chemistry of the hydroxyindoles is discussed. The chemistry of oxindole and 3-indolinone, the stable tautomers of 2- and 3-hydroxyindole, respectively, are also considered.

A. REACTIONS OF OXINDOLES

Sufficient physical evidence has been accumulated to show that 2-hydroxyindole is unstable relative to its carbonyl tautomer **A1** (76). The common name oxindole is very widely used, instead of the more systematic 2-indolinone, for **A1**. Recently, however, *Chemical Abstracts* has begun using the latter nomenclature and its use will doubtlessly increase. The tautomers **A1** and **A2** are conjugate acids of the common anion **A3** and it is this anion which is responsible for much of the chemistry of oxindole. The anion is fully conjugated (indole conjugation) whereas oxindole (**A1**) has the conjugated system of an acylaniline, with C-3 having sp^3 hybridization. The cyclic conjugation, then, would be expected to lower the energy of the anion **A3** relative to a noncyclic analog such as the anion **A4**. Formation of an anion such as **A4**, of course, requires a quite strong base. Reactions involving the formation of the anion **A3**, however, can be catalyzed effectively by much weaker bases. The chemistry of oxindole is similar to that of a typical lactam when reactions of the carbonyl carbon or oxygen are involved. Base-catalyzed reactions which involve abstraction of a proton from C-3 are, however, extraordinarily facile when compared with those of lactams in which the anions are not stabilized by the structural features unique to the

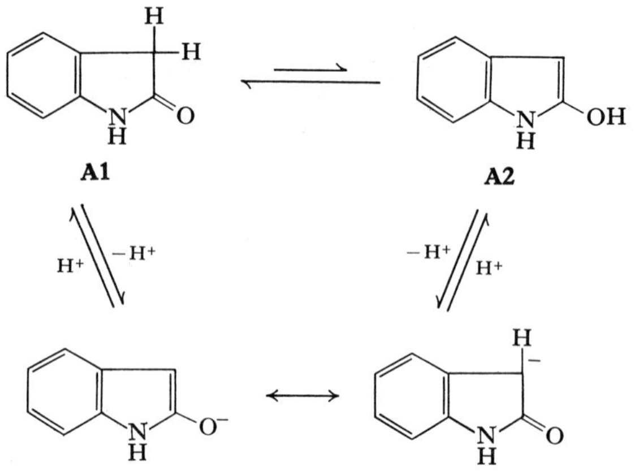

A1 A2 A3a A3b 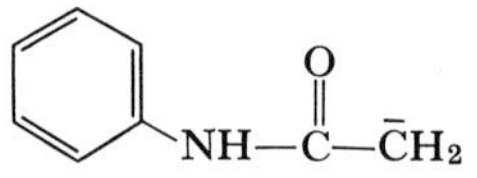 A4

oxindole system. There appears to be no quantitative information on the acidity of oxindole, but sodium carbonate is a sufficiently strong base to promote ionization of 1-acyloxindoles (18).

1. Alkylation of Oxindole

The ambient anion **A3** offers the possibility of carbon or oxygen alkylation. Base-catalyzed alkylations of oxindoles with alkyl halides have consistently led to carbon alkylation. Table XLV records a number of typical alkylations. In a particularly interesting example, it has been shown that treatment of 3-(2-bromethyl)oxindole (**A5**) with base gives the carbon alkylation product **A6** in preference to the O-alkylation product **A7** (164).

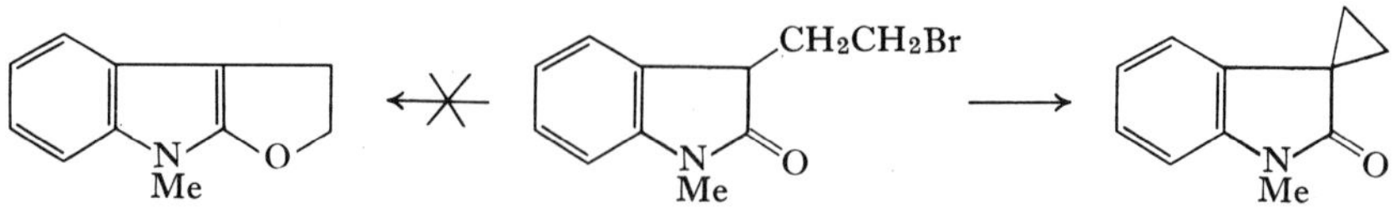

A7 A5 A6

The presence of an alkyl group at C-3 favors further alkylation so that in certain cases only the dialkylated product can be isolated even with the use

TABLE XLV □ ALKYLATION OF OXINDOLES

Oxindole substituent	Alkylating agent	Base	Site of alkylation	Yield (%)	Ref.
None	Benzyl chloride	Sodium ethoxide	3[a]	?	(45)
None	Triethyloxonium fluoroborate	None	Oxygen	91	(49, 127)
1-Me	Triethyloxonium fluoroborate	None	Oxygen	?	(49)
1-Me	Ethyl bromo-acetate	Sodium ethoxide	3[a]	76	(162)
1-Me	β-Dimethylamino-ethyl chloride	Sodium hydride	3	50	(162)
1-Methyl-3-ethyl	β-Dimethylamino-ethyl chloride	Sodium amide	3	65	(58)
1-Methyl-3-benzyl	β-Dimethylamino-ethyl chloride	Sodium amide	3	?	(160)
1-Methyl-3-phenyl	1,2-Dibromo-ethane	Sodium hydride	3	87	(171)
1,3-Dimethyl-5-methoxy	Methyl iodide	Sodium ethoxide	3	87	(84)
1,3-Dimethyl-5-methoxy	Ethylene oxide	Sodium ethoxide	3	?	(91)
1-Methyl-3-ethyl-5-methoxy	Methyl iodide	Sodium ethoxide	3	90	(84)
1-Methyl-5-ethoxy	*N,N*-bis(2-chloro-ethyl)methyl-amine	Sodium amide	3[b]	38–48	(85)
3,3-diMe	Triethyloxonium fluoroborate	None	Oxygen	?	(49)
3-Ph	Methyl iodide	Sodium carbonate	3	63	(18)
3-Ph	Benzyl iodide	Sodium carbonate	3	83	(18)

[a] Dialkylation occurs.
[b] Dialkylation results in cyclization.

of a deficiency of alkyl halide (162). Use of preformed enolate anion can circumvent this difficulty and give monoalkylation product (162).

Alkylation of 3-phenyloxindole with 1-bromo-2-chloroethane gives a 25% yield of the C-alkylation product **A9** but a 5% yield of **A10**, in which initial alkylation on either oxygen or nitrogen has occurred, is also obtained (171). The *N*-methyl derivative of **A8** is cleanly alkylated on carbon (87% yield) by 1,2-dibromoethane. A high yield of O-alkylation is observed in the case of **A11** where C-alkylation is sterically prohibited (171).

The alkylation of oxindole with triethyloxonium fluoroborate gives O-alkylation (49, 127). Few substituted oxindoles have been subjected to

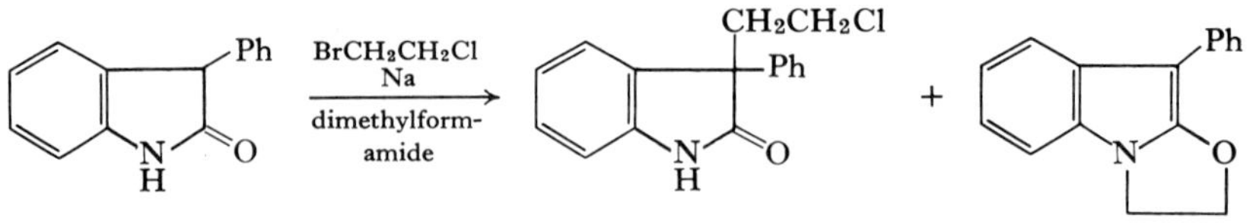

A8 $\xrightarrow[\text{dimethylformamide}]{BrCH_2CH_2Cl,\ Na}$ A9 + A10

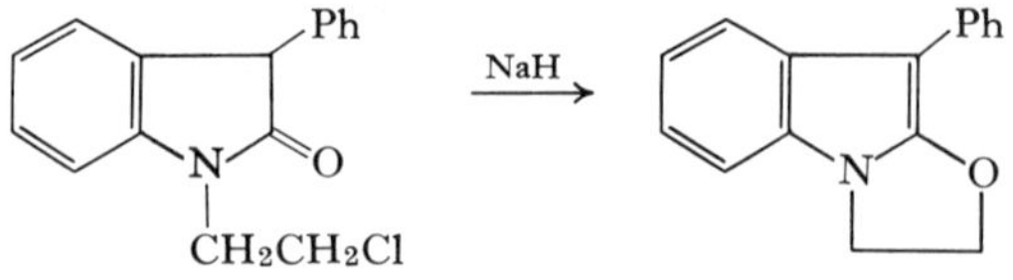

A11 $\xrightarrow{NaH}$ A12

alkylation with this reagent, but both 1-methyl and 3,3-dimethyloxindole have also been found to give O-alkylation (49). The contrasting behavior of oxindoles toward alkylation by alkyl halides versus trialkyloxonium ions is in line with results obtained in other ambident systems. Base-catalyzed alkylation involves the conjugate base of the oxindole. Carbon alkylation tends to be dominant when the transition state for alkylation resembles the products and therefore reflects the stability of the carbon alkylate relative to the oxygen alkylation product. In alkylation with triethyloxonium fluoroborate, it is the neutral molecule which is alkylated. The carbon atom is sp^3 hybridized and not a nucleophilic site. In general, it would be expected that O-alkylation would be the preferred reaction course for alkylating agents which are sufficiently reactive to alkylate the neutral molecule.

The conjugate bases of 3-acyl and 3-formyloxindoles are nucleophiles which present three potential sites for alkylation. Early reports indicated O-alkylation occurred at the 2-oxygen of **A13** but the structure of the product was later shown to be **A14** (163). Alkylation of these enolic

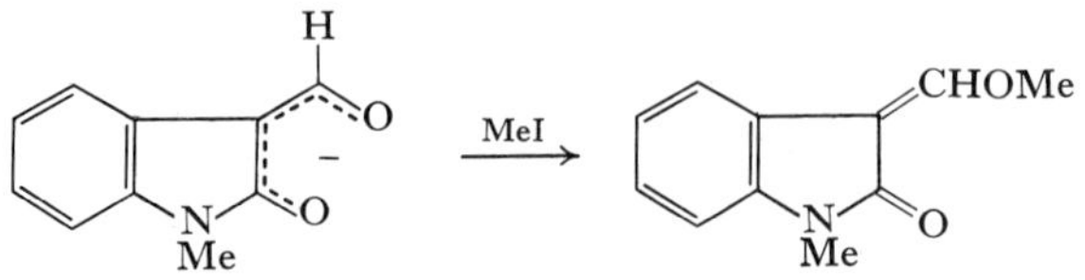

A13 $\xrightarrow{MeI}$ A14

systems with diazomethane gives products resulting from alkylation at both oxygens (162).

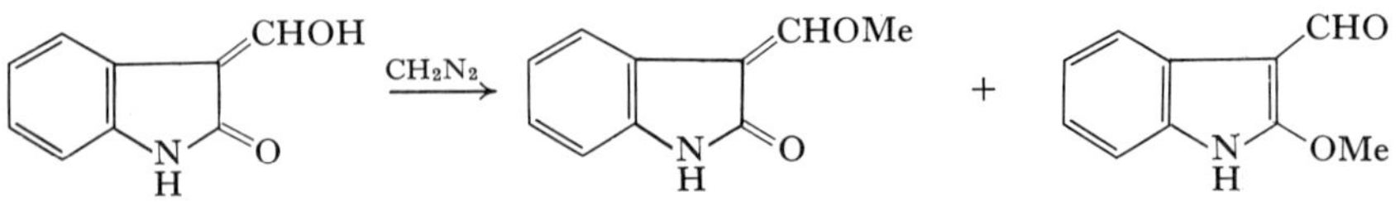

A15 $\xrightarrow{CH_2N_2}$ A16 + A17

2. Condensation and Addition Reactions at C-3 of Oxindole

The anion derived from oxindole is nucleophilic toward unsaturated centers and both aldol-type condensation reactions and Michael addition of the oxindole ring to electron-deficient olefins are well-established reaction types. Base-catalyzed reactions of oxindoles with carbonyl compounds normally lead to 3-alkylideneoxindoles. Rate constants for the condensation of benzaldehyde with several substituted oxindoles have been measured and correlated by the Hammett equation (30a). Table XLVI records a number of typical examples of aldol type condensation reactions.

TABLE XLVI ☐ CONDENSATION REACTIONS OF OXINDOLES WITH ALDEHYDES AND KETONES

Oxindole substituent	Carbonyl component	Catalyst	Yield (%)	Ref.
None	Benzaldehyde	Piperidine	100	(149)
None	Benzaldehyde	Acetic anhydride potassium acetate	81	(148)
None	*p*-Fluorobenzaldehyde	Piperidine	?	(160)
None	Cinnamaldehyde	Piperidine	71	(34)
None	Furan-2-carboxaldehyde	Piperidine	82	(34)
None	Pyridine-2-carboxaldehyde	Pyrrolidine	91	(160)
None	Cyclohexanone (as pyrrolidine enamine)		88	(160)
None	1-Methyl-4-piperidone	Ammonia	72	(160)
1-Ph	*o*-Nitrobenzaldehyde	Piperidine	70	(1)
5-Br	Benzaldehyde	None	50	(1)
5-Br	*o*-Nitrobenzaldehyde	None	90	(1)
5,6-Dimethoxy	3,4-Dimethoxybenzaldehyde	Piperidine	65	(159)
5,6-Dimethoxy	Pyridine-3-carboxaldehyde	Pyrrolidine	80	(160)

A particularly informative paper from the point of view of optimum conditions for condensation of various types of carbonyl compounds with oxindoles has been published by Walker and co-workers (160). These workers found pyrrolidine to be a superior catalyst for many condensations and also noted that the preformed enamines of ketones reacted smoothly with oxindoles to give the desired condensation products. It was suggested that enamines and iminium salts are generally intermediates in the amine-catalyzed reactions of oxindoles with carbonyl compounds.

Base-catalyzed acylation by esters is another typical reaction of the heterocyclic enolate. Hydroxymethylene groups can be introduced at the 3-position in excellent yield (162). Condensation of oxindole with ethyl piperidinoacetate is effected by sodium ethoxide and gives **A20** (165). Diethyl malonate also can be smoothly condensed with oxindole (72).

A18 + $EtOCOCH_2N$ (piperidine) A19 $\xrightarrow{^{-}OEt}$ A20

Mannich condensations of 1,3-dimethyloxindole have been effected in acidic ethanol with formaldehyde and several secondary amines (58, 111). Oxindole itself does not give a simple Mannich condensation product. Instead, a condensation product derived from two molecules of oxindole is formed (52). The structure **A22** has been assigned to this product. Hinman and Bauman attribute the failure of Mannich reactions with oxindole to

A21 $\xrightarrow{CH_2O}$ A22

rapid decomposition of the Mannich base to methyleneoxindole which then reacts with oxindole to give the observed products which they regard as ill defined (55). 3-(2-Aminoethyl)oxindoles give intramolecular condensation reactions with aldehydes (10, 47, 53, 109). The reactions below are typical.

A23 + $PhCH_2CHO$ (A24) ⟶ A25 Ref. (10)

A26 + PhCHO (A27) ⟶ A28 Ref. (47)

Condensations of oxindoles with ortho esters can be effected in the presence of acetic anhydride (14).

The condensation of oxindole with aromatic nitroso compounds has been reported (81).

The anion of oxindole has been alkylated with several electron-deficient olefins. Cyanoethylation of 1-methyl-3-ethyloxindole gives a 71% yield of **A34**. Oxindole itself is dialkylated by ethyl acrylate with both alkylations taking place at C-3 (72).

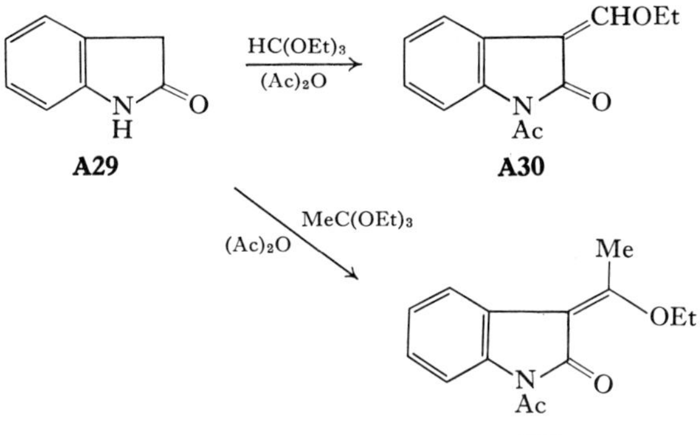

A29 A30 A31

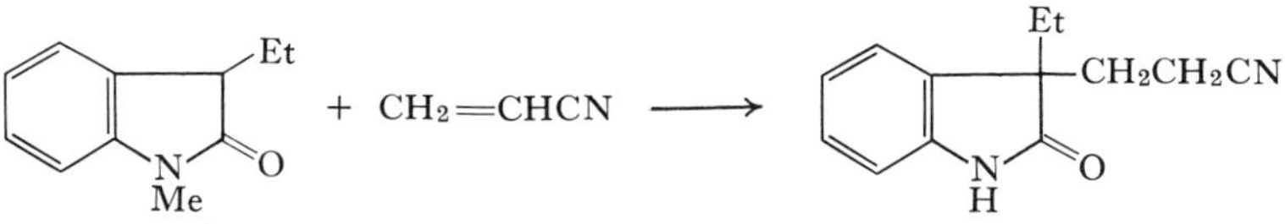

A32 A33 A34

The 2-tetrahydropyranyl derivative of 3-hydroxyoxindole has been successfully alkylated by ethyl acrylate and diethyl methoxymethylmalonate

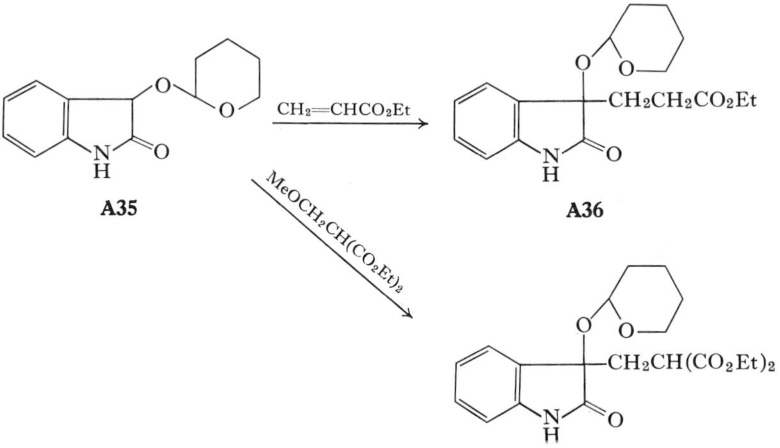

A35 A36 A37

(65, 73). Sodium methoxide catalyzed the addition of 1-methyloxindole to dimethyl acetylenedicarboxylate. The product is the 3-alkylideneoxindole **A41** (8).

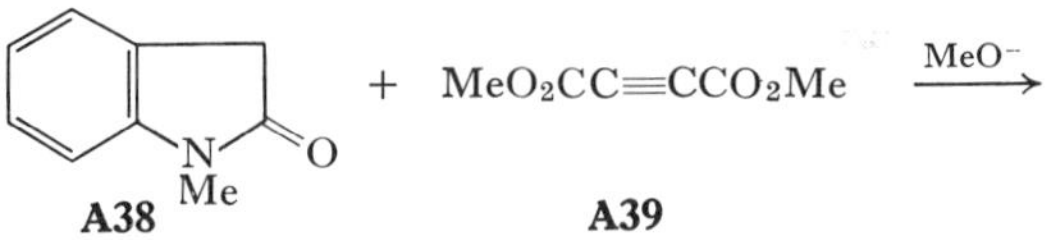

A38 + $MeO_2CC{\equiv}CCO_2Me$ A39 $\xrightarrow{MeO^-}$ [A40] → A41

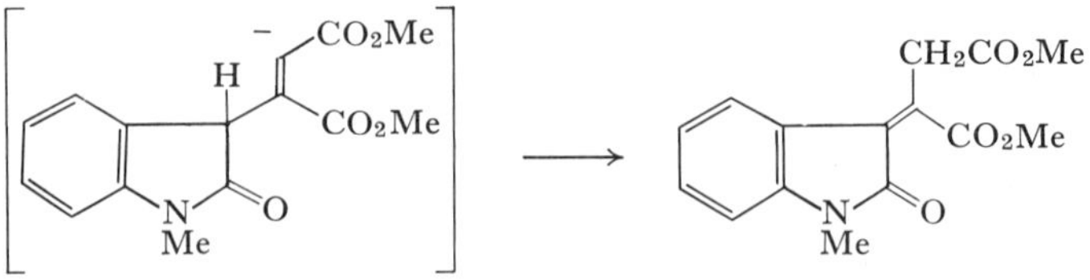

A series of interesting cycloaddition reactions of *N*-acetyl-3-(acetylmethylene)oxindole have been observed. For example, the reaction with dimethyl acetylenedicarboxylate is reported to give **A43** (9). These transformations each require a final aromatization step. Certain features of this reaction are curious. The reaction proceeds much better in acetic anhydride

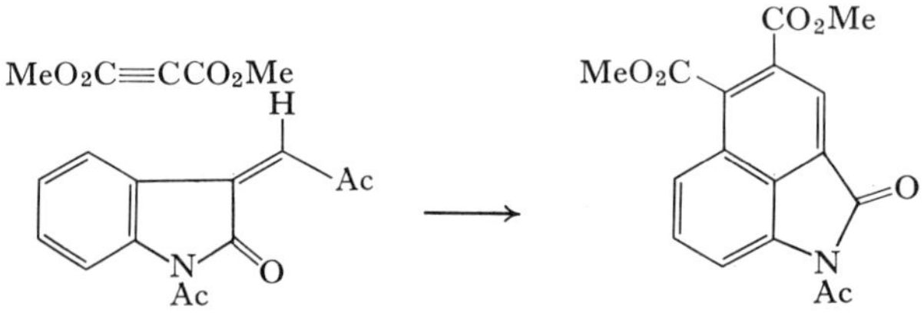

$MeO_2C{\equiv}CCO_2Me$ + A42 → A43

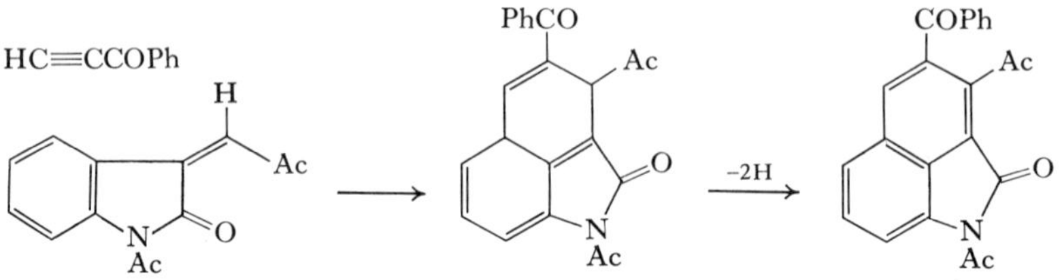

$HC{\equiv}CCOPh$ + A44 → A45 $\xrightarrow{-2H}$ A46

than in other solvents which were investigated, although Diels–Alder reactions usually are not expected to show strong solvent sensitivity. 1-Acetyl-3-phenacylideneoxindole behaves as a dienophile toward cyclopentadiene, cyclohexadiene, and butadiene (75a).

3. Miscellaneous Reactions of Oxindole Derivatives

The stability of the oxindole anion leads to facile cleavage of 3-formyl and 3-acyloxindoles by base (66, 69). The base-catalyzed conversion of **A47** has been interpreted as occurring through the carbene **A48** (33).

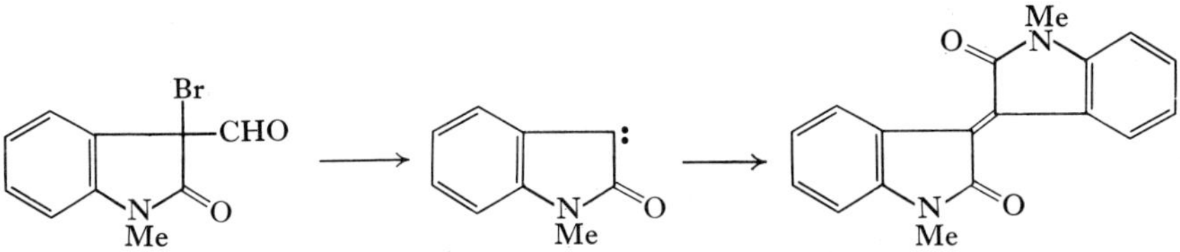

A47 A48 A49

Electronegative substituents at C-3 of the oxindole ring are subject to facile nucleophilic displacement. 3-Bromo-3-methyloxindole has been successfully subjected to the nucleophilic displacements outlined below (55). Elimination of halide ion from 3-halooxindoles is also facile. Treatment of 3-bromoxindole-3-acetic acid with base leads to 3-methyleneoxindole. This rapid decarboxylative elimination is promoted by the ease of

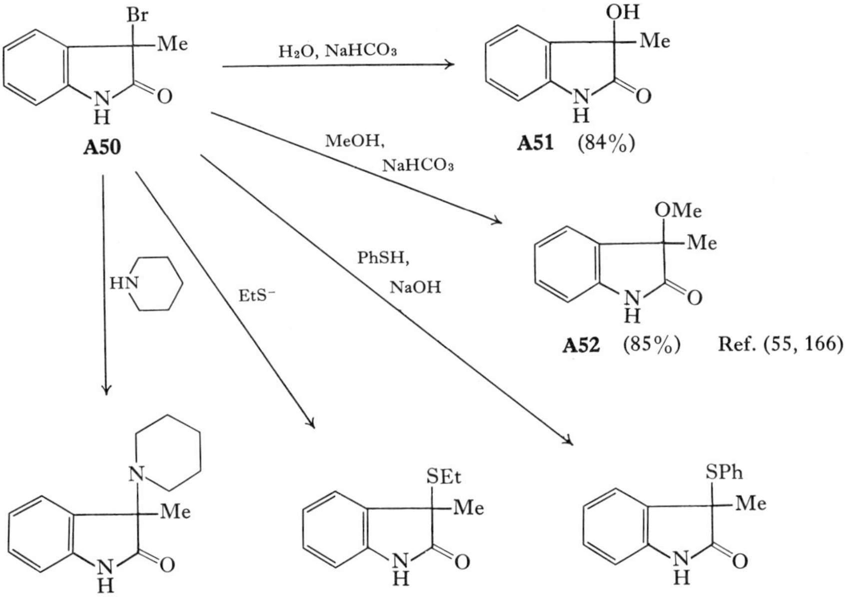

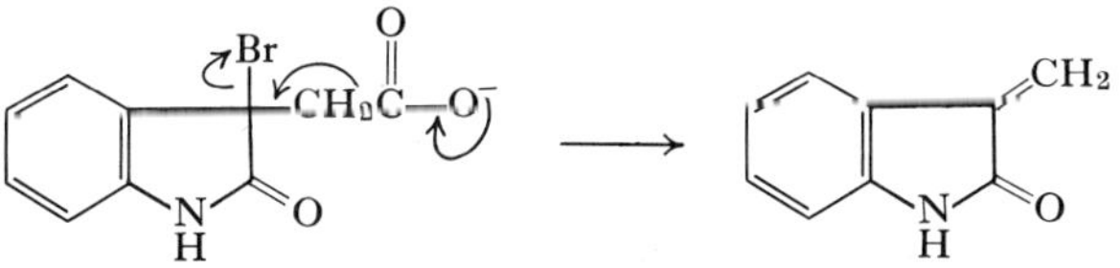

A50 → (H_2O, $NaHCO_3$) A51 (84%)

A50 → (MeOH, $NaHCO_3$) A52 (85%) Ref. (55, 166)

A50 → (PhSH, NaOH) A53 (48%)

A50 → (EtS^-) A54 (80%)

A50 → (piperidine) A55 (91%)

A56 → A57

ionization of the C–Br bond. When decarboxylation is prevented, as in the *t*-butyl ester **A58**, simple elimination occurs. The high reactivity of these

A58 —$(Et)_3N$→ A59

3-halooxindoles can probably be attributed to the fact that the halogen atom is tertiary and also at a benzylic position. The resonance interaction of the nitrogen atom is also favorable for ionization of the halide ion.

3-Methyleneoxindole reacts with nucleophiles to give products of conjugate addition as illustrated by conversion of **A60** to **A61**. The adducts are very sensitive to base.

A60 + PhSH → A61

3-Methyleneoxindoles having a displaceable substituent such as 3-hydroxymethyleneoxindole or 3-chloromethyleneoxindole are subject to easy exchange of the substituent on the methylene group. The exchange may often proceed by addition–elimination mechanisms. A number of nucleophilic displacement reactions of derivatives of 3-methyleneoxindole are summarized in Table XLVII. An interesting case is the reversible addition of succinimide to the ethoxymethyleneoxindole **A63** (33).

TABLE XLVII □ NUCLEOPHILIC DISPLACEMENT REACTIONS OF 3-METHYLENEOXINDOLES

Oxindole substituent	Nucleophilic reagent	Yield (%)	Ref.
3-Hydroxymethylene	Ethyl glycinate	?	(165)
3-Methoxymethylene	Methylamine	88	(165)
3-Chloromethylene	Ammonia	89	(165)
1-Acetyl-3-ethoxymethylene	Diethylamine	?	(14)
1-Acetyl-3-ethoxymethylene	Aniline	?	(14)
1-Acetyl-3-ethoxymethylene	*p*-Tolyl mercaptan	100	(13)
1-Methyl-3-hydroxymethylene	Chloride (thionyl chloride)	94	(13, 129)
1-Methyl-3-chloromethylene	2-Carbethoxycyclohexanone	?	(129)
1-Methyl-3-hydroxymethylene-5-methoxy	Chloride (thionyl chloride)	60	(129)
1-Benzyl-3-hydroxymethylene	Diethylamine	?	(112)

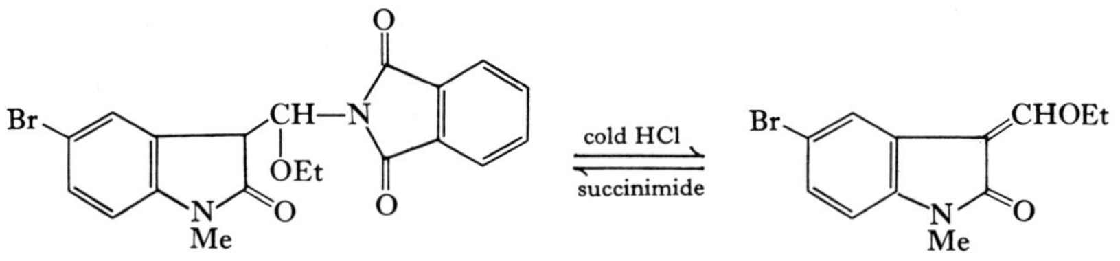

A62 ⇌ A63 (cold HCl / succinimide)

A curious reaction has been reported by Behringer and Weissauer (13). The thiomethyleneoxindole **A64** is said to condense with malonic ester to give **A66**. The structure proof is based on the isolation of the picrate of 2,3-dimethylindole after saponification and decarboxylation of **A67**.

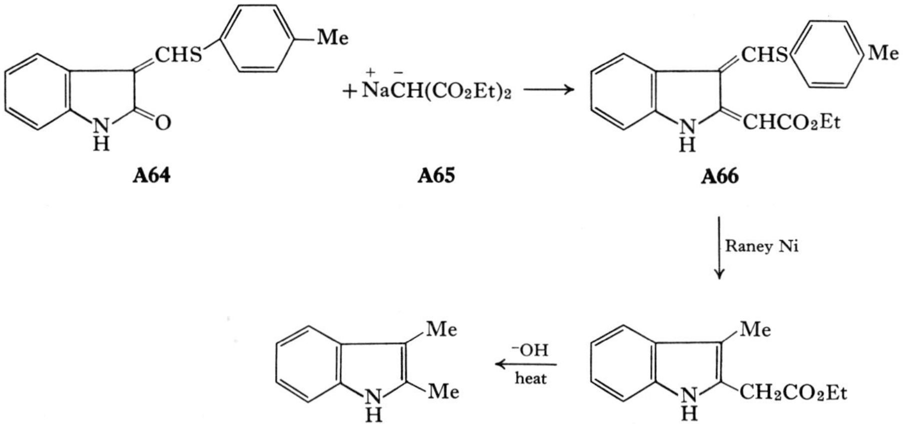

A64 + $\overset{+}{Na}\overset{-}{C}H(CO_2Et)_2$ (A65) → A66 —Raney Ni→ A67 —^{-}OH, heat→ A68

3-Acyloxindoles undergo catalytic reduction to 3-alkyloxindoles (66, 70, 72). The reaction is analogous mechanistically to hydrogenolysis of 3-acylindoles (Chapter II, C). 3-Aminomethyleneoxindoles also undergo hydrogenolysis, giving 3-alkyloxindoles (165). However, acetamidomethylene oxindoles are reduced to acetamidomethyloxindoles. The basis for this

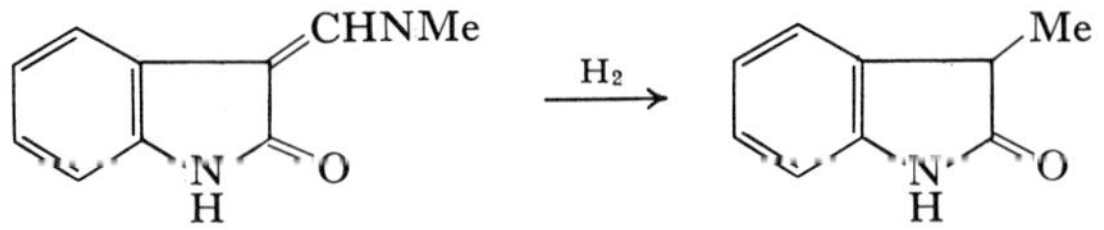

A69 —H_2→ A70

contrasting behavior is not apparent unless protonation of the methylene substituent is required for hydrogenolysis, in which case the more basic

CHNHAc → CH_2NHAc

A71 A72

amino systems would be expected to be more reactive. Reduction of the oximes of 3-acyloxindoles over palladium in ethanol terminates at the aminoalkylideneoxindole oxidation state (161).

Oxidation of oxindoles by oxygen occurs in alkaline solution. Both ring cleavage (3) and formation of 3-hydroxyoxindoles (3, 39, 68, 78, 92) have been observed.

$\xrightarrow[NaH]{O_2}$ Ref. (3)

A73 A74 (77%)

$\xrightarrow[NaH]{O_2}$ Ref. (3)

A75 A76 (71%)

Dioxindole (3-hydroxyoxindole) rings are cleaved to *o*-aminophenyl ketones by alkaline hydrogen peroxide or aqueous potassium ferricyanide

$\xrightarrow[^{-}OH]{K_3Fe(CN)_6}$

A77 A78

(98). Dimeric products result from the oxidation of oxindoles in alkaline solution (48).

$\xrightarrow[^{-}OH]{K_3Fe(CN)_6}$

A79 A80

4. Electrophilic Substitution Reactions of Oxindole and its Derivatives

Electrophilic substitution of oxindoles occurs readily in the 3- and the 5-positions. Examples of both types of reaction have been observed in the case of bromination (54, 89). Hinman and Bauman have discussed the

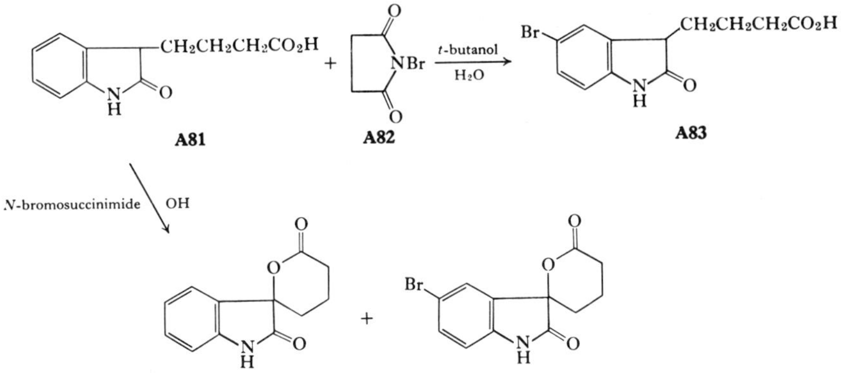

bromination of oxindole derivatives in terms of a competition between aromatic bromination and bromination at C-3 (54). The latter process is catalyzed by base or acid in a manner analogous to halogenation of ketones. The highly reactive anion (or enol) of the oxindole is formed and subsequently rapidly brominated. No precise data on the comparative reactivity of the two sites appear to be available but it is known that in aqueous medium, bromination in position 5 usually dominates (54, 153). It is not surprising that the 5-position is the most reactive of the sites on the carbocyclic ring since it is para to the strongly activating lactam nitrogen.

3-Methyloxindole reacts with benzenesulfenyl chloride to give the 3-substitution product **A87**.

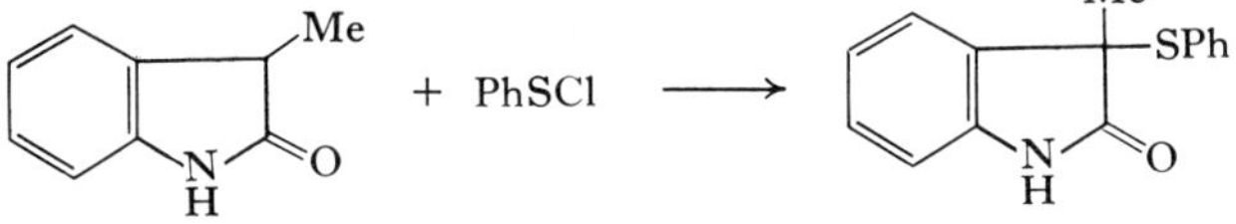

When the 3-position is blocked by substituents, the 5-position is the dominant site of reaction. Nitration of 3-chloromethylene-1-methyloxindole gives the 5-nitro derivative (129).

5. Reactions of 2-Alkoxyindoles

The recent development of synthetic routes to 2-ethoxyindoles (127) has permitted study of the reactions of these oxindole derivatives. The presence of the 2-ethoxy substituents is expected to increase the electron density of the 3-position of the indole ring. The high reactivity of 2-ethoxyindoles toward oxygen bears out this expectation (see Chapter V,A) (131). 2-Ethoxyindole is reported to be in equilibrium with the tautomer **A89** in solution. Indeed, the two tautomers have apparently been isolated as

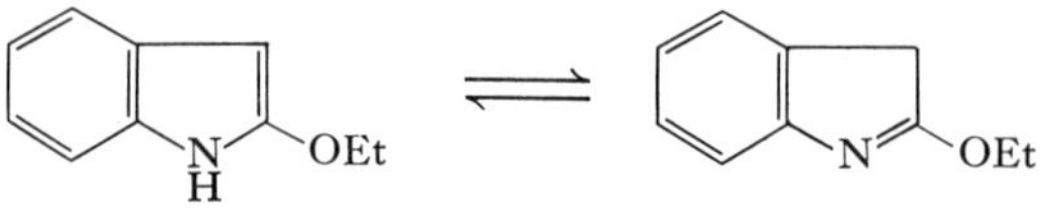

OEt N H **A88** ⇌ N OEt **A89**

distinct crystalline forms (49). Tautomer **A89** is isolated by sublimation and can be reconverted to **A88** by melting.

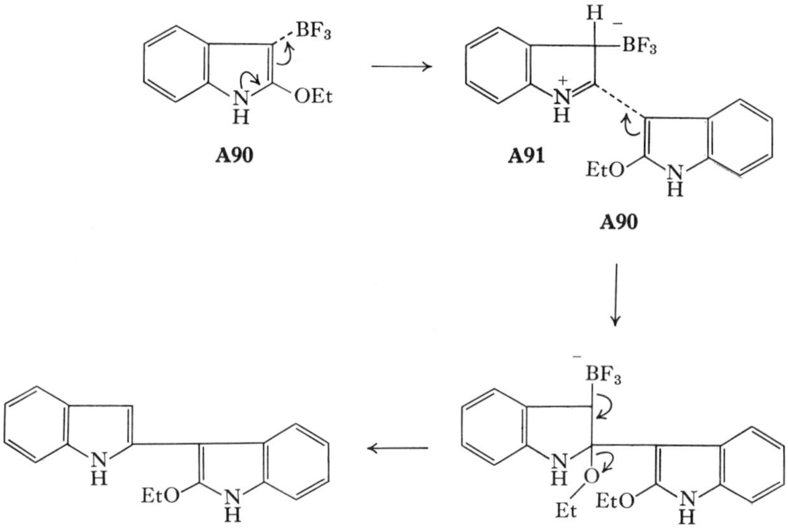

BF_3 N H OEt **A90** → H BF_3 $\overset{+}{N}$ H **A91** EtO N H **A90**

↓

BF_3 N H O Et EtO N H **A92** → N H EtO N H **A93**

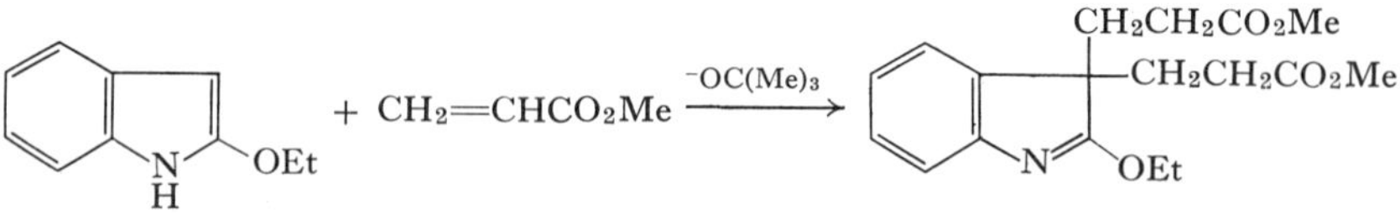

N H OEt **A90** + $CH_2{=}CHCO_2Me$ **A94** $\xrightarrow{^{-}OC(Me)_3}$ $CH_2CH_2CO_2Me$ $CH_2CH_2CO_2Me$ N OEt **A95**

A condensation reaction related to acid-catalyzed dimerization of indole is observed on treatment of 2-ethoxyindole with boron trifluoride (131). 2-Ethoxyindole reacts at the 3-position with maleic anhydride and diethyl azodicarboxylate to give addition products (131). Reaction with methyl acrylate requires a basic catalyst and gives **A95** (131).

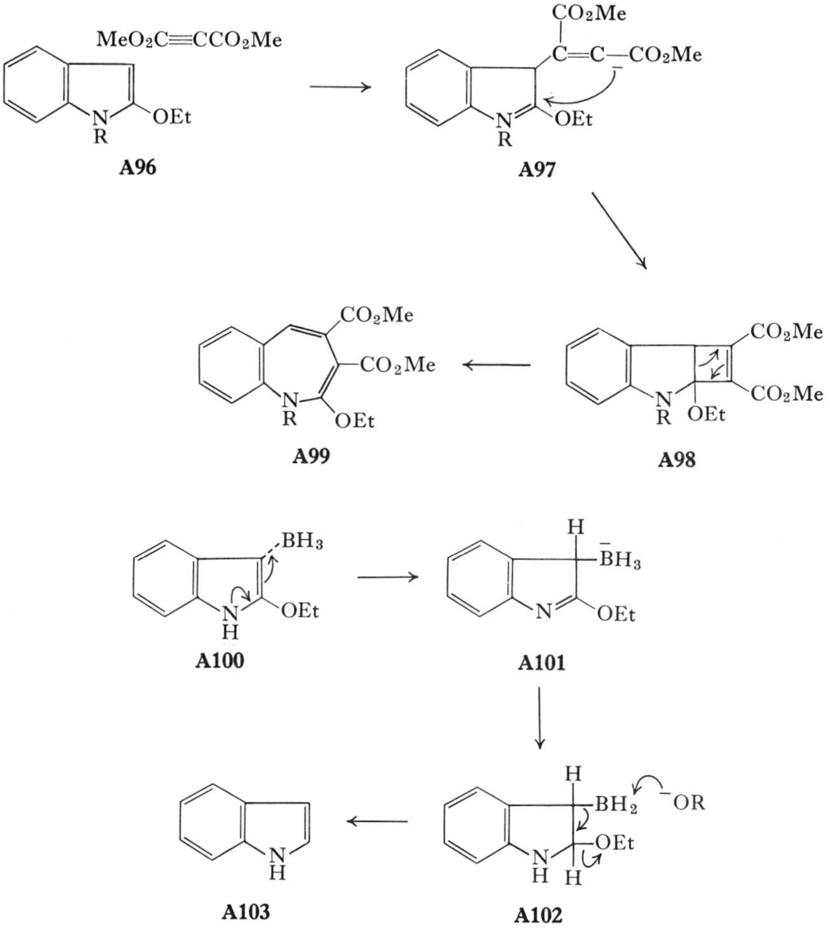

$MeO_2C{\equiv}CCO_2Me$

A96 **A97** **A98** **A99**

A100 **A101** **A102** **A103**

An interesting ring opening has been observed when 2-ethoxyindole and 1-methyl-2-ethoxyindole react with dimethyl acetylenedicarboxylate (132). Normal 3-substituted products as well as the ring-opened product are observed with 2-ethoxyindole. 2-Ethoxyindole is reduced to indole with borane in tetrahydrofuran (128). The electrophilic nature of borane is probably involved in this reduction.

6. Reactions at C-2 of the Oxindole Ring

The reactivity of the oxindole ring toward reagents which would normally attack the carbonyl group is generally that typical of amides. 3-Benzyloxindole and P_4S_{10} apparently give the thio derivative **A105** which can subsequently be reduced to the indole (130). The mercaptoindolenine **A108** has

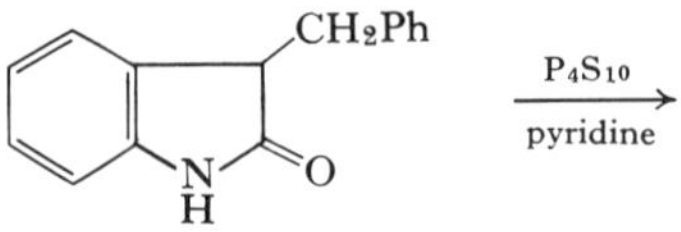

CH₂Ph $\xrightarrow[\text{pyridine}]{P_4S_{10}}$

A104

CH₂Ph S $\xrightarrow{\text{Raney Ni}}$ CH₂Ph

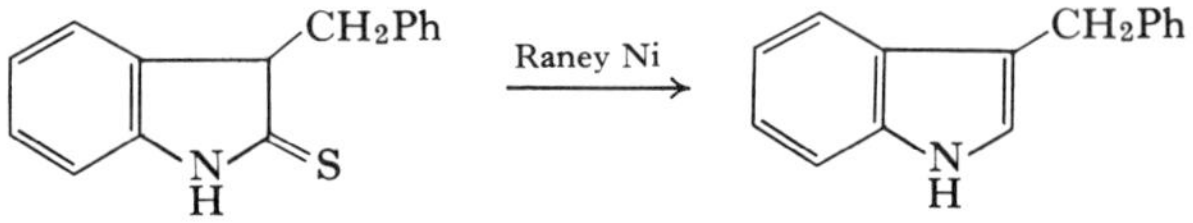

A105 **A106**

been reported to be formed from the reaction of 3,3-dimethyloxindole and P_4S_{10} (79). No basis for rejecting the more likely tautomeric thione structure

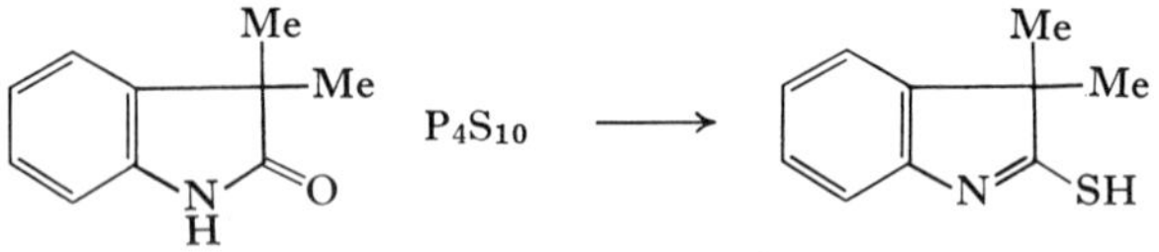

P_4S_{10} $\longrightarrow$

A107 **A108**

is reported. Phosphorus pentachloride gives the chloroindolenine **A109** (36).

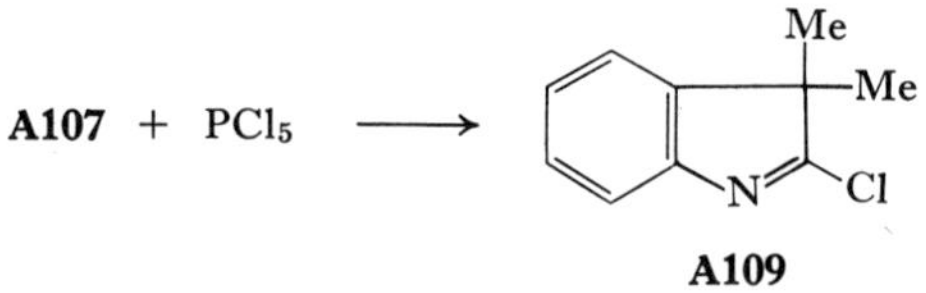

A107 + PCl_5 $\longrightarrow$

A109

Lithium aluminum hydride reduces 1-methyloxindole to a mixture of 1-methylindole and 1-methylindoline (71). This result suggests the intermediacy of the carbinolamine **A111** as its aluminum salt. Partial reduction at C-2 of the oxindole ring has been encountered in several molecules in which a nucleophilic group in the molecule can react with a partially reduced intermediate to give a tricyclic system which is stable towards further reduction.

A110 —$LiAlH_4$→ A111 → A112
A111 → A113
A112 —$^{-}AlH_4$→ A114

A115 —$LiAlH_4$→ A116 Ref. (170)

A117 —Na / PrOH→ A118 Ref. (83)

Several cyclizations of a similar type have been effected using sodium and alcohol as the reducing agent (68). The oxindole ring is susceptible to reduction by diborane although synthetically efficient conditions have not yet been developed. Oxindole gives indoline (15%) on reduction by diborane and *N*-methyloxindole gives a mixture of *N*-methylindoline and *N*-methylindole (128). Sodium in butanol or propanol can also effect reduction of oxindoles to indoles (161).

B. SYNTHESIS OF THE OXINDOLE RING

1. Cyclization of α-Halo and α-Hydroxyacyl Derivatives of Anilines

Treatment of anilides of α-halocarboxylic acids with aluminum chloride effects cyclization to oxindoles. The reaction is presumably of the Friedel–Crafts type. Several typical examples are listed in the equations below. Additional examples can be found in a review by Sumpter (152) and in a paper by Beckett and co-workers (12a).

B1 $\xrightarrow{AlCl_3}$ B2 68–71% Ref. (141)

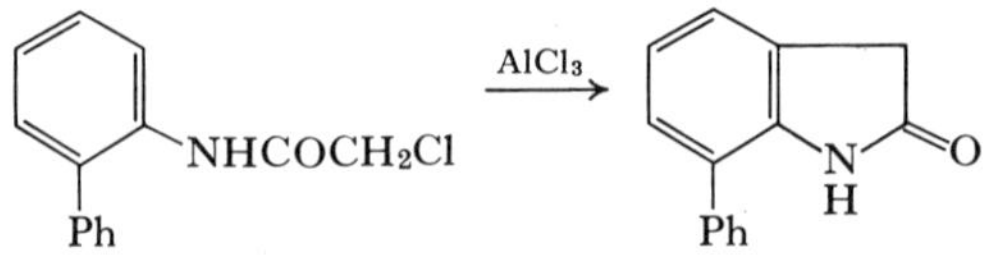

B3 $\xrightarrow{AlCl_3}$ B4 22% Ref. (168)

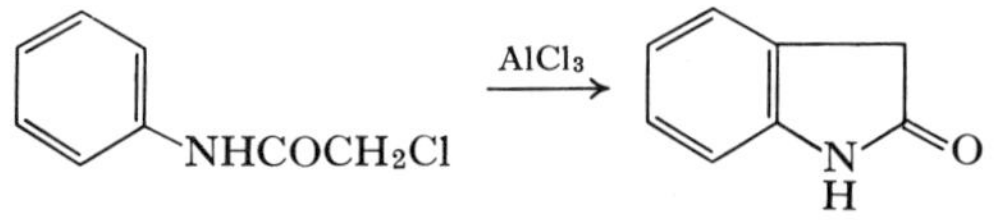

B5 $\xrightarrow{AlCl_3}$ B6 60% Ref. (1)

Petyunin and co-workers have reported the preparation of many 3,3-diphenyloxindoles by cyclization of α,α-diphenyl-α-hydroxyacetanilides (115, 116, 118, 119). The general scheme is outlined below. The cyclization is doubtlessly an intramolecular Friedel–Crafts reaction of the carbonium ion generated by ionization of the tertiary carbinol. α-Chloro-α,α-diphenyl-

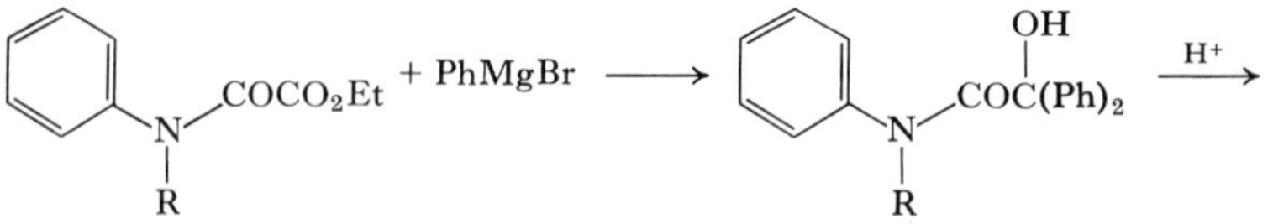

B7 + PhMgBr (B8) ⟶ B9 $\xrightarrow{H^+}$ B10

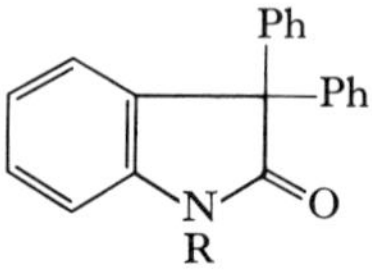

acetanilide and α-methoxy-α,α-diphenylacetanilides are also cyclized by concentrated sulfuric acid (117).

Treatment of α-chloro-α,α-diphenylacetanilide with sodium hydride gives substantial amounts of 3,3-diphenyloxindole. The major product is **B14**, which would appear to be the product of subsequent alkylation of **B12** by **B11** (144). An α-lactam has been suggested (145) to be an intermediate in the formation of **B12** and **B13** but direct evidence for this proposal is lacking. Meyer has reported an exceptionally facile cyclization leading to an oxidole derivative. The hydrazide **B16** apparently cyclized to **B17** in ice-cold ether (97). Meyer has proposed that an intramolecular displacement

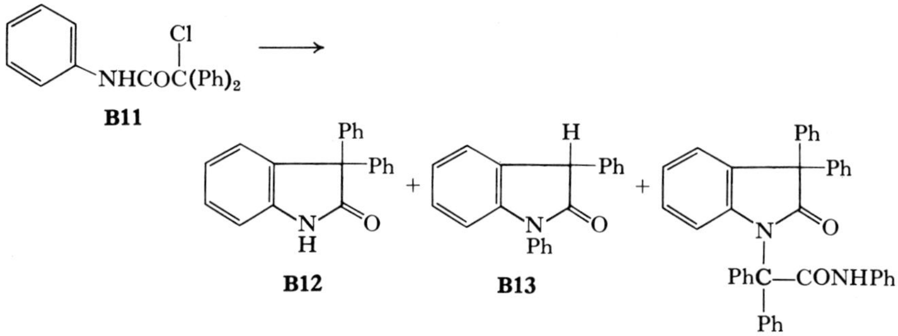

B11 → B12 + B13 + B14

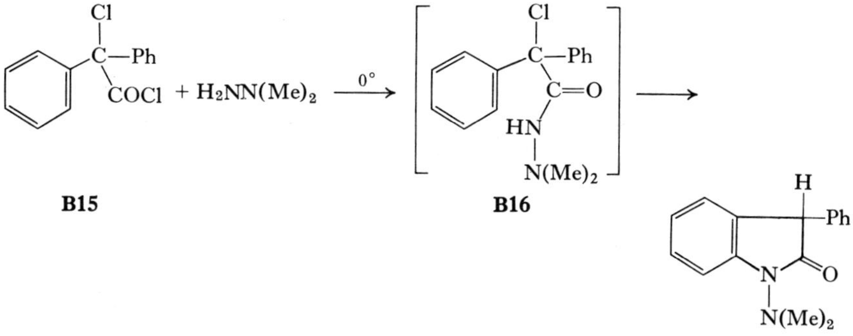

B15 + $H_2NN(Me)_2$ —0°→ [B16] → B17

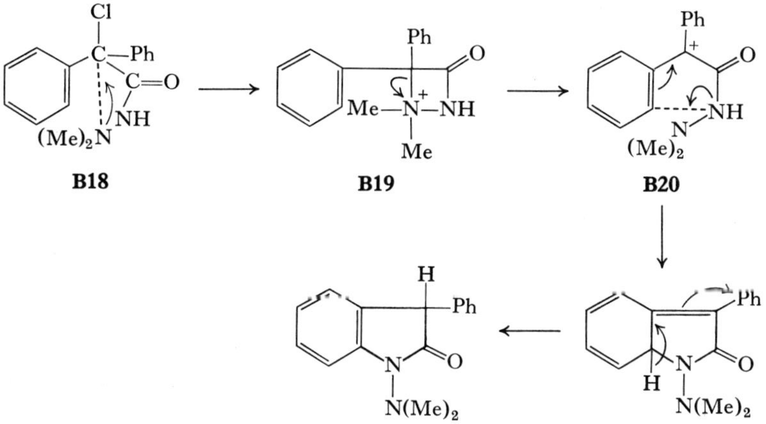

B18 → B19 → B20 → B21 → B22

forming an α-lactam intermediate can account for the cyclization but it seems more likely that the more basic nitrogen of the hydrazide is the source of the increased activity. Analogy for such a scheme can be recognized in the work of Bird (17) in which the sequence **B23** → **B25** → **B26** was described.

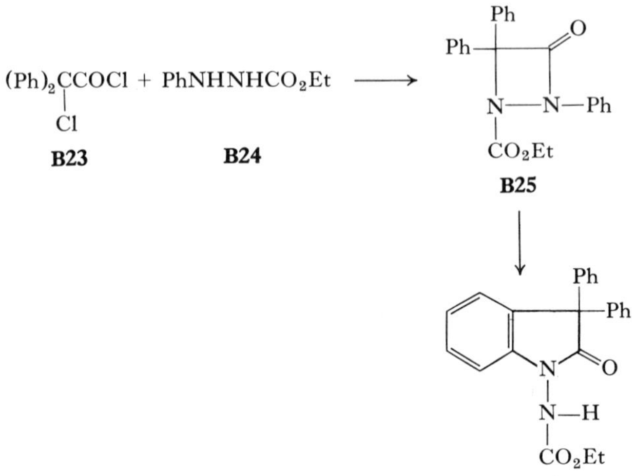

$(Ph)_2C(Cl)COCl$ + $PhNHNHCO_2Et$ ⟶ **B25** ⟶ **B26**

B23 **B24**

2. Cyclization of Derivatives of *o*-Aminophenylacetic Acid

The generation of derivatives of *o*-aminophenylacetic acid often leads to the formation of oxindoles by intramolecular condensation. Thus, catalytic

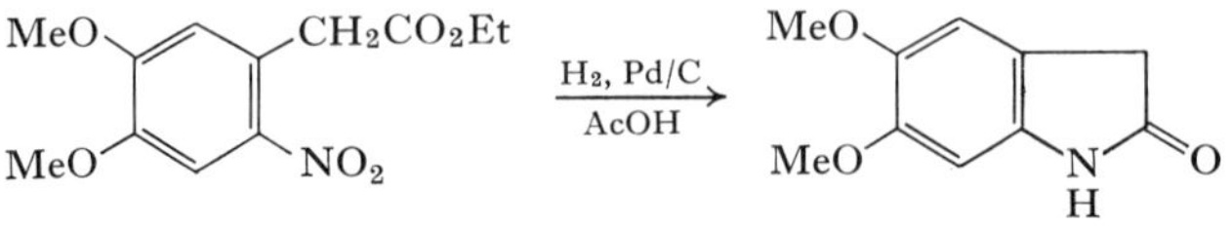

B27 $\xrightarrow[\text{AcOH}]{H_2,\ Pd/C}$ **B28** (75%) Ref. (159)

reduction of **B27** in acetic acid gives **B28**. Reduction of **B29** in aqueous ammonia gives 4-iodooxindole (46). 4-Chlorooxindole has been synthesized by a similar sequence (138). Other examples of preparation of substituted oxindoles by reductive cyclization have been reported by Beckett and coworkers (12a). Neil (105) has reported the synthesis of oxindole in 34–43% yield by treatment of *o*-chlorophenylacetic acid with concentrated ammonium hydroxide and cupric acetate.

B29 → B30 (Fe^{+++}, NH_3, H_2O)

B31 → B32 ($Cu(OAc)_2$, NH_4OH, 160°)

Schulenberg and Archer found that 3-benzoyl-1-(2-carbomethoxyphenyl)oxindole was the product of treatment of **B33** with methoxide ion (143). A very reasonable mechanism has been proposed. The oxindole ring is believed to arise by cyclization of an *o*-aminophenylacetic acid derivative. Later work (142) has provided confirmation of this mechanism by isolation of the enolic form of **B36**. The requirement for a basic catalyst in the final

B33 → B34 (^-OMe) → B35 → B36 → B37

cyclization step was also demonstrated. The formation of oxindole derivatives such as **B42** from the reaction of *p*-substituted anilines with alloxan (24) also involves intramolecular cyclization of an *o*-aminophenylacetic acid system.

B38 B39 B40

B42 B41

3. Cyclization of *N*-Acyl Phenylhydrazides

Heating *N*-acyl phenylhydrazides in the presence of various basic catalysts results in the formation of oxindoles (Brunner reaction) (19). The preparation of 3-methyloxindole is a well-documented example of the reaction. A formal similarity to the Fischer cyclization of arylhydrazones is apparent. It seems likely that the mechanism of the cyclization parallels that of the Fischer reaction.

Table XLVIII records a number of successful examples of this reaction.

TABLE XLVIII □ BASE-CATALYZED CONVERSION OF ACYL PHENYL HYDRAZIDES TO OXINDOLES

Oxindole substitution	Basic catalyst	Temp.	Yield (%)	Ref.
None	Sodium amide	200	?	(149)
1,3,3-triMe	Calcium hydride	210–220	44	(21)
3-Me	Calcium hydride	190	41–44	(35)
3-Me	Sodium, naphthalene	?	75	(167)
3,3-diPh	Sodium, naphthalene	?	42	(147)
3,3-Tetramethylene	Calcium oxide	230–250	40	(101)
5-Methoxy-3,3-pentamethylene	Calcium oxide	230–250	28	(101)

NHNHCOEt $\xrightarrow[190°]{CaH_2}$ (3-methyloxindole) + NH_3 Ref. (35)

B43 **B44** 41–44%

NHNHCOCHR$_2$ $\overset{B^-}{\rightleftharpoons}$ N—N (H H), $C(R)_2=C-O^-$

B45 **B46**

B49 ← **B48** (NH_2, O^-) ← **B47** (CONH$_2$)

B49 **B48** **B47**

4. Miscellaneous Synthetic Approaches to Oxindoles

The reaction of indoles with an equimolar amount of *N*-bromosuccinimide in *t*-butyl alcohol gives moderate yields of the corresponding oxindoles (54). For example, indole-3-butyric acid furnishes a 50% yield of the oxindole. Excess *N*-bromosuccinimide results in bromination of the oxindole product (54, 89). The mechanism of this and related reactions was discussed in Chapter V,C.

Oxindoles have been isolated by treatment of benzophenone anils with chloroform, ethylene oxide, and tetraethylammonium bromide (82). Ethylene oxide and chloroform generate dichlorocarbene and the adduct **B52** is believed to be an intermediate in the formation of the oxindole. An

PhN=C(Ph)$_2$ $\xrightarrow[R_4N^+]{\text{ethylene oxide, } CHCl_3}$ (3,3-diphenyl oxindole, N—CH_2CH_2Cl)

B50 **B51**

authentic sample of **B52** has been shown to react with ethylene oxide to give **B53**. The sequence below can account for the structure of the isolated product, but it is entirely speculative at this point.

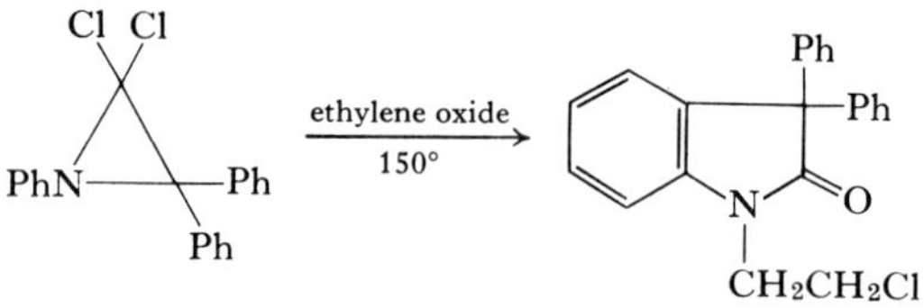

B52 → (ethylene oxide, 150°) → B53

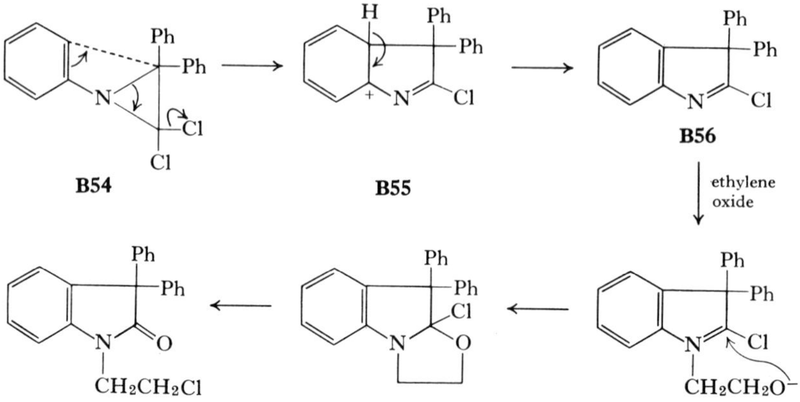

B54 → B55 → B56

B56 → (ethylene oxide) → B57 → B58 → B59

C. DERIVATIVES OF 3-HYDROXYINDOLE AND 3-INDOLINONE (INDOXYLS)

Infrared studies (56) have led to the conclusion that 1-acetyl-3-indolinone exists entirely as the ketonic tautomer in preference to the enolic 1-acetyl-3-hydroxyindole structure. A similar conclusion was reported in the case of 1-methyl-3-indolinone. In contrast, 2-(2-pyridyl)-3-hydroxindole appears

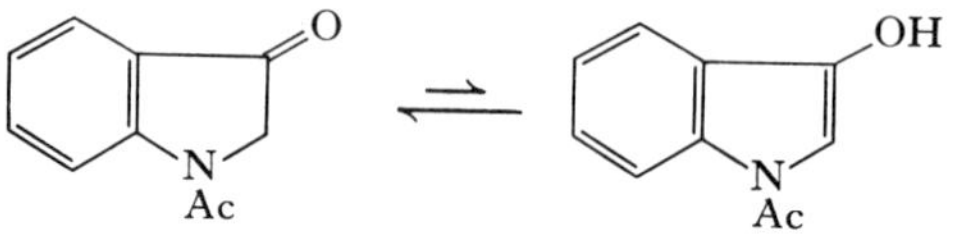

C1 ⇌ C2

to exist in the enolic form (114a). The ketonic and enolic forms of 3-indolinone are, of course, related by the common anion **C4**. Derivatives of 3-indolinone are expected to be especially reactive in base-catalyzed reactions

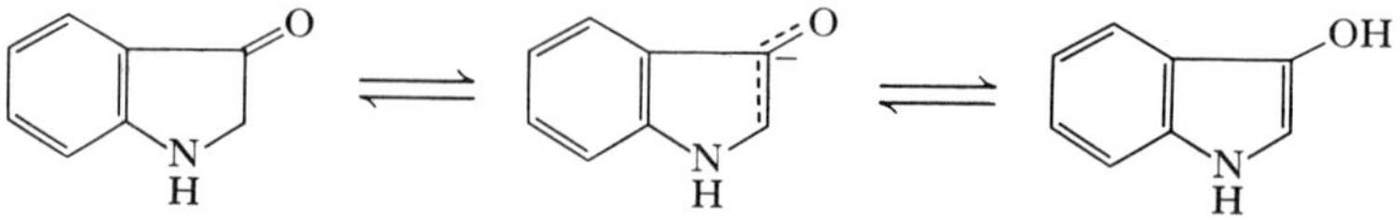

C3 ⇌ C4 ⇌ C5

since abstraction of a proton from the neutral ketone is favored by the additional resonance energy of the aromatic enolate **C4**. This stabilization should make 3-indolinone more acidic than acetophenone, which does not develop a new conjugated ring on ionization. The expectation of high reactivity at C-2 is borne out by considerable chemical experience. While

$$\underset{\textbf{C6a}}{Ph-\overset{O}{\overset{\|}{C}}-\bar{C}H_2} \longleftrightarrow \underset{\textbf{C6b}}{Ph-\overset{^{-}O}{\overset{|}{C}}=CH_2}$$

base-catalyzed alkylation of 3-indolinone with methyl sulfate gives 3-methoxyindole, alkylation with excess methyl iodide gives 2,2-dimethylindolin-3-one (35a). 2-Carbethoxyindolin-3-one is alkylated in the 2-position by allyl chloride and allyl bromide but products of O-alkylation have been isolated from reactions with diazomethane and triethyloxonium fluoroborate (59a, 129a). 3-Indolinone formed from 3-acetoxyindole condenses with benzaldehyde in high yield (1a). Julian, Meyer, and Printy have summarized the early work on base-catalyzed alkylation and condensation reactions of 3-indolinone and its derivatives (67).

3-Indolinones are subject to very easy oxidation. 2-Methyl-3-indolinone is converted by oxygen or hydrogen peroxide to a dimeric oxidation product (41, 50). 2-Phenyl-3-indolinone behaves similarly (41, 50, 74). The oxidation

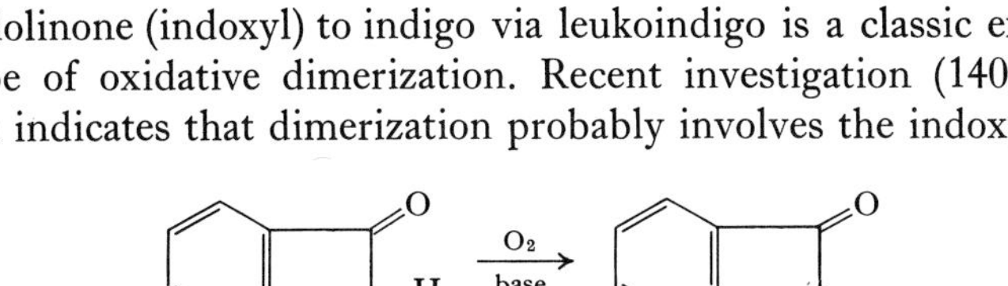

C7 → **C8**

of 3-indolinone (indoxyl) to indigo via leukoindigo is a classic example of this type of oxidative dimerization. Recent investigation (140a) of this reaction indicates that dimerization probably involves the indoxyl radical.

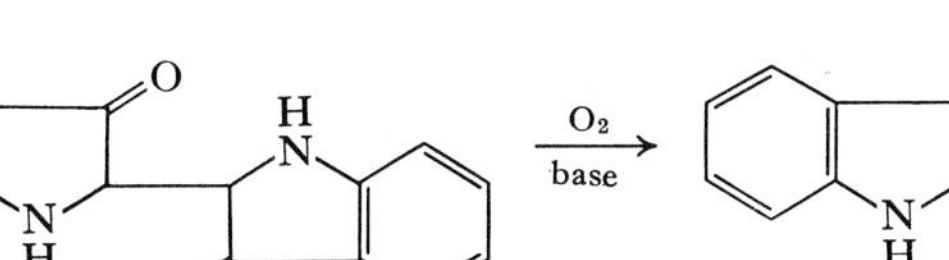

C9 $\xrightarrow[\text{base}]{O_2}$ **C10**

2 **C10** → **C11** $\xrightarrow[\text{base}]{O_2}$ **C12**

C13 $\xrightarrow[D_2O]{H^+}$ C14 $\xrightarrow{LiAlH_4}$ C15

3-Indolinone is reported to give a mixture of 3-hydroxyindoline (5%), indole (25%), and indigo (10%) on reduction with lithium aluminum hydride (43). The indole no doubt arises via dehydration of the indolinol. Advantage has been taken of this reaction to prepare indole specifically labeled with deuterium at C-2 (80). Catalytic reduction of 3-indolinones to indolinols is also possible (64).

C16 $\xrightarrow{H_2, Pt}$ C17

Indolinones behave as typical ketones in certain carbanion-type condensation reactions. 1-Acetyl-3-indolinones condense with cyanoacetic acid in the presence of ammonium acetate (106). Phenol-xylene mixtures were used as the solvent and water was removed by azeotropic distillation. This condensation introduces a 3-cyanomethyl substituent and as such is a potential synthetic route to both substituted tryptamines and indole-3-acetic acids.

C18 + $NCCH_2CO_2H$ (C19) ⟶ [C20] $\xrightarrow{-H_2O, -CO_2}$ C21

C21 $\xrightarrow{^-OH}$ C22 (CH_2CO_2H)

C21 $\xrightarrow{LiAlH_4}$ C23 ($CH_2CH_2NH_2$)

These routes have been little exploited. The main limitation is probably the relative inaccessibility of the required indolinones. Other nucleophilic additions to 3-indolinones include the use of Reformatsky reactions to synthesize indole-3-acetic acids. Dehydration of the intermediate indolinols has been accomplished with phosphorus pentoxide (126, 133).

$$\textbf{C24} + BrCH_2CO_2Et\ (\textbf{C25}) \xrightarrow{Zn} \textbf{C26} \xrightarrow{(1)\ P_2O_5;\ (2)\ ^-OH} \textbf{C27}$$

D. SYNTHESIS OF 3-INDOLINONE DERIVATIVES

Indolinones have usually been prepared via their enol acetates which can, in turn, be obtained from *o*-carboxyphenylglycines. The condensation is usually effected in acetic anhydride using sodium acetate as the catalyst. The products isolated are 1-acetyl-3-acetoxyindoles and subsequent selective hydrolysis (56, 150) affords the 1-acetylindolinone. A number of typical examples are collected in Table XLVIX. Base-catalyzed condensation of

TABLE XLVIX □ SYNTHESIS OF 3-ACETOXYINDOLES FROM *N*-(*o*-CARBOXYPHENYL)GLYCINES

3-Acetoxy-1-acetylindole substituent	Yield (%)	Ref.
None	60	(106)
5-Bromo-4-chloro	?	(75)
5-Bromo-4-chloro-7-methoxy	?	(75)
5-Bromo	42	(56, 150)
5,7-Dibromo	15	(150)
5-Nitro	57	(57)
6-Chloro	?	(56)

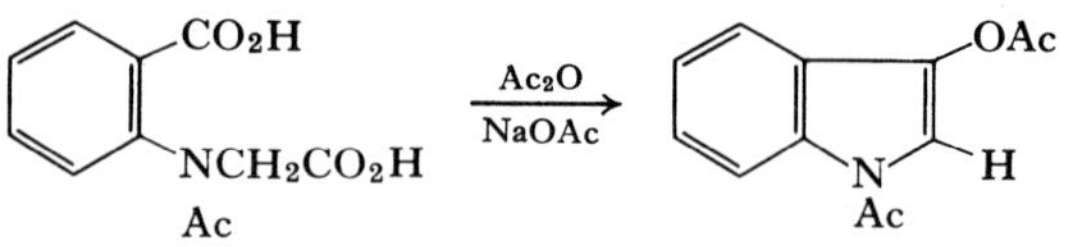

D1 → (Ac_2O, NaOAc) → D2

diesters of *N*-(*o*-carboxyphenyl)glycines also provides the 3-indolinone ring (50). Subsequent oxidation of the very sensitive 3-indolinones often results in the isolation of oxidative coupling products. The loss of the car-

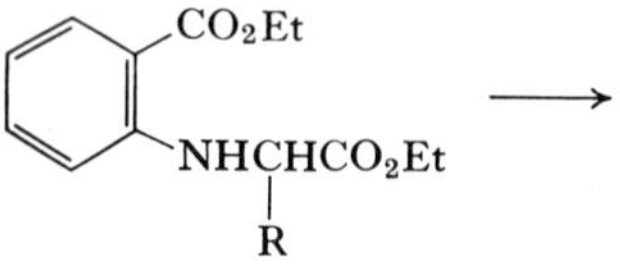

D3 R = Me
D4 R = Ph

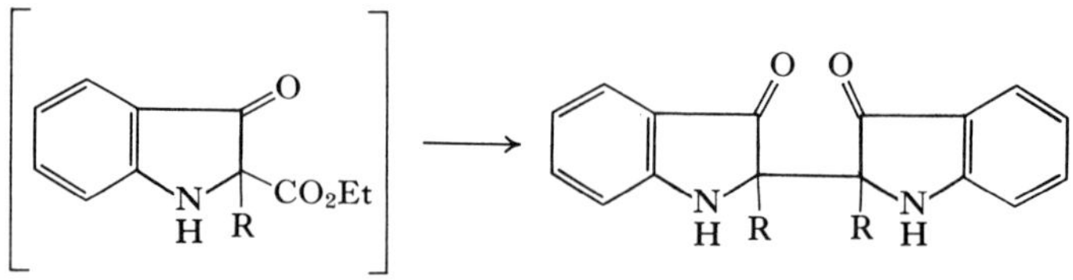

D5 R = Me
D6 R = Ph

D7 R = Me
D8 R = Ph

bethoxy substituent from **D5** and **D6** presumably occurs by a mechanism analogous to base-catalyzed cleavage of α,α-disubstituted β-keto esters and presumably precedes the oxidation.

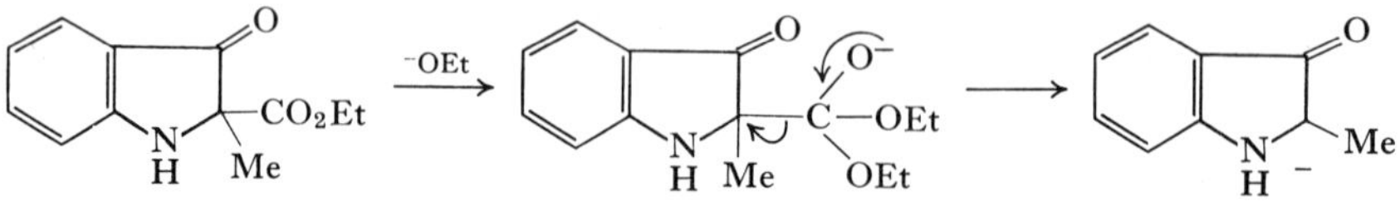

D4 → ($^-$OEt) → **D9** → **D10**

E. INDOLES WITH CARBOCYCLIC HYDROXYL SUBSTITUENTS

Derivatives of indoles with hydroxyl substituents in the benzenoid ring, particularly 5-hydroxy derivatives, have attracted considerable attention since this group is present in such physiologically important derivatives as serotonin. Recently, renewed interest has been sparked by the discovery of the antibiotic properties of certain indole quinones. The special importance of the Nenitzescu synthesis (Chapter III, C) for the synthesis of 5-hydroxyindoles may be recalled. The direct hydroxylation of indole rings was discussed in Chapter V, D. Many aspects of the chemistry of hydroxyindoles are typical of phenols. Mannich reactions of 6-hydroxytetrahydrocarbazoles give adducts in which alkylation has occurred at C-5 (100). 7-Hydroxytetrahydrocarbazole gives the product of substitution at C-8 (100). The Mannich reactions of 5-hydroxyindole and 5-hydroxy-2-methylindole give the respective 4-alkylation products (100). Ethyl 5-hydroxy-2-methylindole-3-carboxylate undergoes Mannich condensation selectively at C-4 (15). Each of the four simple carbocyclic hydroxyindoles gives a Mannich condensation product in which the dimethylaminomethyl group is introduced into the benzenoid ring in preference to C-3 of the indole ring (158). The hydroxy groups have very specific directional effects.

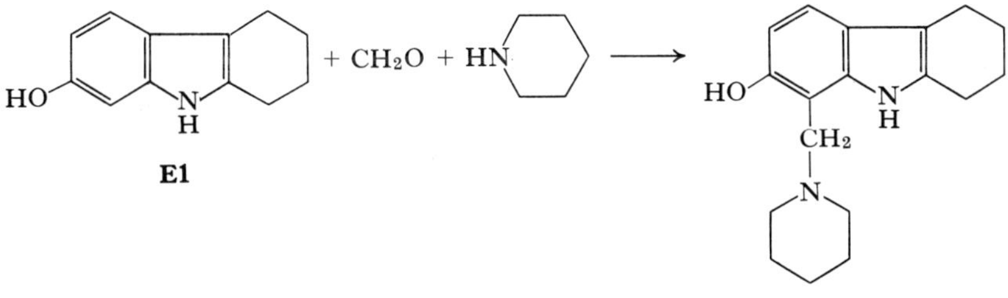

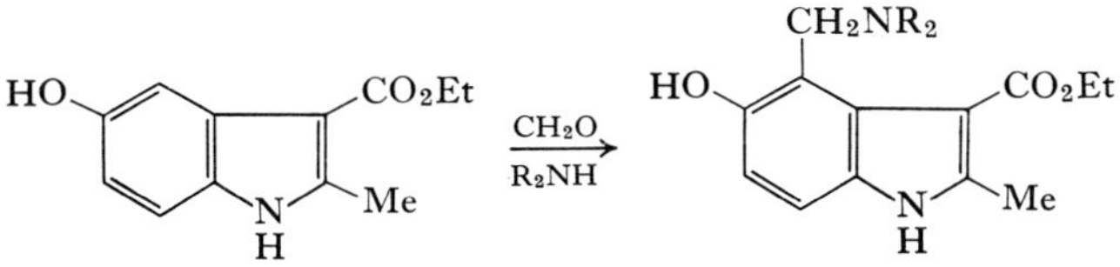

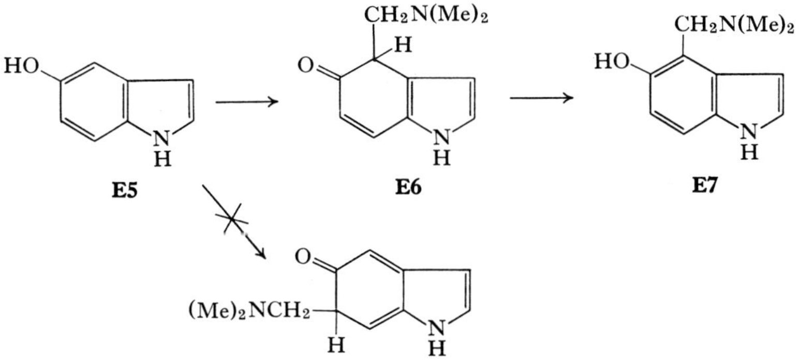

4-Hydroxindole is alkylated at C-5, 5-hydroxyindole at C-4, 6-hydroxyindole at C-7, and 7-hydroxyindole at C-6 (158). This directing effect is reminiscent of directional effects in the naphthalene system. The intermediate **E6** is preferred to **E8** in the case of 5-hydroxyindole. The remaining conjugated residue in **E8** is expected to be considerably less stable than that in **E6** and reaction at C-4 predominates. The preferential exchange of C-4 protons in 5-hydroxyindoles (Chapter I, B) can be attributed to similar factors.

As is true of other aromatic phenols (60), ring alkylation competes with oxygen alkylation under certain conditions. Suehiro has observed extensive C-alkylation of **E9** in xylene (151). In the presence of excess base and halide, dialkylation occurs at C-4.

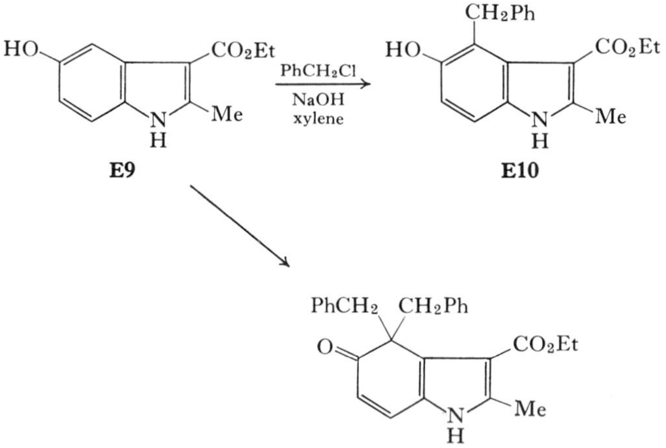

$PhCH_2Cl$, NaOH, xylene

E9 → **E10** + **E11**

Quinone derivatives of indoles can be obtained by oxidation of hydroxyindoles. Much of the recent work in the area of oxidation of hydroxindoles has been done by Teuber and co-workers. These workers have investigated particularly the oxidation of hydroxyindoles with potassium nitrosodisulfonate. 5-Hydroxy-3-methylindole is oxidized in 95% yield to the quinone **E13** (156). 5-Hydroxy-2-phenylindole is oxidized in a similar manner (156). 6-Hydroxytetrahydrocarbazoles also give quinones on

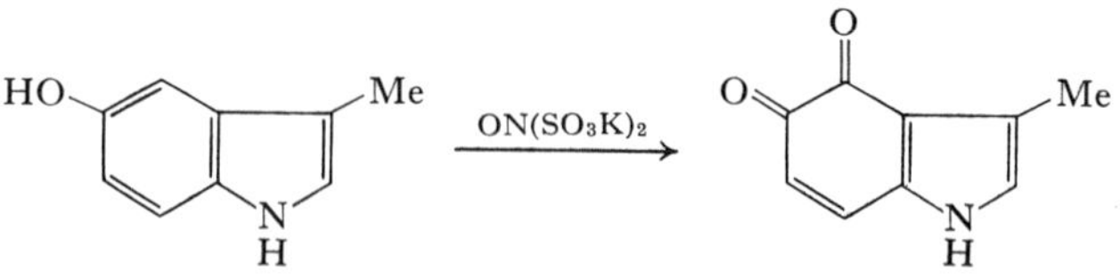

$ON(SO_3K)_2$

E12 → **E13** (95% yield)

oxidation with the reagent (156). The reaction has also been successfully extended to a number of 2-methyl-3-carbethoxy-5-hydroxy indoles and again 4,5-quinones are the products (157). Reduction of these quinones with

$ON(SO_3K)_2$

E9 E14 (85%)

dithionite affords 4,5-dihydroxyindoles (157). The quinones undergo 1,4-addition of hydrogen chloride, giving 7-chloro-4,5-dihydroxyindoles (157).

Potassium nitrosodisulfonate can effect hydroxylation and oxidation of indolines, giving hydroxindoles, which, in turn, are substrate for further oxidation (156).

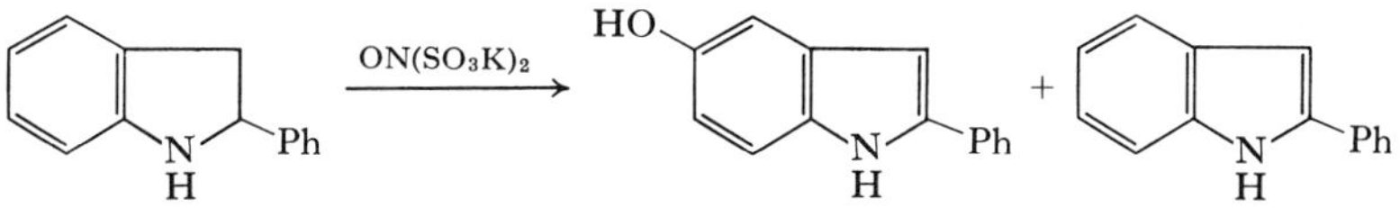

$ON(SO_3K)_2$

E15 E16 (68%) E17 (9%)

$ON(SO_3K)_2$

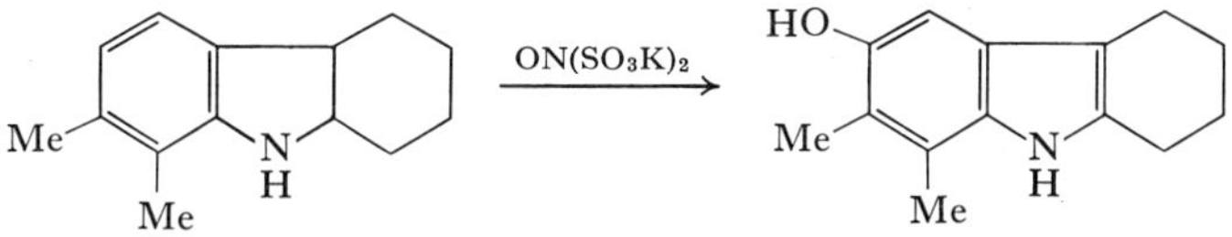

E18 E19 (70%)

1-Ethyl-2,6-dimethyl-4-hydroxindole has been found to give only a low yield of *o*- and *p*-quinones but the corresponding 3-formyl derivative could be oxidized reasonably satisfactorily (139). Remers and Weiss (137) have reported the oxidation of 1-ethyl-4-hydroxy-2-methylindole to the *p*-quinone **E21** in yields of up to 68%. Nucleophilic displacements which occur

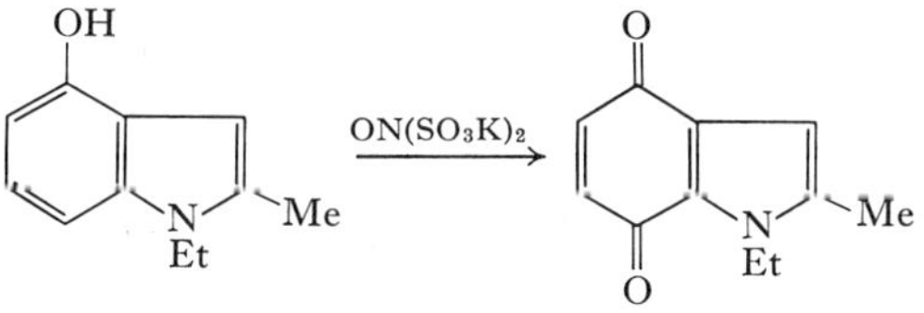

$ON(SO_3K)_2$

E20 E21

by addition–elimination mechanisms are quite facile in indole quinones, as illustrated by the conversion **E22** → **E23** (137). Other examples of syntheses

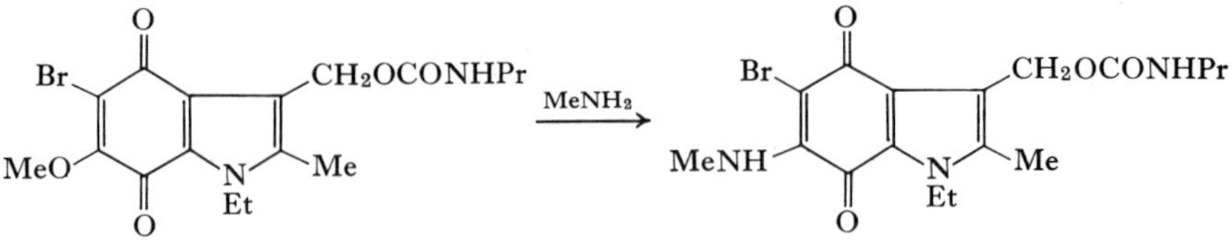

Br, MeO, O, O, $CH_2OCONHPr$, N, Et, Me — E22 $\xrightarrow{MeNH_2}$ E23 (Br, MeNH, O, O, $CH_2OCONHPr$, N, Et, Me)

and reactions of indole quinones are discussed in Chapter X in connection with synthetic work in the area of the mitomycin antibiotics.

F. 1-HYDROXYINDOLES

The relatively inaccessible 1-hydroxyindoles have received scant attention until very recently. Probably the most easily obtained representative of this class is 1-hydroxy-2-phenylindole which can readily be prepared by cyclization of the oxime of benzoin (37, 38). The principal features of

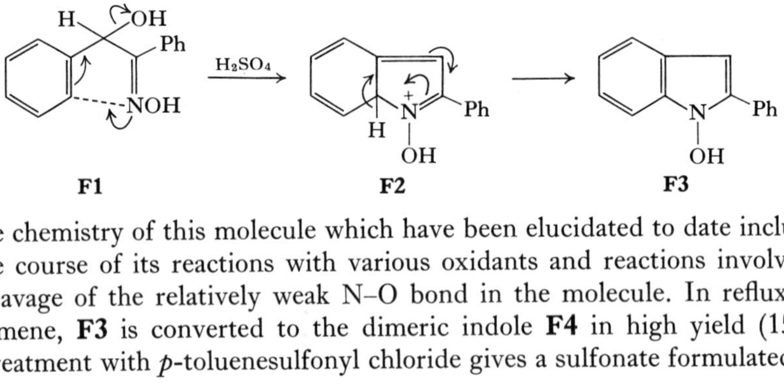

F1 $\xrightarrow{H_2SO_4}$ F2 $\longrightarrow$ F3

the chemistry of this molecule which have been elucidated to date include the course of its reactions with various oxidants and reactions involving cleavage of the relatively weak N–O bond in the molecule. In refluxing cymene, **F3** is converted to the dimeric indole **F4** in high yield (154). Treatment with *p*-toluenesulfonyl chloride gives a sulfonate formulated as

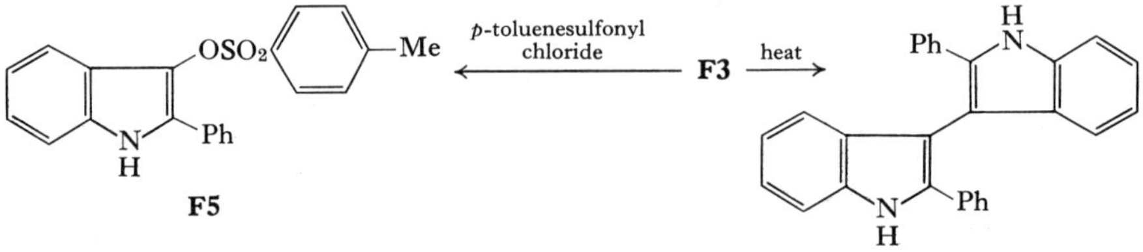

F5 $\xleftarrow{p\text{-toluenesulfonyl chloride}}$ F3 $\xrightarrow{heat}$ F4

F5. A rationale for formation of **F5** could involve the scheme **F3** → **F6** → **F7** → **F5**.

Colonna and co-workers (25–29) have studied the reactions of 1-hydroxy-2-phenylindole with a number of electrophilic and oxidizing reagents. Some of the reported reactions are summarized in the scheme on p. 374.

F3 —(p-toluenesulfonyl chloride)→ F6 → [F7] → F5

The interesting compounds **F11** and **F13** are believed to be diradicals (27) and they have been reported to have electron spin resonance spectra. These radicals have been found to act as oxidizing agents toward such easily oxidized materials as phenylhydrazine and iodide ion. The diradical **F11** is also reported to be the product of oxidation of **F3** with oxygen(photolytic) or *t*-butyl hydroperoxide (25–28). Reactions which are essentially normal electrophilic substitutions at C-3 have also been reported, as with *N*-phenylmaleimide (29) and the phenyldiazonium ion (26).

The reactions and spectral data reported to date seem to indicate that the aromatic system of 1-hydroxy-2-phenylindole is that of a typical indole, but that the presence of the N–OH group adds an additional facet to the chemistry of the molecule by permitting a family of reactions whose closest analogy is to be found in the reactions of arylhydroxylamines

Attempts to prepare the parent *N*-hydroxyindole led instead to a dimer formulated as **F16** (102). The synthetic route employed for attempted preparation of 1-hydroxindole, did, however, give 1-hydroxy-2-methylindole and 1-hydroxy-2-phenylindole (102). Studies of the nmr spectra of 1-hydroxy-2-methylindole have shown that a tautomeric equilibrium exists with the indolenine-*N*-oxide form **F18** (103). The equilibrium is solvent dependent and shifted toward **F18** by acidic solvents. It is reported that 5-bromo-1-hydroxindole-2-carboxylic acid exists entirely in the indolenine-*N*-oxide form in chloroform, and this is attributed to intramolecular hydrogen bonding (2).

Several reports concerning 1-hydroxyindole-2-carboxylic acid (**F20**) can be found in the literature. *o*-Nitrobenzylmalonic acid is converted by hot sodium hydroxide solution into 1-hydroxyindole-2-carboxylic acid (135). In a very similar type of reaction the substituted malonate **F21** gives 3-cyano-1-hydroxyindole-2-carboxylic acid (94). Similarly, treatment of

0.5 mole diethyl azodicarboxylate or 1 mole nitrosobenzene

PhN_2^+

3 moles nitrosobenzene

1 mole diethyl azodicarboxylate

NPh

F3 F8 F9 F10 F12 F13 F11

+ F11

F14 F15 F16

CH_2CHO NO_2

Zn, NH_4Cl

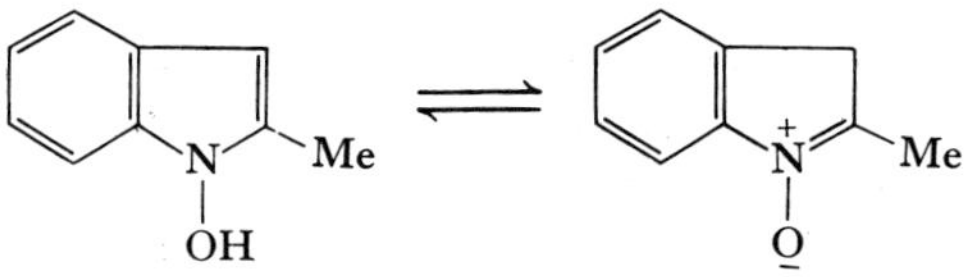

F17 ⇌ F18

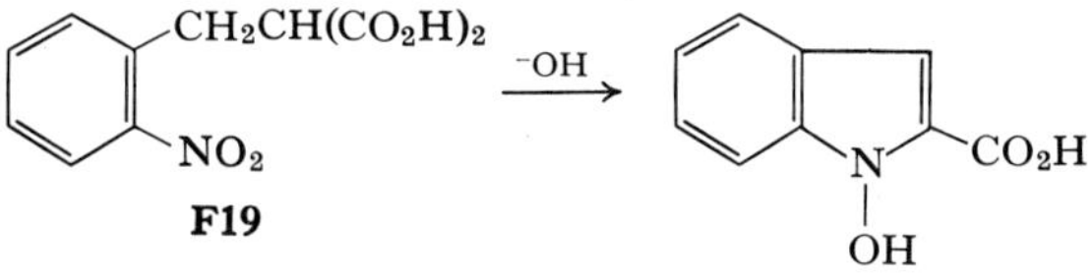

F19 $\xrightarrow{^{-}OH}$ F20

F21 $\xrightarrow{NaHCO_3}$ F22

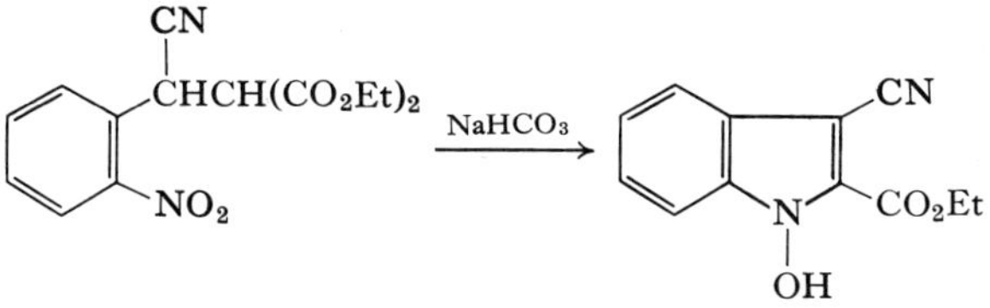

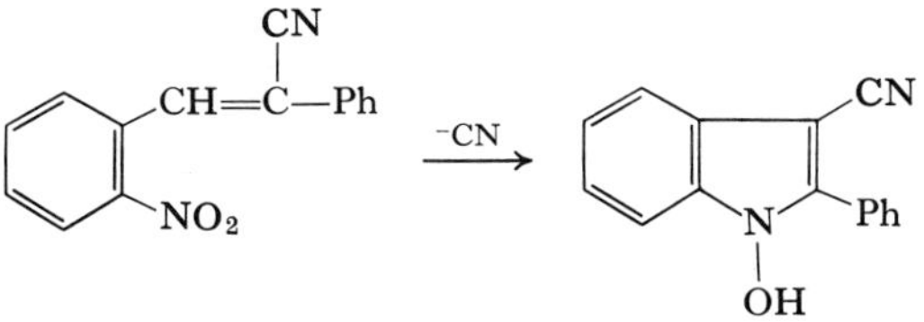

F23 $\xrightarrow{^{-}CN}$ F24

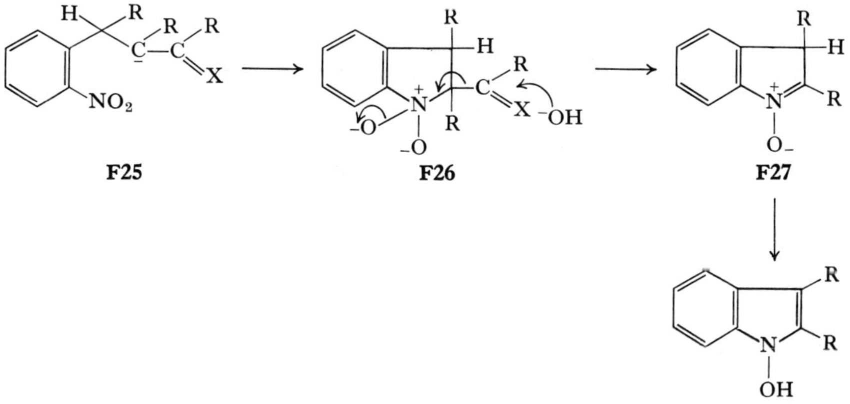

F25 → F26 → F27 → F28

F23 with potassium cyanide gives **F24** as one of the products (93). These cyclizations have been explained in terms of intramolecular condensation reactions of carbanions with nitro groups (93, 94). The condensations also might, however, be initiated by an electron-transfer process from the carbanion to the aromatic nitro compound (140).

Reduction of *o*-nitrophenylpyruvic acid with sodium amalgam provides another route to 1-hydroxyindole-2-carboxylic acid (2, 110, 113, 136). Catalytic reduction of oxime, semicarbazone, or phenylhydrazone derivatives of *o*-nitrophenylpyruvic acid gives varying yields of 1-hydroxyindole-2-carboxylic acid along with indole-2-carboxylic acid (12). The most extensive investigation of the chemistry of 1-hydroxyindole-2-carboxylic acid is that of Gabriel, Gerhard, and Wolter (40). Incompletely characterized bromo and chloro derivatives are reported but, more interestingly, oxidation with ferric chloride to **F30** is reported. Oxidation with potassium dichromate

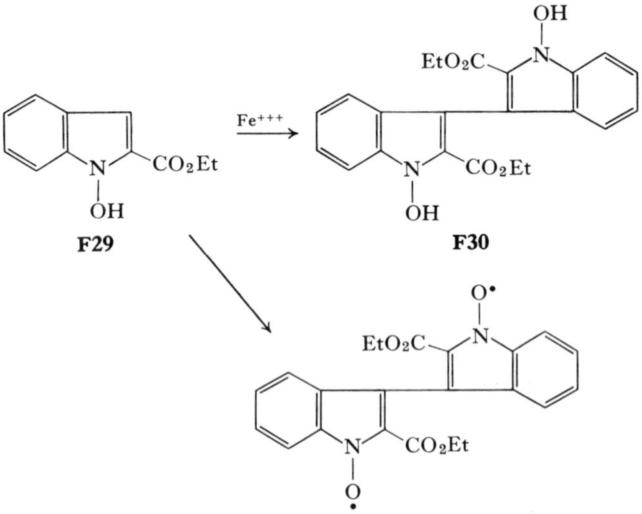

Fe^{+++}

F29

F30

F31

gives a highly colored product which is probably **F31**. To the extent that **F29** has been investigated, its properties seem to parallel those of 1-hydroxy-2-phenylindole.

N-Hydroxindoles can be *O*-alkylated with relative facility (2, 31, 110, 154). Early reports that indolylmagnesium bromide reacts with hydrogen peroxide to give 1-hydroxindole (62) have been discounted (77).

A report (59) that indole-3-acetic acid is oxidized to 1-hydroxyindole-3-acetic acid with ferric chloride in aqueous perchloric acid must be viewed with reservation in view of the tenuous characterization of the product.

A number of 1-hydroxyoxindoles have been prepared by reduction of substituted *o*-nitrophenylacetic acids with zinc and sulfuric acid (169). Several halogen and benzylidene derivatives have been prepared, but other aspects of the chemistry of these substances have not been investigated to date.

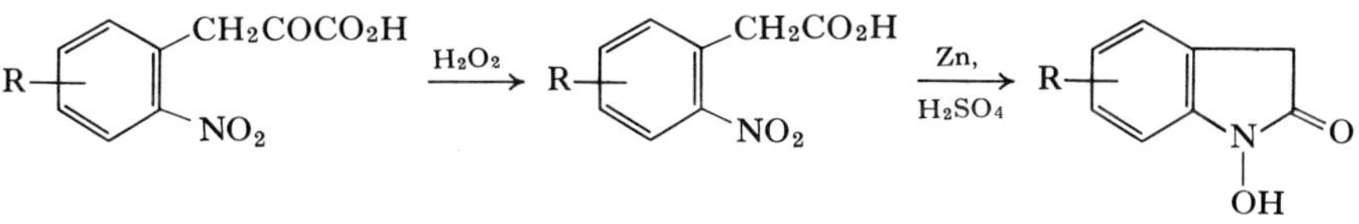

$$\text{R-C}_6\text{H}_3(\text{CH}_2\text{COCO}_2\text{H})(\text{NO}_2) \xrightarrow{H_2O_2} \text{R-C}_6\text{H}_3(\text{CH}_2\text{CO}_2\text{H})(\text{NO}_2) \xrightarrow[H_2SO_4]{Zn,} \text{R-(1-hydroxyoxindole, N-OH)}$$

F32 F33 F34

G. 2,3-INDOLINEDIONE (ISATIN) DERIVATIVES

Isatin, which can formally be considered to be the *o*-quinone of 2,3-dihydroxyindole, has been an important synthetic intermediate for a very long time. The C-3 carbonyl group is strongly electrophilic and isatins are readily converted, by condensation and addition reactions with carbanion-type nucleophiles into 3-substituted oxindoles. These compounds can often subsequently be converted to 3-substituted indoles by reductive processes.

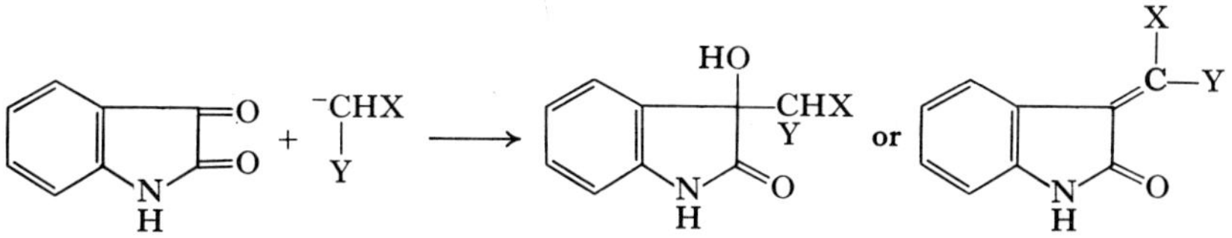

$$\text{G1 (isatin)} + {}^{-}\text{CHX(Y)}\ (\text{G2}) \longrightarrow \text{G3 (3-HO, 3-CHXY oxindole)} \text{ or } \text{G4 (3-(C(X)-Y)ylidene oxindole)}$$

G1 G2 G3 G4

Table L summarizes a number of recent carbanion addition and condensation reactions involving isatin.

Unsymmetrical condensation products can exist as geometric isomers. The configurations of such substances have been assigned only infrequently.

G5 G6

A recent paper by Autrey and Tahk considers in detail assignments for isatylideneacetic acids and related derivatives, principally on the basis of nmr chemical shifts (6). These workers deduce that in systems such as

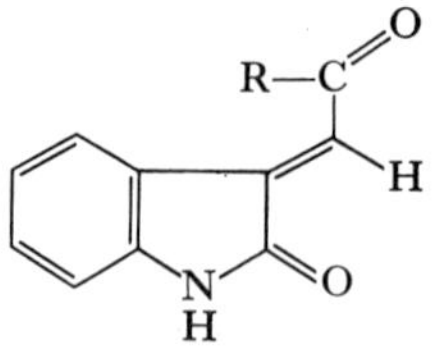

G7

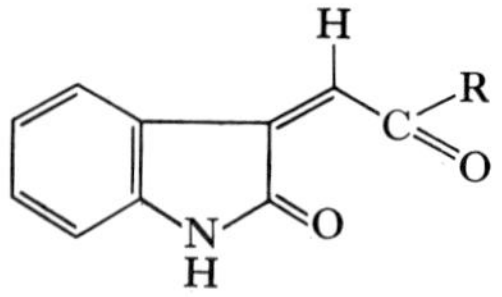

G8

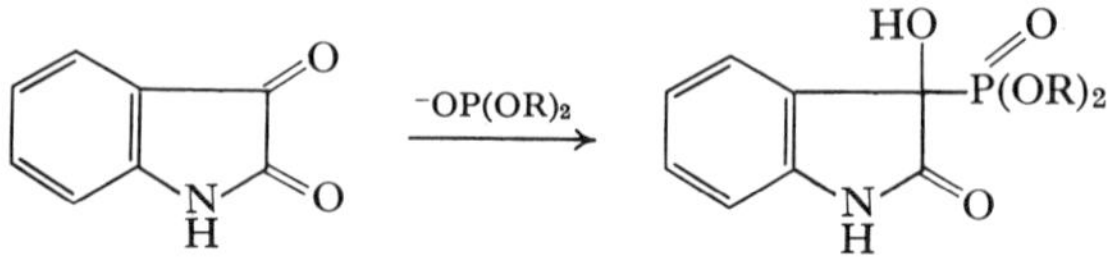

$\xrightarrow{^{-}OP(OR)_2}$

G9

G10

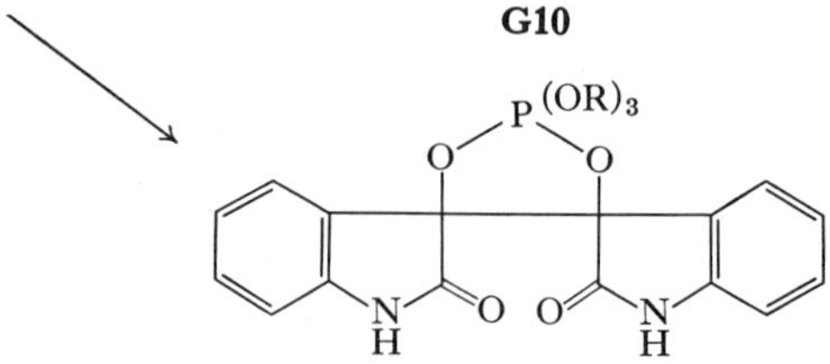

G11

TABLE L □ ADDITION AND CONDENSATION REACTIONS OF ISATINS WITH CARBON NUCLEOPHILES

Isatin substituent	Nucleophile derived from	Catalyst	Yield[a] (%)	Ref.
None	Ethyl pyruvate	Diethylamine	? (C)	(30)
None	Nitromethane	Diethylamine	71 (A)	(155)
None	Nitropropane	Diethylamine	74 (A)	(155)
None	2-Methylcyclohexanone	Piperidine	30 (A)	(5)
None	Acetophenone	Diethylamine	63 (A)	(90)
None	Benzyl cyanide	Diethylamine	85 (A)	(124)
None	Ethyl cyanoacetate	Piperidine	73 (C)	(47)
None	2-Methylpyridine	None	80 (A)	(4)
None	6-Methylnicotinic acid	None	? (C)	(4)
None	4-Phenylbut-3-enoic lactone	Piperidine	? (C)	(11)
None	2-Hydroxy-1-methyl-5-phenylpyrrole	Hydrochloric acid	60 (C)	(7)
1-Me	*p*-Bromopropiophenone	Diethylamine	97 (A)	(95)
1-Me	4-Phenylbut-3-enoic lactone	Acetic acid	75 (C)	(11)
1-Benzyl	Triethyl phosphonoacetate	Sodium hydride	75 (C)	(6)
5-Methoxy	Acetone	Diethylamine	? (A)	(122)
5-Methoxy-1-methyl	Acetophenone	Diethylamine	? (A)	(123)

[a] Notation (A) or (C) indicates addition or condensation product, respectively.

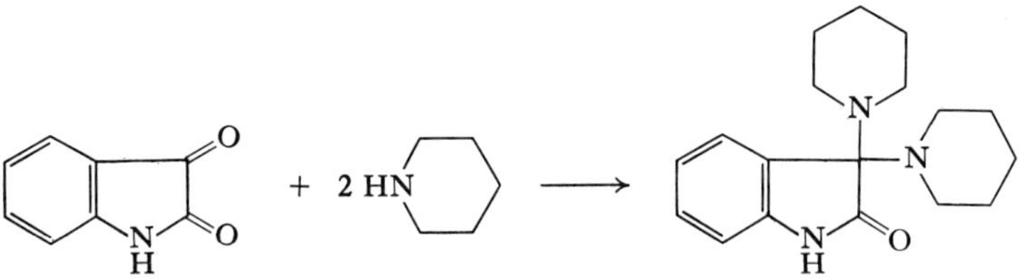

G12 + 2 HN G13 ⟶ G14

G7 and **G8**, a strong dipole-dipole interaction between carbonyl groups favors **G7**.

Nitrogen and phosphorus nucleophiles also react with isatin at C-3. With dialkylphosphites an unexceptional addition product **G10** is formed (104). With trialkyl phosphites the formation of the 2:1 adduct **G11** occurs (104). This reaction is quite characteristic of electron-deficient α-dicarbonyl compounds (134).

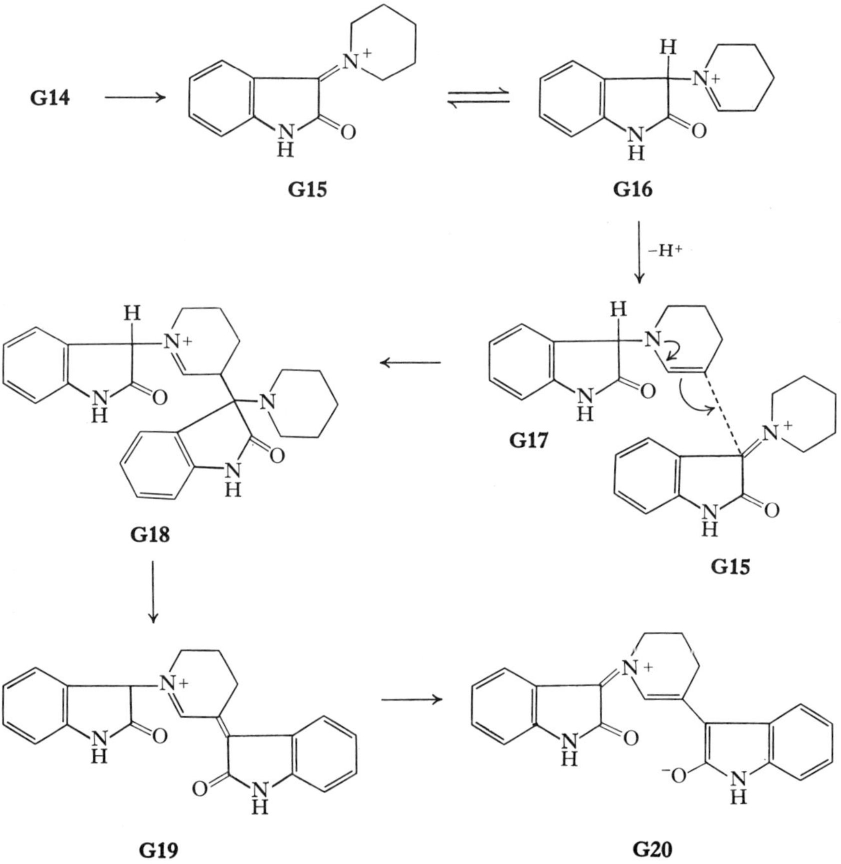

G14 ⟶ G15 ⇌ G16

G16 —$-H^+$→ G17

G17 + G15 ⟶ G18

G18 ⟶ G19 ⟶ G20

N-Acetylisatin condenses readily with aromatic amines at the C-3 carbonyl to give 3-iminooxindoles (114). Secondary amines such as piperidine and indoline react with isatin to give 3,3-diaminooxindoles (63). These condensation products react in hot acetic anhydride to give colored condensation products such as **G20**, probably by a mechanism such as that outlined above.

Isatin condenses with *p*-toluenesulfonylhydrazine and the resulting hydrazone on treatment with base gives 3-diazooxindole (22).

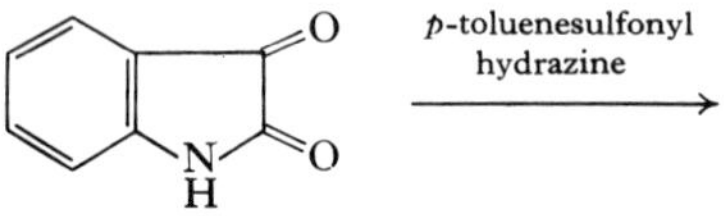

p-toluenesulfonyl hydrazine

G21

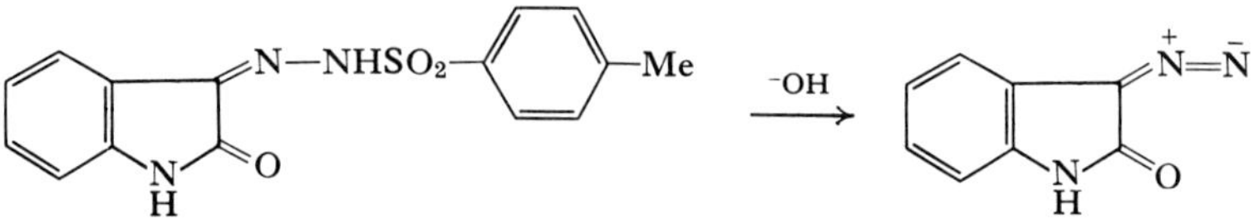

$N{-}NHSO_2{-}C_6H_4{-}Me$ $\xrightarrow{^-OH}$ $\overset{+}{N}{=}\overset{-}{N}$

G22 **G23**

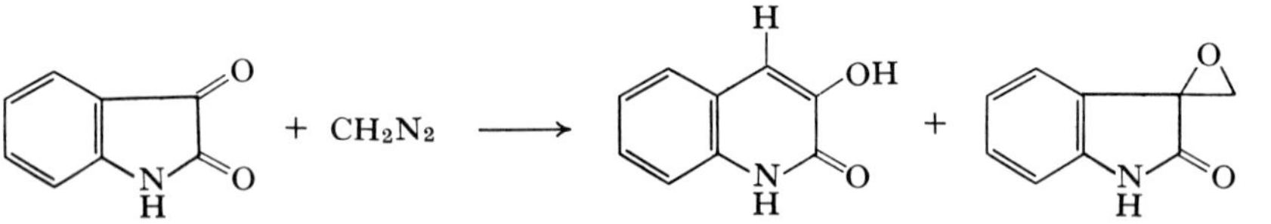

$+ \; CH_2N_2 \longrightarrow$ +

G24 **G25** **G26**

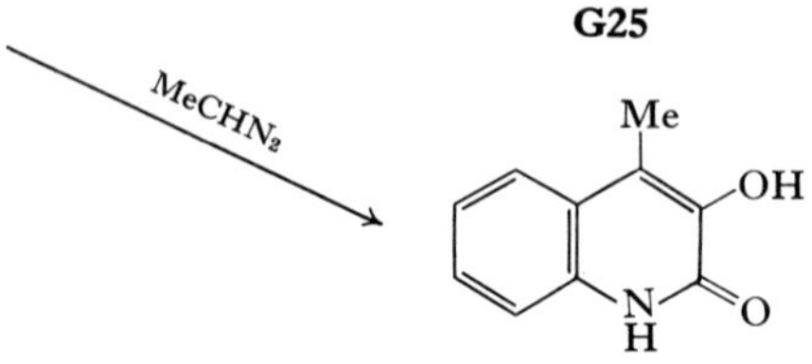

$MeCHN_2$

G27

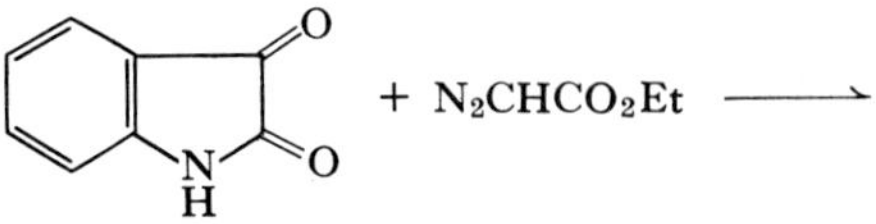

$+ \; N_2CHCO_2Et \longrightarrow$

G28 **G29**

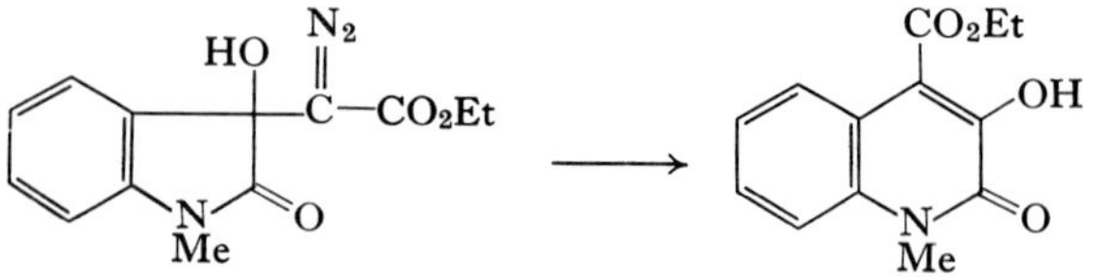

HO N_2 $C{-}CO_2Et$ Me $\longrightarrow$ CO_2Et OH Me

G30 **G31**

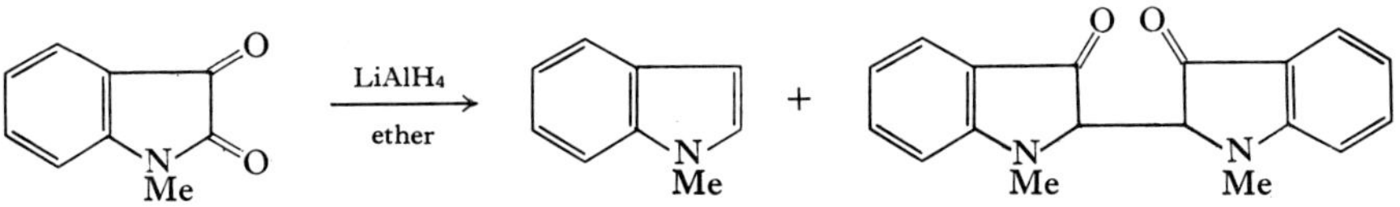

LiAlH4 / ether

G32 → G33 (50%) + G34 (12%)

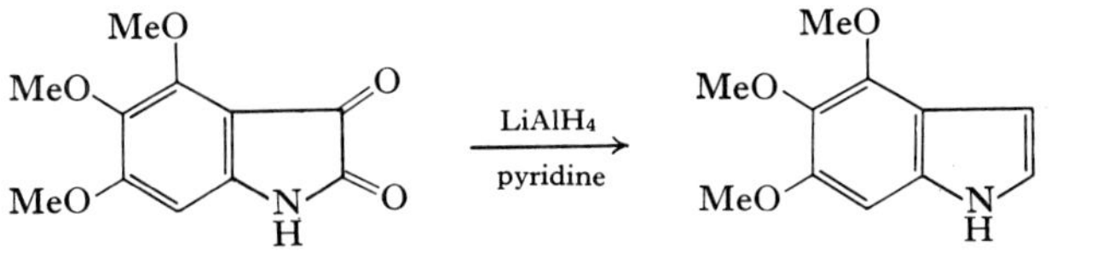

LiAlH4 / pyridine

G35 → G36 (47%) Ref. (20)

Isatins undergo ring expansion with diazoalkanes in a reaction presumably initiated by nucleophilic attack at C-3 (32). An intermediate addition product, presumably **G30**, was isolated in the case of ethyl diazoacetate.

Reduction of isatin derivatives with lithium aluminum hydride usually gives the corresponding indole as the principal product. Some typical results are illustrated in the equations which follow. Coupling to indigo derivatives

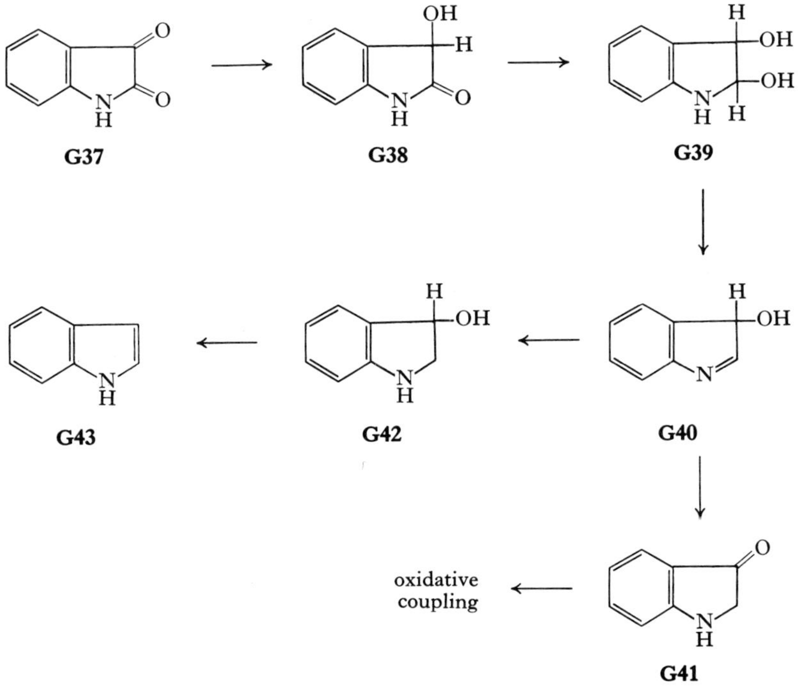

G37 → G38 → G39 → G40 → G42 → G43

G40 → G41 → oxidative coupling

is often noted and under some conditions this becomes the dominant reacton (44, 50). A 21% yield of 3-hydroxyindoline has been reported, under certain conditions, in the reduction of isatin (42). Indole (27%) and coupling products are also found. The formation of each of these types of products can be readily rationalized although no information concerning the sequence of steps is available. Hydrogen sulfide reduces isatin to a dihydrodimer which has been assigned structure **G44** (16).

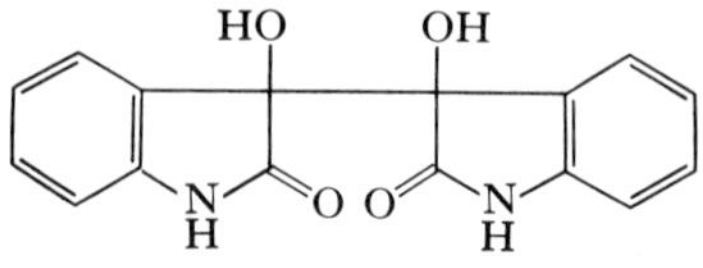
HO
OH
N
H
O
O
N
H

G44

The reaction of anilines with chloral and hydroxylamine to give oximes of glyoxylanilides followed by cyclization with sulfuric acid (96) remains the most versatile synthesis of isatins. Mix and Krause (99) have reported a number of typical examples of this reaction.

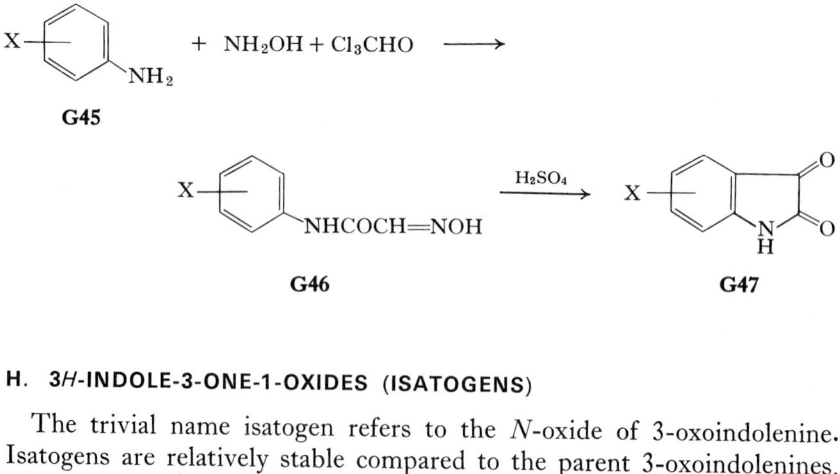
X
NH2
+ NH2OH + Cl3CHO
X
NHCOCH=NOH
H2SO4
X
O
N
H
O

G45

G46

G47

H. 3*H*-INDOLE-3-ONE-1-OXIDES (ISATOGENS)

The trivial name isatogen refers to the *N*-oxide of 3-oxoindolenine. Isatogens are relatively stable compared to the parent 3-oxoindolenines. This stability can be attributed to resonance involving the oxide substituent.

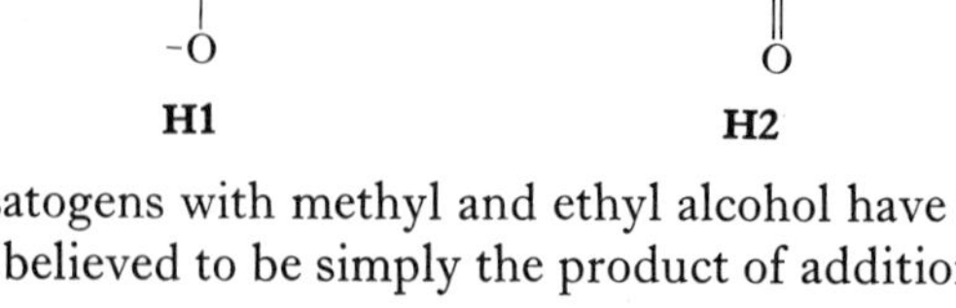
O
+
N
R
-O
O⁻
+
N
R
O

H1 **H2**

Adducts of isatogens with methyl and ethyl alcohol have been described (51). These are believed to be simply the product of addition of the alcohol to the $C{=}\overset{+}{N}{-}\overset{-}{O}$ system (51). A ring-expansion effected by ammonia may be initiated by a similar addition (107). Isatogens are converted by acids into

H3
NH_3
O
Ph
N
NH_2
H—O
H4
O
Ph
NH_2
N
O
H5
O
Ph
N—N—H
HO
H6
$-H_2$
OH
Ph
$\overset{+}{N}$=N
^-O
H7

anthranil derivatives (125). A logical reaction mechanism begins with nucleophilic addition of water to C-2 of the isatogen ring.

COPh
H^+, H_2O
H3
O
N
H8
O
Ph
N
OH
H—O
H9
O
Ph
O
NH
HO
H10
HO
CO—Ph
O
N
H
H11

Isatogens are reported to react with ethyl cyanoacetate to give isoxazoline rings (19a). Again, the initial step of the reaction would appear to involve nucleophilic attack at C-2 of the ring.

Several reactions of isatogens with unsaturated compounds appear to proceed by 1,3-dipolar cycloaddition. The reactions of 2-phenylisatogen with acrylonitrile and nitroethylene fit this pattern (108). Under solvolytic conditions the adducts undergo ring opening leading to 3-indolinones.

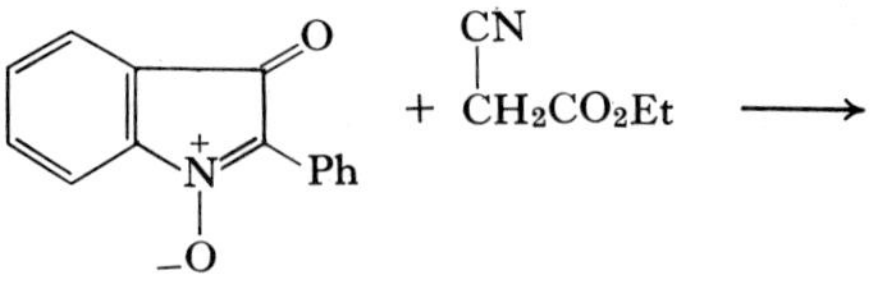

O, $\overset{+}{N}$, ^{-}O, Ph + CN, CH_2CO_2Et ⟶

H3

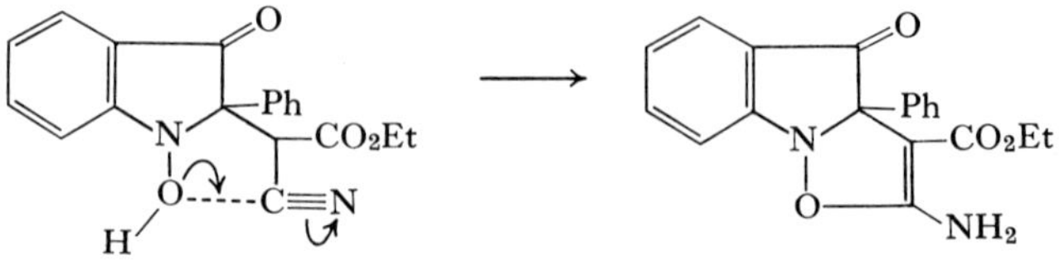

O, Ph, CO_2Et, N, O, H, C≡N ⟶ O, Ph, CO_2Et, N, O, NH_2

H12 H13

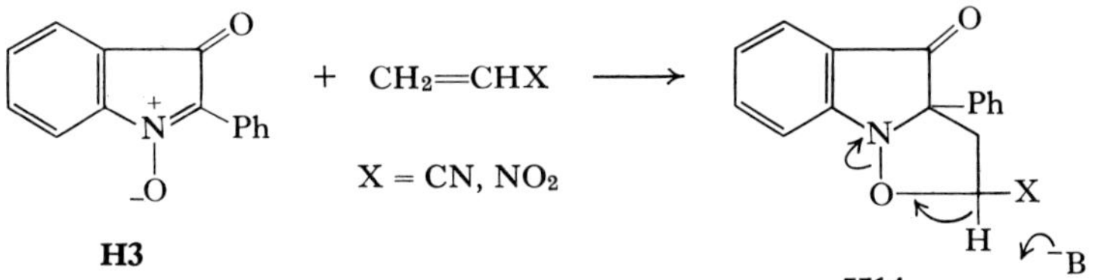

O, $\overset{+}{N}$, ^{-}O, Ph + $CH_2{=}CHX$ ⟶ O, Ph, N, O, X, H, ^{-}B

X = CN, NO_2

H3 H14

↓

O, Ph, N, H, CH_2CO_2R $\xleftarrow{ROH}$ O, Ph, N, H, $CH_2C{-}X$, O

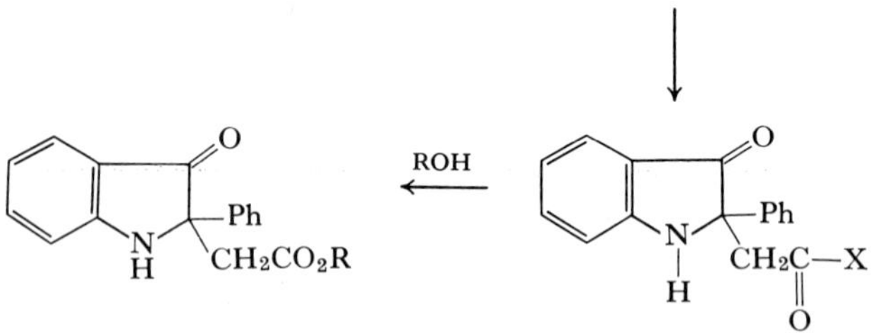

H16 H15

The reaction of 2-phenylisatogen with tetracyanoethylene or trichloroacetonitrile leads to a ring expansion and formation of 3,4-dihydro-2-phenyl-4-quinazolone (107). A related carbon insertion reaction occurs with arylacetylenes (107), leading, for example, to formation of 3-phenyl-4-quinolinol from 2-phenylisatogen and phenylacetylene (107). There is evidence that these ring expansions occur via intermediates formed by dipolar cycloadditions (99a).

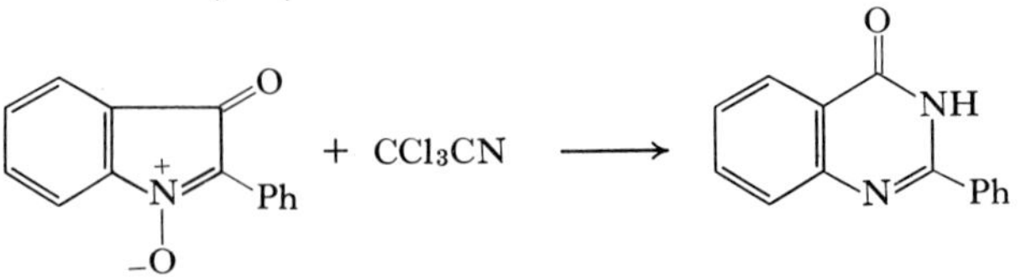

O, $\overset{+}{N}$, ^{-}O, Ph + CCl_3CN ⟶ O, NH, N, Ph

H3 H17

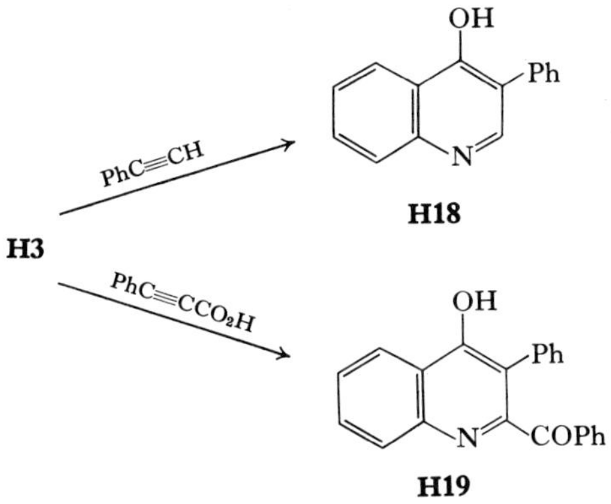

Isatogens have most often been prepared from *o*-substituted nitro compounds by base-catalyzed or photochemical condensation reactions involving the nitro group. One of the most general procedures is that of Kröhnke (86). Benzylpyridinium salts are condensed with *o*-nitrobenzaldehyde. Aqueous solutions of weakly basic catalysts such as diethylamine catalyze the cyclization. The conjugate addition of hydroxide to the vinylpyridinium

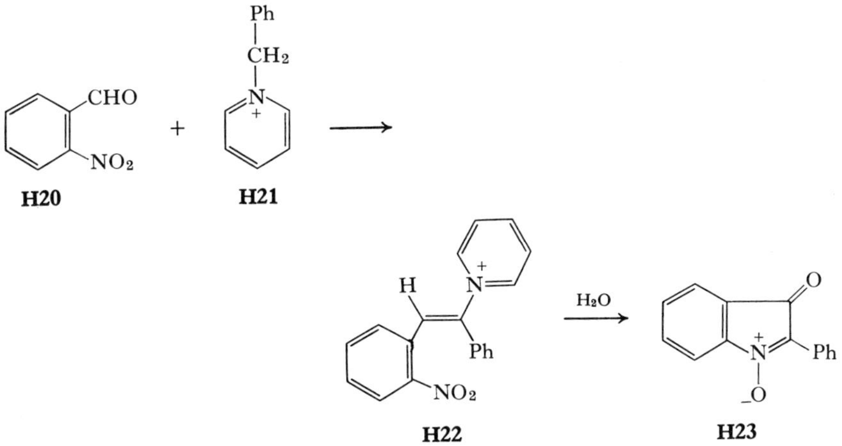

salt could possibly initiate intramolecular condensation with the nitro group. The fact that the aldol condensation product **H26** reverts to starting materials rather than cyclizing on treatment with base argues against this mechanism. Alternatively, the conjugate base of **H22** may be the intermediate which initiates cyclization (61, 86). The Kröhnke procedure has been extended to 2-vinyl and 2-styrylisatogens by use of allyl and cinnamylpyridinium salts as starting materials (88).

H22 → [OH, H, N+, $^-$O, N, $_-$O, Ph] H24 → [OH, N, Ph, $_-$O, $_-$O] H25 → H23

H22 →(?) [HO, H, C, N+, H, Ph, NO_2] H26 → H20 + H21

H22 → [N+, Ph, NO_2] H27 → [Ph, N+, O, N:, $^-$O] H28 → [$^-$O, N+, Ph, N, O] H29 →(– pyridine) H23

[C≡CPh, O_2N, NO_2] H30 →($h\nu$) [O, O_2N, N, Ph, $_-$O] H31 Ref. (121)

[Ph, Me, CHBr—CHBr, NC, NO_2] H32 →(pyridine) [Me, O, NC, N, Ph, $_-$O] H33 Ref. (23)

[HO, H, N, CH, Ph, NO_2] H34 →($h\nu$) [O, N, Ph, $_-$O] H35 Ref. (87)

It has been known for a long time that cyclization of *o*-nitrophenylacetylenes to isatogens can be effected by light (120). Systems which can, by elimination, generate *o*-nitrophenylacetylenes are also generally susceptible to this reaction but the available evidence suggests that the acetylenes are not necessarily intermediates (87). The examples on p. 386 are typical.

The following scheme can rationalize the formation of isatogens in photochemical reactions:

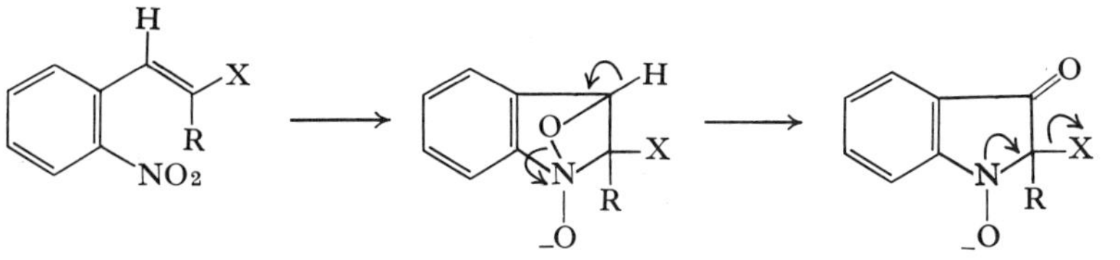

H36 H37 H38 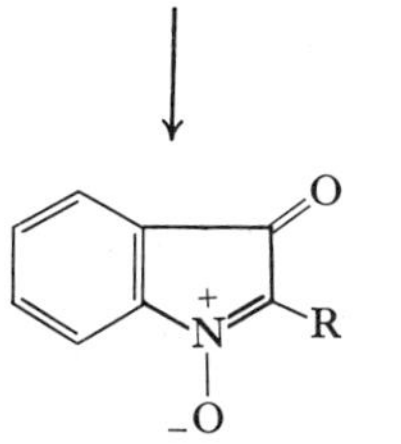 H39

It may be that the *o*-nitrophenylacetylene systems must undergo prior nucleophilic addition since an intermediate corresponding to **H37** cannot be formed directly from a nitrophenylacetylene without violation of Bredt's rule.

Isatogens are also formed when certain *o*-nitrostilbenes are irradiated (146). The minimum requirement for isatogen formation is the presence of an electron-releasing substituent in the other phenyl ring. The existence of intermediates in this reaction has been demonstrated, but a fully satisfactory mechanism has not yet been advanced.

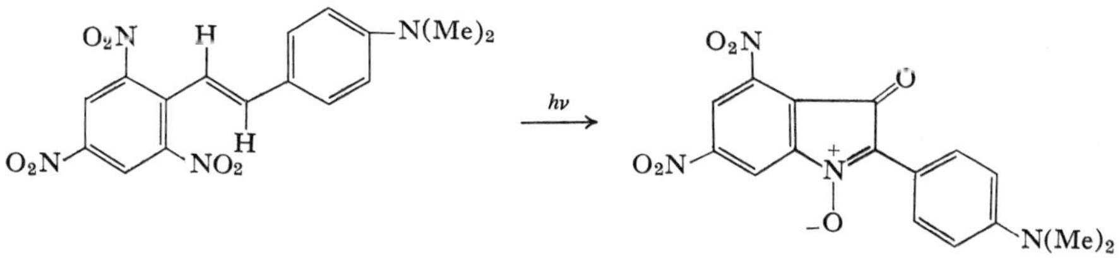

H40 $\xrightarrow{h\nu}$ H41

The parent 3*H*-indole-3-one ring system is quite unstable. 2-Aryl substituents stabilize the system and several 2-aryl-3*H*-indole-3-one derivatives have been prepared by oxidation of 3-aminoindoles to the corresponding imines followed by hydrolysis (23a, 137a). Studies on the chemistry of these compounds are just beginning to appear (23a, 137a)

H42 → **H43** → **H44**

REFERENCES

1. Abramovitch, R. A., and Hey, D. H., *J. Chem. Soc.* p. 1697 (1954).
1a. Abramovitch, R. A., and Marko, A. M., *Can. J. Chem.* **38**, 131 (1960).
2. Acheson, R. M., Brookes, C. J. Q., Dearnaley, D. P., and Quest, B., *J. Chem. Soc., C* p. 504 (1968).
3. Aeberli, P., and Houlihan, W. J., *J. Org. Chem.* **33**, 1640 (1968).
4. Akkerman, A. M., and Veldstra, H., *Rec. Trav. Chim.* **73**, 629 (1954).
5. Angyal, S. J., Bullock, E., Hanger, W. G., Howell, W. C., and Johnson, A. W., *J. Chem. Soc.* p. 1592 (1957).
6. Autrey, R. L., and Tahk, F. C., *Tetrahedron* **23**, 901 (1967).
7. Ballantine, J. A., Beer, R. J. S., Crutchley, D. J., Dodd, G. M., and Palmer, D. R., *J. Chem. Soc.* p. 2292 (1960).
8. Ballantine, J. A., Beer, R. J. S., and Robertson, A., *J. Chem. Soc.* p. 4779 (1958).
9. Bamfield, P., Johnson, A. W., and Datner, A. S., *J. Chem. Soc. C* p. 1028 (1966).
10. Ban, Y., and Oishi, T., *Chem. & Pharm. Bull (Tokyo)* **11**, 441 (1963).
11. Barrett, C. B., Beer, R. J. S., Dodd, G. M., and Robertson, A., *J. Chem. Soc.* p. 4810 (1957).
12. Baxter, I., and Swan, G. A., *J. Chem. Soc., C*, p. 2446 (1967).
12a. Beckett, A. H., Daisley, R. W., and Walker, J., *Tetrahedron* **24**, 6093 (1968).
13. Behringer, H., and Weissauer, H., *Chem. Ber.* **85**, 743 (1952).
14. Behringer, H., and Weissauer, H., *Chem. Ber.* **85**, 774 (1952).
15. Bell, M. R., Oesterlin, R., Beyler, A. L., Harding, H. R., and Potts, G. O., *J. Med. Chem.* **10**, 264 (1967).
16. Bergmann, E. D., *J. Am. Chem. Soc.* **77**, 1549 (1955).
17. Bird, C. W., *J. Chem. Soc.* p. 674 (1963).
18. Bruce, J. M., and Sutcliffe, F. K., *J. Chem. Soc.* p. 4789 (1957).
19. Brunner, K., *Monatsh. Chem.* **17**, 479 (1896).
19a. Bunney, J. E., and Hooper, M., *Tetrahedron Letters* p. 3857 (1966).
20. Carlsson, A., Corrodi, H., and Magnusson, T., *Helv. Chim. Acta* **46**, 1231 (1963).
21. Carson, D. F., and Mann, F. G., *J. Chem. Soc.* p. 5819 (1965).
22. Cava, M. P., Little, R. L., and Napier, D. R., *J. Am. Chem. Soc.* **80**. 2257 (1958).
23. Chardonnens, L., and Kramer, W. J., *J. Am. Chem. Soc.* **79**, 4955 (1957).
23a. Ch'ng, H. S., and Hooper, M., *Tetrahedron Letters* p. 1527 (1969).

24. Clark-Lewis, J. W., and Edgar, J. A., *J. Chem. Soc.* p. 5551 (1965).
25. Colonna, M., and Bruni, P., *Ric. Sci. Rend* [2] **A4**, 151 (1964); *Chem. Abstr.* **61**, 633 (1964).
26. Colonna, M., and Bruni, P., *Gazz. Chim. Ital.* **94**, 1448 (1964).
27. Colonna, M., and Bruni, P., *Gazz. Chim. Ital.* **95**, 1172 (1965).
28. Colonna, M., and De Martino, U., *Gazz. Chim. Ital.* **93**, 1183 (1963).
29. Colonna, M., and Monti, A., *Gazz. Chim. Ital.* **92**, 1401 (1962).
30. Cornforth, J. W., Cornforth, R. H., Dalgliesh, C. E., and Neuberger, A., *Biochem. J.* **48**, 591 (1951); *Chem. Abstr.* **46**, 104 (1952).
30a. Daisley, R. W., and Walker, J., *J. Chem. Soc. B* p. 146 (1969).
31. de Stevens, G., Brown, A. B., Rose, D., Chernov, H. I., and Plummer, A. J., *J. Med. Chem.* **10**, 211 (1967).
32. Eistert, B., and Selzer, H., *Chem. Ber.* **96**, 1234 (1963).
33. Elderfield, R. C., and Rembges, H. H., *J. Org. Chem.* **32**, 3809 (1967).
34. Elliott, I. W., and Rivers, P., *J. Org. Chem.* **29**, 2438 (1964).
35. Endler, A. S., and Becker, E. I., *Org. Syn.* **37**, 60 (1957).
35a. Étienne, A., *Bull. Soc. Chim. France* p. 651 (1948).
36. Ficken, G. E., and Kendall, J. D., *J. Chem. Soc.* p. 3988 (1959).
37. Fischer, E., *Chem. Ber.* **29**, 2062 (1896).
38. Fischer, E., and Hütz, H., *Chem. Ber.* **28**, 585 (1895).
39. Freter, K., Weissbach, H., Redfield, B., Udenfriend, S., and Witkop, B., *J. Am. Chem. Soc.* **80**, 983 (1958).
40. Gabriel, S., Gerhard, W., and Wolter, R., *Chem. Ber.* **56**, 1024 (1923).
41. Giovannini, E., Farkas, F., and Rosales, J., *Helv. Chim. Acta* **46**, 1326 (1963).
42. Giovannini, E., and Lorenz, T., *Helv. Chim. Acta* **40**, 1553 (1957).
43. Giovannini, E., and Lorenz, T., *Helv. Chim. Acta* **40**, 2287 (1957).
44. Giovannini, E., and Lorenz, T., *Helv. Chim. Acta* **41**, 113 (1958).
45. Gruda, I., *Roczniki Chem.* **40**, 1323 (1966); *Chem. Abstr.* **66**, 65352 (1967).
46. Hardegger, H., and Corrodi, H., *Helv. Chim. Acta* **39**, 514 (1956).
47. Harley-Mason, J., and Ingleby, R. F. J., *J. Chem. Soc.* p. 3639 (1958).
48. Harley-Mason, J., and Ingleby, R. F. J., *J. Chem. Soc.* p. 4782 (1958).
49. Harley-Mason, J., and Leeney, T. J., *Proc. Chem. Soc.* p. 368 (1964).
50. Hassner, A., and Haddadin, M. J., *J. Org. Chem.* **28**, 224 (1963).
51. Heller, G., and Boessneck, W., *Chem. Ber.* **55**, 474 (1922).
52. Hellmann, H., and Renz, E., *Chem. Ber.* **84**, 901 (1951).
53. Hendrickson, J. B., and Silva, R. A., *J. Am. Chem. Soc.* **84**, 643 (1962).
54. Hinman, R. L., and Bauman, C. P., *J. Org. Chem.* **29**, 1206 (1964).
55. Hinman, R. L., and Bauman, C. P., *J. Org. Chem.* **29**, 2431 (1964).
56. Holt, S. J., Kellie, A. E., O'Sullivan, D. G., and Sadler, P. W., *J. Chem. Soc.* p.1217 (1958).
57. Holt, S. J., and Petrow, V., *J. Chem. Soc.* p. 607 (1947).
58. Horning, E. C., and Rutenberg, M. W., *J. Am. Chem. Soc.* **72**, 3534 (1950).
59. Houff, W. H., Hinsvark, O. N., Weller, L. E., Wittwer, S. H., and Sell, H. M., *J. Am. Chem. Soc.* **76**, 5654 (1954).
59a. Houghton, E., and Saxton, J. E., *J. Chem. Soc., C* p. 595 (1969).
60. House, H. O., "Modern Synthetic Reactions," pp. 124–125. Benjamin, New York 1965.
61. Huisgen, R., *Angew. Chem.* **75**, 604 (1963).
62. Ingraffia, F., *Gazz. Chim. Ital.* **63**, 175 (1933).
63. Johnson, A. W., and McCaldin, D. J., *J. Chem. Soc.* p. 3470 (1957).

64. Johnson, J. R., and Andreen, J. H., *J. Am. Chem. Soc.* **72**, 2862 (1950).
65. Julian, P. L., Dailey, E. E., Printy, H. C., Cohen, H. L., and Hamashige, S., *J. Am. Chem. Soc.* **78**, 3503 (1956).
66. Julian, P. L., Magnani, A., Pikl, J., and Karpel, W. J., *J. Am. Chem. Soc.* **70**, 174 (1948).
67. Julian, P. L., Meyer, E. W., and Printy, H. C., *in* "Heterocyclic Compounds" (R. C. Elderfield, ed.), Vol. 3, p. 186. Wiley, New York, 1952.
68. Julian, P. L., and Pikl, J., *J. Am. Chem. Soc.* **57**, 539 563, and 755 (1935).
69. Julian, P. L., Pikl, J., and Boggess, D., *J. Am. Chem. Soc.* **56**, 1797 (1934).
70. Julian, P. L., Pikl, J., and Wantz, F. E., *J. Am. Chem. Soc.* **57**, 2026 (1935).
71. Julian, P. L., and Printy, H. C., *J. Am. Chem. Soc.* **71**, 3206 (1949).
72. Julian, P. L., and Printy, H. C., *J. Am. Chem. Soc.* **75**, 5301 (1953).
73. Julian, P. L., Printy, H. C., and Dailey, E. E., *J. Am. Chem. Soc.* **78**, 3501 (1956).
74. Kalb, L., and Bayer, J., *Chem. Ber.* **45**, 2150 (1912).
75. Kambli, E., *Helv. Chim. Acta* **47**, 2155 (1964).
75a. Kato, T., Yamanaka, H., and Ichikawa, H., *Chem. & Pharm. Bull.* (*Tokyo*) **17**, 481 (1969).
76. Katritzsky, A. R., and Lagowski, J. M., *Advan. Heterocyclic Chem.* **2**, 1 (1963).
77. Kawana, M., Yoshioka, M., Miyaji, S., Kataoka, H., Omote, Y., and Sugiyama, N., *Nippon Kagaku Zasshi* **86**, 526 (1965); *Chem. Abstr.* **63**, 11479 (1965).
78. Kendall, E. C., and Osterberg, A. E., *J. Am. Chem. Soc.* **49**, 2047 (1927).
79. Kendall, J. D., and Ficken, G. E., British Patent 811,876 (1959); *Chem. Abstr.* **53**, 21306 (1959).
80. Kirby, G. W., and Shah, S. W., *Chem. Commun.* p. 381 (1965).
81. Kisteneva, M. S., *Zh. Obshch. Khim.* **26**, 1169 and 2019 (1956); *Chem. Abstr.* **50**, 16747 (1956); **51**, 5044 (1957).
82. Klamann, D., Wache, H., Ulm, K., and Nerdel, F., *Chem. Ber.* **100**, 1870 (1967).
83. Kolosov, M. N., Metreveli, L. I., and Preobrazhenskaya, N. A., *Zh. Obshch. Khim.* **23**, 2027 (1953); *Chem. Abstr.* **49**, 3208 (1955).
84. Kolosov, M. N., and Preobrazhenskaya, N. A., *Zh. Obshch. Khim.* **23**, 1779 (1953); *Chem. Abstr.* **49**, 295 (1955).
85. Kretz, E., Müller, J. M., and Schlittler, E., *Helv. Chim. Acta* **35**, 520 (1952).
86. Kröhnke, F., and Meyer-Delius, M., *Chem. Ber.* **84**, 932 (1951).
87. Kröhnke, F., and Vogt, I., *Chem. Ber.* **85**, 376 (1952).
88. Kröhnke, F., Kröhnke, G., and Vogt, I., *Chem. Ber.* **86**, 1500 (1953).
89. Lawson, W. B., Patchornik, A., and Witkop, B., *J. Am. Chem. Soc.* **82**, 5918 (1960).
90. Lindwall, H. G., and Maclennan, J. S., *J. Am. Chem. Soc.* **54**, 4739 (1932).
91. Longmore, R. B., and Robinson, B., *Chem. & Ind.* (*London*) p. 1297 (1965).
92. Longmore, R. B., and Robinson, B., *Collection Czech. Chem. Commun.* **32**, 2184 (1967).
93. Loudon, J. D., and Tennant, G., *J. Chem. Soc.* p. 3466 (1960).
94. Loudon, J. D., and Wellings, I., *J. Chem. Soc.* p. 3462 (1960).
95. Lutz, R. E., and Clark, C. T., *J. Org. Chem.* **25**, 193 (1960).
96. Marvel, C. S., and Hieirs, G. S., "Organic Syntheses" (H. Gilman and A. H. Blatt, eds.), 2nd ed., Coll. Vol. I, pp. 327–330. Wiley, New York, 1952.
97. Meyer, R. F., *J. Org. Chem.* **30**, 3451 (1965).
98. Mills, B., and Schofield, K., *J. Chem. Soc.* p. 5558 (1961).
99. Mix, H., and Krause, H. W., *Chem. Ber.* **89**, 2630 (1956).
99a. Modler, R. F., Ph.D. Thesis with W. E. Noland, University of Minnesota (1965); *Dissertation Abstr.* **26**, 4244 (1966).
100. Monti, S. A., Johnson, W. O., and White, D. H., *Tetrahedron Letters* p. 4459 (1966).
101. Moore, R. F., and Plant, S. G. P., *J. Chem. Soc.* p. 3475 (1951).

102. Mousseron-Canet, M., and Boca, J.-P., *Bull. Soc. Chim. France* p. 1296 (1967).
103. Mousseron-Canet, M., Boca, J.-P., and Tabacik, V., *Spectrochim. Acta* **23A**, 717 (1967).
104. Mustafa, A., Sidky, M. M., and Soliman, F. M., *Tetrahedron* **22**, 393 (1966).
105. Neil, A. B., *J. Am. Chem. Soc.* **75**, 1508 (1953).
106. Nenitzescu, C. D., and Răileanu, D., *Chem. Ber.* **91**, 1141 (1958).
107. Noland, W. E., and Jones, D. A., *J. Org. Chem.* **27**, 341 (1962); Noland, W. E., and Modler, R. F., *J. Am. Chem. Soc.* **86**, 2086 (1964).
108. Noland, W. E., and Jones, D. A., *Chem. & Ind. (London)* p. 363 (1962).
109. Oishi, T., Nagai, M., and Ban, Y., *Tetrahedron Letters* p. 491 (1968).
110. Omote, Y., Fukada, N., and Sugiyama, N., *Bull. Chem. Soc. Japan* **40**, 2703 (1967).
111. Palazzo, G., and Rosnati, V., *Gazz. Chim. Ital.* **82**, 584 (1952).
112. Palazzo, G., and Rosnati, V., *Gazz. Chim. Ital.* **83**, 211 (1953).
113. Pappalardo, G., and Vitali, T., *Gazz. Chim. Ital.* **88**, 574 (1958).
114. Parisi, F., *J. Am. Chem. Soc.* **75**, 3848 (1953).
114a. Patterson, D. A., and Wibberley, D. G., *J. Chem. Soc.* p. 1706 (1965).
115. Petyunin, P. A., *Zh. Obshch. Khim.* **22**, 296 (1952); *Chem. Abstr.* **46**, 11161 (1952).
116. Petyunin, P. A., and Berdinskaya, I. S., *Zh. Obshch. Khim.* **21**, 1703, 1859, and 2016 (1951); *Chem. Abstr.* **46**, 6638 and 8070 (1952).
117. Petyunin, P. A., Berdindskaya, I. S., and Panferova, N. G., *Zh. Obshch. Khim.* **27**, 1901 (1957); *Chem. Abstr.* **52**, 4647 (1958).
118. Petyunin, P. A., and Pesis, A. S., *Zh. Obshch. Khim.* **26**, 223 (1956); *Chem. Abstr.* **50**, 13935 (1956).
119. Petyunin, P. A., and Shklyaev, V. S., *Zh. Obshch. Khim.* **27**, 731 (1957); *Chem. Abstr.* **51**, 15523 (1957).
120. Pfeiffer, P., *Ann. Chem.* **411**, 72 (1916).
121. Pfeiffer, P., and Kramer, E., *Chem. Ber.* **46**, 3655 (1913).
122. Pietra, S., and Tacconi, G., *Farmaco (Pavia), Ed. Sci.* **13**, 893 (1958); *Chem. Abstr.* **53**, 21875 (1959).
123. Pietra, S., and Tacconi, G., *Farmaco (Pavia), Ed. Sci.* **14**, 854 (1959); *Chem. Abstr.* **54**, 9880 (1960).
124. Pietra, S., and Tacconi, G., *Gazz. Chim. Ital.* **89**, 2304 (1959); *Chem. Abstr.* **55**, 5452 (1961).
125. Pinkus, J. L., Woodyard, G. G., and Cohen, T., *J. Org. Chem.* **30**, 1104 (1965).
126. Piper, J. R., and Stevens, F. J., *J. Org. Chem.* **27**, 3134 (1962).
127. Plieninger, H., and Bauer, H., *Angew. Chem.* **73**, 433 (1961).
128. Plieninger, H., Bauer, H., Bühler, W., Kurze, J., and Lerch, U., *Ann. Chem.* **680**, 69 (1964).
129. Plieninger, H., and Castro, C. E., *Chem. Ber.* **87**, 1760 (1954).
129a. Plieninger, H., and Herzog, H., *Monatsh. Chem.* **98**, 807 (1967).
130. Plieninger, H., and Werst, G., *Angew. Chem.* **70**, 272 (1958).
131. Plieninger, H., and Wild, D., *Chem. Ber.* **99**, 3063 (1966).
132. Plieninger, H., and Wild, D., *Chem. Ber.* **99**, 3070 (1966).
133. Pretka, J. E., and Lindwall, H. G., *J. Org. Chem.* **19**, 1080 (1954).
134. Ramirez, F., *Pure Appl. Chem.* **9**, 337 (1964).
135. Reissert, A., *Chem. Ber.* **29**, 639 (1896).
136. Reissert, A., *Chem. Ber.* **30**, 1030 (1897).
137. Remers, W. A., and Weiss, M. J., *J. Am. Chem. Soc.* **88**, 804 (1966).
137a. Richman, R. J., and Hassner, A., *J. Org. Chem.* **33**, 2548 (1968).
138. Romeo, A., Corrodi, H., and Hardegger, E., *Helv. Chim. Acta* **38**, 463 (1955).

139. Roth, R. H., Remers, W. A., and Weiss, M. J., *J. Org. Chem.* **31**, 1012 (1966).
140. Russell, G. A., and Janzen, E. G., *J. Am. Chem. Soc.* **89**, 300 (1967).
140a. Russell, G. A., and Kaupp, G., *J. Am. Chem. Soc.* **91**, 3851 (1969).
141. Rutenberg, M. W., and Horning, E. C., *Org. Syn.* **30**, 62 (1950).
142. Schulenberg, J. W., *J. Am. Chem. Soc.* **90**, 1367 and 7008 (1968).
143. Schulenberg, J. W., and Archer, S., *J. Am. Chem. Soc.* **83**, 3091 (1961).
144. Sheehan, J. C., and Beeson, J. H., *J. Org. Chem.* **31**, 1637 (1966).
145. Sheehan, J. C., and Frankenfeld, J. W., *J. Am. Chem. Soc.* **83**, 4792 (1961).
146. Splitter, J. S., and Calvin, M., *J. Org. Chem.* **20**, 1086 (1955).
147. Staněk, J., *Chem. Listy* **37**, 161 (1943); *Chem. Abstr.* **44**, 5869 (1950).
148. Staněk, J., and Horák, M., *Collection Czech. Chem. Commun.* **15**, 1037 (1951); *Chem. Abstr.* **46**, 7100 (1952).
149. Staněk, J., and Rybar, D., *Chem. Listy* **40**, 173 (1946); *Chem. Abstr.* **45**, 5147 (1951).
150. Su, H. C. F., and Tsou, K. C., *J. Am. Chem. Soc.* **82**, 1187 (1960).
151. Suehiro, T., *Chem. Ber.* **100**, 905 (1967).
152. Sumpter, W. C., *Chem. Rev.* **37**, 443 (1945).
153. Sumpter, W. C., Miller, M., and Hendrick, L. N., *J. Am. Chem. Soc.* **67**, 1656 (1945).
154. Sundberg, R. J., *J. Org. Chem.* **30**, 3604 (1965).
155. Tacconi, G., and Pietra, S., *Farmaco (Pavia), Ed. Sci.* **18**, 409 (1963); *Chem. Abstr.* **59**, 9954 (1963).
156. Teuber, H.-J., and Staiger, G., *Chem. Ber.* **89**, 489 (1956).
157. Teuber, H.-J., and Thaler, G., *Chem. Ber.* **91**, 2253 (1958).
158. Troxler, F., Bormann, G., and Seeman, F., *Helv. Chim. Acta* **51**, 1203 (1968).
159. Walker, G. N., *J. Am. Chem. Soc.* **77**, 3844 (1955).
160. Walker, G. N., Smith, R. T., and Weaver, B. N., *J. Med. Chem.* **8**, 626 (1965).
161. Wenkert, E., Bernstein, B. S., and Udelhofen, J. H., *J. Am. Chem. Soc.* **80**, 4899 (1958).
162. Wenkert, E., Bhattacharyya, N. K., Reid, T. L., and Stevens, T. S., *J. Am. Chem. Soc.* **78**, 797 (1956).
163. Wenkert, E., Bose, A. K., and Reid, T. L., *J. Am. Chem. Soc.* **75**, 5514 (1953).
164. Wenkert, E., and Reid, T. L., *Chem. & Ind. (London)* p. 1390 (1953).
165. Wenkert, E., Udelhofen, J. H., and Bhattacharyya, N. K., *J. Am. Chem. Soc.* **81**, 3763 (1959).
166. Wieland, T., and Grimm, D., *Chem. Ber.* **98**, 1727 (1965).
167. Wieland, T., Weiberg, O., Fischer, E., and Hörlein, G., *Ann. Chem.* **587**, 146 (1954).
168. Wiesner, K., Valenta, Z., Manson, A. J., and Stonner, F. W., *J. Am. Chem. Soc.* **77**, 675 (1955).
169. Wright, W. B., Jr., and Collins, K. H., *J. Am. Chem. Soc.* **78**, 221 (1956).
170. Yamada, S., Hino, T., and Ogawa, K., *Chem. & Pharm. Bull. (Tokyo)* **11**, 674 (1965).
171. Zaugg, H. E., and De Net, R. W., *J. Am. Chem. Soc.* **84**, 4574 (1962).

VIII
AMINOINDOLES

Relatively little work has been done on indoles bearing an amino group directly substituted to one of the ring positions. Treatment of *o*-aminobenzyl cyanide with sodium ethoxide gives "2-aminoindole" (18). The same

CH_2CN NH_2 (1) → NH_2 N H (2) ⇌ H H N NH_2 (3) ⇌ N H NH (4)

compound can be prepared from indole-2-carboxylic acid via a Curtius rearrangement (19). Hydrolysis gives oxindole (19). It has been shown that 2-aminoindole (**2**) is unstable relative to 2-amino-3*H*-indole (**3**) (2, 12, 24). Tautomer **3** is also more stable than 2-iminoindoline (**4**).

The apparent preference for tautomer **3** over tautomer **4** indicates that conjugation of the phenyl ring with the imino nitrogen of the amidine system is energetically more favorable than conjugation with the amino nitrogen. It will be recalled that in oxindole the tautomer corresponding to **4** is stable. In the case of oxindole the stability of the amido system relative to the alternative imido ether system is no doubt the governing factor in determining the position of the equilibrium. Kebrle and Hoffmann (12) have considered the structure of 1-methyl-2-aminoindole in which the two tautomers **5** and **6** are possible. They have concluded on the basis of ultraviolet

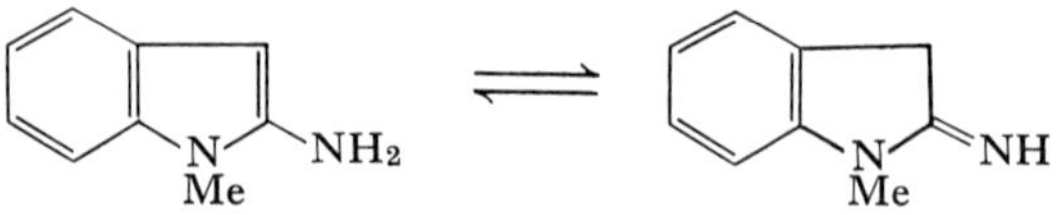

N NH2 Me ⇌ N NH Me

5 **6**

spectral data that in ether solution the tautomer **5** is favored but in polar solvents the equilibrium is shifted toward **6** and the conjugate acid **7**.

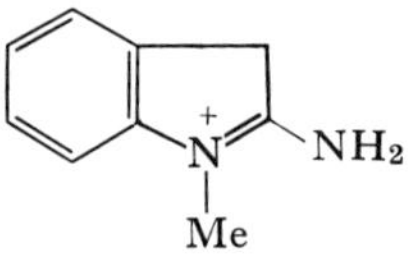

$\overset{+}{N}$ NH_2 Me

7

Kebrle and Hoffmann have assigned reasonable structures to several acetyl derivatives of simple 2-aminoindoles (12). Bromination of 3-methylindole in pyridine gives **9** (9, 12a). This compound can be hydrolyzed to 3-methyloxindole (12a) or reduced to a piperidine derivative which is probably **10** (9). As is characteristic of 2- and 3-aminoindoles, the compound is subject

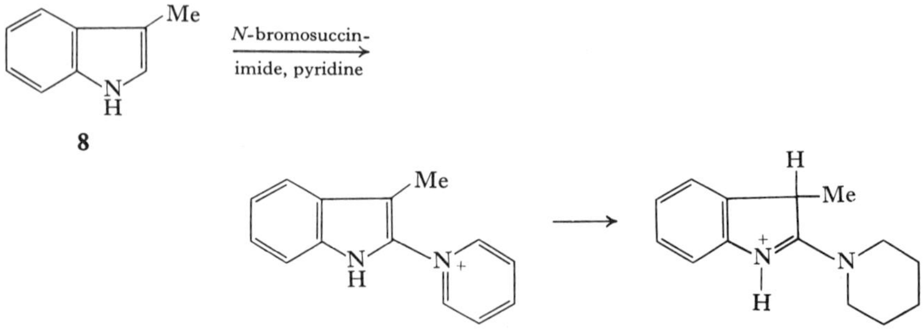

Me N H → *N*-bromosuccinimide, pyridine

8

Me N H N^+ → H Me $\overset{+}{N}$ H N

9 **10**

CO_2Et CHCN NO_2 → Fe AcOH → CO_2Et N H NH_2 ⇌ H CO_2Et N NH_2

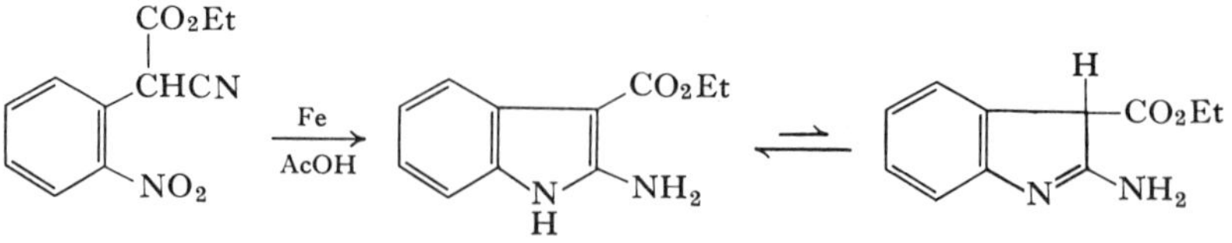

11 **12** **13**

to very facile air oxidation (see Chapter V,A). Grob and Weissbach have characterized the reduction product of **11** as 2-amino-3-carbethoxyindole (**12**) (6). The stability of the 2-aminoindole tautomer in this instance can be attributed to conjugation of the 2-amino substituent with the carbonyl substituent at C-3 in tautomer **12**. In general, it seems that the energies of the three possible tautomeric forms of 2-aminoindole must be rather closely balanced and can be shifted predominantly to any of the three forms by modification of the substitution pattern on the ring.

3-Aminoindoles have most often been prepared by reduction of 3-oximinoindolenines, which can be prepared by nitrosation of indoles. The reactions below are typical.

AcOH, $NaNO_2$ → NOH, OMe

14 **15** (88%)

NH_2, OMe

16 (97%) Ref. (16)

BuONO, EtO^- → NaS_2O_4 → NH_2, Me

17 **18** Ref. (15)

AcOH, $NaNO_2$ → NO, Ph → H_2, Pt → NH_2, Ph

19 **20** (96%) **21** Ref. (10)

Hester has prepared the cyclic 3-aminoindole **24** via a Beckmann rearrangement followed by reduction of the lactam **23** (8). A series of Fischer cyclizations leading directly to 3-dialkylaminoindoles has also been reported (25).

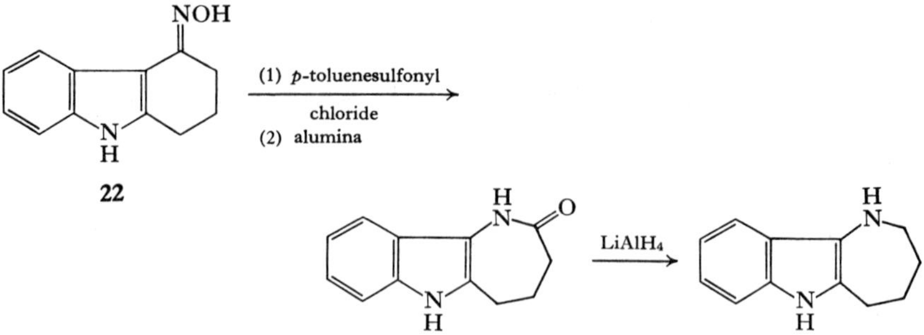

NOH
(1) p-toluenesulfonyl chloride
(2) alumina
22
LiAlH$_4$
23
24

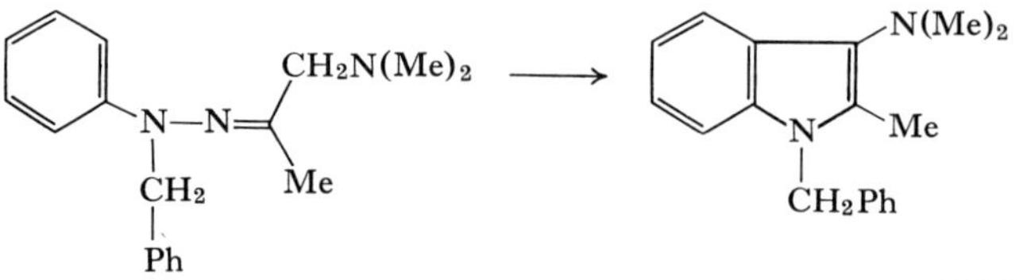

$CH_2N(Me)_2$
N—N=
CH_2
Ph
Me
25
$N(Me)_2$
Me
N
CH_2Ph
26 (24%)
Ref. (25)

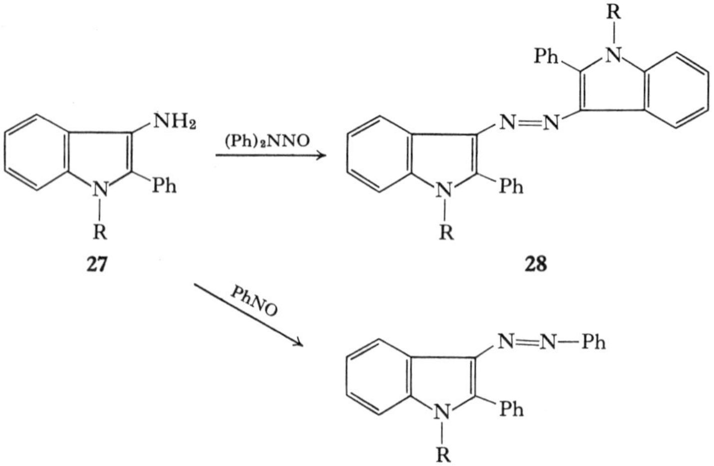

NH_2
Ph
N
R
27
$(Ph)_2NNO$
N=N
Ph
28
PhNO
N=N—Ph
Ph
29

The products of the facile auto-oxidation of 3-aminoindoles were discussed in Chapter V. The 3-aminoindoles can be diazotized to stable, characterizable 3-diazoindolenines (10, 15, 16). As long as precautions to minimize autoxidation are taken, 3-aminoindoles can be successfully condensed with aldehydes and nitrosobenzene. Oxidation with *N*-nitrosodiphenylamine gives azoindoles.

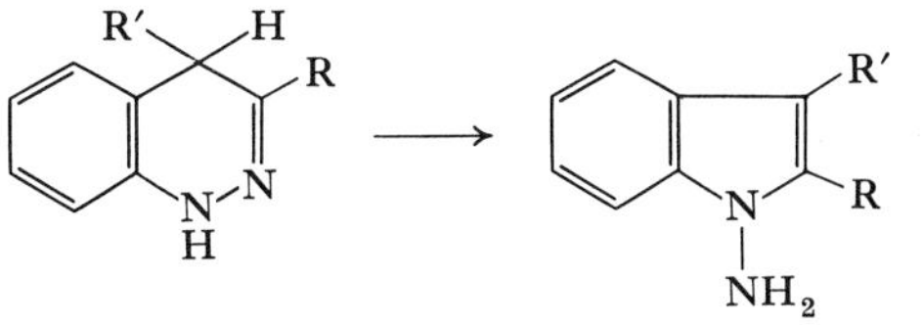

30 → 31

1-Aminoindoles are formed when 1,4-dihydrocinnolines are treated with dilute mineral acid at 100°C (7). A reversible ring opening can account for the rearrangement (7).

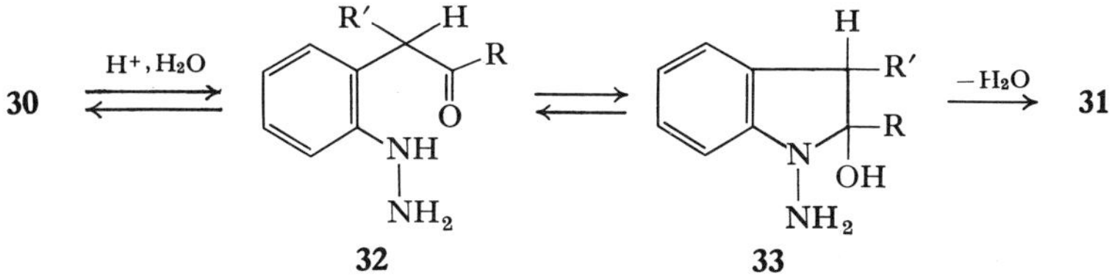

30 $\xrightarrow{H^+, H_2O}$ 32 ⇌ 33 $\xrightarrow{-H_2O}$ 31

Treatment of the vinyl azo compound **34** with acetic acid gives 1-anilino-3-phenylindole (23). No extension of this method to other cases appears to have been reported.

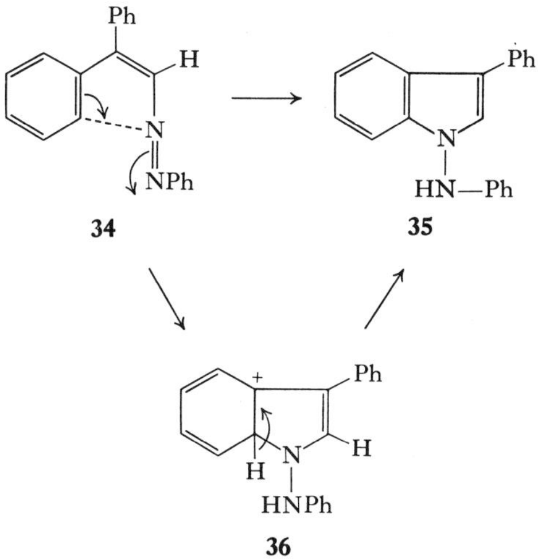

34 → 35; 34 → 36 → 35

Indoles with carbocyclic nitro substituents are potential precursors of 4-, 5-, 6- and 7-aminoindoles. Reduction with hydrazine and Raney nickel has been used with good success as a method of reduction of 6-nitroindole to 6-aminoindole (1). This method has been found preferable to catalytic reduction in the case of certain simple indoles (3, 21). Sodium hydrosulfite has also been used with general success (20).

O_2N ... N Ac — Ni, NaOH / H_2O, 100° → H_2N ... N H — Ref. (11)

37 **38** (53%)

N Ac — Br_2 / AcOH → Br ... N Ac — HNO_3 / H_2SO_4 → Br ... O_2N N Ac — Ref. (5, 11)

39 **40** **41**

Raney Ni / H_2O, NaOH ↓

N H NH_2

42 (49%)

$(Me)_2NSO_2N$= ... =N–SO_2–$N(Me)_2$ + Me ... N →

43 **44**

$(Me)_2NSO_2N$ ... Me ... N–$SO_2N(Me)_2$ — HBr / AcOH → $(Me)_2NSO_2N$ ... Me ... N H

45 **46**

Br ... N H CO_2H — NH_4OH / 220° → NH_2 ... N H

47 **48** (55%)

The "indoline–indole" synthetic sequence (Chapter II, H) is also applicable to the synthesis of indoles with carbocyclic amino substituents. Johnson and Crosby (11) have examined the use of nitroindolines for synthesis of amino-indoles. These workers found that reduction of the nitro group and aromatization of the indoline ring could be effected concurrently in the presence of Raney nickel. The synthesis of 5-aminoindole is typical. Ring halogen substituents are removed under the same conditions and therefore can be used as blocking groups for introduction of the nitro group at ring positions which are not the primary site of electrophilic attack.

As was discussed in connection with the Nenitzescu indole synthesis (Chapter III, C), the reaction of quinone imine derivatives with enamines gives 5-aminoindole derivatives. Domschke and co-workers (4) have recently made application of this approach to the synthesis of several 5-dimethylaminosulfamidoindoles. The reaction has also been investigated briefly by Kuehne (13).

Plieninger and co-workers (17) have reported that 4- and 6-aminoindole can be prepared from the corresponding bromoindole-2-carboxylic acid. It is not known if the decarboxylation is associated directly with the displacement process. This reaction, when applied to 4-chloroindole-2 carboxylic acid, constitutes a satisfactory preparative method for 4-amino-indole (22).

Indoles with carbocyclic carboxy groups have been converted to amino-indoles via Curtius rearrangement (14).

REFERENCES

1. Brown, R. K., and Nelson, N. A., *J. Am. Chem. Soc.* **76**, 5149 (1954).
2. Cohen, L. A., Daly, J. W., Kny, H., and Witkop, B., *J. Am. Chem. Soc.* **82**, 2184 (1960).
3. DeGraw, J. I., *Can. J. Chem.* **44**, 387 (1966).
4. Domschke, G., Heller, G., and Natzeck, U., *Chem. Ber.* **99**, 939 (1966).
5. Gall, W. G., Astill, B. D., and Boekelheide, V., *J. Org. Chem.* **20**, 1538 (1955).
6. Grob, C. A., and Weissbach, O., *Helv. Chim. Acta* **44**, 1748 (1961).
7. Haddlesey, D. I., Mayor, P. A., and Szinai, S. S., *J. Chem. Soc.* p. 5269 (1964).
8. Hester, J. B., Jr., *J. Org. Chem.* **32**, 3804 (1967).
9. Hino, T., Nakagawa, M., Wakatusi, T., Ogawa, K., and Yamada, S., *Tetrahedron* **23**, 1441 (1967).
10. Huang-Hsinmin, and Mann, F. G., *J. Chem. Soc.* p. 2903 (1949).
11. Johnson, H. E., and Crosby, D. C., *J. Org. Chem.* **28**, 2794 (1963).
12. Kebrle, J., and Hoffmann, K., *Helv. Chim. Acta* **39**, 116 (1956).

12a. Kobayashi, T., and Inokuchi, N., *Tetrahedron* **20**, 2055 (1964).

13. Kuehne, M. E., *J. Am. Chem. Soc.* **84**, 837 (1962).
14. Lindwall, H. G., and Mantell, G. J., *J. Org. Chem.* **18**, 345 (1953).
15. Moore, R. G. D., and Woitach, P. T., Jr., British Patent 816,382 (1959); *Chem. Abstr.* **55**, 188 (1961).
16. Patel, H. P., and Tedder, J. M., *J. Chem. Soc.* p. 4593 (1963).

17. Plieninger, H., Suehiro, T., Suhr, K., and Decker, M., *Chem. Ber.* **88**, 370 (1955).
18. Pschorr, R., and Hoppe, G., *Chem. Ber.* **43**, 2543 (1910).
19. Rinderknecht, H., Koechlin, H., and Niemann, C., *J. Org. Chem.* **18**, 971 (1953).
20. Shaw, E., and Woolley, D. W., *J. Am. Chem. Soc.* **75**, 1877 (1953).
21. Vejdelek, Z. J., *Chem. Listy* **51**, 1338 (1957).
22. Walton, E., Holly, F. W., and Jenkins, S. R., *J. Org. Chem.* **33**, 192 (1968).
23. Wasserman, H. H., and Nettleton, H. R., *Tetrahedron Letters* No. 7, p. 33 (1960).
24. Witkop, B., *Experientia* **10**, 420 (1954).
25. Yoneda, F., Miyamae, T., and Nitta, Y., *Chem. & Pharm. Bull.* (*Tokyo*) **15**, 8 (1967).

IX

KETONES, ALDEHYDES, AND CARBOXYLIC ACIDS DERIVED FROM INDOLE

A. KETONES AND ALDEHYDES DERIVED FROM INDOLE

1. General

The ketones and aldehydes derived from the indole ring system have certain distinguishing properties relative to other aromatic ketones and aldehydes. This is especially true of ketones and aldehydes in which the carbonyl substituent is attached to C-2 or C-3 of the indole ring. The unique properties of such compounds can usually be attributed to resonance interaction between the electron-releasing indole ring and the carbonyl

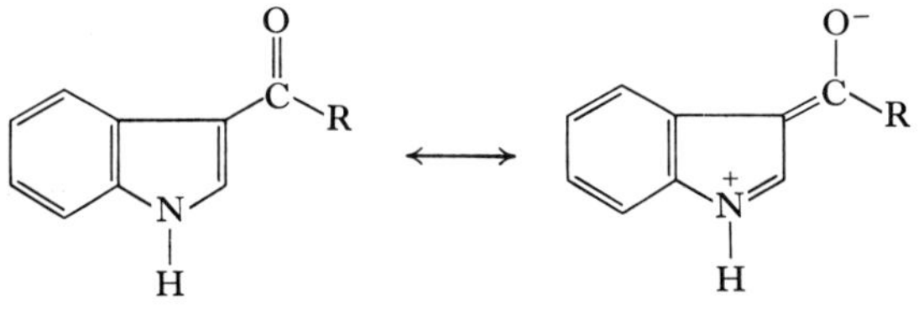

A1a ⟷ A1b

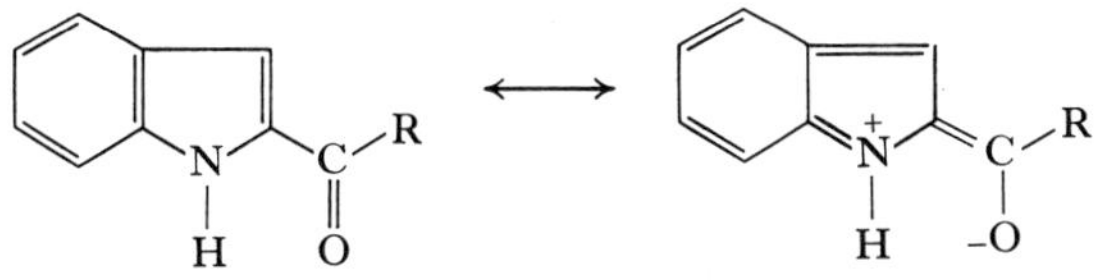

A2a ⟷ A2b

group. The resonance interaction is most pronounced in the case of 3-acylindoles.

The conjugation effects the carbonyl frequency of indolyl ketones and aldehydes, shifting the carbonyl absorption toward longer wavelengths.

For example, the 2-carboxaldehyde absorbs at 1675 cm^{-1} versus a split band at 1634 and 1618 cm^{-1} for the 3-isomer. The methyl esters of the 2- and 3-carboxylic acids absorb at 1689 and 1669 cm^{-1}, respectively (69).

Indole-3-carboxaldehyde has been referred to as a vinylogous amide (106, 116). The important contribution of canonical form **A3b** accounts for the high polarity of the molecule. The polarity, as well as effective hydrogen bonding in the crystal state, can account for the high melting point (195°C, compared to 33°C for 1-napthaldehyde) of indole-3-carboxaldehyde.

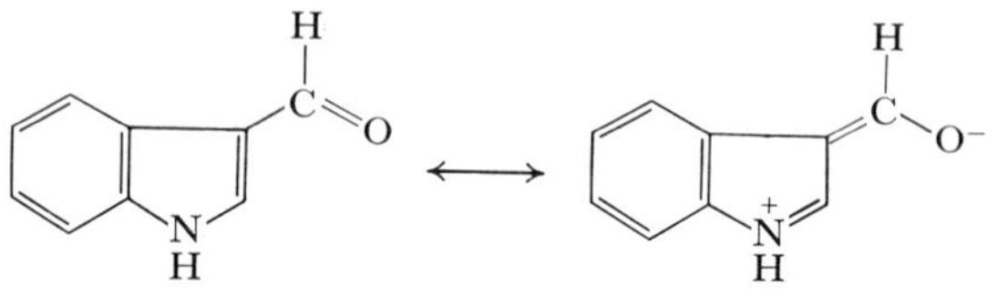

A3a ⟷ **A3b**

The same resonance interaction affects the acidity of the N–H bond. The ultraviolet maxima of indole-3-carboxaldehyde shifts in alkaline solution because of anion formation and indole-3-carboxaldehyde is soluble in strong base (116). While the pK_a of indole as an acid is estimated as 16.97, 3-acetylindole and indole-3-carboxaldehyde have pK_a values of 12.36 and 12.99, respectively (125). The effect can also be recognized qualitatively in the ease with which 3-acylindoles are alkylated. 3-Formylindole is converted to the *N*-methyl derivative in 92% yield by methyl iodide and potassium carbonate in refluxing acetone. Much stronger bases are required for rapid alkylation of indole or its C-alkyl derivatives (49). Indeed, the alkylation of 2-methyl-3-formylindole in the absence of base has been reported (34).

Byrn and Calvin (20) have attributed the relatively low rate of carbonyl oxygen exchange in indole-3-carboxaldehyde (one-third that of benzaldehyde) to the diminished reactivity of the conjugate acid of the aldehyde toward nucleophilic addition. The electrophilicity of the carbonyl group is reduced by the conjugation with the electron-rich ring.

2. Reduction of Indolyl Ketones and Indolecarboxaldehydes

The special features of metal hydride reductions of ketones and aldehydes in which the carbonyl substituent is attached to C-2 or C-3 of the indole ring were considered in Chapter II,C. It will be recalled that 3-carbonyl groups are often easily reduced to methylene groups whereas 2-carbonyl substituents undergo hydrogenolysis only under more vigorous conditions. The availability of the hydride hydrogenolysis reaction has diminished the need for application of the more usual carbonyl to methylene reduction procedures, although scattered examples are available. Wolff–Kishner reduction of **A4** gives indoleacetic acid (71). Wolff–Kishner reduction of 3-benzoyl-

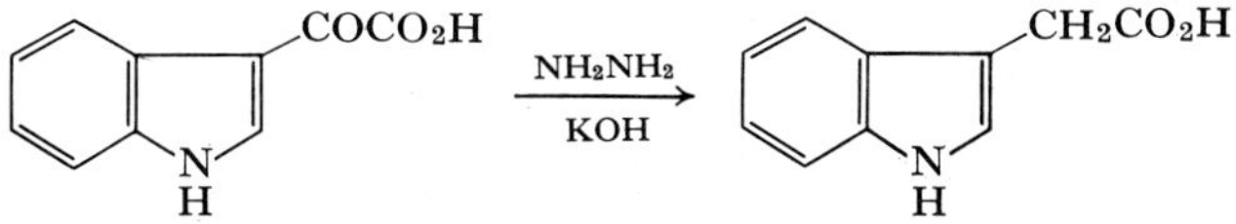

$$\text{A4 (3-}COCO_2H\text{-indole)} \xrightarrow[\text{KOH}]{NH_2NH_2} \text{A5 (3-}CH_2CO_2H\text{-indole)} \quad (95\%)$$

indole to 3-benzylindole proceeds smoothly (101). Various 2- and 3-acetylindoles have also been successfully reduced under Wolff–Kishner conditions (1a). The pyridazinone obtained by treatment of 4-oxo-4-(3-indolyl)-butyric acid with hydrazine is decomposed to 4-(3-indolyl)-butyric acid by potassium hydroxide (57). Similarly, 3,4-dihydro-β-carbolines, which are internal Schiff bases of 2-acylindoles, are reduced to 2-alkyl-3-aminoethylindoles under the conditions of the Huang–Minlon modification of the Wolff–Kishner reaction (36). The reduction of 4-oxo-4-(3-indolyl)butyric acid

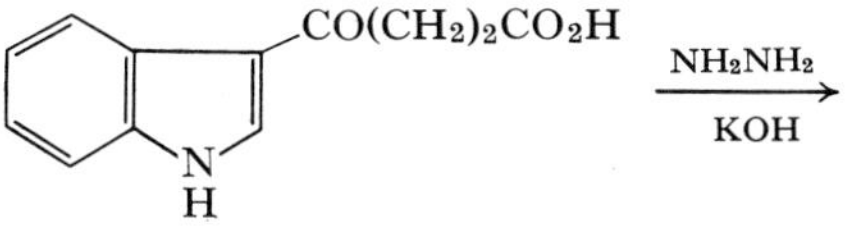

$$\text{A6 (3-}CO(CH_2)_2CO_2H\text{-indole)} \xrightarrow[\text{KOH}]{NH_2NH_2} \text{A7} \xrightarrow[160\text{–}180^\circ]{\text{KOH}} \text{A8 (3-}(CH_2)_3CO_2H\text{-indole)}$$

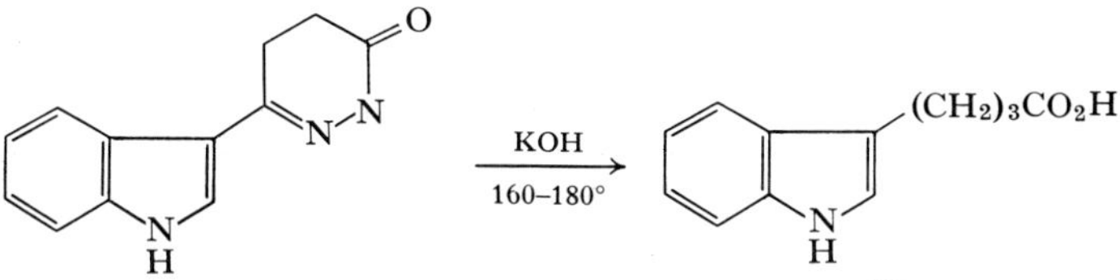

has also been effected by the Clemmensen method, as has the transformation **A9** → **A10** (57). Plieninger and Muller (76) have successfully used the

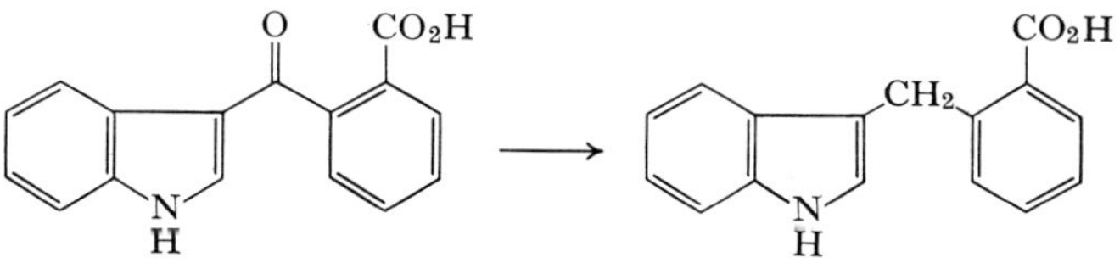

$$\text{A9} \longrightarrow \text{A10 (}CH_2\text{, }CO_2H\text{)}$$

thioketal desulfurization technique in the reduction of the ethyl indole-3-glyoxylate **A11** to **A13**.

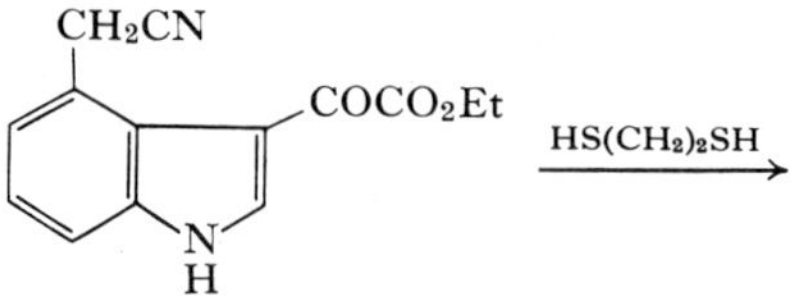

Reeve and co-workers have reported the reduction of a series of methyl 1-acetylindole-3-glyoxylates to the corresponding hydroxy compounds with aluminum amalgam (82). Earlier examples of this type of reduction were reported by Baker (5). This method is very effective for reduction of alkyl glyoxylates but it is doubtful that it is suitable for simple 3-acylindoles.

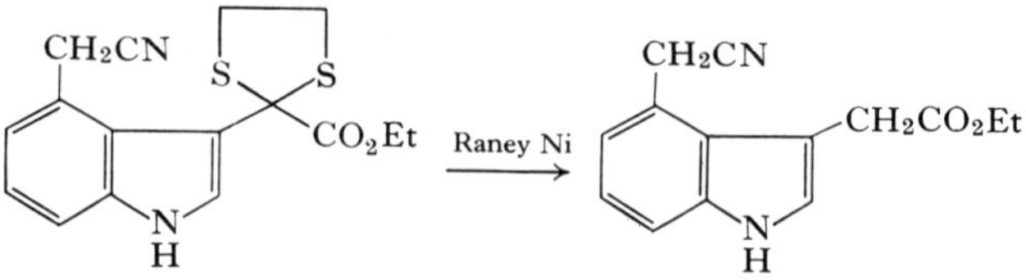

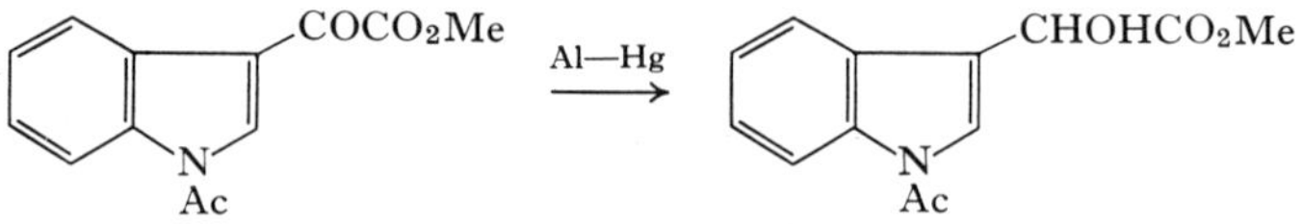

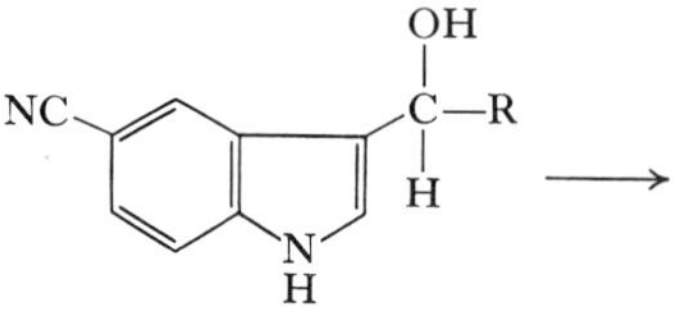

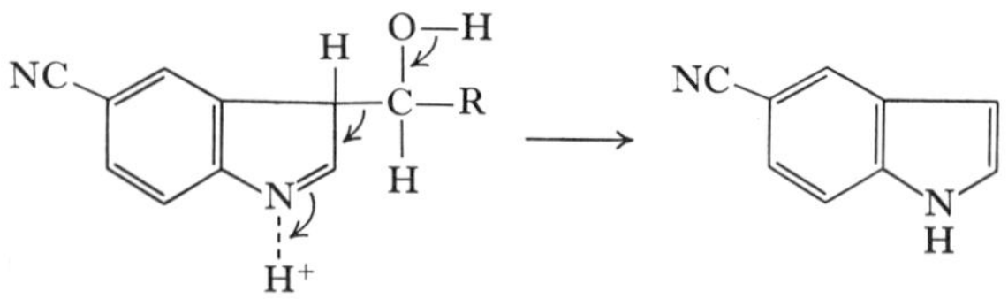

De Graw and co-workers have noticed an interesting cleavage reaction which occurs during sodium borohydride reduction of certain 3-acetyl-5-cyanoindoles (26a). The side-chain cleavage is attributed to base-catalyzed fragmentation of the intermediate indolylcarbinol. This mechanism is the reverse of an indole-aldehyde condensation. The electron withdrawing, cyano substituent may be a factor in facilitating the elimination of the indole from the intermediate **A17**.

3. Condensation and Addition Reactions of Indolyl Carbonyl Compounds with Nucleophiles

The conjugation of the indole nitrogen atom with C-2 and C-3 carbonyl groups would be expected to diminish the electrophilicity of the carbonyl and diminish reactivity, at least when the addition step is rate-controlling, in condensation and addition reactions. Rodinov and Veselovskaya (87) have made a qualitative comparison of the reactivity of indole-3-carboxyaldehyde and its 1- and 2-methyl derivatives and 1,2-dimethylindole-3-carboxaldehyde toward such potential carbon nucleophiles as malonic acid and ethyl cyanoacetate. They conclude that the order of relative reactivities of the various indole-3-carboxaldehydes is 1-Me > H > 2-Me. The reactions below are among the condensations examined.

CHO, N, H
A19
+ $CH_2(CO_2H)_2$ $\xrightarrow{\text{pyridine}}$
CH=CHCO₂H, N, H
A20 (31%)

CHO, N, H, Me
A21
+ $CH_2(CO_2H)_2$ $\xrightarrow{\text{pyridine}}$ no reaction

CHO, N, Me
A22
+ EtO_2CCH_2CN $\xrightarrow{\text{pyridine}}$
$CH{=}C(CO_2Et)(CN)$, N, Me
A23 (94%)

Moffatt (70) has achieved successful condensation with malonic acid using 1-acetylindole-3-carboxaldehyde. Other workers have reported conditions suitable for formation of **A26** in 50% yield directly from indole-3-carboxaldehyde and malonic acid (92). It has been concluded on the basis of nmr studies that the acrylic acid has the expected trans stereochemistry (81). It is possible to obtain 2-methylindole-3-acrylic acid in 30% yield from 2-methylindole-3-carboxaldehyde and malonic acid via 1-acetyl-2-methylindole-3-carboxaldehyde (46).

CHO
N
Ac
A24

$CH_2(CO_2H)_2$

pyridine, 40°

CH=C$(CO_2H)_2$
N
H
A25

pyridine, 80°

CH=CHCO_2H
N
H
A26

The synthetically important condensation of indole-3-carboxaldehydes with nitronate anions was discussed in Chapter IV, B, 3. Indole-2-carboxaldehyde also condenses readily with nitromethane (39). Other carbanion nucleophiles which have been successfully condensed with indole-3-carboxaldehyde include 1,2-dimethylpyridinium ions and β-diketones, as illustrated by the transformations shown below (35, 51a, 123). A successful

CHO
N
H
A27
+
MeN+
Me
A28
⟶
CH=CH
N
H
N^+
Me
A29

CHO
N
H
A30
+
O
O
A31
⟶
H O
N
H
O
A32

application of the Wittig reaction involving the 2-acylindole **A33** has been described (124).

Me
N
N
H
O
A33

$(PH)_3P{=}CH_2$

Me
N
N
H
CH_2
A34

The reactions of 3-acylindoles with organometallic compounds are complicated by the fact that the initial products are very sensitive 3-indolyl carbinols (see Chapter II, B). It is reported (11) that *t*-butyl-3-indolylcarbinol can be obtained in 66% yield from indole-3-carboxaldehyde and *t*-butylmagnesium chloride. It is stated, without detailed experimental support, that similar carbinols from other than *t*-alkylmagnesium halides are unstable. The most extensive recent study of the reactions of organometallic reagents with 3-acylindoles has been the work of Szmuszkovicz. Reaction of 1-methyl-3-benzoylindole with phenylmagnesium bromide gave 1-methyl-2-phenyl-3-benzoylindoline (107). An analogous reaction occur-

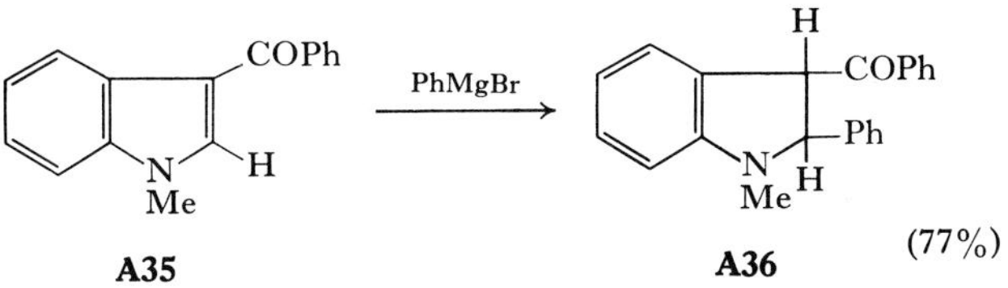

COPh, H, N, Me → (PhMgBr) → H, COPh, Ph, N, Me, H

A35 **A36** (77%)

red with the 2-methyl derivative. With phenyllithium, **A39** was also a product after contact of the crude product with methanol in the presence of alumina. This ether is probably formed by a facile nucleophilic substitution reaction involving methanol and the carbonyl addition product. The 3-

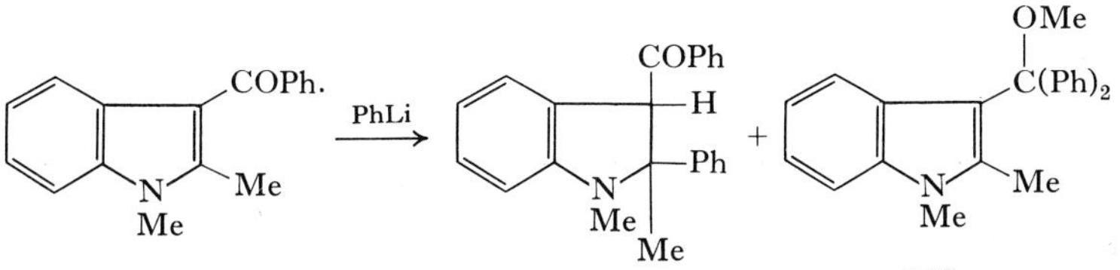

COPh., Me, N, Me → (PhLi) → COPh, H, Ph, N, Me, Me + OMe, C(Ph)$_2$, Me, N, Me

A37 **A38** **A39**

acetylindole **A40** reacts with phenylmagnesium bromide to give **A41** by addition at the carbonyl followed by elimination. The failure of phenylmag-

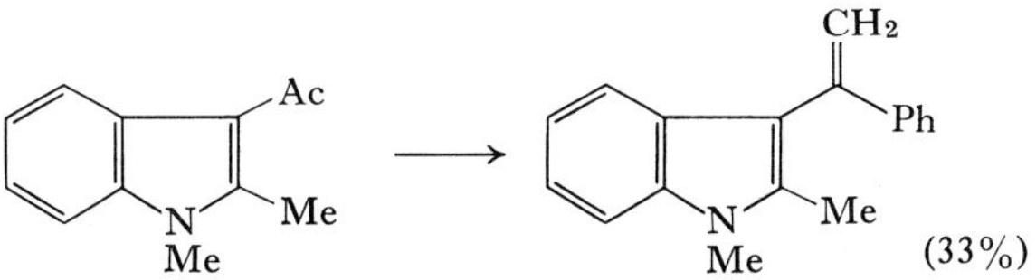

Ac, Me, N, Me → CH$_2$, Ph, Me, N, Me (33%)

A40 **A41**

nesium bromide to react with **A35** at the carbonyl group is attributed to steric hindrance. Nucleophilic attack at C-2 of an indole ring is, of course, very unusual and perhaps a complexed species such as **A42** is a better description of the species which is actually attacked. Szmuszkovicz has also

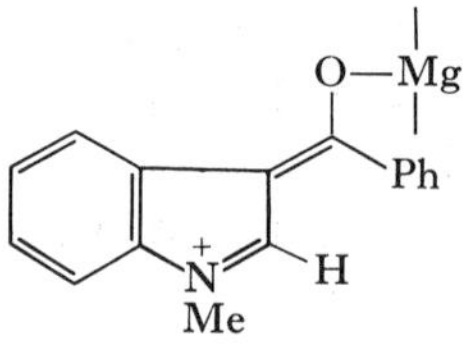

A42

given considerable attention to the reaction of organometallic reagents with indole-3-glyoxylic acid derivatives. Ethyl indole-3-glyoxylate gives **A44** or **A45** depending upon the amount of Grignard reagent employed (108).

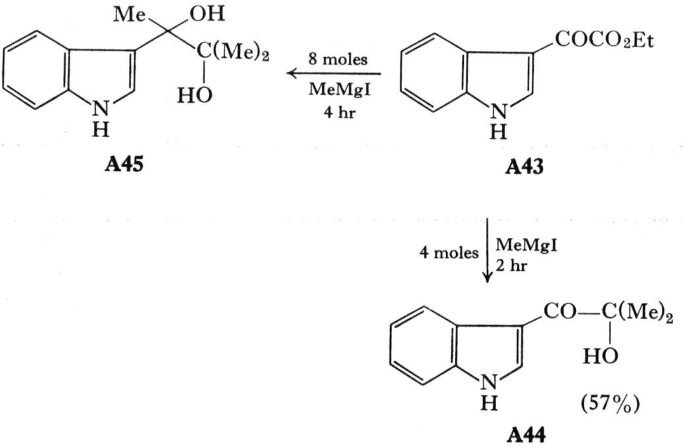

A45 ← (8 moles MeMgI, 4 hr) — A43 — (4 moles MeMgI, 2 hr) → A44 (57%)

One anticipates that the first reaction would be the formation of the anion **A46**. The diminished reactivity of the ketone carbonyl in the conjugate base **A46** can account for preferential reaction at the ester carbonyl. Reaction

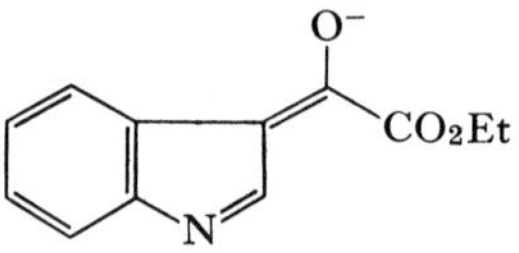

A46

of **A43** with phenylmagnesium bromide gives **A47** and the reaction could not be forced beyond this stage with excess phenylmagnesium bromide (109). The *N*-methyl derivative of **A47** reacts normally with methyllithium to give the diol **A49**. With methylmagnesium iodide, **A50** is formed. The latter reaction is another example of nucleophilic addition to the indole ring.

Schiff bases derived from indole-3-carboxaldehyde can be obtained by heating the aldehyde with a primary amine in refluxing toluene with

A43 $\xrightarrow{PhMgBr}$ A47 $\xrightarrow[MeO^-]{MeI}$ A48

A48 $\xrightarrow{MeMgI}$ A50

A48 $\xrightarrow{MeLi}$ A49

A51 $\xrightarrow{RNH_2}$ A52 (CH=NR) $\xrightarrow{NaBH_4}$ A53 (CH_2NHR)

A54 $\rightleftharpoons$ A55

provisions for azeotropic removal of water (121). The resulting imines have been reduced successfully to substituted 3-indolylmethylamines (121). Intramolecular interaction of basic nitrogen atoms and carbonyl groups have been reported in certain of the 2-acylindole alkaloids. The ultraviolet spectrum of **A54** is strongly solvent dependent (30). In ether, the spectrum is that of **A54**, showing conjugation of the indole ring with a carbonyl group, but in ethanol a spectrum compatible with structure **A55** is observed. These results parallel data from other medium-ring amino ketones (62).

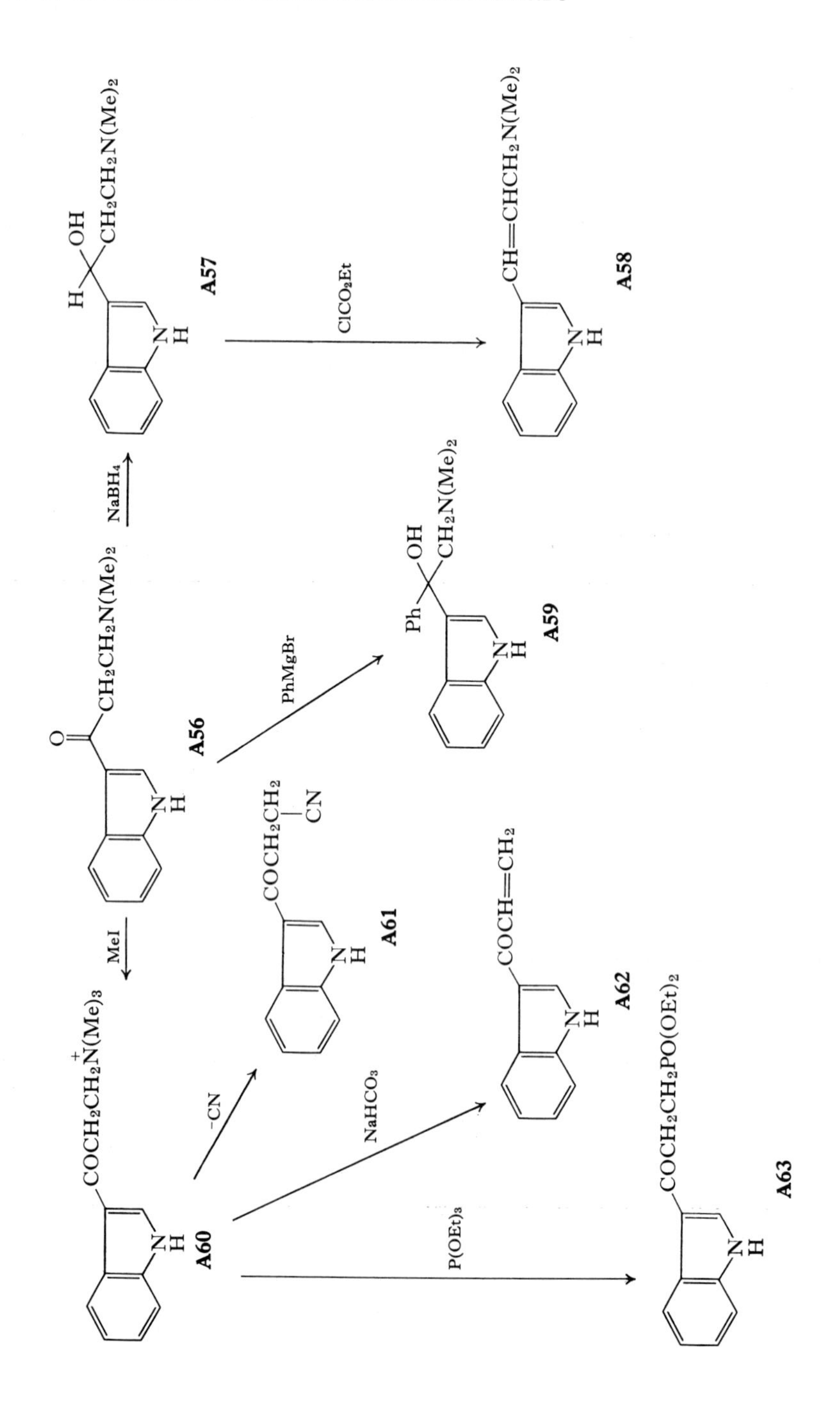
A56
A57
A58
A59
A60
A61
A62
A63
NaBH4
ClCO2Et
PhMgBr
MeI
-CN
NaHCO3
P(OEt)3

4. Reactions of 2- and 3-Acylindoles Involving the α-Carbon

In this section reactions involving the α-carbon of 2-indolyl and 3-indolyl ketones are discussed. In a large part these reactions are typical of the reactions of aromatic ketones. 3-Acylindoles undergo the Mannich base condensation (106). 3-Acetylindole, for example, is converted to 3-(β-dimethylamino)propionylindole in 78% yield. With 3-propionylindole, however, condensation takes place at both the α-carbon and the indolic nitrogen. The condensation of 3-acetylindole takes this course when excess formaldehyde and dimethylamine are used. The Mannich bases are potentially versatile synthetic intermediates and some of the transformations which have reported to date are summarized in the scheme on p. 410 (105, 106, 110, 111).

$COCH_2CH_2$ — NCO_2CH_2Ph

A64

$COCHCH_2$ — NCO_2CH_2Ph

Br

(74%)

A65

(1) HBr
(2) ^{-}OH

A66

A number of successful brominations of 3-acylindoles have been reported. Trimethylphenylammonium tribromide effects successful bromination of the ketone **A64** (27). After removal of the N-protecting group the bromide undergoes intramolecular nucleophilic substitution. 3-Acetylindole can be

A67 $\xrightarrow[\text{pyridine}]{I_2}$ A68 $\xrightarrow[H_2O]{^-OH}$ A69

A70 $\xrightarrow{SeO_2}$ A71

A72 $\xrightarrow{SeO_2}$ A73

brominated readily with bromine in methanol (11). The bromine is readily displaced by secondary amines. 2-Phenyl-3-acetylindole, however, apparently is brominated in the 5-position of the indole ring in preference to the acetyl substituent (16). The phenyl substituent presumably contributes a steric effect which diminishes reactivity at the acetyl group. 3-Acetylindoles react with iodine and pyridine to give pyridinium salts (41). These derivatives undergo cleavage to give indole-3-carboxylic acids when subjected to alkaline hydrolysis.

3-Acylindoles have been oxidized to the corresponding α-dicarbonyl derivatives (98, 112).

N-Substituted 2-acylindoles are reported to undergo the Willgerodt reaction successfully. In contrast, it is reported that compounds unsubstituted at nitrogen do not give normal products (7).

B. SYNTHESIS OF INDOLYL KETONES AND INDOLECARBOXALDEHYDES

1. Synthesis of 2- and 3-Acylindoles by Acylation

The acylation of indoles provides the most versatile approach to 3-acylindoles. In the normal case, the 3-position is strongly preferred as the site of electrophilic attack (Chapter I, Sections A and O) and problems resulting from competing acylation in other positions are minimal. Acylation of indoles and magnesium derivatives of indoles by carboxylic acid chlorides are important synthetic routes to 3-acylindoles. Acid anhydrides have occasionally been used as acylating agents (58). Table LI outlines a number of examples. Discussion of the mechanism of the reaction can be found in Chapter I, I.

TABLE LI □ ACYLATIONS OF INDOLES AND MAGNESIO INDOLES WITH CARBOXYLIC ACID CHLORIDES AND ANHYDRIDES

Indole substituent	Acylating agent	Position of acylation	Yield (%)	Ref.
A. Acylations without added catalyst				
None	Acetic anhydride	1, 3	55–66	(40, 89)
None	Oxalyl chloride	3	92	(92)
1,2-diMe	Acetic anhydride	3	?	(13)
2-Me	Acetic anhydride	3	40	(13)
2-Ph	Acetic anhydride	3	60	(13)
5-Cyano	Trifluoroacetic anhydride	3	88	(26a)
B. Acylations catalyzed by Lewis acids				
None	$Si(O_2CCH_2CH_2CO_2CH_3)_4$, stannous chloride	3	30	(58)
1,3-diMe	Acetyl chloride, zinc chloride	2	?	(32)
1-Methyl-3-carbomethoxymethyl	Acetyl chloride, zinc chloride	2	89	(53)
3-(2-Carbomethoxyethyl)	Acetyl chloride, zinc chloride	2	74	(51)
5-Cyano	Acetyl chloride, stannic chloride	3	70	(26a)
C. Acylations of magnesio indoles				
None	Ethyl chloroformate	1 ; 3	31 ; 32[a]	(52)
None	Chloroacetyl chloride	1, 3	20	(3)
None	3-Carbomethoxypropionyl chloride	3	40	(6)
None	Succinic anhydride	3	54	(58)
None	*N*-Carbobenzyloxyglycinyl chloride	3	?	(3)
None	Ethyl 2-ethylacetoacetate	3	?	(42)
None	Ethyl furan-2-carboxylate	3	?	(42)
None	Phthalic anhydride	3	45	(58)
None	*p*-Methoxyphenylacetyl chloride	3	60–70	(17)

Continued on page 414

TABLE LI □ ACYLATIONS OF INDOLES AND MAGNESIO INDOLES WITH CARBOXYLIC ACID CHLORIDES AND ANHYDRIDES-*continued*

Indole substituent	Acylating agent	Position of acylation	Yield (%)	Ref.
None	3-(*N*-Carbobenzyloxy-4-piperidyl)-propionyl chloride	3	47	(27)
2-Me	Benzoyl chloride	3	50	(18)
2-Me	Phenylacetyl chloride	3	60–70	(17)
2-Me	5-Chlorothiophene-2-carbonyl chloride	3	60	(19)
3-Ph	Benzoyl chloride	1; 2	92; 2	(22)
D. Miscellaneous acylations				
None	Acetic anhydride, perchloric acid	3	28	(31)
None	Acetic acid, silicon tetrachloride	3	21	(127)
2-Me	Benzoyl cyanide, pyridine	3	?	(56)

[a] At 10°; the product ratio has been shown to be temperature dependent.

As can be judged from Table LI, a wide variety of 3-acylindoles can be prepared by the various acylation procedures. The acylation of an indole at C-2 is a more difficult synthetic problem. With 1,3-disubstituted indoles, 2-acylation is observed (13, 32, 53). If only a 3-substituent is present, acylation on nitrogen may be the dominant reaction as in the case of 3-phenylindole (22). However, ethyl indole-3-propionate is easily acylated at C-2 (51). So far, there have been no reports of studies of the mechanism of acylation of 3-substituted indoles as detailed as the studies Jackson has carried out on alkylation reactions (Chapter I, N). If acylation of 3-substituted indoles were to occur initially at C-3, migration of the 3-substituent might follow. There is no evidence that such migrations occur. It would

B1 + R′COCl ⟶ **B2** ⟶ **B3**

thus appear that acylation may occur by direct electrophilic attack at C-2.

Cyclization to a pyrone accompanies acylation of indole-3-acetic acid in the 2-position (77). The pyrone ring is readily reopened by nucleophiles.

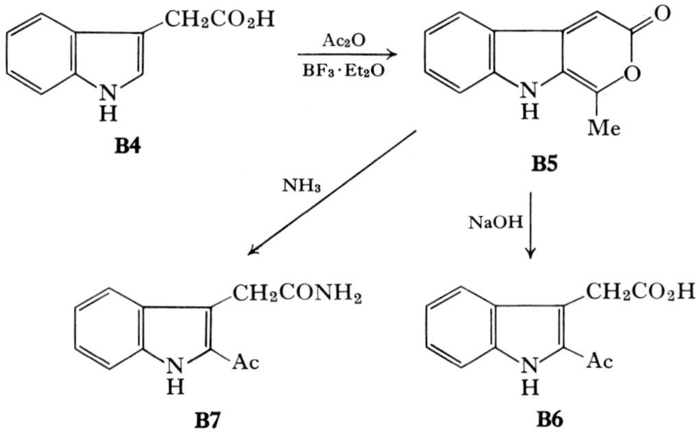

CH_2CO_2H — Ac_2O, $BF_3 \cdot Et_2O$ → **B5** (Me); B5 — NH_3 → **B7** (CH_2CONH_2, Ac); B5 — NaOH → **B6** (CH_2CO_2H, Ac)

B4 **B5** **B7** **B6**

Intramolecular acylations at C-2 of 3-indolealkanonic acids have been reported. The cyclization of **B8** to **B9** proceeds in 11% yield, when effected with phosphorus pentoxide refluxing xylene. The most successful reagent for cyclizations of this sort is polyphosphoric acid. Ishizumi and co-workers (47) have investigated the cyclization of **B8**, and several homologs and their *N*-methyl derivatives with this reagent. The reaction was successful for formation of rings of up to 8 atoms. This reaction has recently seen application in the synthesis of 2-acylindole alkaloids. The cyclization was applied first by Yamada and Shioiri (126) to the synthesis of 1-methyl-16-demethoxycarbonyl-20-desethylidenevobasine (**B12**). Ziegler and Zoretic applied

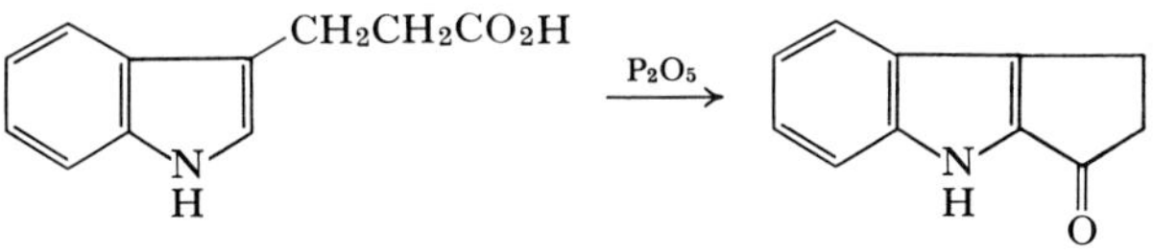

$CH_2CH_2CO_2H$ — P_2O_5 →

B8 **B9**

the reaction to synthesis of the quebrachamine skeleton (128). Dolby and Biere (29) also used the cyclization to excellent advantage in their synthesis of dasycarpidone and epidasycarpidone (**B16**). Other intramolecular acylations accomplished with good efficiency by means of polyphosphoric acid have been reported by Collington and Jones (24).

The Vilsmeier–Haack acylation of indoles using *N*,*N*-dialkylamides and phosphorus oxychloride has been most often employed for the synthesis of 3-formyl derivatives but is also effective for introduction of other acyl groups into the 3-position (4). Table LII summarizes representative synthetic procedures. The reaction has not been extensively investigated for

CH_2CO_2H HN N Me O

polyphosphoric acid →

B10

HN O N Me O

B11

(1) $LiAlH_4$
(2) CrO_3
(3) H_2CO, HCO_2H

N Me N Me O

B12

3-substituted indoles, but 1- and 2- formylation appear to be competitive (123). Formylation occurs at the nitrogen atom in 2,3-disubstituted indoles and in the carbocyclic ring when positions 1, 2, and 3 are all substituted (59).

Formylation and acylations of indole have also been accomplished using acid-catalyzed reactions of hydrogen cyanide and nitriles. This procedure seems to have been largely supplanted by the Vilsmeier–Haack reaction but typical examples are shown in Table LIII.

The introduction of acyl groups at C-5 of the indole ring can be accomplished indirectly via indolines (114). Vilsmeier–Haack formylation of 1-methylindoline gives the 5-formyl derivative in 52% yield and some 1-methyl-5,7-diformylindoline is also apparently formed. Dehydrogenation

O N N H HO_2CCH_2

polyphosphoric acid →

B13

O N N H O

B14

Me N N H CO_2H

polyphosphoric acid →

B15

Me N N H O

B16

TABLE LII □ VILSMEIER–HAACK ACYLATION OF INDOLES

Indole substituent	Amide	Position of substitution	Yield (%)	Ref.
None	Dimethylformamide	3	95	(97)
None	*N*-Methylformanilide	3	53	(91)
None	*N,N*-Dimethylbenzamide	3	51	(4)
2-Me	*N,N*-Dimethylacetamide	3	91	(4)
2-Carbethoxy	Dimethylformamide	3	?	(104)
2-Carbethoxy	*N*-Methylformanilide	3	99	(91)
2-Piperidinocarbonyl	*N*-Methylformanilide	3	65	(14)
2,3-diMe	*N,N*-Dimethylformamide	1	30	(59)
3-Me	*N,N*-Dimethylformamide	1 ; 2	55 ; 20	(123)

of the indoline with chloranil affords a 64% yield of 1-methylindole-5-carboxaldehyde. Acylation can also be accomplished by Friedel–Crafts procedures. For example, 1-acetylindoline reacts with acetyl chloride in the presence of aluminum chloride to give a quantitative yield of 1,5-diacetylindoline (115). Again, dehydrogenation gives 5-acetylindole after hydrolytic cleavage of the *N*-acyl group.

TABLE LIII □ FORMYLATION AND ACYLATION OF INDOLES WITH HYDROGEN CYANIDE AND NITRILES

Indole substituent	Acylating agent	Position of acylation	Yield (%)	Ref.
None	Trifluoroacetonitrile	3	40	(122)
2-Me	Hydrogen cyanide	3	87	(90)
2-Me	Acetonitrile	3	33	(90)
2-Me	Trichloroacetonitrile	3	90	(44)
2-Me	Benzonitrile	3	75	(90)
2-Me	4-Cyanopyridine	3	50	(100)
2-Ethyl	Benzonitrile	3	52	(61)
2-Methyl-5-methoxy	Phenylacetonitrile	3	60	(6)
3-Me	Trifluoroacetonitrile	2	30	(122)
7-Me	Trifluoroacetonitrile	3	35	(122)

2. Synthesis by Means Other than Acylation

The Fischer cyclization of monophenylhydrazones has been used with modest success for the synthesis of 2-acylindoles. The yields are moderate, but the principal limitation has been the availability of the appropriate monophenylhydrazones. These are usually prepared from β-keto acids via Japp–Klingemann reactions. Recently enamines have been shown to react with diazonium salts, opening an important new route to the monophenylhydrazones (93) (see Chapter III,A,3).

Table LIV records typical examples of the synthesis of 2-acylindoles by direct Fischer cyclization. The early example of Manske, *et al.* (65) is illustrative. The yield is essentially quantitative in this instance.

$$PhCH_2CH(CO_2Et)COMe \ (\textbf{B17}) \xrightarrow[H_2O]{NaOH} PhCH_2CH(CO_2^-)COMe \ (\textbf{B18}) \xrightarrow{PhN_2^+} PhNHN{=}C(CH_2Ph)COMe \ (\textbf{B19}) \xrightarrow{HCl} \textbf{B20}$$

B20: 2-acetyl-3-phenylindole (Ph, Ac, N–H)

TABLE LIV □ SYNTHESIS OF ACYLINDOLES BY FISCHER CYCLIZATION

Indole	Cyclization reagent	Yield (%)	Ref.
2-Isonicotinyl	Phosphoric acid	33	(48)
2-Acetyl-3-phenyl	Hydrochloric acid	100	(65)
2-Benzoyl-3-phenyl	Acetic acid	33	(25)
2-Acetyl-3-(2-phthalimidoethyl)	Hydrochloric acid	15–20	(65)
3-Benzoyl-2-carboxy-1-methyl	Zinc chloride	45	(99)
2-Acetyl-1-methyl-5-nitro	Sulfuric acid	50	(28)
2-Acetyl-5,6-dimethoxy-3-phenyl	Hydrochloric acid	?	(63)
2-Acetyl-5-methoxy-6-methyl	Hydrochloric acid	15	(83)
1-Oxo-1,2,3,4-tetrahydrocarbazole	Sulfuric acid	?	(32)
8-Methoxy-1-oxo-1,2,3,4-tetrahydrocarbazole	Sulfuric acid	?	(32)

Organometallic reagents have been used on occasion for the synthesis of indolyl ketones. Phenyllithium and substituted aryllithiums react with indole-2-carboxylic acid to give 2-indolyl aryl ketones (9). The reaction of indole-2-carbonyl chloride with diphenylcadmium gives 2-benzoylindole in 32% yield (50). Diphenylcadmium reacts with 1-methylindole-2,3-dicarboxylic anhydride to give 3-benzoyl-1-methylindole-2-carboxylic acid as the only reported product (70%). The structure was proven by an independent Fischer cyclization (99). 1-Alkylindole-2-carboxaldehydes can be prepared by metalation of a 1-alkylindole followed by reaction with *N*,*N*-disubstituted formamides (43).

B21 (1-methylindole) $\xrightarrow{\text{RLi}}$ B22 (2-lithio-1-methylindole) $\xrightarrow[\text{(2) } H_2O,\ H^+]{\text{(1) PhN(Me)CHO}}$ B23 (1-methylindole-2-carboxaldehyde)

Indole-2-carboxylic acid derivatives have been converted to the corresponding aldehydes by several procedures. The McFadyen–Stevens procedure involving base-catalyzed decomposition of the *p*-toluenesulfonyl hydrazides has been reported to give good yields of indole-2-carboxaldehydes (26). The McFadyen–Stevens route has also been applied successfully to the synthesis of indoles bearing formyl substituents in the benzenoid ring (15).

Lithium tri-(*t*-butoxy)aluminum hydride has been reported to reduce indole-2-carbonyl chloride to indole-2-carboxaldehyde (88). Indole-2-carbonyl chloride can be converted to diazomethyl 2-indolyl ketone with diazomethane and subsequently to the bromomethyl ketone with hydrogen bromide (66, 72).

Plieninger and co-workers have developed conditions for conversion of 3- and 4-cyanoindole to the corresponding aldehydes by hydrogenating the nitriles in the presence of semicarbazide (75, 78). The reduction of indolyl nitriles to aldehydes with sodium hypophosphite and Raney nickel has also been found to be a very useful synthetic procedure (119).

4-Hydroxymethylindole (75) and 2-hydroxymethylindole (39) have been oxidized to the corresponding aldehyde with manganese dioxide. Potassium permanganate (113) is less satisfactory as an oxidizing agent. Thesing (116, 117) has synthesized indole-3-carboxaldehyde from gramine via *N*-(3-indolylmethyl)-*N*-phenylhydroxylamine which is oxidized to the nitrone and finally hydrolyzed.

N-alkyl-*o*-aminophenyl ketones condense with 3-bromo-1-phenyl-1,2-propanedione to give 2-acylindoles of the general structure **B28** (54). A reaction which is related is the conversion of **B29** to the indole **B31** by heating in acid (37). The formation of the indole ring presumably occurs via an intramolecular acid-catalyzed aldol condensation. The general pattern of these reactions, as depicted below (**B32** → **B33**), would seem to have general

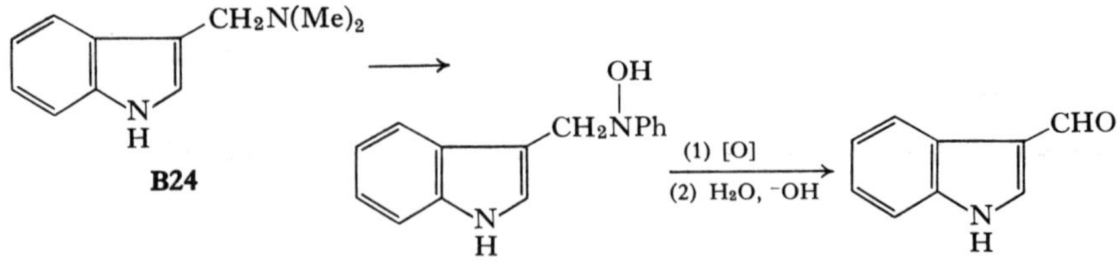

$CH_2N(Me)_2$ **B24** → OH, CH_2NPh **B25** —(1) [O]; (2) H_2O, ^-OH→ CHO **B26**

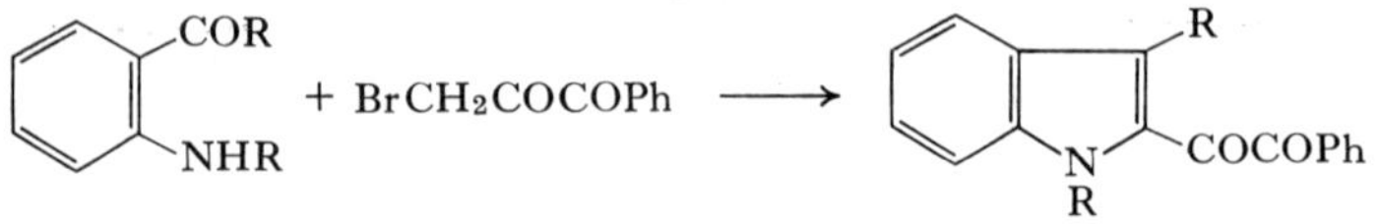

COR, NHR **B27** + $BrCH_2COCOPh$ → R, COCOPh, R **B28**

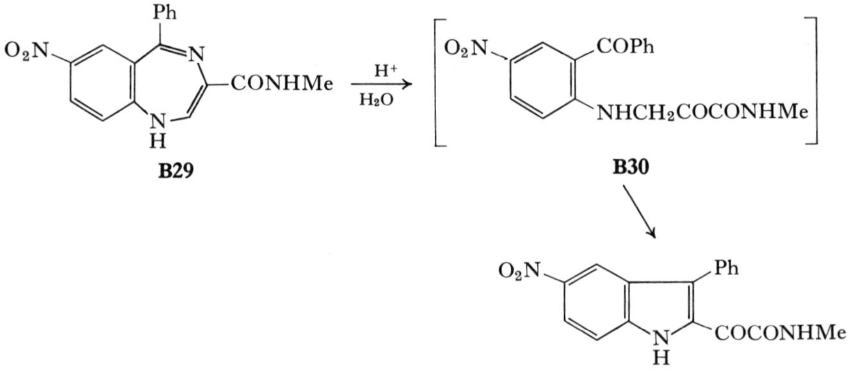

Ph, O_2N, CONHMe **B29** —H^+, H_2O→ [O_2N, COPh, $NHCH_2COCONHMe$] **B30** → O_2N, Ph, COCONHMe **B31**

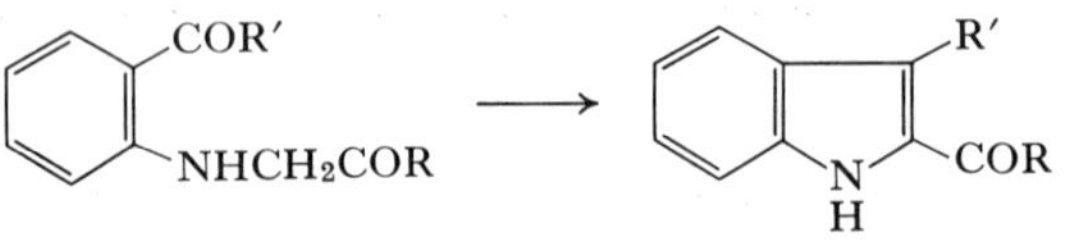

COR′, $NHCH_2COR$ **B32** → R′, COR **B33**

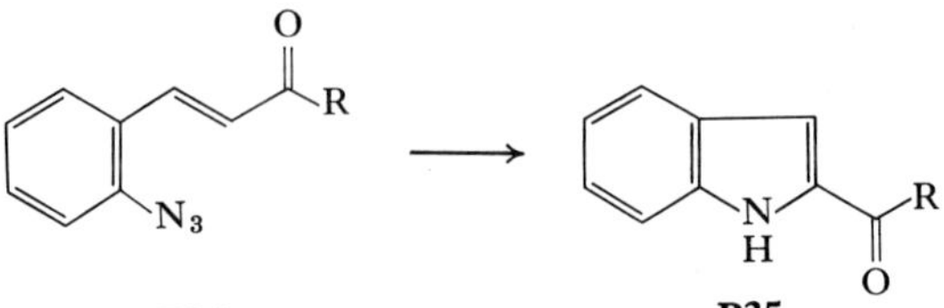

O, R, N_3 **B34** → R, O **B35**

potential for the synthesis of 2-acylindoles but only the two examples cited above seem to have been realized to date.

Thermal decomposition of *o*-azidostyryl ketones affords 2-acylindoles (101a).

C. INDOLECARBOXYLIC ACIDS

The properties of carboxyl groups substituted on the benzenoid ring of the indole nucleus are typical of aromatic carboxylic acids. However, certain features of the chemistry of indole-2-carboxylic acid and indole-3-carboxylic acid are unique. Indole-3-carboxylic acid is a weaker acid than indole-2-carboxylic acid (69). Pyrrole-2-carboxylic acid is also stronger acid than its 3-isomer (55). Since C-3 is the position of highest pi electron density in indole, whereas the opposite is true in pyrrole, it must be concluded that the distribution of electron density in the pi system is not the principal governing factor in acid strength. Intramolecular solvation by hydrogen bonding, involving the indole N–H and the carboxylate anion, and total electron density are also significant. In the case of the indole ring, intramolecular solvation of the 2-carboxylate group and the inductive effect of the nitrogen atom operate in the same direction and make the 2-carboxylic acid ($pK_a = 5.28$) (49a) a considerably stronger acid than the 3-isomer ($pK_a = 7.00$) (67).

1. Decarboxylation

Ease of decarboxylation is probably the most unique feature of the reactivity of indole 2- and 3-carboxylic acid derivatives. Many 3-(2-

$CH_2CH_2\overset{+}{N}H_3$, CO_2H **C1** $\xrightarrow{H^+}$ **C2** $\xrightarrow{-CO_2}$ **C3** $\xrightarrow{-H^+}$ **C4**

aminoethyl)indole-2-carboxylic acids have been decarboxylated in refluxing hydrochloric acid (see the Abramovitch tryptamine synthesis, Chapter III,A,3). The rate of decarboxylation is increased by electron releasing substituents in the benzene ring (1). This order of reactivity has led to the suggestion that decarboxylation proceeds by a mechanism involving protonation of the indole ring. Although the original suggestion involved N-protonation (1), current knowledge indicates that 3-protonation would be involved. The substituent effect then can be attributed to increased basicity of the indole and more extensive formation of the reactive protonated intermediate. In accord with the proposed involvement of 3-protonated indoles in this decarboxylation is the known ease of decarboxylation of 3,3-dialkyl-3*H*-indoles. 3,3-Dimethyl-3*H*-indole-2-carboxylic acid is decarboxylated rapidly at 135–140°C and is also decarboxylated in hot acidic solution (86). These decarboxylations are reminiscent of the facile decarboxylation of pyridine-2-carboxylic acids via ylide intermediates (38a). The exceptional ease of the decarboxylation reaction is illustrated by the conditions reported for the examples below:

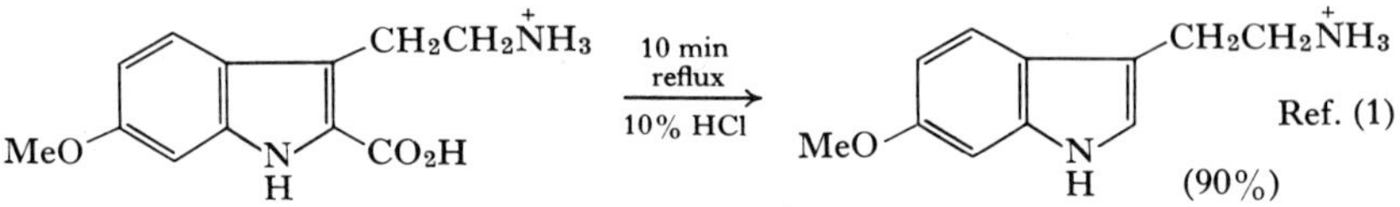

MeO, $CH_2CH_2\overset{+}{N}H_3$, CO_2H, N H — 10 min reflux, 10% HCl → MeO, $CH_2CH_2\overset{+}{N}H_3$, N H (90%) Ref. (1)

C5 **C6**

N, N H, CO_2H — reflux 1 hr, HCl → N H, N Ref. (23)

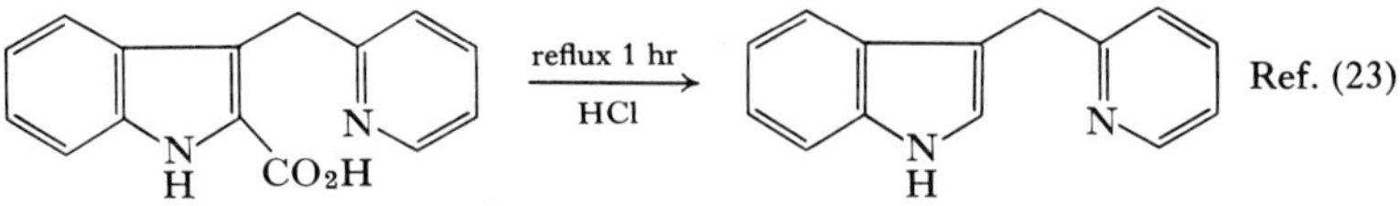

C7 **C8**

An alternative mechanism which could be operating in these decarboxylations would involve decarboxylation via prior 2-protonation.

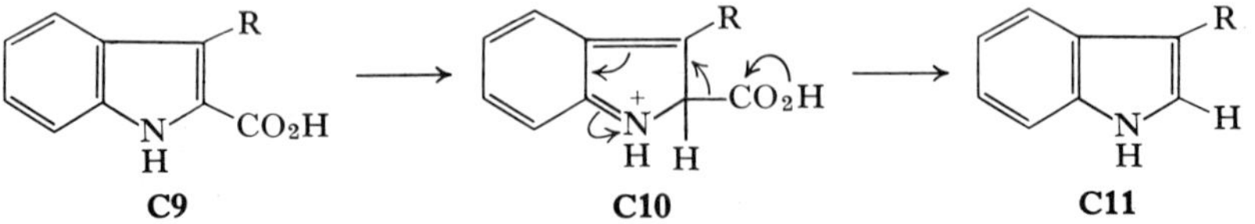

R, N H, CO_2H → R, $\overset{+}{N}$ H H, CO_2H → R, N H, H

C9 **C10** **C11**

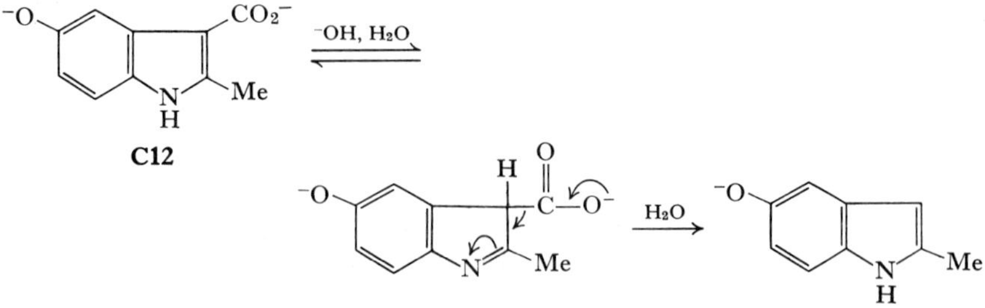

^-O, CO_2^-, N H, Me $\underset{}{\overset{^-OH,\ H_2O}{\rightleftharpoons}}$

C12

^-O, H, O, C—O^-, N, Me — H_2O → ^-O, N H, Me

C13 **C14**

Indole-3-carboxylic acids are decarboxylated when heated in acidic (2) or alkaline aqueous solutions (8). Under these conditions, formation of 3*H*-indoles may be involved in decarboxylation (80).

Hydrolysis of the ethyl trifluoromethylindole-3-carboxylates **C15** and **C17** results not only in ester hydrolysis and decarboxylation but also in removal of the trifluoromethyl group (64). The trifluoromethyl group is presumably lost by hydrolysis to a carboxyl group followed by decarboxylation. Use of the corresponding *t*-butyl esters permits selective cleavage

HO, Cl, CF_3, CO_2Et, Me, NH **C15** —(reflux 1.5 hr, 20% HCl)→ HO, Cl, Me, NH **C16** (56%)

CF_3, HO, CO_2Et, Me, NH **C17** —(reflux, 20% HCl)→ HO, Me, NH **C18** (50%)

of the carbo-*t*-butoxy substituent. Elimination of isobutylene and decarboxylation occurs in refluxing benzene in the presence of *p*-toluenesulfonic acid. These conditions again illustrate the exceptional ease with which indole 3-carboxylic acids decarboxylate. Acid-catalyzed decarboxylation of indole-3-carboxylic acids should be recognized as an example of the general substituent cleavage reaction discussed in Chapter II, D (80). Table LV lists several examples of hydrolytic decarboxylation of indoles.

TABLE LV □ DECARBOXYLATION OF INDOLECARBOXYLIC ACIDS UNDER HYDROLYTIC CONDITIONS

Indole substituent	Reaction conditions	Yield (%)	Ref.
A. Indole-2-carboxylic acids			
3-(2-Pyridylmethyl)	Reflux, hydrochloric acid, 1 hour	80	(23)
6-Methoxy-3-(2-aminoethyl)	Reflux, hydrochloric acid, 10 minutes	90	(1)
B. Indole-3-carboxylic acids			
1-Butyl-5-hydroxy-2,6-dimethyl	Reflux, hydrochloric acid	80	(2)
5-Hydroxy-2-methyl	Reflux, 2 *N* sodium hydroxide	100	(8)
5-Hydroxy-6-methoxy-2-methyl	Reflux, 2 *N* sodium hydroxide	74	(8)
5-Hydroxy-1,2,6-trimethyl	Reflux, hydrochloric acid	100	(2)

Conventional methods for decarboxylation have also been applied to indole carboxylic acids (84) and are, of course, required if the material being decarboxylated or the product is sensitive to hydrolytic conditions. Piers and Brown (74) have reviewed the experience of a number of workers with various copper-catalyzed decarboxylations and have recommended a procedure which involves refluxing the acid with a small amount of its copper salt in quinoline as the method of choice (74). This method was found applicable to acids bearing such substituents as nitro and benzylthio. It is reported to be generally superior to procedures involving use of various copper halides and oxides, copper bronze, or copper chromite. Casini and Goodman (21) found the copper salt procedure of Piers and Brown to be the method of choice for decarboxylation of 7-nitroindole-2-carboxylic acid. *N*,*N*-Dimethylacetamide was found to be a satisfactory solvent.

2. Synthesis of Indole Carboxylic Acids

Indole-2-carboxylic acid derivatives are available either by the Reissert method (Chapter III, D, 1) or by the Fischer cyclization of the arylhydrazones of 2-keto acids, which are usually prepared by the Japp–Klingemann reaction (Chapter III, A, 3). Esters of indole-3-carboxylic acids are formed directly in the Nenitzescu reaction (Chapter III, C) but only a limited number of substitution patterns can be built up directly by use of this reaction and a variety of methods for introduction of a 3-carboxy substituent into preformed indole rings have therefore been investigated. Indole-3-carbonyl chloride can be obtained in 16–23% yield by thermal (115–120°) decarbonylation of the readily available indole-3-glyoxylyl chloride (73). Hydrolysis, of course, gives the acid and esters and amides can be prepared directly from the acid chloride. Indole-3-carboxylic acid can also be obtained by carbonation of indolylmagnesium iodide (33, 67). The reaction of indolylmagnesium iodide with ethyl chloroformate gives ethyl indole-3-carboxylate but the 1-isomer is also formed, the ratio depending upon temperature (see Chapter I, I) (52). The oxidation of 5-nitroindole-3-glyoxylic acid to 5-nitroindole-3-carboxylic acid with alkaline hydrogen peroxide has been reported (67). This may be a generally useful approach in view of the ease of availability of the glyoxylic acids. Several systems have been reported to effect efficient conversion of indole-3-carboxaldehydes into the nitriles. Among these reagents are hydroxylamine followed by acetic anhydride (118), bis(trifluoroacetyl)hydroxylamine (79), ammonium phosphate in acetic acid (10), un*sym*-dimethylhydrazine followed by methyl iodide (38) and dehydration of the oxime with thionyl chloride (33). It is not clear, however, that hydrolysis will afford the acids since decarboxylation via substituent cleavage may intervene (80).

Süs and co-workers have shown that diazo-4-quinolones undergo photochemical ring contraction to give indole-3-carboxylic acids (102). The ring

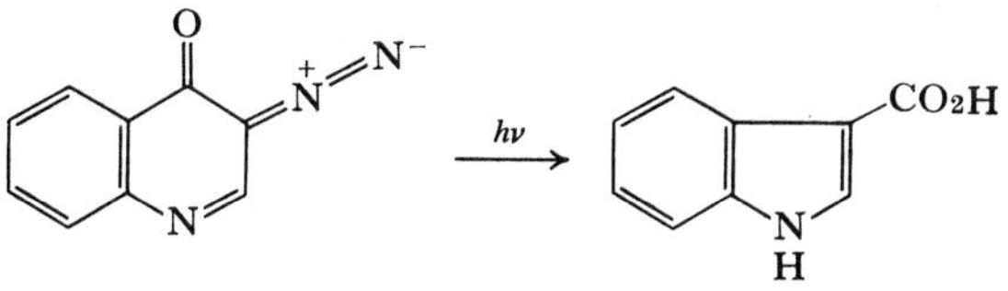

C19 → (hν) C20

contraction is presumably a photochemical Wolff rearrangement. The reaction was shown to be applicable to a number of systems bearing benzen-

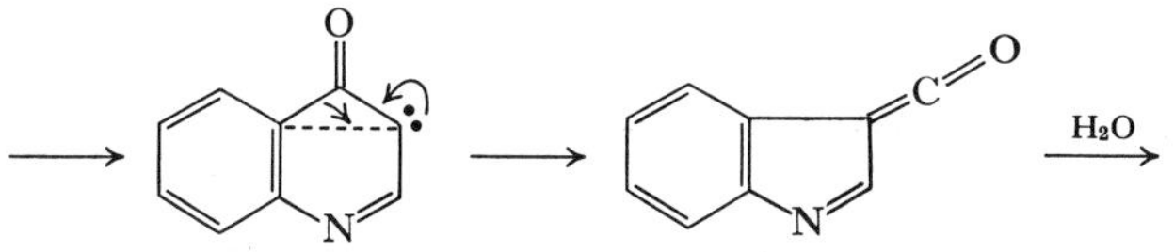

C19 → C21 → C22 → (H_2O) C20

oid substituents (102, 103). This method seems to be quite general, the principal disadvantage being that several steps are involved in the synthesis of the diazoquinolones from anthranilic acids.

A claim that ethyl indole-3-carboxylates can be prepared from anilines and ethyl formylchloroacetate (68) has been disputed (96). The products are, instead, ethyl α-chloro-β-anilinoacrylates and these can apparently be cyclized only with extensive decomposition (96).

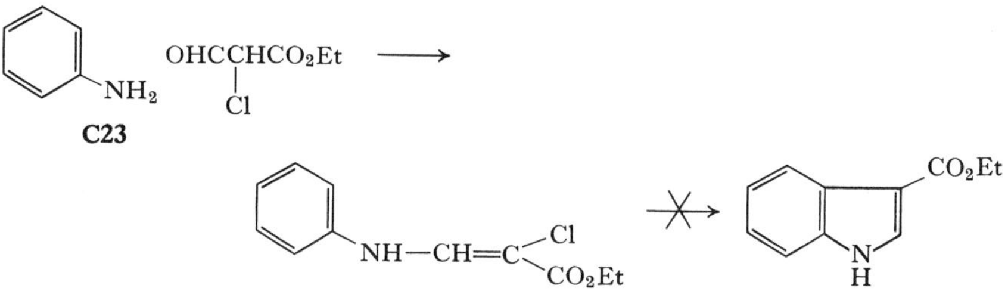

C23 (NH_2) + $OHCCHCO_2Et$ (Cl) →

C24 (NH—CH=C(Cl)CO_2Et) ↛ C25 (CO_2Et)

Indoles bearing carboxyl substituents in the benzenoid ring can be produced directly by several of the general methods of indole ring formation (Chapter III). Indirect introduction has generally involved formation of the appropriate halogen-substituted indole, followed by displacement with cuprous cyanide. Table LVI records a number of typical cyanoindole syntheses. Subsequent hydrolysis affords the carboxylic acids.

An indirect route to cyanoindoles and indolecarboxylic acids starting with aminoindolines has been devised (45). The sequence is illustrated by the synthesis of indole-7-carboxylic acid.

H_2N, N, Ac — **C26** $\xrightarrow[\text{CuCN}]{\text{HCl, }NO_2^-}$ NC, N, Ac — **C27** $\xrightarrow{\text{HCl, }H_2O}$ HO_2C, N, H — **C28** $\xrightarrow{\text{Pd/C}}$ HO_2C, N, H — **C29**

Trifluoromethyl groups on the benzenoid ring of indole are hydrolyzed to carboxyl groups by aqueous base (12) but the reaction has not yet been used for preparative purpose.

TABLE LVI □ SYNTHESIS OF CYANOINDOLES FROM HALOINDOLES

Indole substituent	Reaction conditions	Yield (%)	Ref.
2-Carboxy-4-chloro[a]	Cuprous cyanide, quinoline	51	(120)
2-Carboxy-5-bromo[a]	Cuprous cyanide, quinoline	34	(94)
2-Carboxy-7-chloro[a]	Cuprous cyanide, quinoline	29	(95)
2-Carboxy-5-bromo-1-methyl[a]	Cuprous cyanide, quinoline	40	(60)
5-Benzyloxy-4-chloro	Cuprous cyanide, quinoline	10	(85)
5-Bromo	Cuprous cyanide, *N*-methyl-pyrrolidone	78	(45)

[a] Decarboxylation accompanies displacement.

REFERENCES

1. Abramovitch, R. A., *J. Chem. Soc.* p. 4593 (1956).
1a. Alberti, C., *Gazz. Chim. Ital.* **69**, 568 (1939); *Chem. Abstr.* **34**, 1310 (1940).
2. Allen, G. R., Jr., Pidacks, C., and Weiss, M. J., *J. Am. Chem. Soc.* **88**, 2536 (1966).
3. Ames, D. E., Bowman, R. E., Evans, D. D., and Jones, W. A., *J. Chem. Soc.* p. 1984 (1956).
4. Anthony, W. C., *J. Org. Chem.* **25**, 2049 (1960).
5. Baker, J. W., *J. Chem. Soc.* p. 458 (1940).
6. Ballantine, J. A., Barrett, C. B., Beer, R. J. S., Boggiano, B. G., Eardley, S., Jennings, B. E., and Robertson, A., *J. Chem. Soc.* p. 2227 (1957).
7. Baskakov, Y. A., and Mel'nikov, N. N., *Sb. Statei Obshch. Khim., Akad. Nauk SSSR* **1**, 712 (1953); *Chem. Abstr.* **49**, 1006 (1955).

8. Beer, R. J. S., Clarke, K., Davenport, H. F., and Robertson, A., *J. Chem. Soc.* p. 2029 (1951).
9. Belgian Patent, 637,355 (1964); *Chem. Abstr.* **62**, 7731 (1965).
10. Blatter, H. M., Lukaszewski, H., and deStevens, G., *J. Am. Chem. Soc.* **83**, 2203 (1961); *Org. Syn.* **43**, 58 (1963).
11. Bodendorf, K., and Walk, A., *Arch. Pharm.* **294**, 484 (1961).
12. Bornstein, J., Leone, S. A., Sullivan, W. F., and Bennett, O. F., *J. Am. Chem. Soc.* **79**, 1745 (1957).
13. Borsche, W., and Groth, H., *Ann. Chem.* **549**, 238 (1941).
14. Brehm, W. J., *J. Am. Chem. Soc.* **71**, 3541 (1949).
15. Brown, U. M., Carter, P. H., and Tomlinson, M., *J. Chem. Soc.* p. 1843 (1958).
16. Buchmann, G., and Rossner, D., *J. Prakt. Chem.* [4] **25**, 117 (1964).
17. Buu-Hoi, N. P., Bisagni, E., Royer, R., and Routier, C., *J. Chem. Soc.* p. 625 (1957).
18. Buu-Hoi, N. P., Hoan, N., and Khoi, N. H., *J. Org. Chem.* **15**, 131 (1950).
19. Buu-Hoi, N. P., Hoan, N., and Xuong, N. D., *Rec. Trav. Chim.* **69**, 1083 (1950).
20. Byrn, M., and Calvin, M., *J. Am. Chem. Soc.* **88**, 1916 (1966).
21. Casini, G., and Goodman, L., *Can. J. Chem.* **42**, 1235 (1964).
22. Chen, F. Y., and Leete, E., *Tetrahedron Letters* p. 2013 (1963).
23. Clemo, G. R., and Seaton, J. C., *J. Chem. Soc.* p. 2582 (1954).
24. Collington, E. W., and Jones, G., *Tetrahedron Letters* p. 1935 (1968).
25. Curtin, D. Y., and Poutsma, M. L., *J. Am. Chem. Soc.* **84**, 4887 (1962).
26. Dambal, S. B., and Siddappa, S., *J. Indian Chem. Soc.* **42**, 112 (1965).
26a. De Graw, J. I., and Kennedy, J. G., *J. Heterocyclic Chem.* **3**, 9 (1966).
27. De Graw, J. I., and Kennedy, J. G., *J. Heterocyclic Chem.* **3**, 90 (1966).
28. Diels, O., and Dürst, W., *Chem. Ber.* **47**, 284 (1914).
29. Dolby, L. J., and Biere, H., *J. Am. Chem. Soc.* **90**, 2699 (1968).
30. Dolby, L. J., and Sakai, S., *Tetrahedron* **23**, 1 (1967).
31. Dorofeenko, G. N., *Zh. Veses. Khim. Obshchestva im. D. I. Mendeleeva* **5**, 354 (1960); *Chem. Abstr.* **54**, 22563 (1960).
32. Douglas, B., Kirkpatrick, J. L., Moore, B. P., and Weisbach, J. A., *Australian J. Chem.* **17**, 246 (1964).
33. Doyle, F. P., Ferrier, W., Holland, D. O., Mehta, M. D., and Nayler, J. H. C., *J. Chem. Soc.* p. 2853 (1956).
34. Eistert, B., German Patent 855,563 (1952); *Chem. Abstr.* **52**, 15592 (1958).
35. Finkelstein, J., and Lee, J., U.S. Patent 2,695,290 (1954); *Chem. Abstr.* **49**, 15978 (1955).
36. Fleming, I., and Harley-Mason, J., *J. Chem. Soc., C* p. 425 (1966).
37. Fryer, R. I., Earley, J. V., and Sternbach, L. H., *J. Org. Chem.* **32**, 3798 (1967).
38. Grandberg, I. I., *J. Gen. Chem. USSR (English Transl.)* **34**, 570 (1964).
38a. Haake, P., and Mantecon, J., *J. Am. Chem. Soc.* **86**, 5230 (1964).
39. Harley-Mason, J., and Pavri, E. H., *J. Chem. Soc.* p. 2565 (1963).
40. Hart, G., Liljegren, D. R., and Potts, K. T., *J. Chem. Soc.* p. 4267 (1961).
41. Hart, G., and Potts, K. T., *J. Org. Chem.* **27**, 2940 (1962).
42. Hishida, S., *J. Chem. Soc. Japan, Pure Chem. Sect.* **72**, 312 (1951); *Chem. Abstr.* **46**, 5038 (1952).
43. Hoffmann, K., Rossi, A., and Kebrle, J., German Patent 1,093,565 (1958); *Chem. Abstr.* **56**, 4735 (1962).
44. Houben, J., and Fischer, W., *Chem. Ber.* **64**, 2645 (1931).
45. Ikan, R., and Rapaport, E., *Tetrahedron* **23**, 3823 (1967).
46. Inhoffen, H. H., Nordsiek, K.-H., and Schäfer, H., *Ann. Chem.* **668**, 104 (1963).

47. Ishizumi, K., Shioiri, T., and Yamada, S., *Chem. & Pharm. Bull.* (*Tokyo*) **15**, 863 (1967).
48. Jackson, A., and Joule, J. A., *Chem. Commun.* p. 459 (1967).
49. Jackson, A. H., and Smith, A. E., *J. Chem. Soc.* p. 5510 (1964).
49a. Jaffé, H. H., and Jones, H. L., *Advan. Heterocyclic. Chem.*, **3**, 209 (1964).
50. Jardine, R. V., and Brown, R. K., *Can. J. Chem.* **41**, 2067 (1963).
51. Jennings, K. F., *J. Chem. Soc.* p. 497 (1957).
51a. Johnson, R. A., Ph.D. Thesis with W. E. Noland, University of Minnesota, Minneapolis, Minnesota, 1965; *Dissertation Abstr.* **26**, 5719 (1966).
52. Kašpárek, S., and Heacock, R. A., *Can. J. Chem.* **44**, 2805 (1966).
53. Katritzky, A. R., *J. Chem. Soc.* p. 2581 (1955).
54. Kempter, G., and Schiewald, E., *J. Prakt. Chem.* [4] **28**, 169 (1965).
55. Khan, M. K. A., and Morgan, K. J., *Tetrahedron* **21**, 2197 (1965).
56. Kiang, A. K., and Mann, F. G., *J. Chem. Soc.* p. 594 (1953).
57. Kost, A. N., Eraksina, V. N., and Vingradova, E. V., *J. Org. Chem. USSR* (*English Transl.*) **1**, 126 (1965).
58. Kost, A. N., Mitropol'skaya, V. N., Portnova, S. L., and Krasnova, V. A., *J. Gen. Chem. USSR* (*English Transl.*) **34**, 3025 (1964).
59. Kucherova, N. F., Evdakov, V. P., and Kochetkov, N. K., *J. Gen. Chem. USSR* (*English Transl.*) **27**, 1131 (1957).
60. Kunori, M., *Nippon Kagaku Zasshi* **83**, 839 (1962); *Chem. Abstr.* **59**, 1573 (1963).
61. Kunori, M., *Nippon Kagaku Zasshi* **83**, 841 (1962); *Chem. Abstr.* **59**, 1573 (1963).
62. Leonard, N. J., Adamacik, J. A., Djerassi, C., and Halpern, O., *J. Am. Chem. Soc.* **80**, 4858 (1958).
63. Lions, F., and Spruson, M. J., *J. Proc. Roy. Soc. N. S. Wales* **66**, 171 (1932); *Chem. Abstr.* **27**, 291 (1933).
64. Littell, R., and Allen, G. R., Jr., *J. Org. Chem.* **33**, 2064 (1968).
65. Manske, R. H. F., Perkin, W. H., Jr., and Robinson, R., *J. Chem. Soc.* p. 1 (1927).
66. Matell, M., *Arkiv Kemi* **10**, 179 (1956); *Chem. Abstr.* **51**, 8069 (1957).
67. Melzer, M. S., *J. Org. Chem.* **27**, 496 (1962).
68. Mentzer, C., *Compt. Rend.* **322**, 1176 (1946); *Chem. Abstr.* **41**, 444 (1947).
69. Millich, F., and Becker, E. I., *J. Org. Chem.* **23**, 1096 (1958).
70. Moffatt, J. S., *J. Chem. Soc.* p. 1442 (1957).
71. Nenitzescu, C. D., and Răileanu, D., *Acad. Rep. Populare Romine, Studii Cercetari Chim.* **7**, 243 (1959); *Chem. Abstr.* **54**, 7681 (1960).
72. Nogrady, T., Doyle, T. W., and Morris, L., *J. Med. Chem.* **8**, 656 (1965).
73. Peterson, P. E., Wolf, J. P., III, and Niemann, C., *J. Org. Chem.* **23**, 303 (1958).
74. Piers, E., and Brown, R. K., *Can. J. Chem.* **40**, 559 (1962).
75. Plieninger, H., Höbel, M., and Liede, V., *Chem. Ber.* **96**, 1618 (1963).
76. Plieninger, H., and Müller, W., *Chem. Ber.* **93**, 2029 (1960).
77. Plieninger, H., Müller, W., and Weinerth, K., *Chem. Ber.* **97**, 667 (1964).
78. Plieninger, H., and Werst, G., *Chem. Ber.* **88**, 1956 (1955).
79. Pomeroy, J. H., and Craig, C. A., *J. Am. Chem. Soc.* **81**, 6340 (1959).
80. Powers, J. C., *Tetrahedron Letters* p. 655 (1965).
81. Rappe, C., *Acta Chem. Scand.* **18**, 818 (1964).
82. Reeve, W., Hudson, R. S., and Woods, C. W., *Tetrahedron* **19**, 1243 (1963).
83. Remers, W. A., Roth, R. H., and Weiss, M. J., *J. Am. Chem. Soc.* **86**, 4612 (1964).
84. Rensen, M., *Bull. Soc. Chim. Belges* **68**, 258 (1959); *Chem. Abstr.* **54**, 494 (1960).
85. Robinson, P., and Slaytor, M., *Australian J. Chem.* **14**, 606 (1961).
86. Robinson, R., and Suginome, H., *J. Chem. Soc.* p. 298 (1932).

87. Rodinov, V. M., and Veselovskaya, T. K., *Zh. Obshch. Khim.* **20**, 2202 (1950); *Chem. Abstr.* **45**, 7106 (1951).
88. Sato, Y., and Masumoto, Y., *Takamine Kenkyusho Nempo* **11**, 33 (1959); *Chem. Abstr.* **55**, 5456 (1961).
89. Saxton, J. E., *J. Chem. Soc.* p. 3592 (1952).
90. Seka, R., *Chem. Ber.* **56**, 2058 (1923).
91. Shabica, A. C., Howe, E. E., Ziegler, J. B., and Tishler, M., *J. Am. Chem. Soc.* **68**, 1156 (1946).
92. Shaw, K. N. F., McMillan, A., Gudmundson, A. G., and Armstrong, M. D., *J. Org. Chem.* **23**, 1171 (1958).
93. Shvedov, V. I., Altukhova, L. B., and Grinev, A. N., *J. Org. Chem. USSR (English Transl.)* **1**, 882 (1965); **2**, 1586 (1966).
94. Singer, H., and Shive, W., *J. Org. Chem.* **20** 1458 (1955).
95. Singer, H., and Shive, W., *J. Am. Chem. Soc.* **77**, 5700 (1955).
96. Smith, G. F., *J. Chem. Soc.* p. 1637 (1950).
97. Smith, G. F., *J. Chem. Soc.* p. 3842 (1954).
98. Sprio, V., and Madonia, P., *Gazz. Chim. Ital.* **87**, 454 (1957); *Chem. Abstr.* **52**, 4601 (1958).
99. Staunton, R. S., and Topham, A., *J. Chem. Soc.* p. 1889 (1953).
100. Strell, M., and Kopp, E., *Chem. Ber.* **91**, 1621 (1958).
101. Sundberg, R. J., *J. Org. Chem.* **33**, 487 (1968).
101a. Sundberg, R. J., Lin, L. S., and Blackburn, D. E., *J. Heterocyclic Chem.* **6**, 441 (1969).
102. Süs, O., Glos, M., Möller, K., and Eberhardt, H.-D., *Ann. Chem.* **583**, 150 (1953).
103. Süs, O., and Möller, K., *Ann. Chem.* **593**, 91 (1955).
104. Suvorov, N. N., Ovchinnikova, Z. D., and Sheinker, Y. N., *J. Gen. Chem. USSR (English Transl.)* **31**, 2174 (1961).
105. Szmuszkovicz, J., *J. Am. Chem. Soc.* **80**, 3782 (1958).
106. Szmuszkovicz, J., *J. Am. Chem. Soc.* **82**, 1180 (1960).
107. Szmuszkovicz, J., *J. Org. Chem.* **27**, 511 (1962).
108. Szmuszkovicz, J., *J. Org. Chem.* **27**, 515 (1962).
109. Szmuszkovicz, J., *J. Org. Chem.* **27**, 1582 (1962).
110. Szmuszkovicz, J., U.S. Patent 2,991,291 (1961); *Chem. Abstr.* **56**, 3458 (1962).
111. Szmuszkovicz, J., and Anthony, W. C., U.S. Patent 2,984,670 (1961); *Chem. Abstr.* **55**, 25984 (1961).
112. Takagi, S., Sugii, A., and Machida, K., *Pharm. Bull. (Tokyo)* **5**, 617 (1957); *Chem. Abstr.* **52**, 16331 (1958).
113. Taylor, W. I., *Helv. Chim. Acta* **33**, 164 (1950).
114. Terent'ev, A. P., Ban-Lun, G., and Preobrazhenskaya, M. N., *J. Gen. Chem. USSR (English Transl.)* **32**, 1311 (1968).
115. Terent'ev, A. P., Preobrazhenskaya, M. N., and Sorokina, G. M., *J. Gen. Chem. USSR (English Transl.)* **29**, 2835 (1959).
116. Thesing, J., *Chem. Ber.* **87**, 507 (1954).
117. Thesing, J., Müller, A., and Michel, G., *Chem. Ber.* **88**, 1027 (1955).
118. Trabert, C. H., *Arch. Pharm.* **294**, 246 (1961).
119. Troxler, F., Harnisch, A., Bormann, G., Seeman, F., and Szabo, L., *Helv. Chim. Acta* **51**, 1616 (1968).
120. Uhle, F. C., *J. Am. Chem. Soc.* **71**, 761 (1949).
121. Walker, G. N., and Moore, M. M., *J. Org. Chem.* **26**, 432 (1961).
122. Whalley, W. B., *J. Chem. Soc.* p. 1651 (1954).
123. Whittle, C. W., and Castle, R. N., *J. Pharm. Sci.* **52**, 645 (1963); *Chem. Abstr.* **61**, 13276 (1964).

124. Wilson, N. D. V., Jackson, A., Gaskell, A. J., and Joule, J. A., *Chem. Commun.* p. 584 (1968).
125. Yagil, G., *Tetrahedron* **23**, 2855 (1967).
126. Yamada, S., and Shioiri, T., *Tetrahedron Letters* p. 351 (1967); *Tetrahedron* **24**, 4159 (1968).
127. Yur'ev, Y. K., and Elyakov, G. B., *Zh. Obshch. Khim.* **26**, 2350 (1956); *Chem. Abstr.* **51**, 5042 (1957).
128. Ziegler, F. E., and Zoretic, P. A., *Tetrahedron Letters* p. 2639 (1968); *J. Am. Chem. Soc.* **91**, 2342 (1969).

X

NATURALLY OCCURRING DERIVATIVES OF INDOLE AND INDOLES OF PHYSIOLOGICAL AND MEDICINAL SIGNIFICANCE

A. INTRODUCTION

The most thoroughly studied class of naturally occurring indole derivatives is the indole alkaloids. While certain aspects of the synthesis, biosynthesis, and chemistry of indole alkaloids have been discussed in earlier chapters, it is clearly beyond the scope of this book to attempt to review the vast literature concerning structural elucidation, stereochemical characterization, and chemical transformations and synthesis of the indole alkaloids. Authoritative discussions (35) and compilations (23) are available. It seems similarly inappropriate to initiate a discussion of proteins although it is amusing to claim all tryptophan-containing proteins as "naturally occurring derivatives of indole." There are, however, well-characterized, smaller groups of indoles including antibiotics and other biologically and medicinally interesting molecules, which can usefully be considered in this chapter. Indole-derived antibiotics are discussed first.

B. INDOLE-DERIVED ANTIBIOTICS

The mitomycin family of antibiotics is represented by structural formula **B1**. These compounds, which are isolated (34) from *Streptomyces* cultures, are indoline quinones and members of this family have been found to be

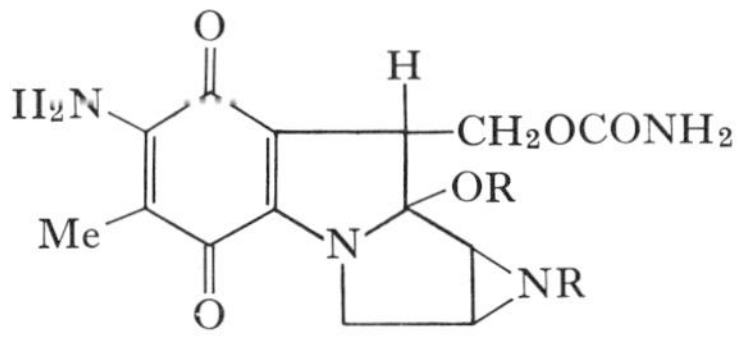

B1

MeO, Me, CO_2Et, N H — **B2**

$CH_2{=}CHCO_2Me$ / $OC(Me)_3$ →

[MeO, Me, CO_2Et, N—$CH_2CH_2CO_2Me$ — **B3**] →

MeO, Me, N, O, CO_2Me — **B4**

H^+, H_2O →

MeO, Me, N, O — **B5**

NH_2NH_2 / $\overline{O}H$ →

MeO, Me, N — **B6**

(1) $HCON(Me)_2$ / $POCl_3$ (2) $AlCl_3$ →

HO, Me, CHO, N — **B7**

$NO(SO_3K)_2$ →

O, O, Me, CHO, N — **B8**

Ac_2O / BF_3 →

AcO, OAc, Me, AcO, CHO, N — **B9**

(1) NaOH (2) O_2 →

HO, O, Me, O, CHO, N — **B10**

CH_2N_2 →

MeO, O, Me, O, CHO, N — **B11**

(1) $NaBH_4$ 2) PhO_2CCl →

MeO, O, Me, O, CH_2OCO_2Ph, N — **B12**

NH_3 →

MeO, O, Me, O, CH_2OCONH_2, N — **B13**

active against bacteria and in cancer chemotherapy. Active research on the mechanism of action of these substances is under way. Much attention has been focused on the interaction of the mitomycins with DNA (13). Structure **B1** ($R = H$, $R' = CH_3$) was established for mitomycin C by chemical evidence (57) reported in 1962. Additional degradative work was described in 1964 (55). An X-ray structure determination on the *p*-bromobenzene-sulfonyl derivative of mitomycin A (**B1**, $R = O_2S\ C_6H_4Br$, $R^1 = CH_3$) has confirmed the chemical work and established the stereochemistry (56). Several other closely related compounds have been isolated and characterized.

Considerable effort has been invested in the synthesis of mitomycin derivatives. Although none of the naturally occurring systems has yet been synthesized, a number of active antibacterials have been prepared by degradative (42) or synthetic approaches (1–5, 43, 46–48).

Synthetic approaches reported to date have accomplished the synthesis of the reduced pyrrolo[1,2-a]indole ring system (2, 4, 47). The scheme on p. 432 illustrates a successful approach to this skeleton (2).

A group of Japanese workers has succeeded in establishing the pyrrolo-[1,2-a]indole ring substituted with the fused aziridine ring (24). Their approach is outlined on p. 434.

Structure **B21** has been established (51, 52) for the antibiotic indolmycin which is also isolated from certain *Streptomyces* cultures. The total synthesis of indolmycin has been accomplished (44, 52) by the schemes outlined on p. 435.

Several strains of *Streptomyces* have also been found to yield a highly vesiscant substance, teleocidin B. This substance is toxic to aquatic animals. Chemical studies permitted the postulation of partial structure **B27** for the dihydro derivative of teleocidin B (40). Concurrent X-ray studies of the bromoacetyl hydrobromide derivative confirmed the structural postulate by establishing structure **B28** (21, 50) for the bromoacetyl derivative.

Gliotoxin is an antibiotic (28) of fungal origin which contains a tetrahydroindole ring and a bicyclic disulfide unit. The structure **B29** was proposed in 1958 (10) and this was confirmed and the absolute configuration was established by a recent X-ray study (9). No major efforts toward the synthesis of gliotoxin have come to the author's attention.

Recently a good deal of interest has developed in metabolites of the fungus *Arachmiotus avereus* (Eidam) Schroeter. Structural investigation has shown that one of the metabolites, called aranotin, has the structure **B30** (39). The major metabolite is the closely related compound **B31**. X-ray methods (36) have established the stereochemistry of **B31**. X-ray structure determination has also been carried out on the diacetyl derivative of **B30**, which was independently isolated from *Aspergillus terreus* (14). Although biological

MeO, Me, N H, CHO (**B14**) + $CH_2{=}CHP^+(Ph)_3$ (**B15**) ⟶ MeO, Me, N (**B16**)

$\xrightarrow{(MeO)_2CO,\ ^-OC(Me)_3}$ MeO, Me, N, CO_2Me (**B17**)

$\xrightarrow{IN_3}$ MeO, Me, N, CO_2Me, N_3, I (**B18**)

$\xrightarrow{(1)\ H_2\ (2)\ MeO^-}$ MeO, Me, N, CO_2Me, NH (**B19**)

$\xrightarrow{ClCO_2Et}$ MeO, Me, N, CO_2Me, NCO_2Me (**B20**)

B21

B22 $\xrightarrow[(Me)_2CHNH_2]{MeCHO}$ B23

B23 $\xrightarrow{(1)\ AcOCH(CO_2Et)_2;\ (2)\ NaOH;\ (3)\ heat}$ B24

B24 $\xrightarrow{CH_2N_2}$ B25

B25 $\xrightarrow{HN{=}C(NHMe)_2}$ B21

B22 + B26 → B25

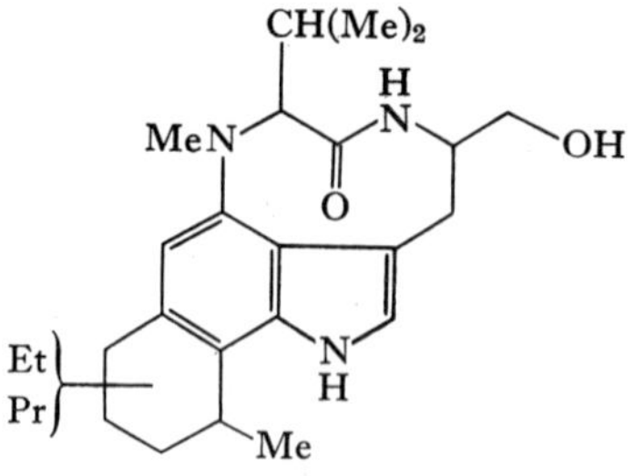

CH(Me)2 H MeN N OH O Et Pr N H Me

B27

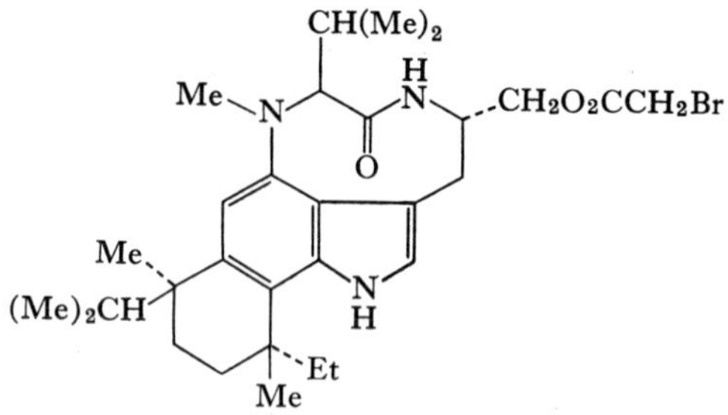

CH(Me)2 H Me N N CH2O2CCH2Br O Me (Me)2CH N H Et Me

B28

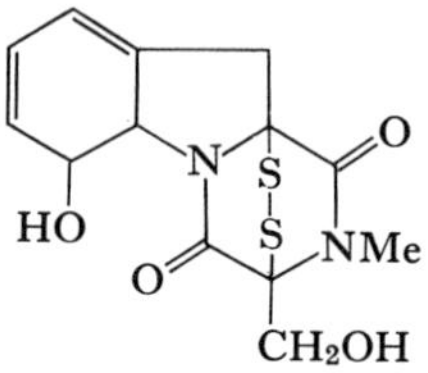

N S S O NMe HO O CH2OH

B29

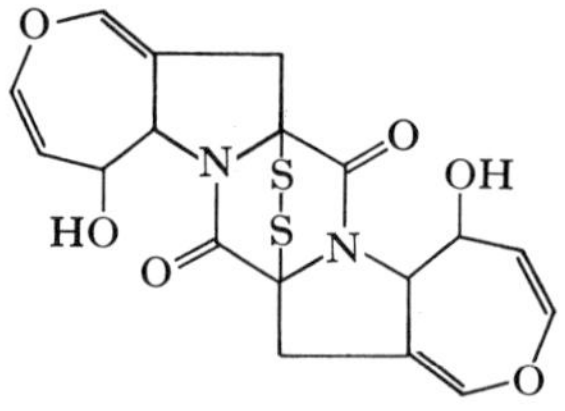

O N S S O OH HO O N O

B30

O S Me N O H OAc AcO H N O S Me O

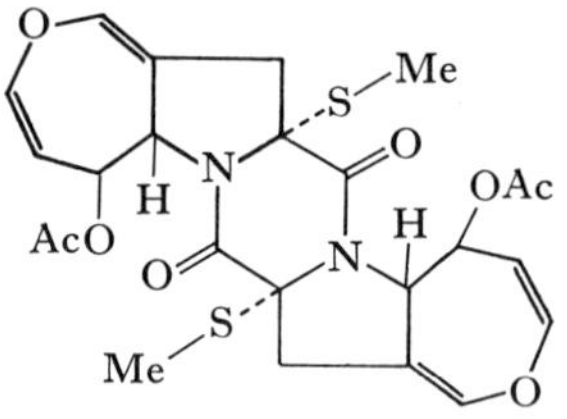

B31

characterization of this family of compounds has not been published in detail, they are reported (14, 39) to exhibit antiviral activity and are therefore of great interest. Justification for inclusion of these compounds as indole-derived natural products comes from their structural relationship to gliotoxin (**B29**) and apoaranotin (**B32**) in which reduced indole rings can be recognized. The structural relationships suggest that the aranotin family is biogenetically related to the tetrahydroindoles gliotoxin and apoaranotin

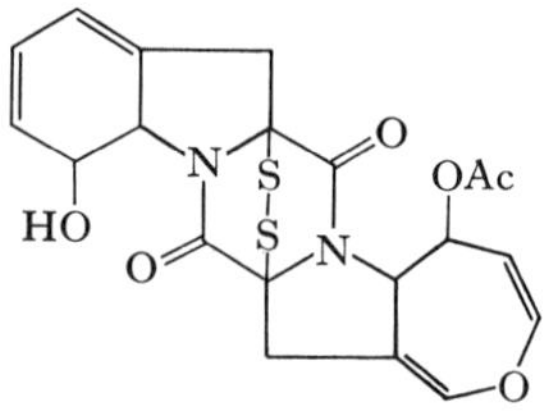

N S S O OAc HO O N O

B32

The suggestion has been made (41) that benzene oxides (see Chapter V, D) may be the key to the biogenetic connection between these compounds.

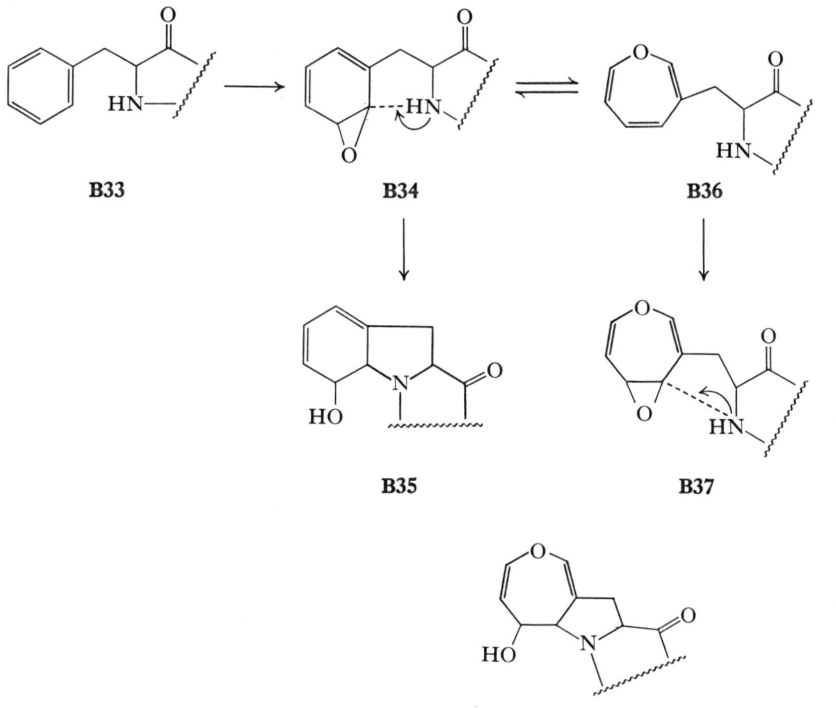

B33

B34

B36

B35

B37

B38

The sulfur-bridged diketopiperazine ring found in aranotin and gliotoxin is also found, although in somewhat different orientation, in sporidesmin (**B39**). Sporidesmin is the toxic agent of the fungus *Pithomyces chartarum* (formerly *Sporidesmium bakeri syd.*) which is responsible for widespread disease in sheep. Sporidesmin has been subjected to X-ray structure analysis which determined the structure to be that depicted (18).

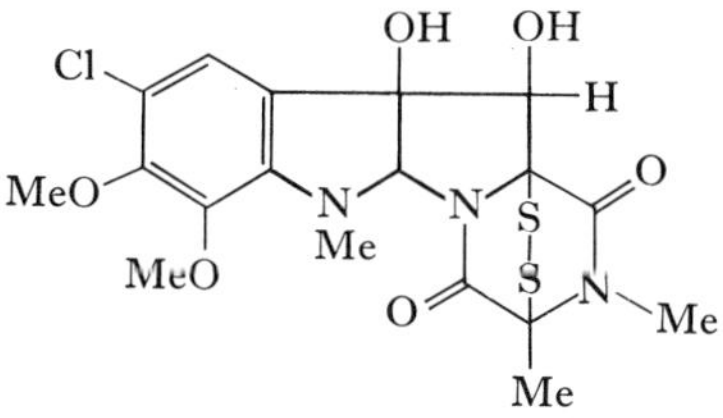

B39

Chemical evidence for this structure has also accumulated (25, 26, 49).

C. INDOLES OF PHYSIOLOGICAL SIGNIFICANCE: TRYPTOPHAN, SEROTONIN, AND MELATONIN

L-Tryptophan (**C1**) is one of the essential amino acids and is, of course, a protein structural unit. The formation of serotonin (**C2**, 5-hydroxytryptamine) from tryptophan by hydroxylation and decarboxylation was discussed in Chapter V, D. These two compounds are the indoles whose physiological action has been subjected to the widest study. Serotonin is a vasoconstrictor and regulates gastric secretion and intestinal peristalsis. It is believed that serotonin may also play a role in the functioning of the central nervous system and this problem is under intensive investigation. Reviews on the

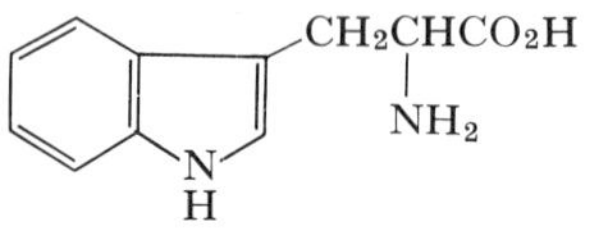

CH_2CHCO_2H / NH_2 / N / H

C1

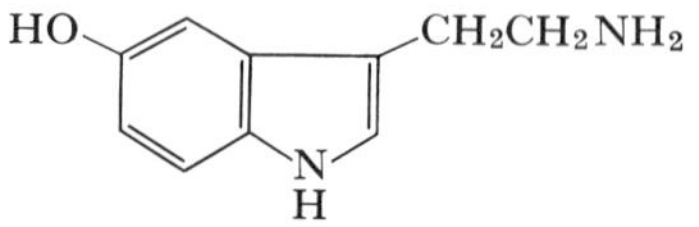

HO / $CH_2CH_2NH_2$ / N / H

C2

enormous amount of work on the distribution, pharmacology, and physiology of the compound are available (17, 19). Much of this work has been stimulated by the theory that defects in tryptophan metabolism and the resulting disruption in serotonin levels might be the cause of certain mental disorders (59). As was pointed out in Chapter V, D, the major pathway for metabolism of tryptophan involves cleavage of the indole ring between C-2 and C-3. This pathway leads directly to formylkynurenine but eventually to nicotinic acid as depicted in the scheme below. Because of this metabolic pathway, tryptophan can substitute nutritionally for nicotinic acid in higher animals (12).

Interest in melatonin (**C9**) is of much more recent origin than the extensive study of serotonin and tryptophan. Its close structural relationship to serotonin is obvious. Melatonin is synthesized from *N*-acetylserotonin in the pineal gland (60). Current interest in this hormone centers around the role it may play in controlling diurnal variations in biological functions. It is believed that direct nerve connections to the pineal gland permit its rate of synthesis to be controlled by variations in light intensity. Melatonin has been shown to inhibit several physiological functions and it may play a key role in the cyclical control of many biological functions (60).

Several of the biologically important hydroxylated phenylethylamines are known to undergo oxidation to products containing the indole nucleus such as adrenochrome (**C10**). Similar oxidative cyclizations lead to the general class of pigments called melanins and, although the structure and mechanism of formation of melanins is still under investigation, indole derivatives, particularly 5,6-dihydroxyindole, have been implicated (41a).

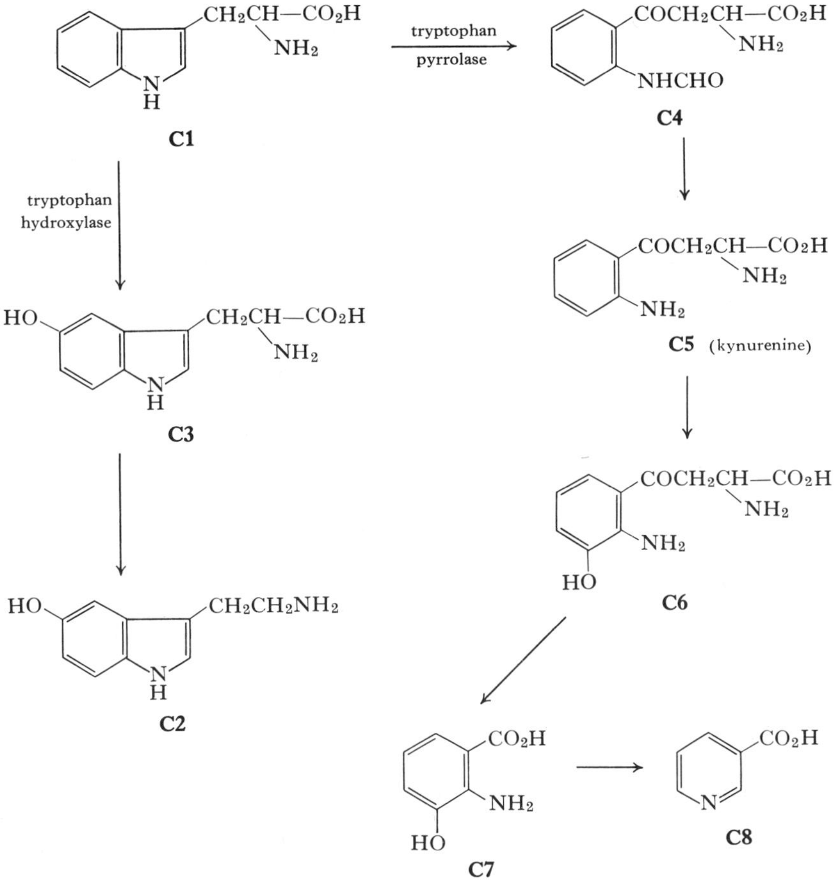

$CH_2CH—CO_2H$, NH_2, N, H

C1

tryptophan pyrrolase

$COCH_2CH—CO_2H$, NH_2, NHCHO

C4

tryptophan hydroxylase

HO, $CH_2CH—CO_2H$, NH_2, N, H

C3

$COCH_2CH—CO_2H$, NH_2, NH_2

C5 (kynurenine)

$COCH_2CH—CO_2H$, NH_2, NH_2, HO

C6

HO, $CH_2CH_2NH_2$, N, H

C2

CO_2H, NH_2, HO

C7

CO_2H, N

C8

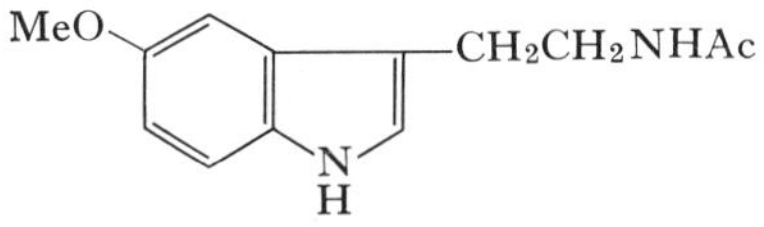

MeO, CH_2CH_2NHAc, N, H

C9

O, OH, ^{-}O, $\overset{+}{N}$, Me

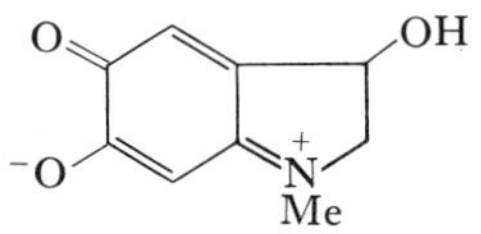

C10

D. INDOLE DERIVATIVES AS HALLUCINOGENS

The psychotomimetic activity of lysergic acid diethylamide (**D1**, LSD) is widely recognized. The simpler indoles 5-hydroxy-*N*,*N*-dimethyltryptamine (bufotenine), *N*,*N*-dimethyltryptamine, and 5-methoxy-*N*,*N*-

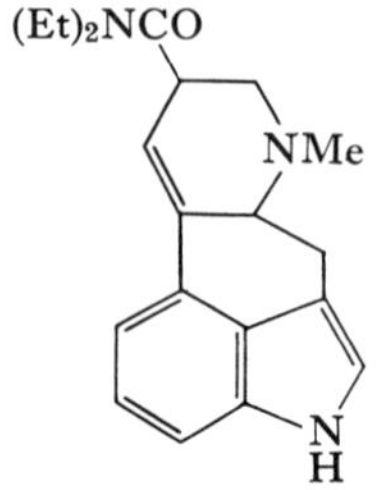

D1

dimethyltryptamine are also hallucinogenic and have been identified in natural extracts which have long been used to induce hallucinations as part of religious rites (27, 53). Only slightly more complex is the zwitterionic tryptamine psilocybin (**D2**) which is also strongly hallucinogenic. Hofmann

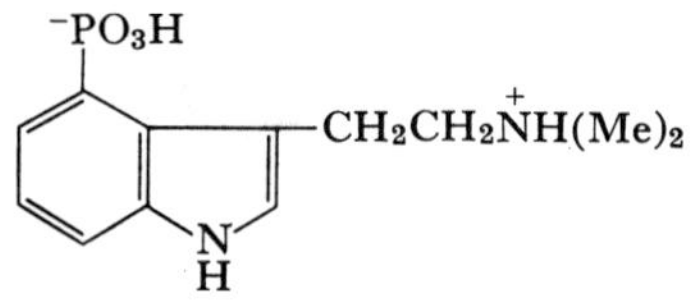

D2

(27) has reviewed the discovery, structure, and some aspects of the pharmacology of psychotomimetic agents. A review outlining synthetic work carried out on these compounds prior to 1962 is also available (16).

E. MEDICINALLY SIGNIFICANT INDOLES

A number of indole derivatives have come to have significant clinical use. Derivatives with antibiotic activity were mentioned earlier in this chapter. Several of the indole alkaloids are particularly important. Reserpine (**E1**) and certain of its derivatives are important tranquilizers and sedatives. Yohimbine, aspidospermine, and strychnine are among the other alkaloids which have found occasional medicinal use.

The dimeric indole alkaloids vincristine (**E2**) and vinblastine (**E3**) are antineoplastic agents and have found clinical use in treatment of choriocarcinoma and certain forms of leukemia and Hodgkins disease (37, 54).

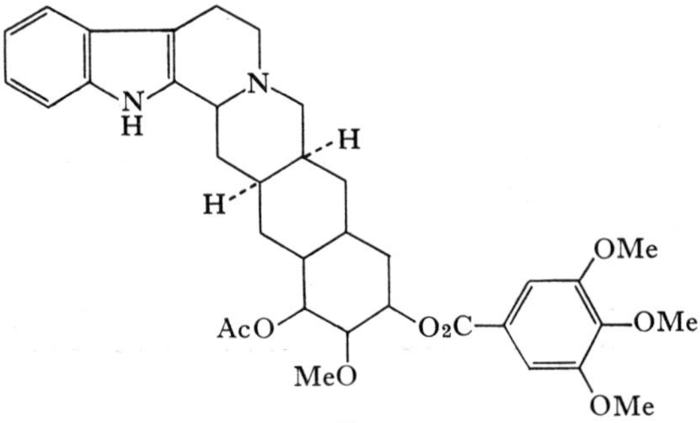

E1

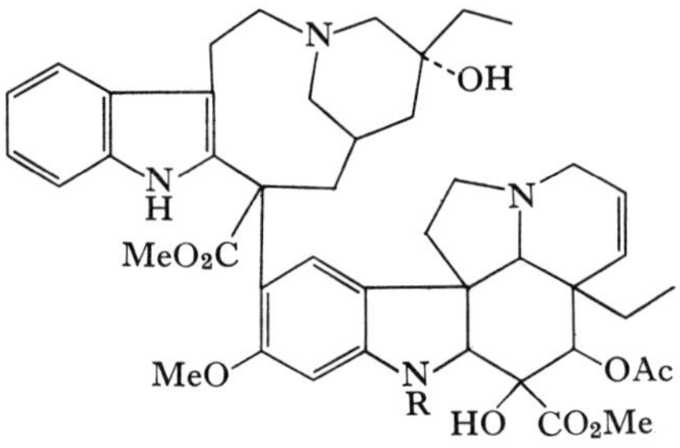

E2 R = H
E3 R = CHO

The lysergic acid derivative ergotamine (**E4**), which occurs naturally in ergot is used as the tartrate salt as a vasoconstrictor in relief of migraine (8). Other amides of lysergic acid are also medicinally useful. The *N*-[2-(1-

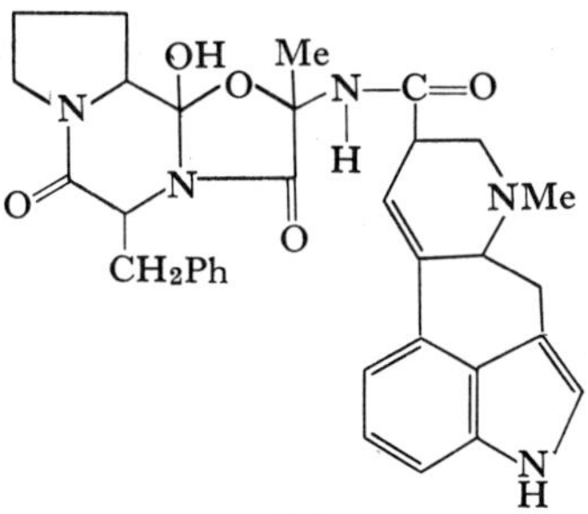

E1

hydroxy)propyl]amide (ergonovine) and the *N*-[2-(1-hydroxy)propyl]-amide (methylergonovine) are oxytocic agents (8).

A great many synthetic indoles have been submitted to pharmacological investigation. Although many show pronounced physiological activity of various types, relatively few have yet been widely used in clinical medicine. Recently, much interest has centered around 1-(*p*-chlorobenzoyl)-5-methoxy-2-methylindole-3-acetic acid. This substance, known as indomethacin, is under evaluation in the treatment of rheumatoid arthritis and related conditions (29, 45, 58). Many synthetic analogs of tryptamine have been synthesized and studied. 3-(2-Aminobutyl)indole (etryptamine) is an antidepressant (11). 1-Benzyl-5-methoxy-2-methyltryptamine (benanserin hydrochloride) is a serotonin antagonist (15). An authoritative review of the chemistry and pharmacological properties of these and other tryptamine derivatives has been published (22).

F. LUCIFERINS DERIVED FROM INDOLE

The determination of the structure of the luciferin responsible for light emission by *Cypridina hilgendorfti* has shown it to be **F1** (30, 31). The

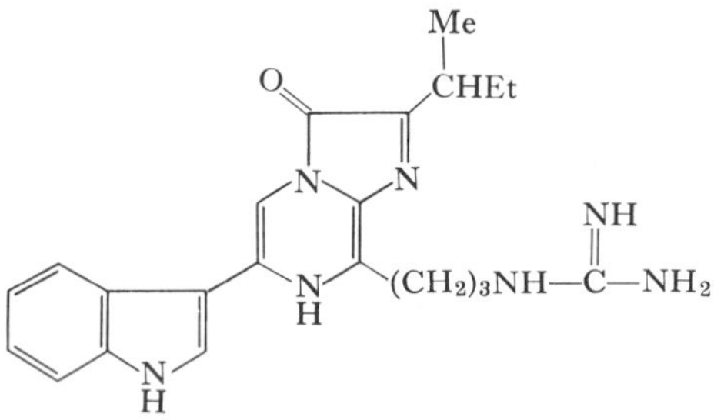

F1

structure has been confirmed by the total synthesis shown below (32):

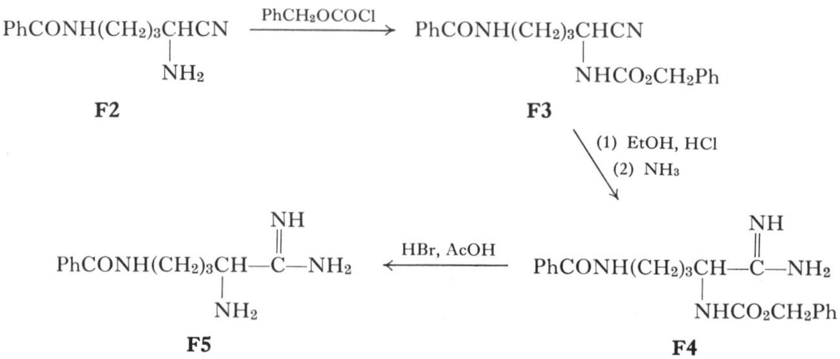

$PhCONH(CH_2)_3CH(NH_2)CN$ (**F2**) $\xrightarrow{PhCH_2OCOCl}$ $PhCONH(CH_2)_3CH(NHCO_2CH_2Ph)CN$ (**F3**)

F3 $\xrightarrow{(1)\ EtOH,\ HCl;\ (2)\ NH_3}$ $PhCONH(CH_2)_3CH(NHCO_2CH_2Ph){-}C({=}NH){-}NH_2$ (**F4**)

F4 $\xrightarrow{HBr,\ AcOH}$ $PhCONH(CH_2)_3CH(NH_2){-}C({=}NH){-}NH_2$ (**F5**)

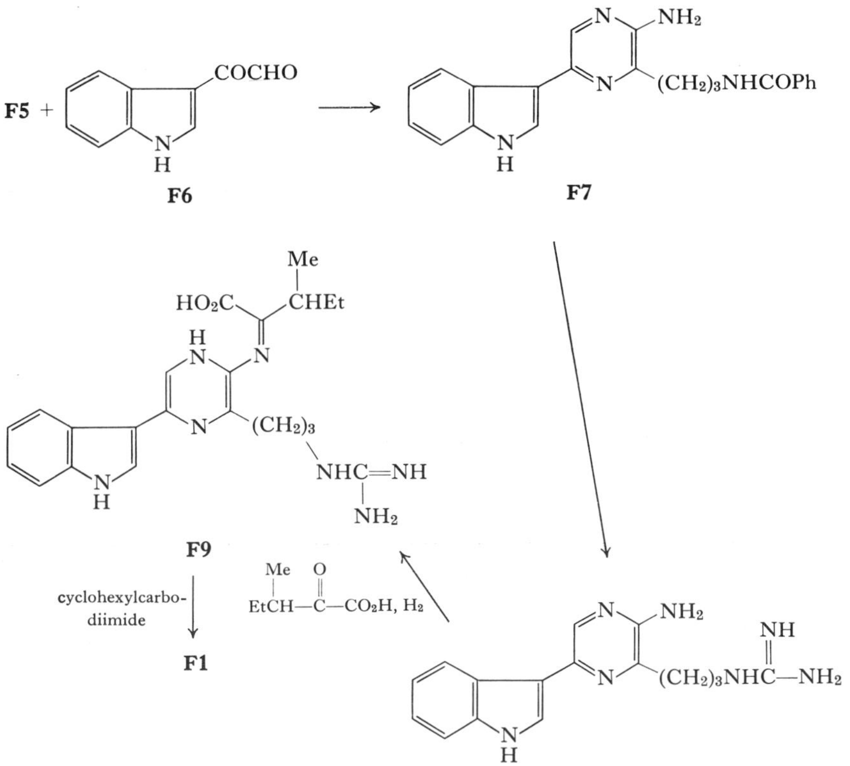

F5 + F6 → F7 → F8 → F9 → F1

COCHO

F6

NH_2

$(CH_2)_3NHCOPh$

F7

Me

HO_2C CHEt

$(CH_2)_3$

NHC=NH

NH_2

F9

cyclohexylcarbo-diimide

F1

Me O

EtCH—C—CO_2H, H_2

NH_2

NH

$(CH_2)_3NHC$—NH_2

F8

The oxidation associated with the chemiluminescent reaction involves the imidazolopyrazine portion of the molecule rather than the indole ring (30).

G. MISCELLANEOUS INDOLE CONJUGATES AND METABOLITES

The name ascorbigen has been given to the condensation product of 3-hydroxymethylindole and ascorbic acid. This compound arises in plants

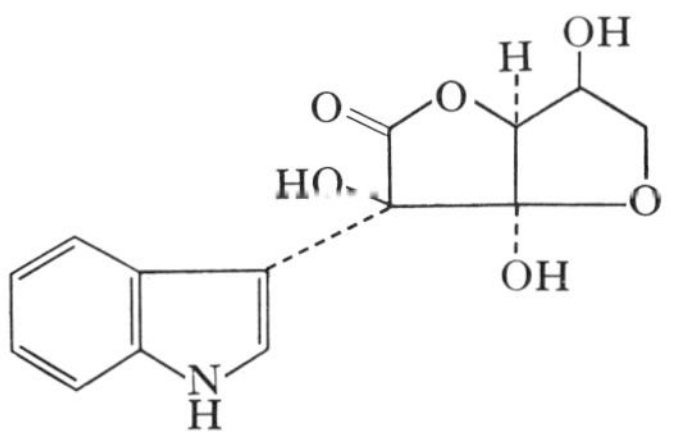

G1

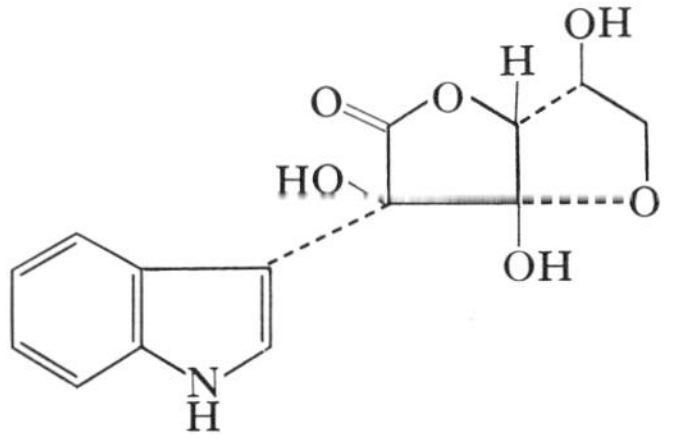

G2

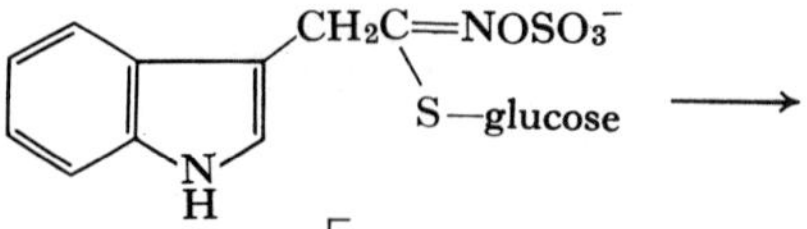

$CH_2C{=}NOSO_3^-$

S—glucose

G3

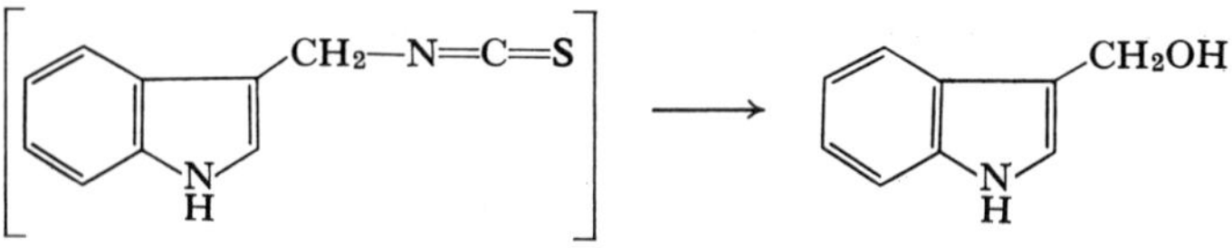

$CH_2{-}N{=}C{=}S$

G4

CH_2OH

G5

ascorbic acid

G1 + G2

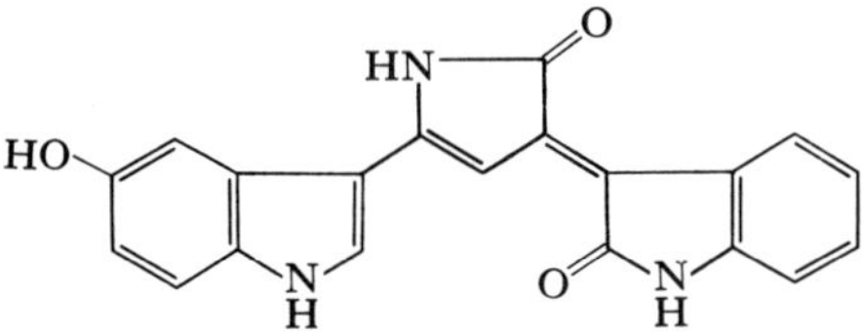

HN O HO N H O N H

G6

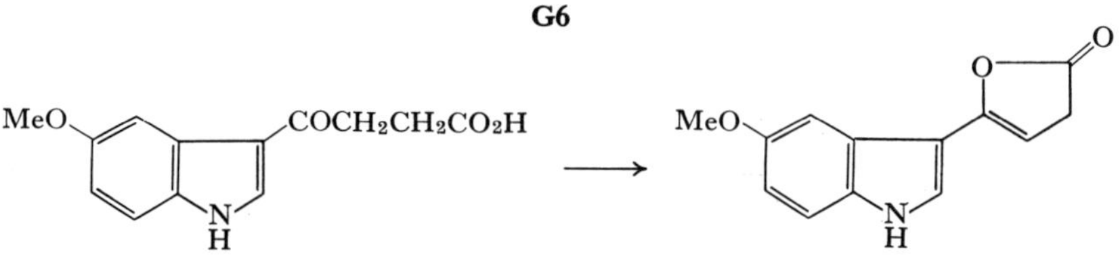

MeO $COCH_2CH_2CO_2H$ N H

G7

MeO O O N H

G8

EtOH, pyridine isatin

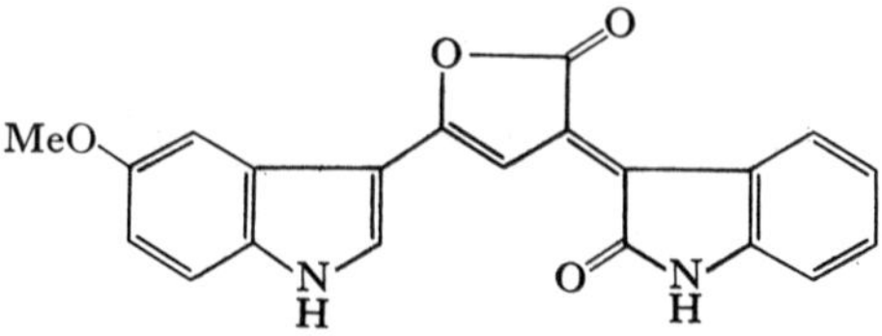

O O MeO N H O N H

G9

NH_3

HN O MeO N H O N H

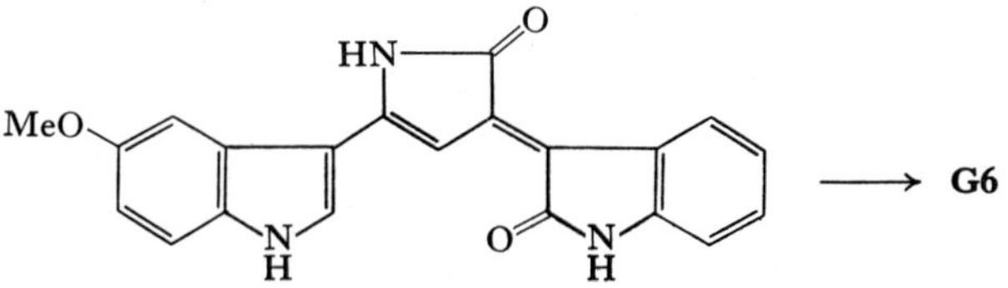

G10

→ G6

by reaction of ascorbic acid with 3-hydroxymethylindole derived from unstable precursors. The stereoisomeric ascorbigens A and B have been assigned structures **G1** and **G2** (33). They are, then, carbon alkylation products of the ascorbate ion. Ascorbigen is believed to arise from hydrolysis of glucobrassicin (**G3**), a derivative of indole-3-acetic acid. An *N*-methoxy analog of **G3**, called neoglucobrassicin has been reported (20). The location of the methoxy group on nitrogen has been deduced from the fact that neoglucobrassicin gives tryptamine and 3-methylindole on treatment with Raney nickel. Isolation of *N*-methoxyindoles from natural sources is unusual, although evidence that 1-methoxy-*N*,*N*-dimethyltyptamine is a component of *Lespedeza bicolor* has recently been reported (38).

Violacein is a pigment elaborated by *Chromobacterium violaceum*. Structure studies led to the proposal that violacein possesses structure **G6** (6). The structure was confirmed by a synthesis reported in 1960 (7).

REFERENCES

1. Allen, G. R., Jr., Binovi, L. J., and Weiss, M. J., *J. Med. Chem.* **10**, 7 (1967).
2. Allen, G. R., Jr., Poletto, J. F., and Weiss, M. J., *J. Org. Chem.* **30**, 2897 (1965).
3. Allen, G. R., Jr., Poletto, J. F., and Weiss, M. J., *J. Med. Chem.* **10**, 14 (1967).
4. Allen, G. R., Jr., and Weiss, M. J., *J. Org. Chem.* **30**, 2904 (1965).
5. Allen, G. R., Jr., and Weiss, M. J., *J. Med. Chem.* **10**, 1, 23 (1967).
6. Ballantine, J. A., Barrett, C. B., Beer, R. J. S., Eardley, S., Robertson, A., Shaw, B. L., and Simpson, T. H., *J. Chem. Soc.* p. 755 (1958).
7. Ballantine, J. A., Beer, R. J. S., Crutchley, D. J., Dodd, G. M., and Palmer, D. R., *J. Chem. Soc.* p. 2292 (1960).
8. Backman, H., "Drugs." Saunders, Philadelphia, Pennsylvania, 1958.
9. Beecham, A. F., Fridrichsons, J., and Mathieson, A. M., *Tetrahedron Letters* p. 3131 (1966).
10. Bell, M. R., Johnson, J. R., Wildi, B. S., and Woodward, R. B., *J. Am. Chem. Soc.* **80**, 1001 (1958).
11. Biel, J. H., *in* "Drugs Affecting the Central Nervous System" (A. Burger, ed.), p. 67. Marcel Dekker, New York, 1968.
12. Cerletti, A., *Progr. Drug Res.* **2**, 227 (1960).
13. Collins, J. F., *Brit. Med. Bull.* **21**, 223 (1965).
14. Cosulich, D. B., Nelson, N. R., and van den Hende, J. H., *J. Am. Chem. Soc.* **90**, 6519 (1968).
15. Desci, L., *Progr. Drug Res.* **8**, 53 (1965).
16. Downing, D. F., *Quart. Rev.* (*London*) **16**, 133 (1962).
17. Erspamer, V., ed., "5-Hydroxytryptamine and Related Indolealkylamines." Springer, Berlin, 1966.
18. Fridrichsons, J., and Mathieson, A. M., *Acta Cryst.* **18**, 1043 (1965).
19. Garattini, S., and Valzelli, L., "Serotonin." Elsevier, Amsterdam, 1965.
20. Gmelin, R., and Virtanen, A. I., *Acta Chem. Scand.* **16**, 1378 (1962).
21. Harada, H., Sakabe, N., Hirata, Y., Tomiie, Y., and Nitta, I., *Bull. Chem. Soc. Japan* **39**, 1773 (1966).
22. Heinzelman, R. V., and Szmuszkovicz, J., *Progr. Drug Res.* **6**, 75 (1963).

23. Hesse, M., "Indolealkaloide in Tabellen." Springer, Berlin, 1964.
24. Hirata, T., Yamada, Y., and Matsui, M., *Tetrahedron Letters* p. 19, 4107 (1969).
25. Hodges, R., Ronaldson, J. W., Shannon, J. S., Taylor, A., and White, E. P., *J. Chem. Soc.* p. 26 (1964).
26. Hodges, R., Ronaldson, J. W., Taylor, A., and White, E. P., *J. Chem. Soc.* p. 5332 (1963).
27. Hofmann, A., *in* "Drugs Affecting the Central Nervous System" (A. Burger, ed.), pp. 169–235. Marcel Dekker, New York, 1968.
28. Johnson, J. R., Bruce, W. F., and Dutcher, J. D., *J. Am. Chem. Soc.* **65**, 2005 (1943).
29. Kelly, M., *Med. J. Australia* **51**, Part II, 541 (1964).
30. Kishi, Y., Goto, T., Eguchi, S., Hirata, Y., Watanabe, E., and Aoyama, T., *Tetrahedron Letters* p. 3437 (1966).
31. Kishi, Y., Goto, T., Hirata, Y., Shimomura, O., and Johnson, F. H., *Tetrahedron Letters* p. 3427 (1966).
32. Kishi, Y., Goto, T., Inoue, S., Sugiura, S., and Kishimoto, H., *Tetrahedron Letters* p. 3445 (1966).
33. Kiss, G., and Neukom, H., *Helv. Chim. Acta* **49**, 989 (1966).
34. Lefemine, D. V., Dann, M., Barbatschi, F., Hausmann, W. K., Zbinovsky, V., Monnikendam, P., Adam, J., and Bohonos, N., *J. Am. Chem. Soc.* **84**, 3184 (1962).
35. Manske, R. H. F., ed., "The Alkaloids," Vol. 8. Academic Press, New York, 1965; Vol. 11, Academic Press, New York, 1968.
36. Moncrief, J. W., *J. Am. Chem. Soc.* **90**, 6517 (1968).
37. Montgomery, J. A., *Progr. Drug Res.* **8**, 431 (1965).
38. Morimoto, H., and Oshio, H., *Ann. Chem.* **682**, 212 (1965).
39. Nagarajan, R., Huckstep, L. L., Lively, D. H., DeLong, D. C., Marsh, M. M., and Neuss, N., *J. Am. Chem. Soc.* **90**, 2980 (1968).
40. Nakata, H., Harada, H., and Hirata, Y., *Tetrahedron Letters* p. 2515 (1966).
41. Neuss, N., Nagarajan, R., Molloy, B. B., and Huckstep, L. L., *Tetrahedron Letters* p. 4467 (1968).
41a. Nicolaus, R. A., "Melanins." Hermann, Paris, 1968.
42. Patrick, J. B., Williams, R. P., Meyer, W. E., Fulmor, W., Cosulich, D. B., Broschard, R. W., and Webb, J. S., *J. Am. Chem. Soc.* **86**, 1889 (1964).
43. Poletto, J. F., Allen, G. R., Jr., and Weiss, M. J., *J. Med. Chem.* **10**, 95 (1967).
44. Preobrazhenskaya, M. N., Balashova, E. G., Turchin, K. F., Padieskaya, E. N., Uvarova, N. V., Pershin, G. N., and Suvorov, N. N., *Tetrahedron* **24**, 6131 (1968).
45. Ragan, C., *Clin. Pharmacol. Therap.* **8**, 11 (1967); *Chem. Abstr.* **66**, 45401 (1967).
46. Remers, W. A., Roth, R. H., and Weiss, M. J., *J. Am. Chem. Soc.* **86**, 4612 (1964).
47. Remers, W. A., Roth, R. H., and Weiss, M. J., *J. Org. Chem.* **30**, 2910 (1965).
48. Remers, W. A., and Weiss, M. J., *J. Am. Chem. Soc.* **88**, 804 (1966).
49. Ronaldson, J. W., Taylor, A., White, E. P., and Abraham, R. J., *J. Chem. Soc.* p. 3172 (1963).
50. Sakabe, N., Harada, H., Hirata, Y., Tomiie, Y., and Nitta, I., *Tetrahedron Letters* p. 2523 (1966).
51. Schach von Wittenau, M., and Els, H., *J. Am. Chem. Soc.* **83**, 4678 (1961).
52. Schach von Wittenau, M., and Els, H., *J. Am. Chem. Soc.* **85**, 3425 (1963).
53. Schultes, R. E., *Science* **163**, 245 (1969).
54. Shimkin, M. B., *in* "Topics in Medicinal Chemistry" (J. L. Rabinowitz and R. M. Myerson, eds.), pp. 96–105. Wiley (Interscience), New York, 1967.
55. Stevens, C. L., Taylor, K. G., Munk, M. E., Marshall, W. S., Noll, K., Shah, G. D., Shah, L. G., and Uzu, K., *J. Med. Chem.* **8**, 1 (1964).

56. Tulinsky, A., *J. Am. Chem. Soc.* **84**, 3188 (1962).
57. Webb, J. S., Cosulich, D. B., Mowat, J. H., Patrick, J. B., Broschard, R. W., Meyer, W. E., Williams, R. P., Wolf, C. F., Fulmor, W., Pidacks, C., and Lancaster, J. E., *J. Am. Chem. Soc.* **84**, 3185, 3187 (1962).
58. Whitehouse, M. W., *Progr. Drug Res.* **8**, 321 (1965).
59. Woolley, D. W., "The Biochemical Bases of Psychoses." Wiley, New York, 1962.
60. Wurtman, R. J., and Axelrod, J., *Probl. Actuels Endocrinol. Nutr.* **10**, 189 (1966).

AUTHOR INDEX

Numbers in parentheses are reference numbers and indicate that an author's work is referred to although his name is not cited in the text. Numbers in italics show the page on which the complete reference is listed.

A

Abdurahman, N., 111 (78), *138*
Abraham, R. J., 437 (49), *446*
Abramovitch, R. A., 150 (1, 2, 3), 157 (2, 5, 7), 159, 186 (3, 4, 6), *206*, 241 (1), *275*, 345 (1), 358 (1), 365 (1a), *388*, 421 (1), 422 (1), 423 (1), *388*
Acheson, R. M., 69 (1), 70 (1), *85*, 226 (2), *275*, 310 (1), *312*, 330 (1, 2), *339*, 373 (2), 376, *388*
Adam, J., 431 (34), *446*
Adamacik, J. A., 409 (62), *428*
Adams, K. A. H., 150 (3), 186 (3), *206*
Adams, R., 173 (9), 174 (9), 175 (8, 9), *206*
Adkins, H., 129 (1), *137*
Adlerova, E., 154 (10), 157 (10), *206*, 246 (141), *279*
Advani, B. G., 110 (111), *139*
Aeberli, P., 352 (3), *388*
Agosta, W. C., 201 (10a), *206*
Ahmad, A., 111 (83), *139*
Ahmad, I., 58 (120), *88*
Ahmad, Y., 186 (4), *206*
Ahmed, M., 121 (1a), *137*
Akaboshi, S., 291 (44), *313*
Akkerman, A. M., 57 (2), *85*, 378 (4), *385*
Akopyran, Z. G., 157 (263), *213*
Aksanova, L. A., 158 (11), (148), *206*, *210*
Aksormvitaya, R., 291 (83), 314
Akutagawa, M., 290 (89), *314*
Alberti, C., 30 (183), *89*, 335 (3), *339*, 403 (1a), *426*
Albertson, N. F., 97 (2), 98 (2), *137*, 231 (3), 233 (3), *275*
Albright, J. D., 94, 95 (3), 102 (2a), *137*
Ali Quershi, A., 57 (120), *88*
Aldrich, P. E., 263 (185), *280*
Allais, A., 178 (269), *213*
Allemann, T., 246 (151), *279*
Allen, C. F. H., 148 (13), 189 (12), *206*
Allen, F. L., 178 (14), *206*
Allen, G. R. Jr., 114 (4), *137*, 171 (17, 151), 173 (15, 17, 157), *206*, *210*, 423 (2, 64), *426*, *428*, 433 (1, 2, 3, 4, 5, 43), *445*
Allessandri, L., 35 (3), *85*
Almirante, L., 243 (108, 109), *278*
Altukhova, L. B., 159 (229), *212*, 418 (93), *429*
Ames, A. F., 220 (4), 221 (4), 224 (4), *275*
Ames, D. E., 34 (4), *85*, 110 (5), 113 (5), *137*, 220 (4), 221 (4), 224 (4), *275*, 413 (3), *426*
Amorosa, M., 217 (5, 6,) *275*, *276*
Anderson, E. L., 263 (192), *280*
Andratschke, P., 166 (139), *209*
Andreen, J. H., 366 (64), *390*
Ang, S. K., 264 (115), *278*
Anger, D. G., 38 (289), *92*
Angyal, S. J., 378 (5), *388*
Aniline, O., 196 (195), *211*

Anthony, W. C., 22(91), 23(91), 37, 38 (289), *85*, *87*, *92*, 113(6), *137*, 219 (165), 220(174), 221(165), 222(74), 224(74), 225(74), 232(74), *277*, *280*, 337, *339*, 411(111), 415(4), 417(4), *426*, *429*
Aoyama, T., 442(30), 443(30), *446*
Arai, I., 233(7), *276*
Archer, S., 97(2), 98(2), *137*, 190(158), *210*, 231(3), 233(3), *275*, 361, *392*
Archer, W. L., 40(85), 71(85), 74(85, 86), *87*
Arifuddin, M., 291(36), *313*
Arigoni, D., 269(64, 107), 270(64), 271 (107), 272(106), 273(106), *277*, *278*
Armen, A., 95(45), 97(45), 98(45), 102(45), *138*
Armstrong, M. D., 221(156), *279*, 405(92), 413(92), *429*
Arnall, F., 13(6), *85*
Arnold, R. D., 14(7), 15(7), *85*
Ash, A. S. F., 224(8), 225(8), *276*
Ashby, E. C., 217(167), *280*
Asma, W. J., 224(98), *278*
Astill, B. D., 398(5), *399*
Atkinson, C. M., 202(18), *206*, 298(2), *312*, 328(5), *339*
Atta-ur-Rahman, 98(51), *138*, 256(69), 261 (70), *277*
Augusti, K. T., 291(36), *313*
Autrey, R. L., 377(6), 378(6), *388*
Avakian, S., 214(96), 233(96), *278*
Axelrod, J., 310(9), *312*, 438(60), *447*

B

Backman, H., 441(8), *445*
Bader, A. R., 194(19, 113), *206*, *209*
Bader, F. E., 241(208), 263(208), *281*
Bader, H., 46, *85*, 114(7), *137*
Badger, G. M., 215(9), *276*
Bagot, J., 165(125, 126), *209*
Bailey, A. S., 17(9), *85*
Bajwa, G. S., 145, *206*
Baker, J. W., 404, *426*
Baker, M. B., 182(187), *211*
Balashova, E. G., 433(44), *446*
Balderman, D., 221(87), 227(87), *278*
Ballach, P., 61(259), *91*, 133(146), *140*
Ballantine, J. A., 34(10), 39(10), *85*, 347(8, 11), *388*, 413(6), 417(6), *426*, 445(6, 7), *445*
Bamfield, P., 348(9), *388*
Ban, Y., 161(136), *209*, 220(90), 245(91), 246(91), *278*, 346(109), *388*, *391*
Ban-Lun Ge, 132(140), 135(140), *140*, 220 (176), *280*, 416(114), *429*
Barbatschi, F., 431(34), *446*
Barltrop, J. A., 129(72), *138*
Barman, T. E., 122(8), *137*
Barnett, B. F., 188(25), *213*
Barrett, C. B., 34(10), 39(10), *85*, 378(11), *388*, 413(6), 417(6), *426*, 445(6), *445*
Bartlett, M. F., 20(11), *85*, 118(9), 122(9a), *137*, 263(10), *276*, 292(3), *312*
Barton, J. E. D., 259(11), *276*
Bärwald, L., 120(50a), *138*
Basangoudar, L. D., 77(12), *85*
Bashkirova, A. Ya., 22(141), 23(14), *88*
Baskakov, Y. A., 412(7), *426*
Battersby, A. R., 269(12, 13, 14, 15), 270 (14, 15a), 271(14, 15), 272(15a), 273 (15a, 16, 17), *276*
Baude, F. J., 73(160), 83(160), *89*, 178(175), 179(176), *210*, 322, *340*
Bauer, H., 291(78), *314*, 343(127), 354(127), 355(128), *391*
Bauman, C. P., 4(95), 15(95a), 16(95a), *87*, 304, *313*, 346, 349(55), 353(54), 363 (54), *389*
Baumgarten, H. E., 202(21), *206*
Baxter, I., 182, *206*, 376(12), *388*
Bayer, J., 365(74), *390*
Beaudet, C., 217(118), *278*
Becher, J., 203(43), *207*
Beck, D., 127, *137*, 162(23), *206*
Becker, E. I., 362(35), 363(35), *389*, 402 (69), 421(69), *428*
Beckett, A. H., 308(4), *312*, 357, 360, *388*
Beecham, A. F., 433(9), *445*
Beer, R. J. S., 34(10), 39(10), *85*, 114(11), *137*, 173(24), 177(25), 178(25), *206*, 282(7), 283(5, 6, 7), *312*, 347(8), 378 (7, 11), *388*, 413(6), 417(6), 423(8), *426*, *427*, 445(6, 7), *445*
Beeson, J. H., 358(144), *392*
Behringer, H., 235(18), *276*, 346(14), 350 (13, 14), 351, *388*
Beilfuss, H. R., 178(235), *212*, 235(158), *279*
Beitz, H., 217(19), *276*
Belen'kiĭ, L. I., 15(251), *91*

Bell, J. B., Jr., 77(13), *85*
Bell, M. R., 369(15), *388*, 433(10), *445*
Bellamy, A. J., 143(25a), *206*
Belleau, B., 248(20), *276*, 325(5a), *339*
Belsky, I., 133(12), 135(12), *137*
Benington, F., 22(14), *85*, 183(26, 170), *206*, *210*, 221(21), *276*
Bennett, O. F., 426(12), *427*
Benson, H. D., 176(105), *208*
Bentov, M., 183(27, 28), *206*
Benzing, E., 178(108), *208*
Berdinskaya, I. S., 358(116, 117), *391*
Bergmann, E. D., 135(59), *138*, 154(29), 157(192), *206*, *211*, 382(16), *388*
Berguer, Y., 295(65, 66), *313*
Bernstein, B. S., 335(36), *340*, 352(161), 357(161), *392*
Bernstein, S., 133(22), *137*, 188(53), *207*
Berti, G., 2(18), 4(18), 5(18), 9(16), 11, 12(17), 13(17), *85*, 115(13), *137*, 295(8), *312*
Besford, L. S., 202(29a), *206*
Betkerur, S. N., 173(30), *206*
Beugelmans, R., 101(14), *137*
Beyler, A. L., 369(15), *388*
Bhramaramba, A., 40(264), *91*
Bhattacharyya, N. K., 343(162), 344(162), 345(162, 165), 351(165), *392*
Biasotti, J. B., 33(187a), *90*
Bickel, H., 122(14a), *137*, 241(208), 263(208), *281*
Biel, J. H., 442(11), *445*
Biemann, K., 122(123a), *140*, 261(123), *279*
Biere, H., 415, *427*
Binger, P., 58(258), *91*
Binks, J. H., 17(19), *85*
Binovi, L. J., 433(1), *445*
Birch, A. J., 122(14b), *137*
Bird, C. W., 290(8a), *312*, 360(388)
Birkofer, L., 35(20), *85*
Bisagni, E., 34(40), *86*, 413(17), 414(17), *427*
Biswas, K. M., 41(21), 80(21a), *85*, *86*, 109(15), 113(15), 131(15), *137*, 220(22), *276*
Blackburn, D. E., 187(253), *212*, 420(101a), *429*
Blackhall, A., 164(31), 166(31), *206*
Blades, C. E., 169(32), *207*
Blaikie, K. G., 178(33), *207*
Blair, J., 178(34), *207*
Blake, J., 331(6), *339*
Blatter, H. M., 224(44), *277*, 424(10), *427*
Blicke, F. F., 59(22), *86*
Bloor, J. E., 1(23), 2(23), 84(23), *86*
Blume, H., 200(96), *208*
Boca, J.-P., 293(68a), *313*, 373(102, 103), *391*
Bodendorf, K., 407(11), 412(11), *427*
Boekelheide, V., 398(5), *399*
Boessneck, W., 382(51), 383(51), *389*
Boggess, P., 348(69), *390*
Boggiano, B. G., 34(10), 39(10), *85*, 413(6), 417(6), *426*
Böhlke, H., 57(291), *92*
Bohonos, N., 431(34), *446*
Bokarev, K. S., 215(122), *278*
Boriavia, A., 150(201), *211*
Bonthrone, W., 331(6), *339*
Booth, D. L., 108(29), 111, 113(29), *137*, 300(19), 301(19), *312*
Booth, S. R. G., 330(1), *339*
Bormann, G., 60(270), *92*, 369(158), 370(158), *392*, 419(119), *429*
Bornstein, J., 201(110), *208*, 426(12), *427*
Borsche, W., 34(24), 36, *86*, 413(13), 414(13), *427*
Bory, M., 217(118), *278*
Bose, A. K., 344(163), *392*
Botimer, L. W., 30(202), *90*
Bowden, K., 85(25), *86*
Bowersox, W., 196(195), *211*
Bowman, R. E., 34(4), *85*, 110(5), 113(5), *137*, 413(3), *426*
Boyd, S. D., 196(195), *211*
Boyd-Barrett, H. S., 319(7), *339*
Boyer, J. H., 187(35), *207*
Bradsher, C. K., 53(26), *86*
Bramley, R. K., 24(26a), *86*
Breen, D. L., 1(23), 2(23), 84(23), *86*
Brehm, W. J., 37(27), 57(28), *86*, 417(14), *427*
Breslow, R., 19(29), *86*
Bretherick, L., 155(36), *207*
Brewster, J. H., 94, *137*
Bridgwater, R. J., 194(19), *206*
Brieskorn, C., 16(30), *86*
Broadhurst, T., 283(5), *312*
Brodie, B. B., 310(9), *312*
Brooke, G., 197(39), *207*
Brookes, C. J. Q., 373(2), 376(2), *388*

Broschard, R. W., 433(57, 42), *446*
Brown, A. B., 376(31), *389*
Brown, F., 166(40), 168(40), *207*
Brown, H. C., 219 (93), *278*
Brown, J. B., 19(32), *86*
Brown, K., 12, 13(33), *86*
Brown, R. D., 3(12, 34, 83, 84), *86*
Brown, R. K., 5(111), 14(189), 15(140, 189), 77(112), *88*, *90*, 145, 154(122), 162(217), 178(197), *206*, *209*, *211*, 397(1), *399*, 419(50), 424, *428*
Brown, R. T., 255(101), 259(100), 269(13, 14), 270(14), 271(14, 15), *276*, *278*
Brown, U. M., 155(41), *207*, 419(15), *427*
Bruce, J. M., 15(35), 72(36), 73(36), 75 (36), *86*, 202(29a), *206*, 342(18), 343 (18), *388*
Bruce, W. F., 433(28), *446*
Bruice, T. C., 114(57), *138*
Brundage, R. P., 173(240), *212*
Bruni, P., 66(49, 50), 69(51), 75(51), *86*, 372(25, 26, 27), 373(25, 26, 27), *389*
Brunner, K., 362, *388*
Brunton, J. C., 178(14), *206*
Brydowna, W., 149(145), *209*
Buchardt, O., 203(42, 43), *207*
Büchi, G., 99(21), 102(19, 20), 103(20), 104(18), 107(19), *137*, 253, 261(123), 262(27), *276*, *279*
Buchmann, G., 28(37), *86*, 412(16), *427*
Bucourt, R., 227(29), *276*
Budylin, V. A., 15(132), *88*
Bühler, W., 291(78), *314*, 355(128), *391*
Bullock, E., 378(5), *388*
Bullock, M. W., 216(58), 217(57, 30, 31, 32), *276*, *277*
Bu'Lock, J. D., 72(38), 73(38), 75(38), *86*, 192(48), *207*
Bunnett, J. F., 197, *207*
Bunney, J. E., 383(19a), *388*
Burgess, J. F., 203(46), *207*
Burgstahler, A. W., 263(185), *280*
Burks, R. E. Jr., 129(1), *137*
Burnett, A. R., 270(15a), 272(15a), 273 (15a), *276*
Burr, G. O., 42(39), 43, *86*
Buu-Hoi, N. P., 34(40), *86*, 149(48, 49, 50), 150(49), 153(47), *207*, 318(8), 413(17), 414(17, 18, 19), *427*
Buzas, A., 243(33), *276*
Byrne, J. C., 273(16), *276*
Byrn, M., 402, *427*

C

Cadogan, J. I. G., 145(51), 184(51), 187 (51), *207*
Calvin, M., 387(146), *392*, 402, *427*
Cameron-Wood, M., 145(51), 184(51), 187 (51), *207*
Campaigne, E., 164(52), 166(52), *207*
Cantrall, E. W., 133(22), *137*, 188(53), *207*
Capaldi, E. C., 192, *210*
Caprio, L., 243(108, 109), *278*
Cardani, C., 17(41), *86*, 297(10), 301(10), *312*
Cardillo, B., 19(42), 20, 21, *86*
Carithers, R., 203(46), *207*
Carlin, R. B., 142, 145(54, 56, 57, 59), 147 (53a), 148, *207*
Carlson, D. P., 145(54, 56), *207*
Carlsson, A., 381(20), *388*
Carnahan, R. E., 97(130), *140*
Caro, A. N., 27(152), *89*
Carpenter, W., 77(43), *86*
Carson, D. F., 121, *137*, 362(21), *388*
Carter, P. H., 155(41), *207*, 419(15), *427*
Casini, G., 233(34), *276*, 424(21), *427*
Casnati, G., 17(41), 19(42), 20(42), 21(42), *86*, 153(60), 189(60), *207*
Castle, R. N., 406(123), 416(123), 417(123), *429*
Castro, C. E., 207(61, 242), *207*, *212*, 350 (129), 353(129), *391*
Caubére, P., 24(118), *88*
Cava, M. P., 237(210), 267(209, 210), *281*, 380(22), *388*
Cecere, M., 298(74), *314*
Cerletti, A., 438(12), *445*
Cerutti, P., 131(157), *141*
Čěrvinka, O., 200(62), *207*
Cetenko, W., 251(120), *278*
Chaikin, S. W., 113(126), *140*
Challis, B. C., 6, *86*
Chan, K. K., 256(102), *278*
Chang, Y. C., 289(63), *313*
Chapman, N. B., 220(36), 221(35), *276*
Chardonnens, L., 386(23), *388*
Chastrette, M., 169, *207*

Chen, F. Y., 35, *86*, 283, 286, 300, 301 (11), *312*, 414 (22), *427*
Chernov, H. I., 376 (31), *389*
Ch'ng, H. S., 388 (23a), *388*
Chibata, I., 161 (280), *213*
Chiu, Y., 155 (147), 156 (147), *210*
Christensen, G. M., 69 (161), 75 (161), *89*, 226 (124), *279*
Christie, B. J., 215 (9), *276*
Clark, C. T., 378 (95), *390*
Clark, L. C., Jr., 22 (14), *85*, 183 (26, 170), *206*, *210*, 221 (21), *276*
Clarke, G., Jr., 130 (101), *139*
Clarke, K., 114 (11), 130 (101), *137*, 173 (24), *206*, 220 (36), 221 (35), *276*, 423 (8), *427*
Clark-Lewis, J. W., 361 (24), *389*
Clemen, J., 40 (276), *92*
Clemo, G. R., 155 (64), *207*, 422 (23), 423 (23), *427*
Clerc-Bory, G., 295 (12), *312*
Clerc-Bory, M., 295 (12, 13), *312*
Clifford, B., 168 (65), *207*
Closs, G. L., 31 (46), *86*
Closson, W. D., 24 (47), *86*
Clusius, K., 148 (66), *207*
Cockerill, D. A., 68 (48), 75 (48), *86*
Coffen, D. L., 104 (18), *137*, 262 (27), *276*
Cohen, A., 130 (24), *137*, 226 (37), *276*
Cohen, H. L., 347 (65), *390*
Cohen, L. A., 393 (2), *399*
Cohen, L. H., 310 (36a), *313*
Cohen, M. P., 97 (151), *140*, 157 (226, 271, 272), *212*, *213*
Cohen, T., 383 (125), *391*
Coker, J. N., 97 (25), *137*
Coller, B. A. W., 3 (34), 83, 84, *86*
Collington, E. W., 415, *427*
Collins, J. F., 433 (13), *445*
Collins, K. H., 377(169), *392*
Colonna, M., 66 (49, 50), 69 (51), 75 (51, 52), 76 (53), *86*, 372, 373 (25, 26, 27, 28, 29), *389*
Cone, N. J., 118 (46a), *138*, 261 (123), *278*, *279*
Conrow, R. B., 133 (22), *137*, 188 (53), *207*
Cook, A. G., 153 (188), *211*
Cook, J. W., 233 (38), *276*
Cook, P. L., 98 (128), 99, (128) *140*
Cooper, A. R., 2 (54), *86*
Corbella, A., 168 (67), *207*
Cornelius, D., 337 (32), *340*
Cornforth, J. W., 378 (30), *389*
Cornforth, R. H., 29 (55), *86*, 378 (30), *389*
Corrodi, H., 360 (46, 138), 381 (20), *388*, *389*, *391*
Corsano, S., 242 (39), *276*
Corwin, D. A., 24 (47), *86*
Cosulich, D. B., 433 (42, 57, 14), 436 (14), *445*, *447*
Cowper, R. M., 166 (68), *207*
Coyne, C. R., 220 (4), 221 (4), 224 (4), *275*
Craig, C. A., 424 (79), *428*
Crandall, D. I., 311 (14), *312*
Crawford, N., 22 (250a), 23 (250a), *91*
Cretney, W. J., 124 (79), *139*, 256 (104), 273 (103), *278*
Creveling, C. R., 308 (94), *314*
Cromartie, R. I. T., 192 (69), *207*
Crosby, D. G., 30 (115), *88*, 129 (65), *138*, 398 (11), 399, *399*
Crowne, C. W. P., 2 (54), *86*
Crutchley, P. J., 378 (7), *388*, 445 (7), *445*
Curtin, D. Y., 21 (56), *86*, 418 (25), *427*
Cushley, R. I., 178 (197), *211*

D

Dacons, J. C., 4 (121), *88*
Dadson, B. A., 259 (40), *276*
Daeniker, H. U., 237 (210), 267 (209, 210), *281*
Dagliesh, C. E., 378 (30), *389*
Dahlbom, R., 40 (57), *86*, 97 (26), *137*
Dailey, E. E., 347 (65, 73), *390*
Daisley, R. W., 345, 357 (12a), 360 (12a), *388*, *389*
Dalgliesh, C. E., 295 (16), 308 (45), 310 (15), *312*, *313*
Daly, J. W., 7 (58), *86*, 129 (27), *137*, 182 (99), 183 (99), *208*, 308 (80, 94), 309 (49, 50, 51), *313*, *314*, 393 (2), *399*
Dambal, S. B., 419 (26), *727*
Dann, M., 431 (34), *446*
Darrah, H. K., 203 (248), *212*
DaSettimo, A., 2 (18), 4 (18), 5 (18), 9 (16), 13 (17, 60, 61), *85*, *86*, 115 (13), *137*, 295 (8), *312*
Datner, A. S., 348 (9), *388*
Da Vanzo, J. P., 38 (289), *92*, 220 (194), 221 (194), *280*

Dave, K. G., 61(280, 281), 65(280), *92*, 111(154), *141*, 227(200), 236(197), 248(198, 199), 263(200), *280*
Davenport, H. F., 114(11), *137*, 173(24), *206*, 423(8), *427*
Davenport, R. W., 33(187a), *90*
David, S., 116(28), *137*, 232(41), *276*, 294 (17), *312*
Dearnaley, D. P., 373(2), 376(2), *388*
Decker, M., 399(17), *400*
DeGraw, J. I., 22(62), *86*, 214(42), *276*, 397(3), *399*, 404 (26a), 411(27), 413 (26a), 414(27), *427*
de Jongh, D. K., 57(2), *85*
DeLong, D. C., 433(39), 436(39), *446*
Delvigs, P., 22(250a), 23(250a), *91*, 214 (43), *276*, 308(18), 312
DeMartino, U., 372(28), 373(28), *389*
DeNet, R. W., 373(171), *392*
Desai, H. S., 47(170), *89*
Desci, L., 442(15), *445*
Despuy, M. E., 178(79), *208*
deStevens, G., 224(24), *277*, 376(31), *389*, 424(10), *427*
Dewar, M. J. S., 2(63), 3, 84(63), *86*
Dickel, D. F., 20(11), *85*, 118(9), *137*, 292 (3), *312*
Diels, O., 13(64), 20(70), *86*, *207*, 418(28), *427*
Dietrich, M. A., 77(70), *87*
Dixon, W. B., 153(188), *211*
Djerassi, C., 99(67), *138*, 409(62), *428*
Dobbs, H. E., 33, *87*
Dodd, G. M., 378(11, 7), *388*, 445(7), *445*
Dodo, T., 204(177), *210*
Doering, W. E., 71(66), *87*
Dohan, F. C., 43(129), *88*
Doig, G. C., 233(45), *277*
Dolby. L. J., 100(33), 108(29), 111, 113(29), 125, 129, *137*, 237, 268, *280*, 300(19, 20), 301(19), *312*, 335, *339*, 409(30), 415, *427*
Dolfini, J. E., 151, *212*
Domschke, G., 109(34), *137*, 173(71), *207*, 220(46), *277*, 399, *399*
Donavanik, T., 283(6), *312*
Dorofeenko, G. N., 414(31), *427*
Dostál, V., 40(67), *87*
Dougherty, G., 78(286), *92*, 97(85), *139*, 218(54), *277*
Douglas, B., 263(192), *280*, 413(32), 414 (32), 418(32), *427*
Downing, D. F., 440(16), *445*
Downey, W. E., 217(167), *280*
Doyle, F. P., 424(33), *427*
Doyle, T. W., 109(90), *139*, 419(72), *428*
Drozd, V. N., 27(158), *89*
Dubini, M., 338(28), *340*
Dürst, W., 13(64), *86*, 418(28), *427*
Dutcher, J. D., 433(28), *446*
Dutton, G. G. S., 69(161), 75(161), *89*, 226(124), *279*
Dyke, S. F., 65(68), *87*

E

Eardley, S., 34(10), 39(10), *85*, 413(6), 417(6), *426*, 445(6), *445*
Earley, J. V., 115(42), *138*, 419(37), *427*
Eberhardt, H.-D., 425(102), *429*
Ebersberger, J., 31(243), *91*
Ebnöthner, A., 163(71a), *207*
Edgar, J. A., 361(24), *389*
Edwards, P. N., 117(35), *137*
Egli, C., 264(115), *278*
Eguchi, S., 442(30), 443(30), *446*
Eich, E., 310(21), *312*
Eigtert, B., 381(32), *389*, 402(34), *427*
Eiter, K., 113(36), *137*, 215(48), 216(48), 229(47, 48), *277*
Ek, A., 183(72), *207*, 322(39), 330, *340*
Elderfield, R. C., 241(51), 248(49, 50), *277*, 348(33), 350(33), *389*
Elgersma, R. H. C., 144(72a), *207*
Eliel, E. L., 61(69), *87*, 94, 97(129, 130), *139*, *140*
Elks, J., 153(73), *207*
Elliot, D. F., 153(73), *207*
Elliott, I. W., 345(34), *389*
Els, H., 29(224), *90*, 433(51, 52), *446*
Elyakov, G. B., 414(127), *430*
Endler, A. S., 362(35), 363(35), *389*
Enezian, J., 178(269), *213*
Engel, R. R., 144(238), *212*
Engelhardt, V. A., 71(222), 75(222), *90*
England, D. C., 77(70), *87*
Enslin, P., 295(52), *313*
Eraksina, V. N., 403(57), *428*
Ermakova, V. N., 173(87), *208*
Erner, W. E., 206(74), *207*
Ernest, I., 154(10), 157(10), *206*

Erskova, L. I., 154(256), *212*
Erdtman, H., 77(71), *87*
Erspamer, V., 308(22), *312*
Etienne, A., 365(35a), *389*
Evans, D. D., 34(4), *85*, 110(5), 113(5), *137*, 413(3), *426*
Evans, F. J., 316(10), 317, 318(10), 319(10), 321(10), *339*
Evdakov, V. P., 38(133), *88*, 166(142), *209*, 416(59), 417(59), *428*
Ewins, A. J., 229(52), *277*

F

Failli, A., 256(102), *278*
Farkas, F., 365(41), *389*
Farnsworth, D. W., 155(141), 161(141), *209*
Farrell, P. G., 2(54), *86*
Fedotova, M. V., 183(255), *212*
Feigelson, P., 310(23, 24, 34, 61, 62), *312*, *313*
Feofilaktov, V. V., 160(75), *208*, 218(53), *218*
Felton, D. G. I., 155(64), *207*
Ferrier, W., 424(33), *427*
Festag, W., 250(177), *280*
Feuer, B. I., 188(233), *212*
Ficken, G. E., 356(7a, 36), *389*, *390*
Fiebig, H., 217(19), *276*
Fiedler, H., 282(103), 293(103), *314*
Finch, N., 300(25), 303(25, 26), *312*, 325(13), 328(11, 12), *339*
Findley, S. P., 218(54), *277*
Finger, G. C., 218(55), *277*
Finkelstein, J., 406(35), *427*
Fischer, B. A., 248(49, 50), *277*
Fischer, E., 40(72), 77(282), *87*, *92*, 142(76), 149(76), *208*, 318(14), *339*, 362(167), 372, (37, 38), *389*, *392*
Fischer, F. E., 220(68), *277*
Fischer, J.-C., 232(41), *276*
Fischer, O., 122(40), *138*
Fischer, W., 417(44), *427*
Fisher, E. E., 142, 145(5a), *207*
Fitzgerald, B. M., 144(185), *211*
Fitzgerald, J. A., 144(185), *211*
Fitzpatrick, J. T., 162(77), *208*
Fleming, C. L., 200(80), *208*
Fleming, I., 246(56), *277*, 403(36), *427*
Fletcher, L. T., 173(240), *212*
Florent'ev, V., 173(88), 178(88), *208*
Floyd, M. B., 286, *314*
Foglia, T. A., 306, *312*
Fontana, A., 77(73), *87*
Forbes, E. J., 192(78), *208*
Force, C. G., 188(236), *212*
Fornefeld, E. J., 129(77), *138*, 268(97), *278*
Foster, G. H., 98(37), 100(38), 129, *137*, 259(40), *276*
Foster, R., 2(74, 75), 71(75), *87*
Fox, L. R., 311, *312*
Fox, S. W., 216(58), 217(57, 30), *276*, *277*
Franke, P., 200(96), *208*
Frankenfeld, J. W., 358(145), *392*
Franklin, C. S., 231(59), *277*
Frankus, E., 35(20), *85*
Fraser, R. R., 21(56), *86*
Freed, M. E., 155(213a), *211*
Freeman, P. R., 194(19), *206*
Freifelder, M., 218(61), 220(61), 248(60), *277*
Freter, K. R., 22(76), *87*, 102(39), 106(39), *138*, 352(39), *389*
Fedotova, M. V., 135(106), *139*
Freedland, R. A., 308(29), *312*
Frey, A. J., 241(208), 263(208), *281*
Fridrichsons, J., 433((9), 437(18), *445*
Fritz, H., 107(114), 121(41), 122(40), *138*, *139*, 262(147), *279*
Fritz, H. E., 30(77), 31, *87*
Fromson, J. M., 256(102), *278*
Frydman, B., 178(79), *208*
Fryer, R. I., 115(42), *138*, 335(15), *339*, 419(37), *427*
Fuhlhage, D. W., 61(78), *87*
Fukada, N., 376(110), *391*
Fukui, K., 3(79), 84(79), *87*
Fukumoto, K., 224(89), 269(67), 271(67), *277*, *278*
Fulmor, W., 433(57), (42), *446*, *447*
Funderbunk, C., 149(114), 158(114), 151(114), *209*
Funk, F. H., 332(33), *340*
Furnas, J. L., 202(21), *206*
Fürst, H., 109(34), *137*, 173(71), *207*, 220(46), 221(46), *277*
Furukawa, S., 75(130), *88*, 335, *339*
Fuson, R. C., 200(80), *208*
Fyfe, C. A., 2(74), *87*

G

Gabriel, S., 376, *389*
Gaimster, K., 155 (36), *207*
Gaines, W. A., 229 (157), *279*
Gall, W. G., 398 (5), *399*
Gallagher, M. J., 158 (81), 162 (81), *208*
Galun, A. B., 135 (59), *138*, 182 (104), *208*
Ganellin, C. R., 19 (80), 22 (81), 29 (80), *87*, 227 (62), *277*
Garanttini, S., 308 (30), 310 (30), *312*, 438 (19), *445*
Garcia, E. E., 335 (15), *339*
Garratt, S., 111 (154), *141*
Gasha, T., 290 (89), *314*
Gaskell, A. J., 105, 111, 118 (43), *138*, 159 (118), *209*, 406 (124), *429*
Gaston-Breton, H., 133 (69), *138*, 197 (127), *209*
Gaudion, W. J., 36 (82), *87*, 301 (31), *312*
Gaughan, E. J., 204 (61), *207*
Geisenfelder, H., 3 (290), *92*
Geissman, T. A., 95 (75), 97 (45), 98 (45), 102 (45), *138*
Geller, K. H., 7 (229), 8 (229), *91*
Gemenden, G. W., 300 (25), 303 (25), *312*, 328 (11, 12), *339*
Gerhard, W., 376, *389*
Gertner, D., 133 (12), 135 (12), *137*
Gibs, G. J., 131 (113), *139*
Gill, N. S., 97 (46), *138*
Ginsburg, D., 188 (82), *208*
Giovannini, E., 365 (41), 382 (42, 44), *389*
Gleicher, G. J., 2 (63), 3, 84 (63), *86*
Glenn, E. M., 166 (261), *212*
Gletsos, C., 256 (102), *278*
Glos, M., 425 (102), *429*
Glosauer, O., 191 (265), *213*
Gmelin, R., 445 (20), *445*
Gnewuch, C. T., 237 (197), *280*
Godtfredsen, W. O., 303 (32), *312*
Goeggel, H., 269 (64), 269 (107), 270 (64), 271 (107), *277*
Goltzsche, W., 65 (273), 66 (273), *92*
Golubeva, S. K., 18 (252), *91*
Gompper, R., 26 (83), *87*
Good, N. E., 217 (149), *279*
Goodman, L., 102 (2a), *137*, 214 (42), 233 (34), *276*, 424, *427*
Goodwin, S., 282 (104), 284 (104), *314*
Gore, P. H., 148 (83), *208*
Gorman, M., 118, *138*, 261 (123), *279*
Gortatowski, M. J., 218 (55), *277*
Gortner, R. A., 42 (39), 43, *86*
Goto, T., 442 (30, 31, 32), 443 (30), *446*
Goutarel, R., 101 (47), *138*
Govindachari, T. R., 178 (84), *208*, 238, 241, 244 (207), *280*
Gower, B. G., 96 (48), 109 (48), *138*
Grahame-Smith, D. G., 308 (33), *312*
Grandberg, I. I., 155 (85), *208*, 424 (38), *427*
Grant, M. S., 77 (43, 84), *86*, *87*
Gray, A. P., 40 (85), 71, 75 (85, 86), 77 (86), *87*, 129 (49), *138*
Green, K. H. B., 149 (86), 173 (86), *208*
Greengard, O., 310 (23, 34), *312*
Greig, M. E., 38 (289), *92*
Grey, T. F., 220 (4), 221 (4), 224 (4), *275*
Gribble, G. W., 125 (31), 129, *137*, 300 (20), *312*
Grigg, R., 24 (26a), *86*
Grimm, D., 349 (166), *392*
Grinev, A. N., 159 (229), 173 (87, 88, 89, 89a, 90, 91, 92), *208*, *212*, 418 (93), *429*
Grob, C. A., 186 (93), *208*, 395, *399*
Gross, E., 305 (72a), *314*
Groth, H., 34 (24), 36, *86*, 413 (13), *427*
Groves, L. H., 239 (65), *277*
Gruda, I., 343 (45), *389*
Gudmundson, A. G., 221 (156), *279*, 405 (92), 413 (92), *429*
Guerra, A. M., 66(50), *86*
Guise, G. B., 292 (35), 293 (35), *313*
Gurevich, P. A., 162 (211), *211*
Gurney, J., 130 (50), *138*
Gurvich, S. M., 77 (253), *91*
Gwinani, S., 291 (36), *313*
Guthrie, R. D., 143 (25a), *206*

H

Haake, P., 422 (38a), *427*
Haarstad, V. B., 178 (197), *211*, 331 (34), *340*
Haase, W. H., 178, *211*
Haberland, H., 31 (243), *91*
Hachová, E., 246 (141), *279*
Haddadin, M. J., 365 (50), 368 (50), 382 (50), *389*
Haddlesey, D. I., 397 (7), *399*
Hadfield, J. F., 273 (103), *278*

G

H

Haeck, H. H., 224(98), *278*
Haendel, D., 200(96), *208*
Hagen, P. B., 310(36a), *313*
Haglid, F., 61(280, 281), 65(280), *92*, 248 (198, 199), *280*
Halpern, O., 409(62), *428*
Hahn, G., 120(50a), *138*, 239, 241, *277*
Hahn, W., 31(243, 244), *91*
Hall, E. S., 269(67), 271(67), 273(103), *277*
Hallmann, G., 60(92), *87*
Halmandaris, A., 224(44), *277*
Hamana, M., 66, *87*
Hamann, K., 7(229), 8(229), *91*
Hamashige, S., 347(65), *390*
Hamlin, K. E., 220(68), *277*
Hammer, C. F., 8(163), 9(163), 76, *89*
Hand, J. J., 217(31, 32), *276*
Hands, A. R., 69(1), 70(1), *85*, 226(2), *275*
Hanger, W. G., 378(5), *388*
Hanson, A. W., 2(88), 3(88), *87*
Hanson, J. R., 269(67), 271(67), *277*
Hanson, P., 2(75), 71(75), *87*
Hara, T., 19(287), *92*
Harada, H., 433(21, 40, 50), *445*, *446*
Hardegger, H., 360(46, 138), *389*, *391*
Harder, U., 227(131), *279*
Harding, H. R., 369(15), *388*
Hargrove, W., 261(123), *278*, *279*
Harley-Mason, J., 72(38), 75(38), *86*, *87*, 98(37, 51), 100(38), 118(52), 129, *137*, *138*, 182(94), 192(44, 69, 94, 95), *207*, *208*, 246(56), 256, 259(40), 260 (71), 261(70), 273(72), *276*, *277*, 343 (49), 344(49), 346(47), 352(48), 354 (49), 378(47), *389*, 403(36), 406(39), *427*
Harnisch, A., 419(119), *429*
Harris, L. S., 113(150), *140*
Hart, G., 412(41), 413(40), *427*
Hartman, P. J., 69(164), 75(164), *89*
Hartmann, G., 200(96), *208*
Hartung, W. H., 113(53), *138*
Harvey, D. G., 31, *88*
Hassner, A., 365(50), 368(50), 382(50), 388(137a), *389*, *391*
Hauptmann, S., 200(96), *208*
Hausmann, W. K., 431(34), *446*
Havinga, E., 144(72a), *207*
Haworth, R.C., 338(22), *340*
Hayaishi, O., 308(69), 310(24, 47), 311(37), *312*, *313*
Heacock, R. A., 34(122), *88*, 182(99), 183 (99), 191(97), *208*, 295(39), 310(38), *313*, 413(52), 424(52), *428*
Hearon, W. M., 201(110, 111, 112), *208*, *209*
Hearst, E., 308(90a), *314*
Heath-Brown, B., 28(90), *87*, 130(27), *137*, 145(101), 153(100), 157(100), *208*, 220(73), 222(73), 224(73), 225(73), 226(37), 227(73), *276*, *277*
Heilmann, D., 166(139), *209*
Heller, G., 382(51), 383(51), *389*, 399(4), *399*
Heinzelman, R. V., 22(91), 23(91), 38(289), *87*, *92*, 166(261), *212*, 218(166), 220 (166, 174), 222(74), 224(74, 174), 225(74), 232(74), *277*, *280*, 442(22), *445*
Heitmeier, D. E., 71(86), 75(86), 77(86), *87*
Hellmann, H., 26(93), 60(92, 94), *87*, 346 (52), *389*
Helsley, G. C., 220(194), 221(194), *280*
Hems, B. A., 153(73), *207*
Henbest, H. B., 19(32), *86*
Hendrick, L. N., 353(153), *392*
Hendrickson, J. B., 346(53), 353(53), *389*
Henley, W. O., Jr., 145(56), *207*
Henry, D. W., 97(54), *138*
Herbst, D., 111, *138*
Herbst, K., 202(173), *210*
Herrmann, E., 42(265), 43(265), 44(266), 68(267), *91*, *92*
Hermann, R. B., 84(94a), *87*
Hertz, E., 155(213a), *211*
Herzog, H., 365(129a), *391*
Hesse, M., 431(23), *446*
Hester, J. B., Jr., 166(261), *212*, 243, 263 (182), *277*, *280*, 290(40), 298(96), 301 (96), *313*, *314*, 326(35), *340*, 395, *399*
Hey, D. H., 345(1), 358(1), *388*
Hieirs, G. S., 382(96), *390*
Higginbothom, D. H., 217(168), *280*
Hill, H. M., 194(266), *213*
Hill, R. K., 245(170), *280*
Hinman, R. L., 4(97a), 5(96), 6, 9(98), 11 (96), 12(97), 15(95a), 16(95a), *87*, 304, 311(42), *313*, 346, 349(55), 353(54), 363(54), *389*
Hino, T., 291(43, 44), *313*, 357(170), *392*, 394(9), *399*
Hinsvark, O. N., 376(59), *389*
Hirai, S., 253(121), *278*

Hirata, T., 433 (24), *446*
Hirata, Y., 433 (21, 40, 50), 442 (30, 31), 443, *445*, *446*
Hirohashi, T., 296 (112), *315*
Hiremath, S. P., 153 (102), *208*
Hiriyakkanavar, J. G., 166 (103), *208*
Hiser, R. D., 162 (77), *208*
Hishida, S., 35 (100), *87*, 413 (72), *427*
Hněvsóva, V., 154 (10), 157 (10), *206*
Hoan, N., 414 (18, 19), *427*
Hobbs, C. F., 21 (101), *87*
Höbel, M., 419 (75), *428*
Hodges, R., 437 (25, 26), *446*
Hodson, H. F., 7, *87*, 122 (14b), *137*
Hofmann, A., 178 (246), *212*, 268, *280*, 440 (27), *446*
Hoffmann, C., 243 (33), *276*
Hoffmann, E., 135 (59), *138*, 154 (29), 182, *206*, *208*
Hoffmann, K., 135, *138*, 393 (12), 394, *399*, 419 (43), *427*
Holland, D. O., 235 (76), *277*, 424 (33), *427*
Holly, F. W., 132 (153), 135 (153), *140*, 189 (274), *213*, 399 (22), *400*
Hollyman, D. R., 19 (80), 29 (80), *87*, 227 (62), *277*
Holmes, R. R., 176 (105), *208*
Holt, S. J., 364 (56), 367 (56, 57), *389*
Hood, W. H., 301 (31), *312*
Hook, W. H., 36 (82), *87*
Hoppe, G., 180 (209), *211*, 393 (18), *400*
Hooper, M., 383 (19a), 388 (23a), *388*
Horák, M., 345 (148), *392*
Horden, R. A., 69 (165), 70 (165), 75 (165), *89*
Hörlein, G., 77 (282), *92*, 362 (167), *392*
Hörlein, U., 229 (77), *277*
Horner, L., 73 (103), *87*
Horning, E. C., 308 (45), *313*, 343 (58), 346 (58), 358 (141), *389*, *392*
Hoshino, T., 46, *87*, 216, *278*
Houben, J., 417 (44), *427*
Houff, W. H., 376 (59), *389*
Houlihan, W. J., 352 (3), *388*
Houghton, E., 365 (59a), *389*
House, H. O., 370 (60), *389*
Hovden, R. A., 226 (125), *279*
Howe, E. E., 37 (231), *91*, 233 (78), *277*, 417 (91), *429*
Howell, W. C., 378 (5), *388*
Hrutfiord, B. F., 197, *207*
Hsu, I. H., 300 (25), 303 (25), *312*, 328 (11, 12), *339*
Huang-Hsinmin, 395 (10), 396 (10), *399*
Hübner, H. H., 102 (39), 106 (39), *138*
Huckstep, L. L., 433 (39), 436 (39), 437 (41), *446*
Hudson, R. S., 404 (82), *428*
Huebner, C. F., 183 (106), *208*
Huisgen, R., 385 (61), *389*
Huffman, J. W., 130 (56), *138*, 153 (107), *208*, 252, 255, *277*
Huffman, R. W., 114 (57), *138*
Hughes, B., 299 (46), 301 (46), *313*
Hughes, G. A., 111 (55), *138*
Hughes, G. K., 148 (83), *208*
Hughes, H., 221 (35), *276*
Huisgen, R., 197 (109), *208*
Hunger, A., 237 (210), 267 (209, 210), *281*
Hünig, S., 178 (108), *208*
Hunt, R. R., 132 (58), 133 (58), *138*
Huntress, E. H., 202 (110, 111, 112), *208*, *209*
Hütz, H., 372 (38), *389*
Hutzinger, O., 182 (99), 183 (99), *208*
Hyre, J. E., 194 (113), *209*

I

Ichikawa, H., 348 (75a), *390*
Ichiyama, A., 308 (69), *313*
Iddon, B., 220 (36), *276*
Igolen, H., 29 (116), *88*, 166 (128), 215 (82), *277*
Igolen, J., 166 (128, 129), *209*
Ikan, R., 132 (60), 133 (60), 135 (59, 60), *138*, 182 (104), *208*, 426 (45), *427*
Illy, H., 14a (114), 150 (114), 151 (114), *209*
Imato, E., 133 (73), *138*
Inaba, S., 296 (112), *315*
Ingleby, R. F. J., 346 (47), 352 (48), 378 (47), *389*
Ingraffia, F., 376 (62), *389*
Inhoffen, H. H., 405 (46), *427*
Inokuchi, N., 305 (54a), *313*, 394 (12a), *399*
Inoue, S., 442 (32), *446*
Inove, H., 133 (73), *138*
Ireland, R. E., 200 (115), *209*
Irie, K., 161 (136), *209*
Isaac, W. M., 214 (43), *276*

Karabatsos, G. J., 143 (137a), *209*
Karpel, W. J., 246 (83), *277*, 348 (66), 351 (66), *390*
Karrer, P., 122 (14a), *137*, 295 (52), *313*
Kašpárek, S., 34 (122), *88*, 413 (52), 424 (52), *428*
Kastron, Y. A., 169 (161), 170 (161), *210*
Kataoka, H., 204 (177, 179), *210*, 376 (77), *390*
Kato, T., 348 (75a), *390*
Katritzky, A. R., 12, 13 (33), 34 (123), 35 (124), *86*, *88*, 341 (76), *290*, 413 (53), 414 (53), *428*
Katz, L., 224, 225 (159), *279*
Kaupp, G., 365 (140a), *392*
Kawana, M., 376 (77), *390*
Kawanishi, M., 224 (130), *279*
Kawata, K., 253 (121), *278*
Keasling, H. H., 38 (289), *92*
Kebrle, J., 135, *138*, 393 (12), 394, *399*, 419 (43), *427*
Keller, K., 7 (125), *88*
Kellie, A. E., 364 (56), 367 (56), *389*
Kelly, A. H., 147 (138), 149 (138), 150 (138), 156 (138), 162 (138), *209*
Kelly, M., 442 (29), *446*
Kelly, W., 295 (16), 308 (45), *312*, *313*
Kempler, G., 166 (139), *209*, 419 (54), *428*
Kendall, E. C., 352 (78, 36), *390*
Kendall, J. D., 356 (79), *389*, *390*
Kennedy, J. G., 22 (62), *86*, 404 (26a), 411 (27), 413 (26a), 414 (27), *427*
Kerr, A., 328 (11), *339*
Kershaw, J. W., 298 (2), *312*, 327 (21), 328 (5, 21), *339*, *340*
Khan, M. K. A., 421 (55), *428*
Kharash, M. S., 219(93), *278*
Khoi, N. H., 414 (18), *427*
Kiang, A. K., 39 (126, 127), *88*, 414 (56), *428*
Kiefer, B., 241 (137), 243 (137), *279*
Kierstead, R. W., 241 (208), 263 (208), *281*
Kierzek, L., 149 (145), 209
Kimura, T., 206 (140), *209*, 220 (90), *278*
King, F. E., 129 (72), *138*
King, L. J., 226 (2), *275*, 308 (53), 310 (1), *312*, *313*
Kinoshita, T., 133 (73), *138*
Kirby, G. W., 135 (74), 136 (74), *138*, 366 (80), *390*
Kirkpatrick, J. L., 263 (192), *280*, 413 (32), 414 (32), 418 (32), *427*
Kishi, Y., 442 (30, 31, 32), 443 (30), *446*
Kishimoto, H., 442 (32), *446*
Kiss, G., 445 (33), *446*
Kissman, H. M., 57 (128), *88*, 155 (141), 161 (141), *209*, 233 (94), *278*, 282 (109), 283 (109), 295 (109), 296 (109), *315*
Kisteneva, M. S., 346 (81), *390*
Klamann, D., 363 (82), *390*
Kleigel, W., 57 (291), *92*
Kline, G. B., 129 (77), *138*, 239 (95), 268 (97), *278*
Klug, J. T., 298 (82), 301 (82), *314*, 327 (31), *340*
Klüssendorf, S., 59 (260), *91*, 133 (146), *140*
Knapp, G. G., 61 (272), *92*
Knight, J. A., 269 (13, 15), *276*
Knöller, G., 202 (173), *210*
Knowlton, M., 43 (129), *88*
Knox, W. E., 310 (91), 311 (54), *313*, *314*
Kny, H., 393 (2), *399*
Kobayashi, G., 75 (130), *88*
Kobayashi, S., 182 (117), 187 (117), 188 (117), *209*
Kobayashi, T., 17 (219), *90*, 98 (120), *140*, 305 (54a), *313*, 394 (12a), *399*
Kochetkov, N. K., 38 (133), *88*, 131 (75), *138*, 158 (143, 149), 166 (142), *209*, *210*, 416 (59), 417 (59), *428*
Koechlin, H., 393 (19), *400*
Kocsis, K., 104 (18), *137*, 253 (25), *276*
Koelsch, C. F., 190 (158), *210*, 301 (55, 56), *313*
Kogodovskaya, A. A., 157 (167), *210*
Köhler, E., 97 (116), *139*
Koizumi, M., 4, *88*
Kolosov, M. N., 343 (84), 357 (83), *390*
Kompis, I., 111 (83), *139*
Komzolova, N. N., 162 (144), *209*
König, H., 197 (109), *208*
Königstein, F. J., 97 (116), *139*
Koo, J., 214 (96), 233 (96), *278*
Kopp, E., 417 (100), *429*
Korezynski, A., 149 (145), *209*
Kornfeld, E. C., 98 (76), 129 (77), *138*, 268 (97), *278*
Koshland, D. E., Jr., 122 (8, 85a), *137*, *139*
Kost, A. N., 15 (132), 70 (254), 75 (254), 77 (253, 255), *88*, *91*, 154 (146), 155 (85, 146, 147), 156 (147), *208*, *209*, *210*, 403 (57), 413 (58), *428*

Kotake, M., 15(144), 34(144), *88*
Kralt, T., 224(98), *278*
Kramer, E., 386(121), *391*
Kramer, W. J., 386(23), *388*
Krasnova, V. A., 413(58), *428*
Kraus, H., 71(86), 75(86), 77(86), *87*, 129 (49), *138*
Krause, H. W., 382, *390*
Krausmann, H., 166(139), *209*
Křepinský, J., 168(67), *207*
Kretz, E., 343(85), *390*
Kröhnke, F., 385(86), 386(87), 387(87), *390*
Kröhnke, G., 385(88), *390*
Krueger, W. E., 187(35), *207*
Kuch, H., 221(152), *279*
Kucherova, N. F., 38(133), *88*, 131(75), *138*, 158(11, 135, 143, 144, 148, 149, 150), 162(135, 144), 166(142), *206*, *209*, *210*, 416(59), 417(59), *428*
Kuehne, M. E., 238(99), *278*, *399*
Kühn, H., 57(134), *88*
Kul'bovskaya, N. K., 173(89), *208*
Kulsa, P., 262(26), *276*
Kumadaki, J., 66, *87*
Kunori, M., 15(135, 136), *88*, 417(61), 426 (60), *428*
Kupchan, S. M., 291(83), *314*
Kutter, E., 26(83), *87*
Kuryla, W. C., 7, 71(169), 75(169), *89*, 126(91), 139
Kurze, J., 291(78), *314*, 355(128), 391
Kutney, J. P., 111(78), 124(79), *138*, *139*, 255, 256(102), 259(100, 105), 273(103), *278*
Kwiatkowski, G. T., 24(47), *86*

L

Lagowski, J. M., 248(50), *277*, 341(76), *390*
Lake, R. D., 164(52), 166(52), *207*
Lambert, B. F., 122(9a), *137*
Lancaster, J. E., 433(57), *447*
Landquist, J. K., 165(151), *210*
Lang, J., 4(97a), 5(96), 11(96), 12(96), *87*, 311(42), *313*
Lange, R. F., 71(169), 75(169), *89*, 126(91), *139*, 226(126), *279*
Langella, M. R., 153(60), 189(60, 198a), *207*, *211*, 294, *314*
Lapp, T. W., 196(194), *211*
Larson, G. W., 145(57), *207*
Lauterbach, R., 198(243), 200, *212*
Lawson, W. B., 303(58), 304(57), 305(72a), *313*, *314*, 353(89), 363(89), *390*
Lawton, R. G., 237, 268, *280*
Lee, J. 406(35), *427*
Leeney, T. J., 343(49), 344(49), 354(49), *389*
Leete, E., 23(137), 35, 40(139), 47, *86*, *88*, 96 (48), 97(54, 84), 102(84), 108(80, 84), 109(48, 80), 110(80), 111, 113, (84), 114(84), 129(81), *138*, *139*, 189(152), *210*, 283(11, 59), 285, 286, 288, 300(11), 301(11), *312*, *313*, 414(22), *427*
Lefemine, D. V., 431(34), *446*
Leggetter, B. E., 15(140), *88*
Lehnert, W., 195(204), *211*
LeMen, J., 101(14), *137*
Lenzi, J., 29(116), *88*, 166(131), *209*, 215 (82), *277*
Leonard, N. J., 409(62), *428*
Leone, S. A., 426(12), *427*
Le Quesne, P., 111(78), 124(79), *138*, *139*, 256(104), *278*
Lerch, V., 291(78), *314*, 355(128), *391*
Lesiak, T., 189(154), *210*
Lesslie, T. E., 201(112), *209*
Levine, R. J., 308(60), *313*
Levkoev, I. I., 22(14), 23(14), *88*
Levy, A., 183(28), *206*
Lewis, A. D., 154(155), *210*
Lewis, R. G., 61(281), *92*, 227(200), 248 (199), 249(201), *280*, *281*
Lewis, T., 13(6), 61(281), *85*
Licari, J. J., 97(85), *139*
Liede, V., 419(75), *428*
Lietman, P. S., 114(121), *140*
Liljegren, D. R., 109(102), 113(102), *139*, 249(139), *279*, 413(40), *427*
Lin, L. S., 187(253), *212*, 420(101a), *429*
Lind, C. J., 22(142), 23(142), *88*
Lindsey, R. V., Jr., 77(70), *87*
Lindwall, H. G., 57(28), 77(13), *85*, *86*, 153(156), *210*, 367(133), 378(90), *390*, *391*, 399(14), *399*
Lingens, F., 60(92), *87*
Lions, F., 97(46), *138*, 418(63), *428*
Lipparini, L., 217(5, 6), *275*, *276*
Lippmann, E., 16(143), *88*
Lipton, H. L., 114(121), *140*
Littell, R., 171(157), 173(157), 210, 423(64), *428*

Little, R. L., 380(22), *388*
Litzinger, E. F., Jr., 53(26), *86*
Lively, D. H., 433(39), 436(39), *446*
Livi, O., 11(15), *85*, 115(13), *137*
Lockhart, I. M., 220(4), 221(4), 224(4), *275*
Loew, P., 269(107), 271(107), 272(106), 273(106), *278*
Logeman, W., 243(108, 109), *278*
Lohse, D., 203(43), *207*
Long, F. A., 6, *86*
Longmore, R. B., 343(91), 352(92), *390*
Lorenz, R. R., 190(158), *210*
Lorenz, T., 382(42, 44), *389*
Loudon, G. M., 122(85a), *139*
Loudon, J. D., 233(38), 233(45), *276*, *277*, 373(94), 376(93, 94), *390*
Lovald, R. A., 47(179), *89*
Lovenberg, W., 308(60), *313*
Lukaszewski, H., 224(44), *277*, 424(10), *427*
Lücke, E., 178(108), *208*
Lukton, A., 122(85a), *139*
Lunsford, C. D., 220(194), 221(194), *280*
Lutz, R. E., 378(95), *390*
Lyle, R. E., 149(159), 150(159), 153(159), 156(159), *210*, 249(172), *280*, 316(10), 317, 318(10), 319(10), 321(10), *339*
Lyttle, D. A., 22(91), 23(91), *87*, 219(193), 222(74), 224(74), 225(74), 232(74), 233(110, 193), *277*, *278*, *280*

M

Maas, I., 42(274), 45, *92*
McCaldin, D. J., 380(63), *389*
McCapra, F., 269(116), 271(67), *277*, *278*, 289(63), *313*
McClean, G. W., 43(225), *90*
McCloskey, P., 233(38, 45), *276*, *277*
McCullagh, L., 203(46), *207*
MacDonald, J. A., 70(238), 75(238), *91*
McGovern, J. P., 133(104), *139*
McGrath, L., 177(25), 178(25), *206*, 282(7), 283(7), *312*
McGrew, G., 144(238), *212*
McKague, B., 124(79), *139*
Machida, K., 412(112), *429*
McIntyre, P. S., 151(186a), *211*
McKague, B., 256(104), *278*
McKay, J. B., 154(164), *210*
Mackie, R. K., 145(51), 184(51), 187(51), *207*
McLean, G. W., 114(121), *140*
McLean, J., 147(165), *210*
McLean, S., 147(165), *210*
Maclennan, J. S., 378(90), *390*
McLeod, D. H., 147(138), 149(138), 150(138), 156(138), 162(138), *209*
McMillan, A., 221(156), *279*, 405(92), 413(92), *429*
McMillin, C. K., 21(101), *87*
Madonia, P., 412(98), *429*
Maeda, M., 156(172), *210*
Maeno, H., 310(61, 62), *313*
Maffii, G., 155(262), *213*
Magistro, A. J., 148(58), *207*
Magnani, A., 246(83), 248(84), *277*, 348(66), 351(66), *390*
Magnusson, T., 381(20), *388*
Mahon, M. E., 295(39), 310(38), *313*
Mains, G. J., 148(58), *207*
Majima, R., 15(144), 34(144), *88*, 216, *278*
Mamaev, V. P., 223(112), *278*
Mann, F. A., 71(145), 71(145), *88*
Mann, F. G., 39(126, 127), *88*, 121, *137*, 158(81), 162(81), 166(40), 168(40), *207*, *208*, 338(22), *340*, 362(21), *388*, 395(10), 396(10), *399*, 414(56), *428*
Mann, M. J., 129(77), *138*, 208(97), *278*
Manning, R. E., 102(19, 20), 103(20), 107(19), *137*, 261(123), 262(27), *276*, *279*
Manoury, P., 22(117), *88*, 166(132), 183(132a), *209*
Manske, R. H., 161(120), *209*, 229(113), *277*, *278*, 418, *428*
Manson, A. J., 358(168), *392*
Mantecon, J., 422(38a), *427*
Mantell, G. J., 153(156), *210*, 399(14), *399*
Marchand, B., 153(160), *139*, *210*
Marchant, J. R., 75(221), *90*
Marchant, R. H., 31, *88*
Marchetti, L., 75(52, 147), *86*, *88*
Marchiori, F., 77(73), *87*
Marion, L., 40(139), *88*, 97(84), 102(84), 108(84), 113(84), 114(84), *139*, 189(152), *210*
Marko, A. M., 365(1a), *388*
Marsden, C. J., 165(151), *210*
Marsh, M. M., 433(39), 436(39), *446*
Marshall, W. S., 433(55), *446*
Martin, G. J., 214(96), 233(96), *278*

Martin, J. A., 269(13, 15), 271(15), 272(17), 273(16), *276*
Martin, L. J., 196(194), *211*
Martoskha, B. K., 27(201), *90*
Mathre, O. B., 97(25), *137*
Martynov, V. F., 169, 170(162, 163), *210*
Marvel, C. S., 382(96), *390*
Masamune, S., 264(115), *278*
Massy-Westrop, R. A., 248, 249(201), *280*
Masumoto, Y., 419(88), *429*
Matell, M., 419(66), *428*
Mathieson, A. M., 433(9), 437(18), *445*
Matsubara, S., 25(220), *90*
Matsuda, Y., 75(130), *88*
Matsueda, R., 19(287), *92*
Matsui, M., 433(24), *446*
Matteson, D. S., 57(239), 59, 60, *91*
Mayer, H., 59(260, 263), *91*, 133(146), *140*
Maynard, M. M., 244(189), *280*
Mayor, P. A., 397(7), *399*
Mehler, A., 311(64), *313*
Mehta, M. D., 424(33), *427*
Meier, J., 220(117), *278*
Meimaroglou, C., 14(189), 15(189), *90*
Melby, L. R., 77(70), *87*
Meli, A., 243(108, 109), *278*
Mel'nikov, N. N., 215(122), *278*, 412(7), *426*
Melzer, M. S., 85(148), *89*, 424(67), *428*
Mentzer, C., 217(118), *278*, 295(65, 66), *313*, 425(68), *428*
Mentzger, C., 294(12), *312*, *313*
Merchant, J. R., 189(166), *210*
Merer, J. J., 17(9), *85*
Merica, E. P., 180(236), *212*
Merz, H., 102(39), 106(39), *138*
Metreveli, L. I., 357(83), *390*
Meyer, E. W., 20(119), *88*, 166(134), 169, *209*, 246(93), *277*, 359(97), *390*
Meyer, R. F., 359(97), 365, *390*
Meyer, W. E., 433(42, 57), *446*
Meyer, W. P., 5, 6(200), *90*
Meyer-Delius, M., 385(86), *390*
Michel, G., 419(117), *429*
Middleton, W. J., 71(222), 75(222), *90*
Mietasch, M., 166(139), *209*
Mikol, G. J., 187(35), *207*
Miller, M., 353(153), *392*
Millich, I., 402(69), 421(69), *428*
Milliman, G. E., 144(238), *212*
Mills, B., 295(67), *313*, 352(98), *390*
Mills, G. A., 206(74), *207*
Milne, G. W. A., 193(278), *213*
Mingoia, Q., 46(149), *89*
Misiorny, A., 40(57), *86*, 97(26), *137*
Mitoma, C., 308(68, 79), *313*, *314*
Mitropol'skaya, V. N., 413(58), *428*
Miwa, T., 25(220), *90*
Mix, H., 382, *390*
Miyaji, S., 376(77), *390*
Miyamae, T., 189(281), *213*, 395(25), *400*
Miyashita, K., 161(136), *209*
Mizianty, M. E., 35, *92*
Mkhitaryan, A. V., 157(167), *210*
Mo, L., 273(17), *276*
Modler, R. F., 383(107), 384(107, 99a), *390*
Moe, O. A., 235(119, 188), *278*
Moffatt, J. S., 405, *428*
Möhlav, R., 72(150), 73(150), 75(150), *89*
Mohr, G., 120(147), 135(147), *140*
Mohrbacker, R. J., 105(98), *139*
Moir, R. Y., 153(231), *212*
Mold, H. D., 224(98), *278*
Molho, D., 295(66), *313*
Möller, K., 425(102, 103), *429*
Molloy, B. B., 437(41), *446*
Moncrief, J. W., 433(36), *446*
Money, T., 269(67, 116), 271(67), *277*, *278*
Monnier, J., 294(17), *312*
Monnikendam, P., 431(34), *446*
Montgomery, J. A., 440(37), *446*
Monti, A., 69(51), 75(51), 76(53), *86*
Monti, S. A., 102(19), 107(19), *137*, 262(27), *276*, 369(100), 372(29), 373(29), *389*, *390*
Moore, B., 122(146), *137*
Moore, B. P., 413(32), 414(32), 418(32), *427*
Moore, J. A., 192, *210*
Moore, M. M., 409(121), *429*
Moore, R. F., 362(101), *390*
Moore, R. G. D., 395(15), 396(15), *399*
Mootoo, B. S., 269(67), 271(67), *277*
Morelli, G., 165(169), *210*
Morgan, K. J., 421(55), *428*
Morimoto, H., 445(38), *446*
Morin, R. D., 22(14), *85*, 183(26, 170), *206*, *210*, 221(21), *276*
Moroder, L., 77(73), *87*
Morozovskaya, L. M., 154(256), *212*
Morris, L., 419(72), *428*

Morrison, D. E., 129(77), *138*, 251(120), 268(97), *278*
Morrison, G. C., 27(151, 152), 80, *89*, 251 (120), *278*
Morton, D. M., 308(4), *312*
Mossini, V., 227(187), *280*
Mousseron-Canet, M., 293(68a), *313*, 373 (102, 103), *391*
Mowat, J. H., 433(57), *447*
Muchowski, J. M., 157(5), *206*
Mühlstädt, M., 16(143), *88*
Müller, A., 419(117), *429*
Muller, G., 178(269), *213*
Müller, J. M., 343(85), *390*
Müller, W., 96(99), *139*, 403, 414(77), 419 (117), *428*
Munk, M. E., 433(55), *446*
Murakoshi, I., 190(171), *210*
Murasheva, V. S., 154(257), *212*
Murphy, H. W., 75(153), *89*
Murphy, N. J., 61(69), *87*
Mustafa, A., 379(104), *391*
Myers, T. C., 97(149), *140*

N

Naggi, M., 346(109), *391*
Nagarajan, K., 178(84), *208*, 433(39), 436 (39), 437(41), *446*
Nagata, C., 3(79), 84(79), *87*
Nagata, W., 253(121), *278*
Naidoo, B., 78(109), *88*, 320(18), 321(18), *340*
Nair, M. D., 173(9), 174(9), 175(9), *206*
Nakada, H. I., 311, *312*
Nakagawa, A., 291(43, 44), 295(88), *313*, *314*
Nakagawa, M., 220(90), *278*, 394(9), *399*
Nakamura, S., 308(69), *313*
Nakata, H., 433(40), *446*
Nakatsuka, N., 264(115), *278*
Nakazaki, M., 19(154, 155, 156), 20(155), *89*, 116(88), 122(87), *139*, 156(172), *210*, 318(23, 25), 319, 321(23), *340*
Nametkin, S. S., 215(122), *278*
Napier, D. R., 380(22), *388*
Narasimhan, K., 57(246), *91*
Natzeck, U., 399(4), *399*
Nayler, J. H. C., 235(76), *277*, 424(33), *427*
Neber, P. W., 202(173), *210*
Neil, A. B., 360, *391*
Neklyudov, A. D., 35(157), *89*
Nelson, D. A., 240(190), 244(190), *280*
Nelson, G. E., 256(102), *280*
Nelson, N. A., 397(1), *399*
Nelson, N. R., 433(14), 436(14), *445*
Nelson, V. R., 273(163), *278*
Nenitzescu, C. D., 173(210), *210*, *211*, 366 (106), 367(106), *391*, 403(71), *428*
Nerdel, F., 363(82), *390*
Nesmeyanov, A. N., 27(158), *89*
Nettleton, H. R., 397(23), *400*
Neuberger, A., 378(30), *389*
Neukom, H., 445(33), *446*
Neuss, N., 118(46a), *138*, 261(123), *279*, 433(39), 436(39), 437(41), *446*
Newbold, G. T., 178(34), *207*
Newman, D., 186(4, 6), *206*
Nezval, A., 229(47), *277*
Nicolaus, R. A., 438(41a), *446*
Niemann, C., 235(148), *279*, 393(19), 400, 424(73), *428*
Niklaus, P., 163(71a), *207*
Nitta, I., 433(21, 50), *445*, *446*
Nitta, Y., 189(281), *213*, 395(25), *400*
Nixon, P., 168(65), *207*
Nogrady, T., 109(90), 113(89), *139*, 144 (206), 182, 211, 419(72), *428*
Noland, W. E., 7, 8(163), 9(163), 11, 12 (174, 175, 176, 177), 13(177), 14(174, 176), 40(172, 173, 178, 289a), 41(172), 45(178), 46(174), 47(170, 179, 180, 289a), 53(172), 54(166, 167), 69(161, 165), 70(165), 71(169), 73(160), 75 (161, 164, 165, 169), 76, 80, 82(160), *89*, *92*, 108(92), 113(153a), 114(93, 94), 115(94), 126(66, 91, 95), *138*, *139*, *141*, 178(175), 179(176), *210*, 221(127), 226(124, 125, 126), *279*, 296(70, 71), 297(71), 301(70, 71), *313*, 322, *340*, 383(107, 108), 384(99a, 107), *390*, *391*, 406(51a), *428*
Noll, K., 433(55), *446*
Nordsiek, K.-H., 405(46), *427*
Nordtrom, J. D., 244(191), *280*
Novák, J., 215(128), 246(141), *279*
Novák, L., 154(10), 157(10), *206*
Nozaki, M., 310(47), *313*
Nutt, R. F., 189(274), *213*
Nutter, W. M., 14(7), 15(7), *85*

O

O'Brien, S., 131, *139*
Ochiai, E., 204(177, 178, 179), *210*
Ockenden, D. W., 146, 155(180), 161(180), *210*, 295, 297, 299(72), 300(72), *314*
O'Connor, R., 143(182, 183), *210*
Oddo, B., 7(182), 30(183), 35, 45, 46(184), *89*
Oesterlin, R., 369(15), *388*
Ogareva, D. B., 183(255), *212*
Ogawa, K., 291(43), *313*, 357(170), 392, 394(9), *399*
Ohno, M., 125(97), *139*
Oishi, T., 61(281), *92*, 220(90), 248(199), *278*, *280*, 346(10, 109), *388*, *391*
Okuda, T., 60(186), *89*
Okumura, T., 253(121), *278*
Ollis, W. D., 237(210), 267(209, 210), *281*
Omar, A. M. E., 246(129), *279*
Omote, Y., 289(90), 290(90), *314*, 376(77, 110), *390*, *391*
Onda, M., 224(130), *279*
Opitz, G., 26(93), *87*
Orgareva, O. B., 135(106), *139*
Orlando, G., 97(119), *140*
Orlova, L. M., 96(107), 139, 183(255), *212*
Oroshnik, W., 46, *85*, 114(7), *137*
Orr, A. H., 168(184), *211*
Oshio, H., 445(38), *446*
Osterberg, A. E., 352(78), *391*
O'Sullivan, D. G., 364(56), 367(56), *389*
Otting, W., 2(242), *91*
Ovchinnikova, Z. D., 417(104), *429*
Owellen, R. J., 144(185, 238), 158(239), *211*, *212*
Owsley, D. C., 204(61), *207*

P

Pacheoco, H., 161(186), *211*, 294(12), *312*
Pachter, I. J., 105(98), *139*
Padieskaya, E. N., 433(44), *446*
Page, I. H., 22(250a), 23(250a), *91*
Palazzo, G., 346(111), 350(112), *391*
Palmer, D. R., 378(7), *388*, 445(7), *445*
Palmer, M. H., 151(186a), *211*
Panferova, N. G., 358(117), *391*
Panisheva, E. K., 173(90), *208*
Panizzi, L., 242(39), *276*
Papadopoulos, E. P., 21(101), *87*
Pappalardo, G., 16(187), *89*, 376(113), *391*
Pappo, R., 188(82), *208*
Paragamian, V., 182(187), *211*
Parham, W. E., 33, *90*
Parisi, F., 380(114), *391*
Parke, D. V., 308(53), *313*
Parkhurst, R. M., 154(164), *210*
Parkin, D. C., 85(25), *86*
Parmerter, S. M., 153(188), *211*
Parrick, J., 147(138), 149(138), 150(138), 156(138), 162(138), *209*
Parsons, R. G., 270(15a), 272(15a), 273 (15a), *276*
Parsons, T. G., 5, 6(200), *90*
Passalacqua, V., 75(147), *88*
Patchornik, A., 304(57), 305(72a), *313*, *314*, 353(89), 363(89), *390*
Patel, H. P., 395(16), 396(16), *399*
Patrick, J. B., 282(105, 106, 107, 108, 109, 110), 283(106, 108, 109), 284(110), 285(110), 289, 293(106), 294(105), 295 (108, 109), 296(108, 109), *314*, *315*, 322(40), 324(40, 42), 325(43), 326 (27), 327(41), *340*, 433(42, 57), *446*, *447*
Patterson, D. A., 364(114a), *391*
Pausacker, K. H., 147(190), 148(191), 149 (189), 150(189), *211*
Pavri, E. H., 406(39), *427*
Payne, T. G., 273(16), *276*
Pelah, Z., 221(87), 227(87), *278*
Pelchowicz, Z., 157(192), 183(27, 28), *206*, *211*
Pennington, F. C., 196, *211*
Perekalin, V. V., 36(188), *90*
Périn-Roussel, O., 149(49), 158(49), *207*
Perkin, W. H., Jr., 130(50), *138*, 178(33), *207*, 418(65), *428*
Perotti, L., 46(184), *89*
Pershin, G. N., 433(44), *446*
Pesis, A. S., 358(118), *391*
Pessina, R., 218(86), 227(85), *277*, *278*
Peterson, P. E., 424(73), *428*
Petrow, V., 367(57), *389*
Petruchenko, M. I., 158(143, 150), *209*, *210*
Petyunin, P. A., 358(115, 116, 117, 118, 119), *391*
Peyer, J., 178(246), *212*
Pfaender, P., 121(41), *138*
Pfeiffer, P., 386(121), 387(120), *391*
Pfeil, E., 227(131), *279*

Phillips, G. T., 269(67), 271(67), *277*
Phillips, R. R., 160, *211*
Philpott, P. G., 28(90), *87*, 145(101), 153 (101), 157(100), *208*, 220(73), 222(73), 224(73), 225(73), 227(73), *277*
Piccinni, A., 215, *279*
Pickard, R. L., 132(58), 133(58), *138*
Pidacks, C., 114(4), 131(113), *137*, *139*, 171, 173(15), *206*, 423(2), *426*, 433(57), *447*
Piers, E., 111(78), 124(79), *138*, *139*, 178 (197), *211*, 255(101), 256(104), 259 (100, 105), *278*, 424, *428*
Piers, K., 14(189), 15(189), *90*
Pietra, S., 157(198), *211*, 231(133, 134, 135), *279*, 378(122, 123, 124, 155), *391*, *392*
Pikl, J., 348(66, 69), 351(66, 70), 352(68), *390*
Pillai, P. M., 178(84), *208*
Pinkus, J. L., 383(125), *391*
Piozzi, F., 17(41), *86*, 153(60), 189(60, 198a), *207*, *211*, 294, 297(10), 298(74), 301(10), *312*, *314*, 338(28, 29), *340*
Piper, J. R., 178(199), *211*, 215(136), *279*, 367(129), *391*
Placeway, C., 263(183), *280*
Plancher, G., 149(200), 150(201), *211*
Plant, S. G. P., 13(191), 16(191), 36(82, 190), *87*, *90*, 130(50), *138*, 298(76, 77), 301(31), *314*, 362(101), *390*
Plieninger, H., 19(192), 22(192), 23(192), *90*, 96(99, 100), *139*, 144, 155(202), 195(204), *211*, 215(138), 241(137), 243(137), *279*, 291(78), *314*, 343(127), 350(129), 353(129), 354(127, 131), 355(128, 131, 132), 356(130), 365 (129a), *391*, 399, *400*, 403, 414(77), 419(75, 78), *428*
Plummer, A. J., 376(31), *389*
Plunkett, A. O., 269(13, 14, 15), 270(14), 271(14, 15), *276*
Pochini, A., 19(42), 20(42), 20(42), *86*
Poletto, J. F., 178(16), *206*, 433(2, 3, 43), *445*, *446*
Pomeroy, J. H., 424(79), *428*
Pomykáček, J., 154(10), 157(10), *206*
Pope, W. J., 130(101), *139*
Pople, J. A., 1(193), *90*
Portnova, S. L., 413(58), *428*
Portsmouth, D., 122(85a), *139*
Posner, H., 308(94), *314*
Potier, P., 101(14), *137*
Potts, G. O., 369(15), *388*
Potts, K. T., 19(195, 196), 20(196), 39(194), 65(197), *90*, 97(46), 109(102), 113 (102), *138*, *139*, 248(140), 249(139), *279*, 412(41), 413(40), *427*
Poutsma, M. L., 418(25), *427*
Powers, J. C., 5, 6(200), 15(198), 16(198), 61(199), 65(199), 66(199), 78(198), *90*, 114(103), *139*, 423(80), 424(80), *428*
Pozharskii, A. F., 27(201), *90*
Prasad, K. B., 158(208), *211*
Pratt, E. F., 30(202), 90, 133(104), *139*
Preietzel, H., 26(275), *92*
Prelog, V., 101(47), *138*
Preobrazhenskaya, M. N., 96(107), 132 (108, 140, 141, 143, 144, 145), 133(105, 139), 135(105, 106, 140, 142, 143, 144, 145), *139*, *140*, 157(258), *212*, 220(176), *280*, 416(114), 417(115), *429*, 433(44), *446*
Přeobrazhenskaya, N. A., 343(84), 357(68), *390*
Pretka, J. E., 367(133), *391*
Prins, D. A., 118(115), *139*
Printy, H. C., 20(119), *88*, 166(134), 169, *209*, 345(72), 346(72), 347(65, 73), 351(70), 356(71), *390*
Prior, A. F., 39(127), *88*
Pronina, L. P., 158(143), *209*
Protiva, M., 154(10), 157(10), *206*, 246 (141), *279*
Pryde, C. A., 188, *212*
Pryke, J. M., 215(9), *276*
Pschorr, R., 180(209), *211*, 393(18), *400*
Puma, B. M., 182(187), 205(249), *211*, *212*
Purves, W. K., 311, *312*
Putney, F., 308(90a), *314*

Q

Quan, P. M., 248(142), *279*
Quest, B., 373(2), 376(2), *388*
Quin, L. D., 248(142), *279*
Qureshi, A. A., 116(110), *139*, 273(143, 144, 145), *277*, *279*

R

Radunz, H., 65(284), *92*
Ragan, C., 442(45), *446*

Răileanu, D., 173(210), *211*, 366(106), 367 (106), *391*, 403(71), *428*
Rajagopolan, P., 110(111), *139*
Rajšner, M., 154(10), 157(10), *206*
Ralph, R. S., 220(4), 221(4), 224(4), *275*
Ramachandran, L. K., 135(112), 136(112), *139*
Ramirez, F., 379(134), *391*
Ramloch, H., 226(178), *280*
Ranganathan, S., 57(247, 248), 58, (247, 248), *91*
Rao, C. B. S., 252, 255, *277*
Rapaport, E., 132(60), 133(60), 135(60), *138*, 426(45), *427*
Rapoport, H., 178(79), *208*, 331(6), *339*
Rappe, C., 405(81), *428*
Ratusky, J., 215(128, 146), *279*
Razumov, A. I., 162(211), *211*
Rebstock, T. L., 2(278), *92*
Redfield, B. G., 308(94), *314*, 352(39), *389*
Redlich, A., 72(150), 73(150), 75(150), *89*
Reed, R. I., 147(165), *210*
Rees, C. W., 31(205), 32, 33(205), 54, *90*
Rees, R., 111(55), *138*
Reese, J., 201(70), *207*
Reeve, W., 404(82), *428*
Régent, P., 116(28), *137*
Régnier, G., 243(33), *276*
Reich, C., 108(92), *139*
Reid, R. E., 196(194), *211*
Reid, T. L., 342(164), 343(162), 344(162, 163), 345(162), *392*
Reinecke, M. G., 20(206), *90*
Reinehr, U., 337(32), *340*
Reiners, W., 16(30), *86*
Reissert, A., 176, *211*, 373(135), 376(136), *391*
Rembges, H. H., 348(33), 350(33), *389*
Remers, W. A., 77(207), *90*, 131(113), *139*, 200, *211*, 371(139), *391*, *392*, 418(83), *428*, 433(46, 47, 48), *446*
Renner, V., 107(114), 118(115), *139*, 262 (147), *279*
Rensen, M., 124(84), *428*
Renson, J., 308(80, 81), *314*
Renz, E., 346(52), *389*
Reppe, W., 29(208), 70(210), 77(209), *90*
Ricca, A., 19(42), 20(42), 21(42), *86*, 153 (60), 189(60), *207*
Rice, L. M., 155(213a), *211*
Richards, C. G., 47(170, 180), *89*
Richman, R. J., 388(137a), *391*
Richter, E., 167, *213*
Richter, K., 16(143), *88*
Ridd, J. H., 17(19), 83, *85*, *90*
Ridley, H. F., 19(80), 22(81), 29(80), *87*, 227(62), *277*
Ried, W., 97(116), *139*
Rieke, R. D., 11(171), *89*
Riley, J. G., 335(15), *339*
Rinderkneckt, H., 235(148), *279*, 393(19), *400*
Ritchie, E., 148(83), 149(86), 173(86), *208*, 292(35), 293(35), *313*
Rivers, P., 345(34), *389*
Robertson, A., 34(10), 39(10), *85*, 114(11), *137*, 173(24), 177(25), 178(25), *206*, 282(7), 283(5, 6, 7), 378(11), *388*, 413(6), 417(6), 423(8), *426*, *427*, 445 (6), *445*
Robinson, B., 32, 33(212), 66, 68(213), 80 (1), *90*, 121(1a), *137*, *139*, 142(214), 155(216), 156(215), 163(214), 194 (218), *211*, 343(91), 252(92), *390*
Robinson, D. N., 40(172, 173), 41(172), 53(172), 80(172), *89*
Robinson, F. P., 145, *211*
Robinson, G. C., 194(218), *211*
Robinson, G. M., 142, *211*
Robinson, J. R., 217(149), *279*
Robinson, P., 426(85), *428*
Robinson, R., 30(55), 35(124), 47(214), 52(214, 215, 216), 68(48), 75(216), *86*, *88*, *90*, 162(215, 217), *211*, 218(114), 246(92), 248(140), *278*, *279*, 319(30), *340*, 418(65), 422(86), *428*
Robinson, R. A., 28(217), 68(217), *90*
Rochelmeyer, H., 310(21), *312*
Rodda, H. J., 215(9), *276*
Rodina, O. A., 223(112), *278*
Rodinov, V. M., 405, *428*
Rodzevich, N. E., 173(89a), *208*
Rogers, K. M., 36(190), *90*
Roman, S. A., 24(47), *86*
Romeo, A., 212(39), *276*, 360(138), *391*
Ronaldson, J. W., 437(25, 26, 49), *446*
Rosales, J., 365(41), *389*
Rosati, R. L., 262(26), *276*
Rose, D., 376(31), *389*
Rosemund, P., 178, *211*

Rosenblum, M., 282(110), 284(110), 285 (110), *315*
Rosenbrook, W., Jr., 143(183), *210*
Rosnati, V., 350(112), *391*
Rossi, A., 135(71), *138*, 419(43), *427*
Rossiter, E. D., 108(118), *139*
Rossner, D., 412(16), *427*
Roth, R. H., 371(139), *392*, 418(83), *428*, 433(46, 47), *446*
Roussel, P. A., 135, *140*
Routier, C., 34(40), *86*, 413(17), 414(17), *427*
Royer, R., 34(40), *86*, 149(50), *207*, 413 (17), 414(17), *427*
Rubstov, M. V., 133(156), 134(156), *141*, 197(279), *213*
Rühl, K., 154(277), 155(277), *213*
Runti, C., 45(218), *90*, 97(119), *140*
Rush, K. R., 11(174, 175, 177), 12(174, 175, 177), 13(177), 14(174), 46(149), *89*, 114(93, 94), 115(94), *139*, 296(71), 297(71), 301(71), *313*
Russell, G. A., 365(140a), *392*
Rutenberg, M. W., 343(58), 346(58), 357 (141), *392*
Rutherford, R. J. D., 197(39), *207*
Ruyle, W. V., 229(157), *279*
Rybar, D., 345(149), *392*
Rydon, H. N., 178(221), *211*
Ryskiewicz, E. E., 113(126), *140*

S

Sabet, C. R., 54, *90*
Sadler, P. W., 364(56), 367(56), *389*
Saettone, M. F., 11(60, 61), 12(61), 13(60, 61), *86*
Safrazbekyan, R. R., 157(263), *213*
Saiga, Y., 290(89), *314*
Saikar, S. K., 264(115), *278*
Sainsbury, M., 65(68), *87*
Sakabe, N., 433(21, 50), *445*, *446*
Sakai, S., 100(33), *137*, 409(30), *427*
Sakakibara, H., 17(219), *90*, 98(120), *140*
Sakan, T., 25(220), *90*
Salgar, S. S., 75(221), *90*
Sallay, S. I., 251(150), *279*
Salt, C., 168(65), *207*
Sample, T. F., 193(278), *213*
Samuels, W. P., Jr., 175(18), *206*
Sarel, S., 298(82), 301(82), *314*, 327(31), *340*
Sasamoto, M., 224(130), *279*
Sato, E., 245(91), 246(91), *278*
Sato, Y., 419(88), *429*
Sauer, G. L., 69(161), 75(161), *89*, 226 (124), *279*
Sausen, G. N., 71(222), 75(222), *90*
Saxton, J. E., 19(195, 196), 20(196), 34 (223), 47(214), 52(214, 215, 216), 68 (48), 75(48, 216), *86*, *90*, 108(118), *139*, 365(59a), *389*, 413(89), *429*
Sazonova, V. A., 27(158), *89*
Schäfer, H., 405(46), *427*
Schales, O., 120(50a), *138*
Schellenberg, K. A., 43(225), *90*, 114(121), *140*
Schenker, K., 127, *137*, 162(23), *206*, 237 (210), 267(209, 210), *281*
Schiemenz, G. P., 263(183), *280*
Schiewald, E., 419(54), *428*
Schindler, W., 99(122, 123), *140*
Schleigh, W. R., 291(83), *314*
Schlittler, E., 246(151), *279*, 343(85), *390*
Schlossberger, H. G., 221(152), *279*
Schmid, H., 122(14a), *137*, 259(153), *279*
Schmidt, T., 318(14), *339*
Schmitz-Du Mont, O., 7(228, 229), 8(228, 229), *90*, *91*
Schnoes, H. K., 122(123a), *140*, 291(83), *314*
Schofield, K., 11(226), 13(227), *90*, 146, 149(222), 155(180), 161(180), *210*, *211*, 295(67), 297, 299(72), 300(72), 301(84, 85), *313*, *314*, 352(98), *390*
Schroeder, H. D., 102(39), 106(39), *138*
Schubert, C. I., 147(190), 148(191), *211*
Schülde, F., 214(179), 220(179), 249(172), *280*
Schulenberg, J. W., 191(223), *211*, 361 (142), *392*
Schultes, R. E., 440(53), *446*
Schumann, D., 259(153), *279*
Schwartz, G. M., 31(46), *86*
Scoffone, E., 77(73), *87*
Scolastico, G., 168(67), *207*
Scott, A. I., 116(110), *139*, 269(67, 116), 271(67), 373(143, 144, 145), *277*, *278*, 279
Scott, B. D., 182(99), 183(99), *208*
Searle, R. J. G., 145(51), 184(51), 187(51), *207*

Ishikawa, M., 182(117), 203(116), *209*
Ishimura, Y., 310(24, 47), *312*, *313*
Ishizumi, K., 296(112), *315*, 415, *428*
Isoe, S., 116(88), *139*
Isomura, K., 187(117), 188, *209*
Itaya, T., 19(287), *92*

J

Jackson, A., 159, 192(95), *208*, *209*, 406 (124), 418(48), *429*
Jackson, A. H., 11(106), 19(107, 108), 41 (21, 78), 80(21a, 108), 82(105), 83, *85*, *86*, *88*, 109(15), 113(15), 131(15), *137*, *138*, 236(80), 237(80), *276*, *277*, 320(16, 17, 19), 321(18), *339*, *340*, 402 (49), *428*
Jackson, R. W., 161(120), *209*, 215, 220 (22), *277*
Jacquignon, P., 149(48, 49), 150(49), *207*, 318(8), *339*
Jaffé, H. H., 85(110), *88*
James, K. B., 97(46), *138*
Janetzky, E. F. J., 149(121), *209*
Janot, M.-M., 101(14, 47), *137*, *138*
Jansen, A. B. A., 133(63), *138*
Jardine, R. V., 5(11), 14(189), 15(189), 77 (112), *88*, *90*, 154(122), *209*, 419(50), *428*
Jauer, H., 153(160), *210*
Jellinek, M., 196(193), *211*
Jenisch, G., 149(123), *209*
Jenkins, S. R., 132(153), 135(153), *140*, 189(274), *213*, 399(22), *400*
Jennings, B. E., 34(10), 39(10), *85*, 413(6, 51), 414(51), 417(6), *426*
Jennings, K. F., 414(51), *428*
Jepson, J. B., 308(48), *313*
Jerina, D. M., 309(49, 50, 51), *313*
Jilek, J. O., 154(10), 157(10), *206*, 246(141), *279*
Johnson, A. W., 68(113), 75(113), *88*, 348 (9), 378(5), 380(63), *388*, *389*
Johnson, D. C., 11(176), 12, 13(176), 14 (176), *89*, 296, 301(70), *313*
Johnson, F. H., 442(31), *446*
Johnson, H. E., 30(114, 115), *88*, 129(65), 132(64), 133, *138*, 398(11), 399, *399*
Johnson, H. W., 20(206), *90*
Johnson, J. E., 54(166, 167), *89*
Johnson, J. M., 133(63), *138*
Johnson, J. R., 366(64), *390*, 433(10, 28), *445*, *446*
Johnson, R., 200(80), *208*
Johnson, R. A., 126(66), *138*, 406(51a), *428*
Johnson, W. O., 366(64), 369(100), *390*
Joly, R., 227(28), *276*
Jommi, G., 168(67), *207*
Jones, D. A., 383(107, 108), 387(107), *391*
Jones, E. R. H., 19(32), *86*
Jones, G., 319(20), *340*, 415, *427*
Jones, H. L., 85(110), *88*
Jones, R. G., 129(77), *138*, 268(97), *278*
Jones, W. A., 34(4), *85*, 110(5), 113(5), *137*, 413(3), *426*
Jönsson, A., 77(71), *87*, 165(124), *209*
Joule, J. A., 95(68), 99(67), 105, 111(44), 117, 118(43), *138*, 159, *209*, 418(48), 406(124), *428*, *429*
Juhasz, G. J., 331(6), *339*
Julia, M., 22(117), 24(118), 29(116), *88*, 133(69), *138*, 165(125, 126, 133), 166 (128, 129, 131, 132), 183(132a), 197 (127), *209*, 215(82), *277*
Julian, P. L., 20(119), *88*, 166(134), 169, *209*, 246(83), 248(84), *277*, 345(72), 346(72), 347(65, 73), 348(66, 69), 351 (66, 70, 72), 352(68), 356(71), 365, *390*
Justoni, R., 218(86), 227(85), *277*

K

Kakurina, L. N., 158(135), 162(135), *209*
Kalb, L., 365(74), *390*
Kalir, A., 221(87, 88), 227(87), *277*
Kallianpur, C. S., 189(166), *210*
Kaluszyner, A., 183(27), *206*
Kamal, A., 58(120), *88*
Kambli, E., 367(75), *390*
Kamentani, T., 224(89), *278*
Kamiya, T., 252, 255, *277*
Kamlet, M. J., 4(121), *88*
Kanaoka, Y., 161(136), 203(116, 137), *209*, 220(90), 245(91), 246(91), *278*
Kane, S. S., 219(93), *278*
Kaneko, C., 203(137), *209*
Kao, Ĭ., 246(92), *278*
Kapil, R. S., 268(13, 14), 270(14), 271(14), 273(17), *276*
Kaplan, M., 260(71), *277*

Seaton, J. C., 422(23), 423(23), *427*
Sebastian, J. F., 20(206), *90*
Seemann, F., 60(270), *92*, 369(158), 370 (158), *392*, 419(119), *429*
Segal, G. A., 1(193), *90*
Segnini, D., 2(18), 4(18), 5(18), 9(16), *85*, 295(8), *312*
Seka, R., 39(230), *91*, 417(90), *429*
Sell, H. M., 2(278), *92*, 376(59), *389*
Sellstedt, J. H., 179(176), *210*
Selzer, H., 381(32), *389*
Semenov, A. A., 97(124), *140*
Semenova, N. K., 218(53), *277*
Semler, G., 58(261), *91*, 120(147), 135(147), *140*
Serafin, F., 27(153), 80(152), *89*
Shabica, A. C., 37(231), *91*, 417(91), *429*
Shagalov, L. B., 29(232), *91*, 183(229), *212*
Shah, G. D., 433(55), *446*
Shah, L. G., 433(55), *446*
Shah, S. W., 135(74), 136(74), *138*, 366(80), *390*
Shamma, M., 245(184), 263(185), *280*
Shannon, J. S., 437(25), *446*
Shapiro, D., 157(7), 159), *206*
Shavel, J., Jr., 27(151, 152), 65(292), 80 (152), *89*, *92*, 97(151), *140*, 155(225), 157, 251(120), *278*, 303(86, 113), *314*, *315*, 325, *340*, 445(6), *445*
Shaw, B. L., 445(6), *445*
Shaw, E., 154(227), 156(227), *212*, 217 (154), 221(156), 227(155), *279*, 397 (20), *400*
Shaw, J. T., 37(271), *92*
Shaw, K. N. F., 220(154), *279*, 405(92), 413(92), *429*
Shchelkunov, A. V., 170(163), *210*
Shchukina, L. A., 35(157), *89*
Sheehan, J. C., 358(144, 145), *392*
Sheinker, Y. N., 144(259), 162(259), *212*, 417(104), *429*
Shen, T., 35(233), *91*
Shevdov, V. I., 173(88, 90, 91), *208*
Shimkin, M. B., 440(54), *446*
Shimomura, O., 442((31), *446*
Shin, H., 65(197), *90*
Shine, H. J., 142(228), *212*
Shingu, H., 3(79), 84(79), *87*
Shioiri, T., 19(28), *92*, 415(47), *428*, *430*
Shirley, D. A., 135, *140*
Shirley, R. H., 218(55), *277*
Shive, W., 153(230), *212*, 426(94, 95), *429*
Shklyaev, V. S., 358(119), *391*
Shore, P. A., 310(9), *312*
Shroeder, D. C., 183(106), *208*
Shull, E. R., 9(98), *87*
Shvedov, V. I., 159(229), 173(88, 90, 91, 92), *208*, *212*, 418(93), *429*
Siddappa, S., 77(12), *85*, 153(102), 166 (103), 173(30), *206*, *208*, 419(26), *427*
Sidky, M. M., 379(104), *391*
Siebrasse, K. V., 298(96), 301(96), *314*, 326(35), *340*
Siehnhold, E., 198(244), *212*
Sietzinger, M., 229(157), *279*
Siffert, O., 165(126), *209*
Silva, R. A., 346(53), 353(53), *389*
Silverstein, R. M., 113(126), *140*, 154(164), *210*
Sim, G. A., 328(11), *339*
Siminov, A. M., 27(201), *90*
Simonoff, R., 113(53), *138*
Simpson, J. C. E., 202(18), *206*
Simpson, T. H., 445(6), *445*
Singer, H., 153(230), *212*, 426(94, 95), *429*
Singh, G., 215(171), *280*
Sjoerdsma, S., 308(60), *313*
Skarlos, L., 149(159), 150(159), 153(159), 156(159), *210*
Skinner, W. A., 22(62), *86*, 154(164), *210*
Sklar, M., 224(44), *277*
Slavachevskaya, N. M., 36(188), *90*
Slaytor, M., 426(85), *428*
Sliwa, H., 24(118), *88*
Smit, V. A., 70(254), 75(254), 77(255), *91*
Smith, A., 130(127), *140*
Smith, A. E., 11(106), 19(107), 80, 82 (105), *88*, 121(61), *138*, 236(80), 237 (80), *277*, 320(16, 17), *339*, 402(49), *428*
Smith, C. W., 94(131), 97(131), *140*, 161 (237), *212*, 215(161), 231(160), 233 (160), *279*
Smith, D. C. C., 131, *139*
Smith, G. F., 7(235), 9(237), 37, 66, 68 (213), 80, *87*, *90*, *91*, 95(68), 117(35), 122(146), *137*, *138*, 417(97), 425(96), *429*
Smith, H., 111(55), *138*
Smith, L. R., 11(175, 176, 177), 12(175, 176, 177), 13(177), 14(176), *89*, 114 (94), 115(94), *139*, 296(70, 71), 297 (71), 311(70, 71), *313*

Smith, P., 19(108), 78, 80(108), 82, *88*, 121(62), *138*, 320(16, 17, 19), 321(18), *340*
Smith, R. K., 206(74), *207*
Smith, R. T., 343(160), 345(160), *392*
Smith, W. S., 153(231), *212*
Smithen, C. E., 31(205), 32, 33(205), *90*
Smolinsky, G., 188(233), *212*
Snaith, R. W., 330(2), *339*
Šneberg, V., 215(128), *279*
Snyder, H. R., 57(239), 59, 60, 70(238), 75(238), 77(43, 84), *86*, *87*, 91, 94 (131), 95(3), 97(129, 130, 131), 98 (128), 99(128), 130(158), *137*, *140*, *141*, 161(237), 178(235), 180(236), *212*, 215(161), 224, 225(159), 231(160), 233(78, 160), 235(158), *277*, *279*
Sogn, A. W., 22(142), 23(142), *88*
Soliman, F. M., 379(104), *391*
Sonnet, P. E., 104(18), *137*, 249(172), *276*
Sorm, F., 215(128, 146), *279*
Sorokina, G. M., 132(145), 135(106, 145), *140*, 417(115), *429*
Sorokina, N. P., 29(232), 36(245), *90*, 135 (106), *139*, 144(259), 162(259), 183 (224), *212*
Southwick, P. L., 144(185), 158(239), *211*, *212*
Sova, J., 154(10), 157(10), *206*
Spande, T. F., 122(132), 125(97), *139*, *140*
Speeter, M. E., 22(240), *91*, 113(133), *140*, 216(163), 218(166), 219(165), 220 (166), 221(164, 165), *279*, *280*
Spenser, I. D., 241(1), *275*
Spietschka, W., 73(103), *87*
Splitter, J. S., 387(146), *392*
Sprague, P. W., 227(200), 237(197), 263 (200), *280*
Sprince, H., 43(129), *88*
Sprio, V., 2(98), 4, *429*
Spruson, M. J., 418(63), *428*
Staab, H. A., 2(241, 242), *91*
Staiger, G., 306(93), *314*, 370(156), 371 (156), *392*
Stammer, C. H., 189(274), *213*
Staněk, J., 345(148, 149), 362(141, 147), *392*
Staudinger, H. J., 311(95), *314*
Staunton, R. S., 418(99), 419(99), *429*
Steck, E. A., 173(240), *212*
Stein, M. L., 165(169), *210*
Stein, O., 57(134), *88*
Stephen, H., 180(241), *212*
Stephens, R. D., 204(242), *212*
Stepp, W. L., 14(7), 15(7), *85*
Sternbach, L. H., 115(42), *138*, 419(37), *427*
Stetter, H., 198(200, 243, 244), *212*
Stevens, C. L., 433(55), *446*
Stevens, F. J., 155(245), 178(199), *211*, *212*, 215(136), 217(167, 168), *279*, *280*, 367(126), *391*
Stevens, R. V., 61(281), *92*, 248(199), *280*
Stevens, T. S., 154(245), 166(68), *207*, 319 (20), *340*, 343(162), 344(162), 345 (162), *392*
Stewart, J. M., 215(161), *279*
Stoll, A., 178(246), *212*, 268, *280*
Stoll, W. G., 118(115), *139*
Stonner, F. W., 358(168), *392*
Stork, G., 151, *212*, 215(171), *280*
Streffer, C., 153(160), *210*
Strehlke, P., 65(284), 69(285), *92*, 120(155), *141*
Streith, J., 203(248), *212*
Strell, M., 417(100), *429*
Stroh, H., 217(19), *276*
Stroh, R., 31(244, 243), *91*
Stuber, F., 162(23), *206*
Su, H. C. F., 154(245), 155(245), *212*, 367 (150), *392*
Suehiro, T., 96(100), 113(36), *139*, 295(87, 88), *314*, 370(151), *392*, 399(17), *400*
Suess, R., 163(71a), *207*
Sugii, A., 412(112), *429*
Suginome, H., 319(30), *340*, 422(86), *428*
Sugiura, S., 442(32), *446*
Sugiyama, N., 289(90), 290, *314*, 376(77, 110), *390*, *391*
Sugorova, I. P., 154(146), 155(146), *208*, *209*
Suh, J. T., 205(249), *212*
Suhr, K., 195(207), *211*, 215(138), *279*, 399(17), *400*
Sukayan, R. S., 157(263), *213*
Sullivan, W. F., 426(12), *427*
Sulochana, S., 57(248), 58(248), *91*
Sumpter, W. C., 353(153), 357, *392*
Sundberg, R. J., 100(134), 102(134), 126 (95), *139*, *140*, 184(250, 251, 252), 186 (251, 254), 187(253), 188(252), *212*, 221(127), *279*, 372(154), 376(154), *392*, 403(101), 420(101a), *429*

Supple, J. H., 249(172), *280*
Surtee, J. R., 133(63), *138*
Sûs, O., 425(102, 103), *429*
Suschitzky, H., 178(14), *206*, 299(46), 301 (46), *313*
Sutcliffe, F. K., 342(18), 343(18), *388*
Suter, C. M., 97(2), 98(2), *137*, 231(3), 233 (3), *275*
Suvorov, N. N., 29(232), 35(157), 36(245), *89*, *91*, 96(107), 132(108), 135(106, 109), *139*, 144(259), 154(256, 257), 157(258), 162(259), 183(224, 255), *212*, 417(104), *429*, 433(44), *446*
Svierak, O., 113(361), *137*, 215(48), 216 (48), 229(48), *277*
Swaminathan, S., 57(246, 247, 248), 58 (247, 248), *91*
Swan, G. A., 182, *206*, *211*, 239(65), 243 (173), *277*, *280*, 376(12), *388*
Sweeley, C. C., 308(45), *313*
Swern, D., 306, *312*
Sych, E. D., 156(260), *212*
Szabo, L., 419(119), *429*
Szara, S., 221(88), *278*, 308(90a), *314*
Szinai, S. S., 397(7), *399*
Szmuszkovicz, J., 22(91), 23(91), 38(289), 66(249), 75(249), 78, *87*, *91*, *92*, 109 (135), 113(6, 136), 129(136), *137*, *140*, 166(261), *212*, 220(174), 222(74), 224(74), 225(74), 224(174, 175), 232 (74), *277*, *280*, 402(106), 407(107), 408 (108, 109), 411(105, 106, 110, 111), *429*, 442(22), *445*

T

Tabacik, V., 373(103), *391*
Taborsky, R. G., 22(250a), 23(250a), *91*, 214(43), *206*, 308(18), *312*
Tacconi, G., 157(198), *211*, 231(133, 134, 135), *279*, 278(122, 123, 124, 155), *391*, *392*
Tahk, F. C., 377(6), 378(6), *388*
Takagi, H., 25(220), *90*
Takagi, S., 412(112), *429*
Takahashi, M., 204(177, 178, 179), *210*
Taller, R. A., 143(137a) *209*
Tamai, Y., 204(179), *210*
Tamm, R., 263(185), *280*
Tamura, M., 310(47), *313*
Tanaka, T., 310(91), *314*
Taniguchi, H., 182(117), 187(117), 188 (117), *209*
Tanno, K., 116(88), *139*
Tatevosyan, G. T., 155(264), 157(167, 263), *210*, *213*
Taul, H., 235(18), *276*
Taylor, A., 298(2), *312*, 327(21), 328(521), *339*, *340*, 437(25, 26, 49), *446*
Taylor, C. R., Jr., 220(194), 221(194), *280*
Taylor, J. B., 263(10), 270(14), 271(14), *276*
Taylor, K. G., 433(55), *446*
Taylor, W. C., 292(35), 293(35), *313*
Taylor, W. I., 20(11), *85*, 101(47), 113(138), 118(9), 122(9a), 137, 138, *140*, 220 (194), 221(194), 269(14), 276), 285, 286, 292(3), 300(25), 303(25, 26), 325 (13), 328(11, 12), *339*, 419(113), 429
Tchernoff, G., 165(133), *209*
Tedder, J. M., 395(16), 396(16), *399*
Teichmann, K., 60(94), *87*
Tennant, G., 376(93), *390*
Teotino, U. M., 155(262), *213*
Terashima, M., 61(281), *92*, 220(90), 248 (199), *278*, *280*
Terent'ev, A. P., 15(251), 18(252, 256), 70 (254), 77(253, 255), 132(140, 141, 143, 144, 145), 133(139), 135(140, 142, 143, 144, 145), *140*, 155(85), 173(87, 88, 89, 89a, 92), 220(176), *280*, 416(114), 417 (115), *429*
Terenteva, I. V., 97(124), *140*
Tertzakian, G., 186(6), *206*
Terzyan, A. G., 155(264), 157(167, 263), *210*, *213*
Tetlow, A. J., 71(254), 75(145), *88*
Teuber, H. J., 191(265), *213*
Teuber, H.-J., 306, *314*, 337, *340*, 370(156), 371(156, 157), *392*
Thaler, G., 371(157), *392*
Theobold, R. S., 13(227), *90*, 149(222), *211*, 301(84, 85), *314*
Thesing, J., 40(257), 58(261), 59, 61(259), 80(262), *91*, 120(147), 133(146), 135 (147), *140*, 214(179), 220(179), 226 (178), 249(172), 250(177), *280*, 332 (33), *340*, 402(116), 419, *429*
Thomas, D. W., 291(83), *314*
Thomas, R., 273(180), *280*
Thomas, R. C., 224(175), *280*

Thomson, R. H., 164(31), 166(31), *206*
Thurn, R. D., 196(193), *211*
Tiffeneau, M., 101(147), *140*
Tishler, M., 37(213), *91*, 233(78), *277*, 417(91), *429*
Todd, W. H., 97(25), *137*
Toffoli, C., 45, *89*
Tokunaga, Y., 25(220), *90*
Tokyama, K., 311(54), *313*
Tomiie, Y., 433(21, 50), *446*
Tomlinson, M., 155(41), 168, *207*, *211*, 419(15), *427*
Tomlinson, M. L., 13(191), 16(191), *90*, 298(76), *314*
Topham, A., 39(127), *88*, 418(99), 419(99), *429*
Torralba, A. F., 97(149), *140*
Towne, E. B., 194(266), *213*
Trabert, C. H., 424(118), *429*
Trautmann, P., 28(37), *86*
Treibs, A., 26(268), 27(268), 40(264), 42 (265), 43(265), 44(266), 68(267), *91*, *92*
Tretter, J. R., 331(6), *339*
Trissler, A., 202(173), *210*
Trittle, G. L., 196(195), *211*
Troxell, H. A., 183(106), *208*
Troxler, F., 60(269), *92*, 178(246), *212*, 369(158), 370(158), *392*, 419(119), *429*
Tsou, K. C., 367(150), *392*
Tsurui, R., 161(280), *213*
Tsymbal, L. V., 18(252), *91*
Tulinsky, A., 433(56), *447*
Tullar, B. F., 190(168), *210*
Turchin, K. F., 433(44), *446*
Tweddle, J. C., 178(221), *211*
Tyson, F. T., 37(271), *92*, 189(267), *213*

U

Udelhofen, J. H., 335(36), *340*, 345(165), 350(165), 351(165), 352(161), 357 (161), *392*
Udenfriend, S., 308(79, 94, 48, 68, 80, 81), 309(50, 51), 310(9), *312*, *313*, *314*, 352 (39), *389*
Ueberle, A., 2(242), *91*
Uffer, H., 70(210), *90*
Uhle, F. C., 113(150), *140*, 178(268), *213*, 426(120), *429*
Ulm, K., 363(82), *390*
Ullrich, V., 311(95), *314*
Umani-Ronchi, A., 338(29), *340*
Untch, K. G., 176(105), *208*
Urarova, N. V., 135(106), *139*
Uritskaya, M. Ya., 133(156), 134(156), *141*, 197(279), *213*
Utley, J. H. P., 130(127), *140*
Uvarova, N. Y., 157(258), *212*, 433(44), *446*
Uzu, K., 433(55), *446*

V

Vaghani, D. V., 189(166), *210*
Valenta, Z., 358(168), *392*
Valls, J., 227(28), *276*
Valzelli, L., 308(30), 310(30), *312*, 433(19), *445*
Van Allen, J., 189(12), *206*
van den Hende, J. H., 433(14), 436(14), *445*
Vander-Werf, C. A., 21(101), 61(78), *87*
Vangedal, S., 303(32), *312*
Vejdělek, Z. J., 154(10), 157(10), *206*, 233 (186), 246(141), *279*, *280*, 397(21), *400*
Veldstra, H., 57, *85*, 378(4), *388*
Velluz, L., 178(269), *213*
Venkiteswaran, M. R., 40(178), 45(178), 47(170, 179, 180), *89*
Verkade, P. E., 149(121), *209*
Vernon, J. M., 330(2), *339*
Veselovskaya, T. K., 405, *428*
Vigdorchik, M. M., 135(109), *139*
Vignau, M., 227(29), *276*
Vingradova, E. V., 403(57), *428*
Virtanen, A. I., 445(20), *445*
Viswananathan, N., 178(84), *208*
Vitali, T., 16(187), *89*, 227(187), *280*, 376 (113), *391*
Vlattas, I., 111(78), *138*
Vogt, I., 385(88), 386(87), 387(87), *390*
Voillaume, C., 183(132a), *209*
Voltz, S. E., 206(270), *213*
von Dobeneck, H., 26(275), 42(274), 45, 65(273), 66(273), *92*
von Strandtmann, M., 97(151), *140*, 155 (225), 157(226, 271, 272), *212*, *213*
von Walther, P., 40(276), *92*
von Whittenau Schach, M., 29(224), *90*, 433(51, 52), *446*
Vrotek, E., 173(87), *208*

W

Wache, H., 363(82), *390*
Wahren, M., 26(268), 27, *92*
Waite, R. O., 27(151, 152), 80(152), *89*
Wakasuki, T., 291(43), *313*
Wakatusi, T., 394(9), *399*
Walk, A., 407(11), 412(11), *427*
Walker, D. M., 144(185), *211*
Walker, G. N., 180, *213*, 343(160), 345(159, 160), 357(12a), *392*, 409(121), *429*
Walker, J., 345(30a), 360(12a), *388*, *389*
Wallace, J. G., 145(59), *207*
Walley, R. J., 129(72), *138*
Walls, F., 122(152), *140*
Walton, E., 132(153), 135(153), *140*, 189(274), *213*, 399(22), *400*
Wantz, F. E., 351(70), *390*
Warner, D. T., 235(119), 235(188), *278*, *280*
Warnhoff, E. W., 99(21), *137*
Washida, Y., 75(130), *88*
Wasserman, H. H., 286, *314*, 397(23), *400*
Watanabe, E., 442(30), 443(30), *446*
Waterfield, W. R., 100(38), 118(52), *137*, *138*, 273(72), *277*
Waters, A. E., 9(237), *91*
Watkins, J., 316(10), 318, 319(10), *339*
Wawzonek, S., 240(190), 244(189, 190), *280*
Weaver, R. N., 343(160), 345(160), *392*
Webb, J. S., 433(42, 57), *446*, 447
Weiberg, O., 77(282), *92*, 362(167), *392*
Weil, M., 203(248), *212*
Weil, R. A. N., 71(66), *87*
Weinerth, K., 414(77), *428*
Weinman, J. M., 113(153a), *141*
Weisbach, J. A., 263(192), *280*, 413(32), 414(32), 418(32), *427*
Weisblat, D. I., 218(166), 219(193), 220(166), 224, 233(110, 193), *278*, *280*
Weiss, M. J., 77(207), *90*, 114(4), 131(113), *137*, *139*, 171(17), 173(15, 17), 200, *206*, *211*, 371(139), *391*, *392*, 418(83), 422(2), *426*, *428*, 433(1, 2, 3, 4, 5, 43, 46, 47), *445*, *446*
Weissauer, H., 346(14), 350(13, 14), 351, *385*
Weissbach, H., 308(68, 80, 81), *313*, *314*, 352(39), *389*
Weissbach, O., 395, *399*
Weisser, H. R., 148(66), *207*
Weissgerber, R., 15(277), 16(277), *92*
Weller, L. E., 2(278), *92*, 376(59), *389*
Wellings, I., 373(94), 376(94), *390*
Welsh, D. A., 144(185), *211*
Welstead, W. J., Jr., 220(194), 221(194), *280*
Wenkert, E., 61(279, 280, 281), 65(280), *92*, 111(154), *141*, 188(275), *213*, 227(200, 202, 203), 237(197), 245, 248(196, 198, 199), 249(201), 250, 251, 255, 263(200, 203), 273, *280*, *281*, 335, *340*, 342(164), 373(162), 344(162, 163), 345(162, 165), 350(165), 351(165), 352(161), 357(161), *392*
Werbel, L. M., 173(9), 175(9), 175(9), *206*
Werblood, H. M., 122(9a), *137*
Werner, H., 120(50a), *138*, 239, 241, *277*
Werst, G., 356(130), *391*, 419(78), *428*
Weygand, F., 167, *213*
Whaley, W. M., 238, 241, 244(207), *280*
Whalley, W. B., 417(122), *429*
Whipple, E. B., 4, 6, *87*
Whitaker, W. B., 298(77), *314*
White, D. H., 369(100), *390*
White, E. G., 180(236), *212*
White, E. P., 437(25, 26, 49), *446*
White, R. H., 218(55), 231(59), *277*
Whitehouse, M. W., 442(58), *447*
Whittle, B. A., 222(206), 224(206), *281*
Whittle, C. W., 406(123), 416(123), 417(123), *429*
Wibberley, D. G., 364(114a), *391*
Wickberg, B., 227(202, 203), 245, 250, 251(202), 263(203), *280*
Wieland, T., 77(282), *92*, 154(277), 155(277), *213*, 349(166), 362(167), *392*
Wiesner, K., 358(168), *392*
Wigfield, D. C., 273(103), *278*
Wilchek, M., 122(132), *140*, 193(278), *213*
Wild, D., 354(131), 355(131, 132), *391*
Wildi, B. S., 433(10), *445*
Wilds, A. L., 169(32), *207*
Willersinn, C., 80(262), *91*
Willersinn, C.-H., 226(178), *280*
Williams, J. K., 235(158), *279*
Williams, K. R., 263(192), *280*
Williams, R. P., 433(42), *446*
Williams, R. T., 308(53, 98), *313*, *314*

Williams, W. E., 433 (57), *447*
Williamson, W. R. N., 19 (283), *92*
Wilson, C. V., 148 (13), *206*
Wilson, N. D. V., 159 (118), *209*, 406 (124), *429*
Winterfeldt, E., 65 (284), 69 (285), *92*, 243 (207), *281*, 120 (155), *141*
Witkop, B., 7 (58), 57 (128), *86*, *88*, 122 (132), 125 (97), 129 (27), 131 (157), 135 (112), 136 (112), *137*, *139*, *140*, *141*, 155 (141), 161 (141), 182 (99), 183 (99, 72), 193 (278), *207*, *208*, *209*, *213*, 233 (94), *278*, 282, 283 (101, 106, 108, 109), 284 (104, 110), 285 (110), 289, 293 (101a, 103, 106), 294 (100, 105), 295 (99, 108, 109), 296, 303 (58), 304 (57), 305 (72a, 102), 308 (94, 80), *313*, *314*, *315*, 322 (37, 39, 40), 324 (40, 42), 325 (43), 326 (27), 327 (37, 41), 330, 331, *340*, 352 (39), 353 (89), 363 (89), *389*, *390*, 393 (2, 24), *399*, *400*
Wittwer, S. H., 376 (59), *389*
Woitach, P. T., Jr., 395 (15), 396 (15), *399*
Wolf, C. F., 433 (57), *447*
Wolf, J. P., III., 424 (73), *428*
Wolinsky, J., 263 (185), *280*
Wolter, R., 376, *389*
Wood, J. R., 241 (51), *277*
Woodbridge, R. G., III, 78 (286), *92*
Woddier, A. B., 173 (24), 177 (25), 178 (25), *206*
Woods, C. W., 404 (82), *428*
Woodward, R. B., 129 (77), *138*, 237 (210), 241 (208), 263 (208), 276, (209, 210), 268 (97), *278*, *281*, 433 (10), *445*
Woodyard, G. G., 383 (125), *391*
Woolley, D. W., 154 (227), 156 (227), *212*, 227 (155), *279*, 308 (111), *315*, 397 (20), *400*, 438 (59), *447*
Wragg, W. R., 155 (36), *207*, 224 (8), 225 (8), *276*
Wright, I. G., 263 (183), 269 (116), *278*, *280*
Wright, W. B., Jr., 377 (169), *392*
Wurtman, R. J., 438 (60), *447*

X

Xuong, N. D., 414 (19), *427*

Y

Yagil, G., 19 (286a), *92*, 402 (125), *430*
Yaknontov, L. N., 133 (156), 134 (156), *141*, 197 (279), *213*
Yakubov, A. P., 155 (146), *209*
Yamada, S., 19 (287), *92*, 161 (280), 203 (116, 137), 206 (140), *209*, *213*, 246 (129), *279*, 291 (43), *313*, 357 (170), *392*, 394 (9), *399*, 415 (47), *428*, *430*
Yamada, Y., 433 (24), *446*
Yamagami, K., 318 (25), *340*
Yamamota, K., 318 (25), *340*
Yamamoto, H., 289 (90), 290 (89, 90), 296 (112), *314*, *315*
Yamanaka, H., 348 (75a), *390*
Yamazaki, I., 310 (47), *313*
Yamazaki, T., 186 (254), *212*
Yanovskaya, L. A., 15 (251), 18 (256), *91*
Yaryshev, N. G., 15 (132), *88*
Yasunari, Y., 264 (115), *278*
Yim, N. C., 263 (192), *280*
Yoneda, F., 189 (281), *213*, 395 (25), *400*
Yonemitsu, O., 131 (157), *141*, 161 (136), *209*, 220 (90), *278*
Yonezawa, T., 3 (79), 84 (79), *87*
Yoshiota, M., 376 (77), *390*
Young, D. V., 130 (158), *141*
Young, E. H. P., 222 (206), 224 (206, 211), *281*
Young, T. E., 35, *92*, 178 (282), *213*
Youngdale, G. A., 38 (289), *92*, 166 (261), *212*
Yudin, L. G., 15 (132), *88*, 155 (147), 156 (147), *210*
Yurashevski, N. K., 241 (212), *281*
Yur'ev, Y. K., 414 (127), *430*

Z

Zacharias, D. E., 105 (98), *139*
Zagorevskii, V. A., 158 (11, 135, 148, 150), 162 (135, 144), *206*, *209*, *210*
Zaltzman, P., 308 (48), *313*
Zaltzman-Nirenberg, P., 309 (50, 51), *313*
Zambito, A. J., 233 (78), *277*
Zaugg, H. E., 343 (171), *392*
Zbinovsky, V., 431 (34), *446*
Zee, S. H., 40 (289a), 47 (289a), *92*

Zeig, H., 59(263), *91*
Zeile, K., 102(39), 106(39), *138*
Zhukova, I. G., 131(75), *138*
Ziegler, F. E., 104(18), *137*, 253(25), *276*, 417(91), *430*
Ziegler, J. B., 37(231), *91*, 415, *429*
Zilkha, A., 133(12), 135(12), *137*
Zimmermann, H., 3(290), *92*
Zinner, G., 57(291), *92*
Zinnes, H., 65(292), *92*, 303(86, 113), *314*, *315*, 325, *340*,
Zoretic, P. S., 415, *430*
Zuleski, F. R., 65(292), *92*
Zürcher, R., 162(23), *206*

SUBJECT INDEX

A

4-Acetamido-4,4-dicarbethoxybutyraldehyde, in synthesis of tryptophans, 234–235
Acetic anhydride
 reactions with indoles, 34, 36
  with indolycarbinylamines, 99–102
  with tetrahydro-β-carbolines, 100–102
Acetone
 reactions with indoles, 45, 47–48
  with indolylmagnesium iodide, 47
  with substituted tryptamines, 243–244
Acetyl cyanide, reactions with indoles, 39
Acetylenes, electron deficient, 68–69
Acrylonitrile, reactions with indoles, 70–71
N-Acyl-*o*-alkylanilines, as indole precursors, 189–191
o-Acylaminophenyl ketones, *see o, N*-Diacylanilines
Acylation of indoles and metalloindoles, 33–39, 412–417
 by hydrogen cyanide and nitriles, 39, 416–417
 by Vilsmeier Haack reaction, 36–37, 416–417
2-Acylindole alkaloids, synthetic approaches, 415
Acylindoles, 401–420, *see also* specific acylindoles
 hydrolytic deacylation, 114, 175–176
Acylindoles—*continued*
 organometallic reagents in synthesis of, 418–419
 reduction of, 108–110, 219–222, 402
 spectral properties, 402
 structure of, 401
 synthesis, 33–39, 412–420
1-Acylindoles
 as by-products in C-acylations, 34–35
 hydrolysis of, 34
 oxidation of, 296
 ozonolysis, of, 295
 synthesis, 35
2-Acylindoles
 as oxidation products of indoles, 285–287, 300
 synthesis
  by acylation, 412–417
  from *o*-aminophenyl ketones, 419
  from *o*-azidostyryl ketones, 420
  by Fischer cyclization, 417–418
   intramolecular, 415
  by periodic acid oxidation of 2-alkylindoles, 300
 Wittig reaction of, 406
3-Acylindoles
 bromination, 411
  deacylation
   hydrolytic, 114, 175–176
   during nitration, 14, 115
 Mannich condensation of, 411

3-Acylindoles—*continued*
nucleophilic ring opening with hydrazine, 335–336
oxidation, 412
reactions with organometallic reagents, 407
reduction of, 219–222, 402–405
with diborane, 220
substituent cleavage during, 404
synthesis by acylation of indoles, 33–39, 412–417
by miscellaneous methods, 417–420
by Nenitzescu reaction, 173
by Vilsmeier–Haack reaction, 36–37, 415–417
Adrenochrome, 438
Ajmalicine, synthesis, 263–264
Ajmaline, synthesis, 264, 266
Akuammicine, thermal cleavage of, 117
Alcohols, base-catalyzed reactions with indoles, 30–31
Aldehydes, *see also* Indolecarboxaldehydes
reactions with indoles, 39–56
with indolymagnesium halides, 45–47
Alkaloids
biosynthesis, 269–275
formation of "dimeric", 107, 261–262
oxidation, 291–293
be lead tetraacetate, 300–303
reaction with *t*-butyl hypochlorite, 124–125, 303, 325–326
synthesis, 151, 251–269
Alkylation of indoles, 19–31
alkylations with alcohols, 30–31
with aziridines, 29, 227
with carbonium ions, 26–28
with diazoesters and diazoketones, 27, 215–216
with epoxides, 28–29
with ethylene-trianilinoaluminum, 31
with lactones, 29–30
with Mannich reactions, 56–67
with olefins, 27
with 1-phenyltetrahydro-β-carbolines, 106
intramolecular, 24
in synthesis of alkaloids, 252–255, 266–267
of 3-substituted indoles, 77–82
transannular in alkaloid synthesis, 257–259
Alkylidene-3*H*-indoles
as hydrogen acceptors, 30, 114
as intermediates in alkylation of indoles, 30
in diindolylmethane formation, 40
in reduction reactions, 108–110
in trimerization of indoles, 8
reactions with nucleophiles, 95, 128–129
synthesis, 42–44, 68–69
N-Allylanilines, conversion to indoles, 194
3-Allyl-3*H*-indoles, *see* 3*H*-Indoles
Aluminum chloride, as catalyst for rearrangements, 318, 321
Amberlite IR-120, as catalyst for Fischer cyclizations, 161
4-Aminobutyraldehyde, in synthesis of tryptamines, 229
Aminoindoles, *see also* various aminoindoles
oxidation, 290–291
structure, 393–394
synthesis, 173–176, 180–182, 393–399
from nitroindolines, 132–133, 398–399
by reduction of nitroindoles, 397
1-Aminoindoles, 397
2-Aminoindoles
hydrolysis, 393
oxidation of, 291
structure and derivatives, 393–395
synthesis, 180–182, 393–394
3-Aminoindoles
diazotization, 396
oxidation, 290–291, 396
synthesis, 395–396
o-Aminophenylacetaldehyde derivatives as indole precursors, 195–196
o-Aminophenylacetic acid derivatives, conversion to oxindoles, 360–362
o-Aminophenylacetylenes, as indole precursors, 204–205
α-Anilinoketones, as indole precursors, 164–171
β-Anilinovinyl phenyl ketone, reaction with indoles, 69
Anthranils, from isatogens, 382–383
Antibiotics derived from indole, 431–437
Apoaranotin, structure, 436
Aranotin, structure, 433, 436–437
Arylhydrazones
cyclization
to indoles, 143–163
of unsymmetrical, 148–152

Arylhydrazones—*continued*
from enamines, 159, 418
synthesis by Japp-Klingemann reaction, 159–161, 217–218
1-Aryltriazoles, photochemical conversion to indoles, 203
Ascorbigens, 443–445
Aspidospermidine, synthesis, 260
Aspidospermine, synthesis, 151
Autoxidation, of indoles, 282–295
Azides, reactions with indoles, 17–18
β-Azidostyrenes, as indole precursors, 186–188
o-Azidostyrenes, as indole precursors, 186–188
Aziridines, reactions with indoles, 29, 227

B

Benzoylacetylene, reactions with indoles, 68–69
Benzoyl cyanide, reactions with indole, 39
Biosynthesis of indole alkaloids, 269–275
Bischler indole synthesis, 164–171
mechanism, 166–168
starting materials
α-bromopropionaldehyde diethyl acetal, 169
α-diazoketones, 169
epoxides, 169
α-hydroxy ketones, 166, 168
Bischler–Napieralski cyclization
of N-acyltryptamines, 244–246
in synthesis of ajmalicine, 263
of eburnamonine, 263
of reserpine, 263
of tetrahydro-β-carbolines, 244–246
of thioamides, 245–246
Boron trifluoride
as catalyst for Fischer cyclizations, 161
for rearrangements, 317
Bromination
of 3-acylindoles, 411–412
of indoles, 14–17, 303–305, 363
of oxindoles, 353
2-Bromoindoles, hydrolysis, 304
3-Bromoindole
synthesis by bromination, 14–15
by oxidation of indolylmagnesium bromide, 293
Bromomethyl 2-indolyl ketone, 419
N-Bromosuccinimide
cleavage of peptides by, 305
cyclization of tryptamine and tryptophan derivatives by, 125, 303–305
reactions with indoles, 15–17, 303–305
Bufotenine, *see* 5-Hydroxy-N,N-dimethyltryptamine
t-Butyl hypochlorite
cyclization of tryptamines by, 125
reactions with alkaloids, 124–125, 303, 325, 326
with indoles, 256–257
Butyrolactone, reactions with indoles, 29–30

C

Carbazoles
from indoles and 1,4-diketones, 47, 52
from 2-methylindole and hydroxymethylene compounds, 54–55
Carbenes, reactions with indoles, 31–33
β-Carboline derivatives, *see* Dihydro-β-carbolines, Tetrahydro-β-carbolines
Carbonium ions, alkylations by, 26–28
Chemiluminescence, in indole oxidations, 289–290
Chlorocarbene, reaction with indole, 31–32
2-Chloroindole, hydrolysis, 15
3-Chloroindole, hydrolysis, 15–16, 306–307
β-Chlorovinyl phenyl ketone, reaction with indoles, 69
Cinchonamine, cleavage by acetic anhydride, 101
Cinnolines, conversion to indoles, 202–203
Complexes of indoles
association constants, 2
structure, 3
Coronaridine, synthesis, 255
Corynantheidine, synthesis, 262–263
Cyanoindoles
conversion to indolecarboxaldehydes, 419
hydrolysis, 425–426
synthesis from haloindoles, 425–426
from indole-3-carboxaldehydes, 424
1,3-Cyclohexanedione, reaction with indoles, 43–44
1,4-Cyclohexanedione, reaction with indoles, 44–45
Cyclohexanone, reaction with indoles, 43–45

D

Dasycarpidone, synthesis, 159–160
Deacylation
  hydrolytic, 114, 175–176
  during nitration, 14, 115
Decarboxylation, 421–424
  of indole-2-acetic acid systems, 118–119, 241-242
  of indole-2-carboxylic acids, 159, 421–423
  of indole-3-carboxylic acids, 114, 173, 175, 423
Dehydrogenation of indolines to indoles, 132–135
o,N-Diacylanilines
  from indole oxidations, 283–292, 295, 300
  from ozonolysis of indoles, 295
Diazoalkanes, reactions with isatins, 380–381
Diazo compounds, alkylation of indoles by, 27, 215–216
Diazo ketones, in Bischler indole synthesis, 169
Diazomethyl 2-indolyl ketone, 419
Diazonium ions
  as precursors of arylhydrazones, 159–161, 217–218, 418
  reactions with indoles, 17–18
3-Diazo-4-quinolones, conversion to indole-3-carboxylic acids, 425
Diborane
  reaction with 2-ethoxyindole, 355
  in reduction of acylindoles, 109
    of indole, 131
    of indole-3-glyoxamides, 220
Dibromocarbene, reaction with indole, 32
Dichlorocarbene, reaction with indole, 31–33
Dichloromethyl-3*H*-indoles, *see* 3*H*-Indoles
N,N-Dichlorourethane, reaction with indoles, 306–307
Diethyl azodicarboxylate, reactions with indoles, 76
Difluorocarbene, reaction with indole, 32
Dihydro-β-carboline derivatives, 244–246
  as intermediates in alkaloid synthesis, 263
Dihydrocatharanthine, synthesis, 255
Dihydrocorynantheine, synthesis, 263
Dihydrogambirtannine, synthesis, 248
Dihydropyridines, intramolecular reactions with indoles, 249–250
3,4-Dihydroxyphenylalanine, oxidative cyclization of, 192
Diindolylmethanes
  mechanism of formation, 39–41
  reactions with aldehydes, 44–45
  synthesis, 39–43
Dimethyl acetylenedicarboxylate, indoles from reactions with hydrazobenzene and phenylhydroxylamine, 201–202
p-Dimethylaminobenzaldehyde, reaction with indoles, 42–43
1-Dimethylaminoindole, 57
2-Dimethylaminoindole and derivatives
  displacement reaction of, 97–101, 256–258
  from Mannich condensations, 58
3-Dimethylaminoindole, *see* Gramine and derivatives
N,N-Dimethyltryptamine, as hallucinogen, 440
o,β-Dinitrostyrenes, conversion to indoles, 182–183
Dioxindoles
  oxidation of, 352
  as products of indole oxidations, 286–288, 298

E

Eburnamine, synthesis, 259
Electron density
  effect on reactivity, 83
  of indole, 1–2
Electrophilic substitution, 3–85
  mechanism for 3-substituted indoles, 78–83
  orientation of, 3
  theoretical treatment of, 3, 83–85
Enamines
  acylation and reductive cyclization of, 178–179
  alkylation by gramine derivatives, 97
  conversion to 2-acylindoles, 159
  indoles from, 200–201
  Japp–Klingemann reactions of, 159, 418
  in synthesis of indoles by Nenitzescu reaction, 171–176
Epoxides, reactions with indoles, 28–29, 435
Ergonovine, as oxytocic agent, 441
Ergotamine, as vasoconstrictor, 441
Ergot alkaloids, synthesis, 268–269

1-Ethoxyindole(s), 184–185
2-Ethoxyindole
  reactions, 354–355
  structure, 354
o-Ethylaniline, as indole precursor, 206
Ethyl diazoacetate, alkylation of indoles by, 215–216
Ethylene, alkylation of indole by, in presence of trianilinoaluminum, 31
Exchange, isotopic
  in hydroxyindoles, 7
  in indole and derivatives, 4–7
  in indolylmagnesium bromide, 5–6

F

Fischer cyclization, 142–163
  abnormal, 145, 162–163
  catalysts for, 161, 218
  direction of cyclization in, 148–152
  2,6-disubstituted phenylhydrazones in, 145–147
  intermediates in, 142–147
  mechanism, 142–148
  scope of reaction, 152–162
  synthesis of 2-acylindoles, 417–418
    of ibogamine, 251–252
    of indolealkanoic acids, 217–218
    of indoles
      with heterocyclic substituents, 159, 161
      substituted, 152–162
    of tryptamines, 228–229
    of tryptophans, 234–235
  thermal, 147, 162
Friedel–Crafts acylation
  of indoles, 34, 36, 412–415
  of indolines, 135, 268, 416–417
Friedel–Crafts alkylation of indoles, 26–27
Frontier electron density in prediction of reactivity, 84

G

Geraniol, incorporation into indole alkaloids 269–273
Gliotoxin, structure, 433, 436
Glucobrassicin, 445
Gramine and derivatives, *see also* Indolylcarbinylamines
Gramine and derivatives—*continued*
  alkylations by, 94–97, 214–215, 224–226, 231–234
  displacement reactions of, 94–97, 214–215
  hydrogenolysis of, 113
  methiodide of, synthesis, 95
  reaction with acetic anhydride, 102
  synthesis by Mannich condensation, 56–59
  in synthesis of indole-3-acetic acids, 214–215
    of tryptamines, 224–226
    of tryptophans, 231–234

H

Hallucinogens, indole derivatives as, 440
α-Haloacetonitriles, as alkylating agents, 216
Halogenation of indole derivatives, 14–17, 78, 303–307
Heterocyclic ketones in Fischer cyclization, 158, 161
Hydrogenation of indoles, 129–131
Hydrogenolysis
  of *O*-benzyl groups, 233
  of *N*-benzyl tryptamine derivatives, 221
  of substituted indoles, 108–114
3-Hydroperoxy-3*H*-indole derivatives
  chemiluminescent decomposition of, 289–290
  as intermediates in indole oxidations, 282–295
  reactions of, 283–290
  reduction of, 283
  synthesis, 282–283
    by oxidation of indolylmagnesium halides, 293
5-Hydroxy-*N*,*N*-dimethyltryptamine, as hallucinogen, 440
Hydroxyindole(s), *see also* 1-Hydroxyindoles, Oxindoles, various indolinones
  alkylation of, 370
  carbocyclic hydroxy substituents, 368–372
  formation in hydroxylation reactions, 308–310
  isotopic exchange in, 7
  Mannich condensations of, 369–370

Hydroxyindole(s)—*continued*
oxidation of, 370–371
by oxidative cyclizations of phenylethylamines, 191–193
1-Hydroxyindole and derivatives, 372–377
attempted synthesis of 1-hydroxyindole, 373
1-hydroxyindole-2-carboxylic acid, reactions and synthesis, 373, 376
1-hydroxy-2-methylindole, structure, 373
1-hydroxy-2-phenylindole
as intermediate in deoxygenation of *o*-nitrostilbene, 184–185
oxidation of, 372–374
reactions of, 372–374
synthesis, 372
synthesis of derivatives, 372–376
2-Hydroxyindole, *see* Oxindole
3-Hydroxyindole, *see* 3-Indolinone
5-Hydroxyindole derivatives
alkylation of, 370
Mannich condensations of, 369–370
Nenitzescu synthesis of, 171–176
oxidation of, 370–371
proton exchange in, 7
Hydroxylation
of indoles, 308–310
of tryptophan, 308–310
2-Hydroxy-5-nitrobenzyl bromide (Koshland's reagent), 122–123
3-Hydroxy-1,2,3,4-tetrahydroquinolines, as indole precursors, 196

I

Iboga alkaloids
decarboxylation of derivatives, 118–120
nucleophilic displacement reaction in, 102–103
synthetic approaches, 251–256
Ibogaine, oxidation of, 292
Ibogamine, synthesis, 251–255
Imines and iminium systems, as electrophiles toward indoles, 56–57, 236–244, 246–251
Indigo
derivatives from reduction of isatins, 381–382
mechanism of formation from indoxyl, 365
Indole, *see also* Indole derivatives
acidity of, 19
basicity of, 2–5
dimer of, 7–8
dipole moment of, calculation, 2
electrophilic substitution in, 83–85
isotopic exchange of, 4–7
magnesium derivative, *see* Indolylmagnesium halides
Mannich condensation products, 57
metabolism of, 308
oxidation of, 294
protonation of, 4–8
reaction with mercuric acetate, 135–136
reduction by diborane, 131
resonance energy of, 2–3
sodium salt of, 20–21
theoretical treatment of reactivity, 3, 83–85
trimer of, 7–8
Indole-3-acetamide derivatives, reduction of, 218–220
Indole-3-acetic acid derivatives
hydroxylation of, 308
intramolecular acylation, 415
oxidation by peroxidase, 311–312
synthesis of derivatives, 164–165, 214–218, 366–367
Indole-3-acetonitrile derivatives, reduction, 218–220
Indolecarboxaldehydes, *see also* Indole-2-carboxaldehydes, etc.
condensation reactions, 222–224, 405–406
conversion to cyanoindoles, 424
deformylation during nitration, 115
reduction of, 108–110
resonance in, 401–402
synthesis, 412–420
from cyanoindoles, 419
by McFadyen–Stevens reaction, 419
by oxidation of hydroxymethylindoles, 419
by Vilsmeier–Haack formylation, 36–37, 419
Indole-2-carboxaldehydes, synthesis, 419
Indole-3-carboxaldehydes
acidity of, 402
alkylation of, 402
deformylation of, 114–115
nitration of, 13–14

Indole-3-carboxaldehydes—*continued*
reactivity of, in condensation reactions, 222, 405
resonance in, 402
reduction with diborane, 41, 109
with lithium aluminum hydride, 108
with sodium borohydride, 108, 110
Schiff bases of, 408
synthesis, 37, 416–417, 419
as tryptamine precursors, 222–224
Indolecarboxylic acids, *see also* indole-2-carboxylic acids, etc.
conversion to aminoindoles by Curtius reaction, 393, 399
decarboxylation of, 421–424
dissociation constants of, 85, 421
rates of esterification and hydrolysis of esters, 85
synthesis, 424–426
Indole-2-carboxylic acids
acid dissociation constants, 85
conversion to 2-aminoindoles, 393
to indole-2-carboxaldehydes, 419
hydrolytic decarboxylation, 118, 159, 421–423
reactions with organometallic reagents, 418–419
synthesis by Fischer cyclization, 160–161, 424
by Reissert reaction, 176–178, 424
Indole-3-carboxylic acids
dissociation constants of, 85
esters of, by Nenitzescu synthesis, 171–173
hydrolytic decarboxylation, 118, 173, 421–423
reduction of esters, 108
resonance in, 421
synthesis, 34, 171–173, 412, 424–426
Indole derivatives
acid-catalysed dimerization and trimerization, 7–9
acylation, *see* Acylation
acyl, *see* Acylindoles, 2-Acylindoles, etc.
2-alkoxy
oxidation of, 291
reactions of, 354–355
alkylation, *see* Alkylation
arylation, 27
basicity of, substituent effects, 2–5
bromination, *see* Bromination
Indole derivatives—*continued*
t-butyl, dealkylation, 116
catalytic reduction, 129–130
cyanoethylation, 70–71, 76
cyclizative condensations with ketones, 47–53
deuteration of, 4–7
dimerization of, 7–9
halo, *see* Bromoindoles, Chloroindoles
halogenation, *see* Halogenation
hydroxylation, *see* Hydroxylation
isotopic exchange in, 4–7
magnesium derivatives, *see* Indolylmagnesium halides
Mannich reactions of, 56–59
nitration, *see* Nitration
oxidation, *see* Oxidation
ozonolysis, 295–296
protonation, 3–11
reactions with aldehydes, 39–56
with azides, 17–18
with aziridines, 29
with carbenes, 31–33
with diazonium ions, 17–18
with *N*,*N*-dichlorourethane, 306–307
with iminium bonds, 56–67, 236–244, 246–251
with ketones, 39–56
with nitroethylenes, 69–70
with phosphorus pentachloride, 78
with phthalaldehyde, 54
with 1-piperideine, 61
with pyridine-1-oxides, 66
with pyridinium compounds, 65
with 1-pyrroline, 61
with quinoline-1-oxides, 66
with quinones, 72–74, 83
with sulfenyl chlorides, 77
with tetracyanoethylene, 71–72
with tetrahydropyridines, 61–64, 246–251
with thiocyanogen, 77
with thionyl chloride, 77–78
with vinyl pyridines, 71
rearrangements of, 316–331
reduction
in acidic solution, 130
catalytic, 129–130
hydrogenolysis of substituents, 108–114
in liquid ammonia, 131

Indole derivatives—*continued*
sodium salts, of, 19–21
solvolysis of indolylethyl bromides and tosylates, 24–25
substituent cleavage reactions, 114–120
3-substituted, mechanism of electrophilic substitution, 78–83, 255
sulfonation, 18
synthesis, 143–206
from *N*-acyl-*o*-alkylanilines, 189–191
from *N*-allylanilines, 194
from *o*-aminophenylacetaldehydes, 195–196
from *o*-aminophenylacetylenes, 204–205
from aryltriazoles, 203
from azidostyrenes, 186–188
Bischler cyclization, 164–171
from cinnolines, 202–203
by dehydrogenation of indolines, 134–135, 416–417
by deoxygenation of *o*-nitrostyrenes, 184–186
Fischer cyclization, 152–162
from hydrazobenzenes, 201–202
Madelung reaction, 189–191
Nenitzescu reaction, 171–176
by oxidative cyclization of phenylethylamines, 191–193
from phenylhydroxylamines, 201–202
from quinoline-*N*-oxides, 203–204
by reductive cyclizations, 176–183
3*H*-Indole derivatives, 120–125
3-acetoxy
from lead tetraacetate oxidations, 300–303, 328
reactions of, 300–303, 328–330
3-chloro
reactions, 122–124, 303
rearrangement to oxindoles, 303, 325–326
synthesis, 257, 303, 325
3-diazo, 396
3-dichloromethyl, from indoles and dichlorocarbene, 32–33
electrophilicity of, 10–11, 118
3-hydroperoxy, *see* 3-Hydroperoxy-3*H*-indoles
3-hydroxy, synthesis, 283, 293
as intermediates in electrophilic substitution, 78–83, 236–238
3*H*-Indole derivatives—*continued*
nitration of, 12
nucleophilic addition to, 120–125, 331–339
rearrangement of, 316–322
3-allyl derivatives, 22–24, 320
3,3-dialkyl derivatives, 319–321, 325
migratory aptitude in, 320
1,2,3,3-tetrasubstituted, 319
synthesis of
by alkylation of indoles, 19–21
by Fischer cyclization, 151, 156
trimers, 121
Indole-3-glyoxalates
reaction with organometallic reagents, 408
reduction of, 109, 404
Indole-3-glyoxamides
reduction with diborane, 220
as tryptamine precursors, 219–222
Indole-3-glyoxylic acids, oxidation to indole-3-carboxylic acids, 424
Indole-3-glyoxylyl chlorides
decarbonylation of, 424
as tryptamine precursors, 219–221
Indolenines, *see* 3*H*-Indole derivatives
Indolenyl hydroperoxides, *see* 3-Hydroperoxy-3*H*-indoles
3*H*-Indole-3-one derivatives, 388
3*H*-Indole-3-one-1-oxide, *see* Isatogens
2,3-Indolinediones, *see* Isatins
Indolines
conversion to hydroxyindoles, 371
dehydrogenation to indoles, 132–134
2,3-dihydroxy
by oxidation of indoles, 297–299
rearrangements of, 327–329
Friedel–Crafts acylation of, 417
interconversion with indoles, 129–135
as intermediates in synthesis of indoles, 188, 268–269, 399, 416–417, 426
synthesis from *o*-alkylphenyl azides, 188
Vilsmeier–Haack formylation of, 416
Indolinols, rearrangements of, 322–331
2-Indolinones, *see* Oxindoles
3-Indolinones
alkylation of, 365
anions of, 363–365
oxidation of, 365, 368
as products of rearrangement of 2,3-dihydroxyindolines, 327–329
of 3-acetoxy-3*H*-indoles, 328–330

3-Indolinones—*continued*
reactions with nucleophiles, 366–367
rearrangements of, 322–331
reduction of, 366
structure of, 364
synthesis, 283, 327–330, 367–368, 383–384
Indolmycin
structure, 433
synthesis, 435
Indolylcarbinols, 38–42, 407
Indolylcarbinylamines, *see also* Gramine
3-aminomethylindole
hydrogenolysis by lithium aluminum hydride, 109
reactions and synthesis, 96
2-dimethylaminomethylindole, 97–98
hydrogenolysis of, 109
oxidation of, 296–297
reactivity of, 10, 93–102
synthesis
from indole-3-carboxaldehydes via Schiff bases, 409
by Mannich condensation, 56–58
2-(3-Indolyl)ethyl bromide
in alkylation of amines, 227
solvolysis of, 24–25
1-[2-(3-Indolyl)ethyl]piperidines, oxidation, 250–251
1-[2-(3-Indolyl)ethyl]pyridinium salts, reduction of, 246–250
1-[2-(3-Indolyl)ethyl]tetrahydropyridines, cyclization of, 246–251, 255–259
2-(3-Indolyl)ethyl tosylates, solvolysis of, 24–25
Indolyl ketones, *see* Acylindoles, 2-Acylindoles, etc.
Indolylmagnesium halides
acylation of, 33–39, 413
alkylation of, 19–22, 27–28, 216, 235
oxidation of, 293
protonation of, 5–6
reactions with aldehydes, 45–47
with allyl halides, 80
with epoxides, 30
with esters, 35
with ketones, 45–47
with nitroethylenes, 70
structure of, 20–21
3-Indolylmethanol, *see also* Indolylcarbinols
3-Indolylmethanol—*continued*
reactions of, 40–42, 97, 108
synthesis, 110, 113
Indomethacin, 442
3-Iodoindole, 15
Isatins, 377–382
reactions with amines, 379–80
with diazoalkanes, 380–381
with nucleophiles, 377–381
with phenylmagnesium bromide, 322
with trialkyl phosphites, 378–379
reduction, 381–382
synthesis, 382
Isatogens, 382–387
conversion to anthranils, 383
cycloadditions of, 383–384
nucleophilic addition to, 382–383
photochemical synthesis, 386–387
ring expansion of, 384, 385
synthesis, 385–387
Isogramine, *see* indolylcarbinylamines

J

Japp–Klingemann reaction
in synthesis of arylhydrazones, 159–161
of indolealkanoic acids, 217–218
of monophenylhydrazones of diketones from enamines, 159, 418

K

Ketones
indolyl, *see* Acylindoles, 2-Acylindoles, etc.
reactions with indoles, 39–56
with indolylmagnesium halides, 45–47
with tryptamines, 243–244
unsaturated, reactions with indoles, 67–69
Koshland's reagent, mode of reaction, 122–123

L

Lead tetraacetate
in oxidation of alkaloids, 300–303, 328
of indoles, 300–303
Localization energy, in prediction of reactivity, 83–84
Loganin, in biosynthesis of indole alkaloids, 269–273

Luciferin, of cypridina, 442–443
Lysergic acid
  diethylamide as hallucinogen, 440
  synthesis, 268–269

M

Madelung synthesis of indoles, 189–191
Mannich condensations
  of hydroxyindoles, 60, 369–370
  of indoles, 56–59
  of oxindoles, 346
Melanins, 438
Melatonin, 438
Metabolism of indoles, 308–311
5-Methoxy-*N*,*N*-dimethyltryptamine, as hallucinogen, 440
1-Methylindole, metallation with butyllithium, 135–136, 419
2-Methylindole
  condensation with hydroxymethyleneketones, 54
    with ketones, 47–51
  oxidation of, 289, 293–294
3-Methylindole
  condensation with 1,4-dicarbonyl compounds, 52–53
  dimer of, 9
  Mannich condensations of, 57–58
  oxidation of, 293–295, 304–305
  reactions with *N*-bromosuccinimide, 304
    with perbenzoic acid, 293
    with quinones, 72–74, 83
    with unsaturated ketones, 80–81
  trinitrobenzene complex, structure of, 3
Methyl vinyl ketone, reaction with indoles, 67
Mevalonic acid, as alkaloid precursor, 269–273
Migratory aptitudes, in rearrangement of 3*H*-indole derivatives, 320
Mitomycins
  structure, 431–433
  synthesis, 432–434
Molecular orbital calculations, 1–3, 83–85

N

Nenitzescu synthesis, 171–176
  mechanism, 171–173
  substituent effects, 171–173
Nitration of indole derivatives
  mechanism, 11–14
  orientation in, 11–14
Nitroalkanes
  alkylation by gramine derivatives, 224–225
  condensations with indolecarboxaldehydes, 222, 406
o-Nitrobenzyl ketones, reductive cyclization to indoles, 176–179
o-Nitrobenzyl cyanides, reductive cyclization to indoles, 180–182
Nitroethylenes
  reactions with indoles, 69–70
  as tryptamine precursors, 226
o-Nitrophenylacetaldehydes, reductive cyclization to indoles, 178–179
Nitrosation of indoles, 338–339, 395
o-Nitrostyrenes
  as indole precursors, 194–196
  photochemical conversion to isatogens, 385–387
3-(2-Nitrovinyl)indoles
  nucleophilic addition to, 222–223
  reduction of, 224
  structure of, 222–223
  synthesis of, 222–223
  as tryptamine precursors, 222–224

O

Olefins, electron deficient, reactions with indoles, 67–72
Oxidation, *see also* Autooxidation
  of alkaloids, 291–293, 300–303, 325–326
  of 2-alkoxyindoles, 291
  of 3-aminoindoles, 290–291
  of dioxindoles, 352
  of indole derivatives with N-bromosuccinimide, 303–305
    with *t*-butyl hypochlorite, 303
    with chromic acid, 296–297
    with hydrogen peroxide, 293–294
    with hydrogen peroxide-ammonium molybdate, 295
    with lead tetraacetate, 300–303
    with manganese dioxide, 299
    with nitric acid, 298–299
    with osmium tetroxide, 297–299
    with peracids, 293
    with periodic acid, 300

Oxidation—*continued*
of indole derivatives—*continued*
with potassium nitrosodisulfonate, 306–307, 432
with potassium permanganate, 297
with sodium periodate, 300
Oxidative cyclization
of 1-[2-(3-indolyl)ethyl]piperidines, 250–251
of phenylethylamine derivatives, 191–193
Oxindole derivatives
acidity of, 341
acylation, base-catalyzed, 345–346
acyl derivatives
alkylation, 344
base-catalyzed cleavage, 348–349
conversion to indoles, 335–336
alkylation of, 342–347
3-alkylidene, reactions of, 350–352
3-(2-aminoethyl)-, condensations with aldehydes, 346
bromination, 304–305, 353
condensation reactions with carbonyl compounds, 345–346
3-diazo, 380
Diels–Alder reactions of 3-acylmethylene, 348
electrophilic substitution of, 353
3-formyl, alkylation of, 344
3-halo, reactivity of, 349–350
hydrogenolysis of 3-acyl, 351
of 3-aminomethyl, 351–352
1-hydroxy, 377
hydroxylation of, 308
Mannich condensation of, 346
3-methylene, as metabolite of indole-3-acetic acid, 311–312
Michael alkylation of, 346–348
nitration, 353
oxidation, 352
reactions, 341–357
reduction with lithium aluminum hydride, 356–357
with sodium-alcohol, 356–357
structure of, 341
synthesis, 357–364
from 3-acetoxy-3*H*-indoles, 328–330
from N-acylphenylhydrazides, 362–363
from *o*-aminophenylacetic acids, 360–362
Oxindole derivatives—*continued*
synthesis—*continued*
from 3-chloro-3*H*-indoles, 306–307, 325–326
from indoles, by oxidation, 294–295, 303–307, 363
4-Oxo-4,5,6,7-tetrahydroindoles
as intermediates in synthesis, 198–199
synthesis, 198–200
Ozonides of indoles, 295–296
Ozonolysis of indoles, 295–296

P

Peracids, reactions with indoles, 293–294
Phenacylanilines, conversion to indoles, 164–171
Phenylethylamines, as indole precursors, 191–193
N-Phenylmaleimide, reaction with indoles, 76
Phthalaldehyde, reaction with indoles, 54–56
Picryl azide, reaction with indoles, 17–18
Pictet–Spengler reaction
mechanism, 236–238
in synthesis of alkaloids, 259, 262, 265, 267–268
of tetrahydro-β-carbolines, 236–244, 246–251
Polyphosphate ester
as catalyst for Bischler–Napieralski cyclization, 245
for Fischer cyclization, 161
Polyphosphoric acid
as catalyst for Fischercyclization, 151, 161
for intramolecular acylation of indoles, 415–416
for rearrangement of indoles, 316–318
for ring opening of indole-3-acetamides 332–333
Potassium nitrosodisulfonate, as oxidant, 191, 306–307, 370–371
Propiolactone, reaction with indoles, 29–30
Psilocybin, as hallucinogen, 440
Pyridines, partial reduction of, 61–63, 247–248
Pyridinium compounds
as electrophiles toward indoles, 65–66
reductive cyclization, 246–250
1-Pyrroline, reaction with indole, 61

Pyruvic acids
condensations with tryptamines, 239–242
reactions with indoles, 69–70

Q

Quebrachamine, synthesis, 257
Quinolines, from indoles and carbenes, 31–33
Quinoline-1-oxides
photochemical conversion to indoles, 203–204
reactions with indoles, 66
Quinones
oxidation of indolines to indoles, by 132–133
reactions with indoles, 72–74

R

Rearrangements, 316–331
of 3-allyl-2-methyl-3*H*-indole, 22–24
in cyclization of (β-indolylethyl)isoquinuclidone derivatives, 251–255
of diindole derivatives, 76
in electrophilic substitution of 3-substituted indoles, 78–83, 236–238
of 3*H*-indole derivatives, 317–322
in solvolysis of 2-(3-indolyl)ethyl bromides and tosylates, 24–25
in synthesis of aspidospermidine, 259–261
of iboga alkaloids, 102–104, 252–255
of tetrahydro-β-carbolines, 118
Reductive cyclization, in synthesis of indoles, 176–183
Reissert synthesis, 176–178
Reserpine
equilibration with isoreserpine, 104–105
synthesis, 263–264
as tranquilizer, 440
Resonance energy, of indole, 2
Ring expansion of indoles with carbenes, 31–33
Ring-opening reactions, 126–127, 331–339

S

Serotonin
mechanism of biosynthesis, 308–309
physiological significance, 438
Sporidesmin, structure, 437
Strychnine, synthesis, 266–268
Substituent effects
on acidity of indolecarboxylic acids, 85
on basicity of indole derivatives, 4–5
Hammett correlations of, 85
Sulfenyl chlorides, reactions with indoles, 77
Sulfonation of indoles, 18

T

Tabersonine, acid-catalyzed rearrangements, 116
Teleocidin B, structure, 433, 436
Tetracyanoethylene, reactions with indoles, 71–72
Tetrahydrocarbazole, oxidation, 282–284, 300
Tetrahydro-β-carboline derivatives
acid-catalyzed epimerization of, 104–105
alkaloids, conversion to oxindoles via 3-chloro-3*H*-indoles, 303, 325–326
as intermediates in alkaloid synthesis, 262–265
mechanism of formation from tryptamines, 236–238
mode of formation in alkaloid biosynthesis, 273
oxidation of, 284–285
ring cleavage of
by acid anhydrides, 100–102, 258
by nucleophiles, 256–257
be reduction, 111–112
synthesis from tryptamines, 236–251, 256, 259
4,5,6,7-Tetrahydroindoles
4-oxo derivatives, 198–200
synthesis, 198–200
Tetrahydropyridines
formation, 246–251, 262–263
intramolecular cyclization of *N*-(3-indolyl)ethyl, 246–251, 262–263
by oxidation of piperidines, 250–251
reactions with indoles, 61–64
in synthetic elaboration of indoles, 246–251
p-Toluenesulfonyl azides, reactions with indoles, 17–18
Transannular cyclization in alkaloid synthesis, 256–259
Triethyl orthoformate, reactions with indoles, 26

Triethyloxonium fluoroborate in alkylation of oxindoles, 343–344
Triphenylmethyl chloride, reaction with indoles, 26
Tryptamine and derivatives
1-alkyl, synthesis, 220
basicity of, 11
condensation reactions with ketones, 243–244
conversion to tetrahydro-β-carboline derivatives, 236–251
cyclization of *N*-acyl derivatives, 244–246
by *t*-butyl hypochlorite, 125
of *N*-thioacyl derivatives, 245–246
hallucinogenic derivatives, 440
metabolic transformations of, 308
pharmacological properties, 441–442
protonation of, 10–11
reactions with aldehydes, 236–240
with ketones, 243–244
with pyruvic acids, 240–242
synthesis by means of Abramovitch procedure, 157, 159
of Fischer cyclization, 228–229
of gramine derivatives, 224–226
of 3-(2-haloethyl)indoles, 227
of indole-3-acetic acid derivatives, 218–219
of indole-3-glyoxamides, 219–222
of 3-indolinones, 366
of isatins, 229–230
of 3-(2-nitroethyl)indoles, 226
of 3-(2-nitrovinyl)indoles, 222–224
Tryptophan, *see also* Tryptophan derivatives
alkaloid biosynthesis from, 269
cleavage at, in peptides, 305
hydroxylation of, 308–310
modification of, in peptides, 122–123
oxidation of, 291, 294, 310–311
photochemical reduction with sodium borohydride, 131
physiological significance, 438
reaction with Koshland's reagent, 122–123
synthesis, 230–235
Tryptophan derivatives
conversion to quinolines, 331–332
cyclization by *N*-bromosuccinimide, 125
synthesis, 230–235
Tryptophan pyrrolase, 310–311
Tryptophol, synthesis, 28, 109
Tubifolidine, synthesis, 258–259
Tyrosine, oxidative cyclization, 193

U

Unsaturated ketones, reactions with indoles, 67–69
Uleine, nucleophilic ring opening of, 99–100

V

Velbanamine, synthesis, 262
Vilsmeier–Haack acylation of indoles, 36–38, 415–416
Vinblastine
as antineoplastic agent, 440
structure, 261
Vincadifformine, synthesis, 256–257
Vincadine, synthesis, 256–257
Vincamine, synthesis, 238, 240
Vincristine, as antineoplastic agent, 440
Vindoline, alkylation of, 261–262
Vinylindoles, 125–129
Diels–Alder reactions of, 125–127
nucleophilic additions to, 129
synthesis, 129
by condensation reactions of indoles and carbonyl compounds, 47–51
Vinylpyridines, reactions with indoles, 71
Violacein, 445
Voacamine, 107
Voacangine, oxidation of, 292

W

Wolff–Kishner reduction of acylindoles, 403

Y

Yohimbane and derivatives
ozonolysis, 285
synthesis, 239–241, 245, 248
Yohimbene, ozonolysis, 285
Yohimbine, synthesis, 263, 265

Z

Zinc chloride
as catalyst for Fischer cyclization, 161
for rearrangements, 318